Springer Series in Operations Research

Editor:
Peter Glynn

Springer Science+Business Media, LLC

Springer Series in Operations Research

Altiok: Performance Analysis of Manufacturing Systems
Dantzig and Thapa: Linear Programming 1: Introduction
Drezner (Editor): Facility Location: A Survey of Applications and Methods
Fishman: Monte Carlo: Concepts, Algorithms, and Applications
Olson: Decision Aids for Selection Problems
Yao (Editor): Stochastic Modeling and Analysis of Manufacturing Systems

George B. Dantzig Mukund N. Thapa

Linear Programming

1: Introduction

With 87 Illustrations

Springer

George B. Dantzig
Professor of Operations Research
and Computer Science
Department of Operations Research
Stanford University
Stanford, CA 94305
USA

Mukund N. Thapa
President
Stanford Business Software, Inc.
Suite 304
2680 Bayshore Parkway
Mountain View, CA 94043
and
Consulting Professor of Operations Research
Stanford University
Stanford, CA 94305
USA

Series Editor:
Peter Glynn
Department of Operations Research
Stanford University
Stanford, CA 94305
USA

Library of Congress Cataloging-in-Publication Data
Dantzig, George Bernard, 1914–
Linear programming 1 : introduction / George B. Dantzig & Mukund N. Thapa.
p. cm. – (Springer series in operations research)
Includes bibliographical references and index.

ISBN 978-1-4757-8112-0 ISBN 978-0-387-22633-0 (eBook)
DOI 10.1007/978-0-387-22633-0

1. Linear programming. I. Thapa, Mukund Narain-Dhami. II. Title.
III. Series.
T57.74.D365 1997
619.7'2–dc20 96-36411

Printed on acid-free paper.

Production managed by Timothy Taylor; manufacturing supervised by Joe Quatela.
Photocomposed pages prepared from the authors' LaTeX files.

9 8 7 6 5 4 3 2 1

ISBN 978-1-4757-8112-0 SPIN 10523424

ABOUT THE AUTHORS

George B. Dantzig received the National Medal of Science from the President of the United States "for inventing Linear Programming and for discovering the Simplex Algorithm that led to wide-scale scientific and technical applications to important problems in logistics, scheduling, and network optimization, and to the use of computers in making efficient use of the mathematical theory." He is world famous for his twin discoveries; linear programming and the Simplex Algorithm, which together have enabled mankind for the first time to structure and solve extremely complex optimal allocation and resource problems. Among his other discoveries are the Decomposition Principle (with Philip Wolfe) which makes it possible to decompose and solve extremely large linear programs having special structures, and applications of these techniques with sampling to solving practical problems subject to uncertainty.

Since its discovery in 1947, the field of linear programming, together with its extensions (mathematical programming), has grown by leaps and bounds and is today the most widely used tool in industry for planning and scheduling.

George Dantzig received his master's from Michigan and his doctorate in mathematics from Berkeley in 1946. He worked for the U.S. Bureau of Labor Statistics, served as chief of the Combat Analysts Branch for USAF Headquarters during World War II, research mathematician for RAND Corporation, and professor and head of the Operations Research Center at the University of California, Berkeley. He is currently professor of operations research and computer science at Stanford University. He served as director of the System Optimization Laboratory and the PILOT Energy-Economic Model Project. Professor Dantzig's seminal work has laid the foundation for the field of systems engineering, which is widely used in network design and component design in computer, mechanical, and electrical engineering. His work inspired the formation of the Mathematical Programming Society, a major section of the Society of Industrial and Applied Mathematics, and numerous professional and academic bodies. Generations of Professor Dantzig's students have become leaders in industry and academia.

He is a member of the prestigious National Academy of Science, the American Academy of Arts and Sciences, and the National Academy of Engineering.

Mukund N. Thapa is the president of Stanford Business Software, Inc., as well as a consulting professor of operations research at Stanford University. He received a bachelor of technology degree in metallurgical engineering from the Indian Institute of Technology, Bombay, and M.S. and Ph.D. degrees in operations research from Stanford University. His Ph.D. thesis was concerned with developing specialized algorithms for solving large-scale unconstrained nonlinear minimization problems. By profession he is a software developer who produces commercial software products as well as commercial-quality custom software. Since 1978, Dr. Thapa has been applying the theory of operations research, statistics, and computer science to develop efficient, practical, and usable solutions to a variety of problems.

At Stanford Business Software, Dr. Thapa, ensures that the company produces high-quality turnkey software for clients. His expert knowledge of user-friendly interfaces, data bases, computer science, and modular software design plays an important role in making the software practical and robust. His speciality is the application of numerical analysis methodology to solve mathematical optimization problems. He is also an experienced modeler who is often asked by clients to consult, prepare analyses, and to write position papers. At the Department of Operations Research, Dr. Thapa teaches graduate-level courses in mathematical programming computation and numerical methods of linear programming.

TO

Tobias and Anja Dantzig, my parents, *in memoriam*,
Anne S. Dantzig, my wife, and to
the great pioneers who made this field possible:
Wassily Leontief, Tjalling Koopmans, John von Neumann,
Albert Tucker, William Orchard-Hays, Martin Beale.

— George B. Dantzig

Devi Thapa and Narain S. Thapa, my parents,
and Radhika H. Thapa, my wife.

— Mukund N. Thapa

Contents

List of Figures

List of Tables

FOREWORD

By George B. Dantzig

LINEAR PROGRAMMING

The Story About How It Began: Some legends, a little about its historical significance, and comments about where its many mathematical programming extensions may be headed.

Industrial production, the flow of resources in the economy, the exertion of military effort in a war, the management of finances—all require the coordination of interrelated activities. What these complex undertakings share in common is the *task of constructing a statement of actions to be performed, their timing and quantity (called a program or schedule), that, if implemented, would move the system from a given initial status as much as possible towards some defined goal.*

While differences may exist in the goals to be achieved, the particular processes, and the magnitudes of effort involved, when modeled in mathematical terms these seemingly disparate systems often have a remarkably similar mathematical structure. *The computational task is then to devise for these systems an algorithm for choosing the best schedule of actions from among the possible alternatives.*

The observation, *in particular*, that a number of economic, industrial, financial, and military systems can be modeled (or reasonably approximated) by mathematical systems of *linear inequalities and equations* has given rise to the development of the *linear programming* field.

The first and most fruitful industrial applications of linear programming were to the petroleum industry, including oil extraction, refining, blending, and distribution. The food processing industry is perhaps the second most active user of linear programming, where it was first used to determine shipping of ketchup from a few plants to many warehouses. Meat packers use linear programming to determine the most economical mixture of ingredients for sausages and animal feeds.

In the iron and steel industry, linear programming has been used for evaluating various iron ores. Pelletization of low-grade ores, additions to coke ovens, and shop loading of rolling mills are additional applications. Linear programming is also used to decide what products rolling mills should make in order to maximize profit. Blending of iron ore and scrap to produce steel is another area where it has been used. Metalworking industries use linear programming for shop loading and for determining the choice between producing and buying a part.

Paper mills use it to decrease the amount of trim losses. The optimal design and routing of messages in a communication network, contract award problems, and the routing of aircraft and ships are other examples where linear programming methods are applied. The best program of investment in electric power plants and transmission lines has been developed using linear programming methods.

More recently, linear programming (and its extensions) has found its way into financial management, and Wall Street firms have been hiring mathematical programmers that they call "rocket scientists" for a variety of applications, especially for lease analysis and portfolio analysis.

Linear programming can be viewed as part of a great revolutionary development that has given mankind the ability to state general goals and to lay out a path of detailed decisions to be taken in order to "best" achieve these goals when faced with practical situations of great complexity. Our tools for doing this are ways to formulate real-world problems in detailed mathematical terms (models), techniques for solving the models (algorithms), and engines for executing the steps of algorithms (computers and software).

This ability began in 1947, shortly after World War II, and has been keeping pace ever since with the extraordinary growth of computing power. So rapid have been the advances in decision science that few remember the contributions of the great pioneers that started it all. Some of their names are von Neumann, Kantorovich, Leontief, and Koopmans. The first two were famous mathematicians. The last three received the Nobel Prize in economics for their work.

In the years from the time when it was first proposed in 1947 by the author (in connection with the planning activities of the military), linear programming and its many extensions have come into wide use. In academic circles decision scientists (operations researchers and management scientists), as well as numerical analysts, mathematicians, and economists have written hundreds of books and an uncountable number of articles on the subject.

Curiously, in spite of its wide applicability today to everyday problems, linear programming was unknown prior to 1947. This statement is not quite correct; there were some isolated exceptions. Fourier (of Fourier series fame) in 1823 and the well-known Belgian mathematician de la Vallée Poussin in 1911 each wrote a paper about it, but that was about it. Their work had as much influence on post-1947 developments as would the finding in an Egyptian tomb of an electronic computer built in 3,000 B.C. Leonid Kantorovich's remarkable 1939 monograph on the subject was shelved by the communists for ideological reasons in the U.S.S.R. It was resurrected two decades later after the major developments had already taken place in the West. An excellent paper by Hitchcock in 1941 on the transportation problem went unnoticed until after others in the late 1940s and early 50s had independently rediscovered its properties.

What seems to characterize the pre-1947 era was a lack of any interest in trying to optimize. T. Motzkin in his scholarly thesis written in 1936 cites only 42 papers on linear inequality systems, none of which mentioned an objective function.

The major influences of the pre-1947 era were Leontief's work on the input-output model of the economy (1932), an important paper by von Neumann on game theory (1928), and another by him on steady economic growth (1937).

My own contributions grew out of my World War II experience in the Pentagon. During the war period (1941–45), I had become an expert on programs and planning methods using desk calculators. In 1946 I was mathematical advisor to the U.S. Air Force comptroller in the Pentagon. I had just received my Ph.D. (for research I had done mostly before the war) and was looking for an academic position that would pay better than a low offer I had received from Berkeley. In order to entice me to not take another job, my Pentagon colleagues D. Hitchcock and M. Wood challenged me to see what I could do to mechanize the Air Force planning process. I was asked to find a way to compute more rapidly a time-staged deployment, training, and logistical supply program. In those days "mechanizing" planning meant using analog devices or punch-card equipment. There were no electronic computers.

Consistent with my training as a mathematician, I set out to formulate a model. I was fascinated by the work of Wassily Leontief, who proposed in 1932 a large but simple matrix structure that he called the *Interindustry Input-Output Model of the American Economy.* It was simple in concept and could be implemented in sufficient detail to be useful for practical planning. I greatly admired Leontief for having taken the three steps necessary to achieve a successful application:

1. Formulating the inter-industry model.

2. Collecting the input data during the Great Depression.

3. Convincing policy makers to use the output.

Leontief received the Nobel Prize in 1976 for developing the input-output model.

For the purpose I had in mind, however, I saw that Leontief's model had to be generalized. His was a steady-state model, and what the Air Force wanted was a highly *dynamic model*, one that could change over time. In Leontief's model there was a one-to-one correspondence between the production processes and the items being produced by these processes. What was needed was a model with many *alternative activities.* Finally, it had to be computable. Once the model was formulated, there had to be a practical way to compute what quantities of these activities to engage in consistent with their respective input-output characteristics and with given resources. This would be no mean task since the military application had to be *large scale*, with hundreds and hundreds of items and activities.

The *activity analysis* model I formulated would be described today as a time-staged, dynamic linear program with a staircase matrix structure. *Initially there was no objective function*; broad goals were never stated explicitly in those days because practical planners simply had no way to implement such a concept. Noncomputability was the chief reason, I believe, for the total lack of interest in optimization prior to 1947.

A simple example may serve to illustrate the fundamental difficulty of finding an optimal solution to a planning problem once it is formulated. Consider the problem of assigning 70 men to 70 jobs. Suppose a known value or benefit v_{ij} would result if the ith man is assigned to the jth job. An *activity* consists in assigning the ith man to the jth job. The restrictions are (i) each man must be assigned a job (there are 70 such), and (ii) each job must be filled (also 70). The level of an activity is either 1, meaning it will be used, or 0, meaning it will not. Thus there are 2 × 70 or 140 restrictions, 70 × 70 or 4,900 activities with 4,900 corresponding zero-one decision variables x_{ij}. Unfortunately there are $70! = 70 \times 69 \times 68 \cdots \times 2 \times 1$ different possible solutions or ways to make the assignments x_{ij}. The problem is to compare the 70! solutions with one another and to select the one that results in the largest sum of benefits from the assignments.

Now 70! is a big number, greater than 10^{100}. Suppose we had a computer capable of doing a million calculations per second available at the time of the big bang 15 billion years ago. Would it have been able to look at all the 70! combinations by now? The answer is no! Suppose instead it could perform at nanosecond speed and make 1 billion complete assignments per second? The answer is still no. Even if the earth were filled solid with such computers all working in parallel, the answer would still be no. If, however, there were 10^{40} Earths circling the sun each filled solid with nanosecond-speed computers all programmed in parallel from the time of the big bang until the sun grows cold, then perhaps the answer might be yes.

This easy-to-state example illustrates why up to 1947, and for the most part even to this day, a great gulf exists between man's aspirations and his actions. Man may wish to state his wants in complex situations *in terms of some general objective to be optimized*, but there are so many different ways to go about it, each with its advantages and disadvantages, that it would be impossible to compare all the cases and choose which among them would be the best. Invariably, man in the past has left the decision of which way is best to a leader whose so called "experience" and "mature judgment" would guide the way. Those in charge like to do this by issuing a series of ground rules (edicts) to be executed by those developing the plan.

This was the situation in 1946 before I formulated a model. In place of an explicit goal or objective function, there were a large number of ad hoc ground rules issued by those in authority in the Air Force to guide the selection. Without such rules, there would have been in most cases an astronomical number of feasible solutions to choose from. Incidentally, "Expert System" software, a software tool used today (1996) in artificial intelligence, which is very much in vogue, makes use of this adhoc ground-rule approach.

Impact of linear programming on computers: All that I have related up to now about the early development took place in late 1946 before the advent of the computer, more precisely, before we were aware that it was going to exist. But once we were aware, the computer became a vital tool for our mechanization of the planning process. So vital was the computer going to be for our future progress, that our group successfully persuaded the Pentagon (in the late 1940's) to fund the development of computers.

To digress for a moment, I would like to say a few words about the electronic computer itself. To me, and I suppose to all of us, one of the most startling developments of all time has been the penetration of the computer into almost every phase of human activity. Before a computer can be intelligently used, a model must be formulated and good algorithms developed. To build a model, however, requires the axiomatization of a subject-matter field. In time this axiomatization gives rise to a whole new mathematical discipline that is then studied for its own sake. Thus, with each new penetration of the computer, a new science is born. Von Neumann notes this tendency to axiomatize in his paper on *The General and Logical Theory of Automata.* In it he states that automata have been playing a continuously increasing role in science. He goes on to say

> *Automata have begun to invade certain parts of mathematics too, particularly but not exclusively mathematical physics or applied mathematics. The natural systems (e.g., central nervous system) are of enormous complexity and it is clearly necessary first to subdivide what they represent into several parts that to a certain extent are independent, elementary units. The problem then consists of understanding how these elements are organized as a whole. It is the latter problem which is likely to attract those who have the background and tastes of the mathematician or a logician. With this attitude, he will be inclined to forget the origins and then, after the process of axiomatization is complete, concentrate on the mathematical aspects.*

By mid-1947, I had formulated a model which satisfactorily represented the technological relations usually encountered in practice. I decided that the myriad of adhoc ground rules had to be discarded and replaced by an explicit objective function. I formulated the planning problem in mathematical terms in the form of axioms that stated

1. the total amount of each type of item produced or consumed by the system as a whole is the algebraic sum of the amounts inputted or outputted by the individual activities of the system,

2. the amounts of these items consumed or produced by an activity are proportional to the level of an activity, and

3. these levels are nonnegative.

The resulting mathematical system to be solved was the *minimization of a linear form subject to linear equations and inequalities.* The use (at the time it was proposed) of a linear form as the objective function to be maximized was a novel feature of the model.

Now came the nontrivial question: Can one solve such systems? At first I assumed that the economists had worked on this problem since it was an important special case of the central problem of economics, the optimal allocation of scarce resources. I visited T.C. Koopmans in June 1947 at the Cowles Foundation (which at that time was at the University of Chicago) to learn what I could from the mathematical economists. Koopmans became quite excited. During World War II, he had worked for the Allied Shipping Board on a transportation model and so had the theoretical as well as the practical planning background necessary to appreciate what I was presenting. He saw immediately the implications for general economic planning. From that time on, Koopmans took the lead in bringing the potentialities of linear programming models to the attention of other young economists who were just starting their careers. Some of their names were Kenneth Arrow, Paul Samuelson, Herbert Simon, Robert Dorfman, Leonid Hurwicz, and Herbert Scarf, to name but a few. Some thirty to forty years later the first three and T.C. Koopmans received the Nobel Prize for their research.

Seeing that economists did not have a method of solution, I next decided to try my own luck at finding an algorithm. I owe a great debt to Jerzy Neyman, the leading mathematical statistician of his day, who guided my graduate work at Berkeley. My thesis was on two famous unsolved problems in mathematical statistics that I mistakenly thought were a homework assignment and solved. One of the results, published jointly with Abraham Wald, was on the Neyman-Pearson Lemma. In today's terminology, this part of my thesis was on the existence of Lagrange multipliers (or dual variables) for a semi-infinite linear program whose variables were bounded between zero and one and satisfied linear constraints expressed in the form of Lebesgue integrals. There was also a linear objective to be maximized.

Luckily, the particular geometry used in my thesis was the one associated with the columns of the matrix instead of its rows. This column geometry gave me the insight that led me to believe that the *Simplex Method* would be a very efficient solution technique. I earlier had rejected the method when I viewed it in the row geometry because running around the outside edges seemed so unpromising.

I proposed the Simplex Method in the summer of 1947. But it took nearly a year before my colleagues and I in the Pentagon realized just how powerful the method really was. In the meantime, I decided to consult with the "great" Johnny von Neumann to see what he could suggest in the way of solution techniques. He was considered by many as the leading mathematician in the world. On October 3, 1947, I met him for the first time at the Institute for Advanced Study at Princeton.

John von Neumann made a strong impression on everyone. People came to him for help with their problems because of his great insight. In the initial stages of the development of a new field like linear programming, atomic physics, computers, or whatever, his advice proved to be invaluable. Later, after these fields were developed in greater depth, however, it became much more difficult for him to make the same spectacular contributions. I guess everyone has a finite capacity, and Johnny was no exception.

I remember trying to describe to von Neumann (as I would to an ordinary mortal) the Air Force problem. I began with the formulation of the linear programming model in terms of activities and items, etc. He did something which I believe was not characteristic of him. "Get to the point," he snapped at me impatiently. Having at times a somewhat low kindling point, I said to myself, "O.K., if he wants a *quickie*, then that's what he'll get." In under one minute I slapped on the blackboard a geometric and algebraic version of the problem. Von Neumann stood up and said, "Oh, that!" Then, for the next hour and a half, he proceeded to give me a lecture on the mathematical theory of linear programs.

At one point, seeing me sitting there with my eyes popping and my mouth open (after all, I had searched the literature and found nothing), von Neumann said

> I don't want you to think I am pulling all this out of my sleeve on the spur of the moment like a magician. I have recently completed a book with Oscar Morgenstern on the *theory of games*. What I am doing is conjecturing that the two problems are equivalent. The theory that I am outlining is an analogue to the one we have developed for games.

Thus I learned about *Farkas's Lemma* and about *duality* for the first time. Von Neumann promised to give my computational problem some thought and to contact me in a few weeks, which he did. He proposed an iterative nonlinear interior scheme. Later, Alan Hoffman and his group at the Bureau of Standards (around 1952) tried it out on a number of test problems. They also compared it to the Simplex Method and with some interior proposals of T. Motzkin. The Simplex Method came out a clear winner.

As a result of another visit in June 1948, I met Albert Tucker, who later became the head of mathematics department at Princeton. Soon Tucker and his students Harold Kuhn and David Gale and others like Lloyd Shapley began their historic work on game theory, nonlinear programming, and duality theory. The Princeton group became the focal point among mathematicians doing research in these fields.

The early days were full of intense excitement. Scientists, free at last from wartime pressures, entered the post-war period hungry for new areas of research. The computer came on the scene at just the right time. Economists and mathematicians were intrigued with the possibility that the fundamental problem of optimal allocation of scarce resources could be numerically solved. Not too long after my first meeting with Tucker there was a meeting of the Econometric Society in Wisconsin attended by well-known statisticians and mathematicians like Hotelling and von Neumann, and economists like Koopmans. I was a young unknown and I remember how frightened I was at the idea of presenting for the first time to such a distinguished audience, the concept of linear programming.

After my talk, the chairman called for discussion. For a moment there was the usual dead silence; then a hand was raised. It was Hotelling's. I must hasten to explain that Hotelling was fat. He used to love to swim in the ocean and when he did, it is said that the level of the ocean rose perceptibly. This huge whale of a man stood up in the back of the room, his expressive fat face taking on one of those all-knowing smiles we all know so well. He said: "*But we all know the world is nonlinear.*" Having uttered this devastating criticism of my model, he majestically sat down. And there I was, a virtual unknown, frantically trying to compose a proper reply.

Suddenly another hand in the audience was raised. It was von Neumann. "Mr. Chairman, Mr. Chairman," he said, "if the speaker doesn't mind, I would like to reply for him." Naturally I agreed. Von Neumann said: "The speaker titled his talk 'linear programming' and carefully stated his axioms. If you have an application that satisfies the axioms, well use it. If it does not, then don't," and he sat down. In the final analysis, of course, Hotelling was right. The world is highly nonlinear. Fortunately, systems of linear inequalities (as opposed to equalities) permit us to approximate most of the kinds of nonlinear relations encountered in practical planning.

In 1949, exactly two years from the time linear programming was first conceived, the first conference (sometimes referred to as the Zero Symposium) on mathematical programming was held at the University of Chicago. Tjalling Koopmans, the organizer, later titled the proceedings of the conference *Activity Analysis of Production and Allocation.* Economists like Koopmans, Kenneth Arrow, Paul Samuelson, Leonid Hurwitz, Robert Dorfman, Georgescu-Roegen, and Herbert Simon, academic mathematicians like Albert Tucker, Harold Kuhn, and David Gale, and Air Force types like Marshall Wood, Murray Geisler, and myself all made contributions.

The advent or rather, *the promise*, that the electronic computer would soon exist, the exposure of theoretical mathematicians and economists to real problems during the war, the interest in mechanizing the planning process, and last but not least the availability of money for such applied research all converged during the period 1947–1949. The time was ripe. The research accomplished in exactly two years is, in my opinion, one of the remarkable events of history. The proceedings of the conference remain to this very day an important basic reference, a *classic*!

The Simplex Method turned out to be a powerful theoretical tool for proving theorems as well as a powerful computational tool. To prove theorems it is essential that the algorithm include a way of avoiding degeneracy. Therefore, much of the early research around 1950 by Alex Orden, Philip Wolfe, and myself at the Pentagon, by J.H. Edmondson as a class exercise in 1951, and by A. Charnes in 1952 was concerned with what to do if a degenerate solution is encountered.

In the early 1950, many areas that we collectively call *mathematical programming* began to emerge. These subfields grew rapidly with linear programming, playing a fundamental role in their development. A few words will now be said about each of these.

Nonlinear Programming began around 1951 with the famous Karush, Kuhn-Tucker Conditions, which are related to the Fritz John Conditions (1948). In 1954, Ragnar Frisch (who later received the first Nobel Prize in economics) proposed a nonlinear interior-point method for solving linear programs. Earlier proposals such as those of von Neumann and Motzkin can also be viewed as interior methods. Later, in the 1960s, G. Zoutendijk, R.T. Rockafellar, P. Wolfe, R. Cottle, A. Fiacco, G. McCormick, and others developed the theory of nonlinear programming and extended the notions of duality.

Commercial Applications were begun in 1952 by Charnes, Cooper, and Mellon with their (now classical) optimal blending of petroleum products to make gasoline. Applications quickly spread to other commercial areas and soon eclipsed the military applications that had started the field.

Software—The Role of Orchard-Hays In 1954, William Orchard-Hays of the RAND Corporation wrote the first commercial-grade software for solving linear programs. Many theoretical ideas such as ways to compact the inverse, take advantage of sparsity, and guarantee numerical stability were first implemented in his codes. As a result, his software ideas dominated the field for many decades and made commercial applications possible. The importance of Orchard-Hays's contributions cannot be overstated, for they stimulated the entire development of the field and transformed linear programming and its extensions from an interesting mathematical theory into a powerful tool that changed the way practical planning was done.

Network Flow Theory began to evolve in the early 1950 by Merrill Flood and a little later by Ford and Fulkerson in 1954. Hoffman and Kuhn in 1956 developed its connections to graph theory. Recent research on combinatorial optimization benefited from this early research.

Large-Scale Methods began in 1955 with my paper "Upper Bounds, Block Triangular Systems, and Secondary Constraints." In 1959–60 Wolfe and I published our papers on *the Decomposition Principle.* Its dual form was discovered by Benders in 1962 and first applied to the solution of mixed integer programs. It is now extensively used to solve stochastic programs.

Stochastic Programming began in 1955 with my paper "Linear Programming under Uncertainty" (an approach which has been greatly extended by R. Wets in the 1960s and J. Birge in the 1980s). Independently, at almost the same time in 1955, E.M.L. Beale proposed ways to solve stochastic programs. Important contributions to this field have been made by A. Charnes and W. Cooper in the late 1950s using chance constraints, i.e., constraints that hold with a stated probability. Stochastic programming is one of the most promising fields for future research, one closely tied to large-scale methods. One approach that the author, Peter Glynn, and Gerd Infanger began in 1989 combines Bender's decomposition principle with ideas based on importance sampling, control variables, and the use of parallel processors.

Integer Programming began in 1958 with the work of R. Gomory. Unlike the earlier work on the traveling salesman problem by D.R. Fulkerson, S. Johnson, and Dantzig, Gomory showed how to systematically generate the "cutting" planes. Cuts are extra necessary conditions that when added to an existing system of inequalities guarantee that the optimization solution will solve in integers. Ellis Johnson of I.B.M. extended the ideas of Gomory. Egon Balas and many others have developed clever elimination schemes for solving 0-1 covering problems. *Branch and bound* has turned out to be one of the most successful ways to solve practical integer programs. The most efficient techniques appear to be those that combine cutting planes with branch and bound.

Complementary Pivot Theory was started around 1962–63 by Richard Cottle and Dantzig and greatly extended by Cottle. It was an outgrowth of Wolfe's method for solving quadratic programs. In 1964 Lemke and Howson applied the approach to bimatrix games. In 1965 Lemke extended it to other nonconvex programs. Lemke's results represent a historic breakthrough into the nonconvex domain. In the 1970's, Scarf, Kuhn, and Eaves extended this approach once again to the solving of fixed-point problems.

Computational Complexity. Many classes of computational problems, although they arise from different sources and appear to have quite different mathematical statements can be "reduced" to one another by a sequence of not-too-costly computational steps. Those that can be so reduced are said to belong to the same *equivalence class*. This means that an algorithm that can solve one member of a class can be modified to solve any other in same equivalence class. The *computational complexity* of an equivalence class is a quantity that measures the amount of computational effort required to solve the most difficult problem belonging to the class, i.e., its *worst case*. A nonpolynomial algorithm would be one that requires in the worst-case a number of steps not less than some exponential expression like Ln^m, $n!$, or 100^n, where n and m refer to the row and column dimensions of the problem and L the number of bits needed to store the input data.

Polynomial Time Algorithms. For a long time it was not known whether or not linear programs belonged to a nonpolynomial class called "hard" (such as the one the traveling salesman problem belongs to) or to an "easy" polynomial class (like the one that the shortest path problem belongs to). In 1970, Victor Klee and George Minty created a worst-case example that showed that the classical Simplex Algorithm would require an "exponential" number of steps to solve a worst-case linear program. In 1978, the Russian mathematician, L.G. Khachian developed a polynomial-time algorithm for solving linear programs. It is a *method* that uses *ellipsoids* that contain points in the feasible region. He proved that the computational time is guaranteed to be less than a polynomial expression in the dimensions of the problem and the number of digits of input data. Although polynomial, the bound he established turned out to be too high for his algorithm to be used to solve practical problems.

Karmarkar's algorithm (1984) was an important improvement on the theoretical result of Khachian that a linear program can be solved in polynomial time. Moreover, his algorithm turned out to be one that could be used to solve practical linear programs. As of this writing, interior algorithms are in open competition with variants of the Simplex Method. It appears likely that commercial software for solving linear programs will eventually combine pivot-type moves used in the Simplex Methods with interior type moves, especially for those problems with very few polyhedral facets in the neighborhood of the optimum.

Origins of Certain Terms

Here are some stories about how various linear-programming terms arose. The military refer to their various plans or proposed schedules of training, logistical supply, and deployment of combat units as a *program*. When I had first analyzed the Air Force planning problem and saw that it could be formulated as a system of linear inequalities, I called my first paper *Programming in a Linear Structure*. Note that the term "program" was used for linear programs long before it was used for the set of instructions used by a computer to solve problems. In the early days, these instructions were called *codes*.

In the summer of 1948, Koopmans and I visited the RAND Corporation. One day we took a stroll along the Santa Monica beach. Koopmans said "Why not shorten 'Programming in a Linear Structure' to 'Linear Programming'" I agreed: "That's it! From now on that will be its name." Later that same day I gave a talk at RAND entitled "Linear Programming"; years later Tucker shortened it to *Linear Program*.

The term *mathematical programming* is due to Robert Dorfman of Harvard, who felt as early as 1949 that the term linear programming was too restrictive.

The term *Simplex Method* arose out of a discussion with T. Motzkin, who felt that the approach I was using, when viewed in the geometry of the columns, was best described as a movement from one simplex to a neighboring one. A simplex is the generalization of a pyramid-like geometric figure to higher dimension. Mathematical programming is also responsible for many terms that are now standard in mathematical literature—terms like *Arg-Min, Arg-Max, Lexico-Max, Lexico-Min.* The term *dual* is an old mathematical term. But surprisingly, the term *primal* is new and was first proposed by my father, Tobias Dantzig, around 1954, after William Orchard-Hays stated the need for a shorter phrase to call the "original problem whose dual is..."

Summary of My Own Early Contributions

If I were asked to summarize my early and perhaps my most important contributions to linear programming, I would say they are three:

1. Recognizing (as a result of my wartime years as a practical program planner) that most practical planning relations could be reformulated as a system of linear inequalities.

2. Replacing ground rules for selecting good plans by general objective functions. (Ground rules typically are statements by those in authority of the means for carrying out the objective, not the objective itself.)

3. Inventing the Simplex Method which transformed the rather unsophisticated linear-programming model for expressing economic theory into a powerful tool for practical planning of large complex systems.

The tremendous power of the Simplex Method is a constant surprise to me. To solve by brute force the assignment problem that I mentioned earlier would require a solar system full of nanosecond electronic computers running from the time of the big bang until the time the universe grows cold to scan all the permutations in order to select the one that is best. Yet it takes only a moment to find the optimum solution using a personal computer and standard Simplex Method software.

In retrospect, it is interesting to note that the original class of problems that started my research is beginning to yield—namely the problem of planning or scheduling dynamically over time, particularly when there is uncertainty about the values of coefficients in the equations. If such problems could be successfully solved, it could eventually produce better and better plans and thereby contribute to the well-being and stability of the world.

The area of *planning under uncertainty* or *stochastic programming* has become a very exciting field of research and application, with research taking place in many countries. Some important long term planning problems have already been solved. Progress in this field depends on ideas drawn from many fields. For example, our group at Stanford is working on a solution method that combines the nested decomposition principle, importance sampling, and the use of parallel processors.

Prior to linear programming, it was not of any use to explicitly state general goals for planning systems (since such systems could not be solved) and so objectives were often confused with the ground rules in order to have a way of solving such systems. Ask a military commander what the goal is and he probably will say, "The goal is to win the war." Upon being pressed to be more explicit, a Navy man might say, "The way to win the war is to build battleships," or, if he is an Air Force general, he might say, "The way to win is to build a great fleet of bombers." Thus the *means to attain the objective becomes an objective in itself* which in turn spawns new ground rules as to how to go about attaining the means such as how best to go about building bombers or space shuttles. These *means* in turn become confused with *goals*, etc., down the line.

From 1947 on, the notion of what is meant by a goal has been adjusting to our increasing ability to solve complex problems. As we near the end of the twentieth century, planners are becoming more and more aware that it is possible to optimize a specific objective while at the same time hedging against a great variety of unfavorable contingencies that might happen and taking advantage of any favorable opportunity that might arise.

The ability to state general objectives and then be able to find optimal policy solutions to practical decision problems of great complexity is the revolutionary development I spoke of earlier. We have come a long way down the road to achieving this goal, but much work remains to be done, particularly in the area of uncertainty. The final test will come when we can solve the practical problems under uncertainty that originated the field back in 1947.

PREFACE

Linear programming and its generalization, mathematical programming, can be viewed as part of a great revolutionary development that has given mankind the ability to state general goals and lay out a path of detailed decisions to be taken in order to "best" achieve these goals when faced with practical situations of great complexity. The tools for accomplishing this are the *models* that formulate real-world problems in detailed mathematical terms, the *algorithms* that solve the models, and the *software* that execute the algorithms on computers based on the mathematical theory.

Our goal then is to provide a simple introduction to these various tools in the context of linear programming. Because this is an introduction to the field at the Undergraduate level, no proofs of key ideas are provided except for a few that are easy to follow. We prefer to state the key ideas as theorems and lemmas, rather than as facts or properties, as is done in some introductory texts because we wish to highlight the importance of the mathematical results. Examples are provided to illustrate the main ideas. Proofs of all the theorems and lemmas can be found in *Linear Programming 2*. Selected bibliographical references can be found at the end of each chapter.

We assume that the reader has some knowledge of elementary linear algebra. For those whose knowledge is weak or non-existent, necessary details on linear algebra used in this text are provided in the appendices.

OUTLINE OF CHAPTERS

Chapter 1 (The Linear Programming Problem): This chapter begins with a formal definition of the mathematical programming field and, in particular, formulation of the linear programming problem in mathematical terms so that it can be solved on a computer. A number of examples are formulated. We point out that most text book examples are simple in concept and small, making it is easy to represent them as system of linear inequalities (the *row-oriented approach*) and to solve on a computer. However, most real-life applications tend to be large and complex and are much easier to formulate and update in an activity (*column-oriented*) approach. We point out the advantages of viewing the problem from both the row and column perspectives. This chapter concludes with the assumptions (*axioms*) that the linear programming problem must approximately satisfy in practice.

Chapter 2 (Solving Simple Linear Programs): In the second chapter, we introduce methods for solving very simple linear programs. We start with the graphical solution of a two-variable linear program; here we also introduce the concept of a *dual linear program*, properties of which are developed in Chapter 5. Next we discuss how to solve a two-equation linear program graphically. This is followed by a simple exposition of the *Fourier-Motzkin Elimination* (FME) process for solving linear inequalites. It is a powerful theoretical tool that provides an easy proof of the important *Infeasibility Theorem* of Linear Programming. While the FME process is not practical for solving a large system of inequalities, it can be used to solve systems having a very small number of inequalities and variables.

Chapter 3 (The Simplex Method): Having introduced the basics, we are now ready to describe the *Simplex Algorithm*, which solves a linear program given a starting basic feasible solution for the standard form. The approach is first illustrated through examples. The *Simplex Method* is a two-phase process, with each phase using the Simplex Algorithm. In the first phase, an initial feasible solution, if one exists, is obtained. In the second phase, an optimal solution, if one exists, is obtained, or a class of solutions is obtained whose objective value goes to $+\infty$. Next, the solution method is extended to handle conveniently linear inequality systems with simple upper and lower bounds on the variables. Finally, the Revised Simplex Method, which is the Simplex Method in a form more suitable to large problems, is described.

Chapter 4 (Interior Point Methods): From the 1980s on there has been exponential growth of interest in solving linear programs using *interior-point methods.* We describe the *primal-affine algorithm* for historical reasons and because it is easy to understand. Details on this and other interior-point methods, including a summary of the state of the art as of 1996 in interior-point methods, can be found in *Linear Programming 2.*

Chapter 5 (Duality): The important details about the concept of duality is introduced through several examples. The Tucker diagram and von Neumann primal-dual systems are illustrated. *Weak Duality*, *Strong Duality*, the key theorems of duality, and the central results in duality, are presented and illustrated through examples. Formal proofs can be found in *Linear Programming 2.*

Chapter 6 (Equivalent Formulations): We take a slight detour and spend some time describing how a variety of problems encountered in practice can be reduced to linear programs. For example, if your software cannot handle problems whose variables are *unrestricted in sign*, such problems can be handled by splitting the variables into the difference of two nonnegative variables. Next we show how an *absolute value* in the objective can be modeled. *Goal* *programming* is discussed next followed by a discussion of how to obtain the *minimum of the maximum* of several linear functions. The next topic covered is *curve-fitting.* Finally we discuss how to use *piecewise linear approximations* to model convex functions.

Chapter 7 (Price Mechanism and Sensitivity Analysis: The chapter starts by dicussing the *price mechanism* of the Simplex Method. *Sensitivity analysis* is concerned with measuring the effect of changes in cost coefficients, the right-hand side, the matrix coefficients, or whether or not it is worthwhile to introduce additional rows and columns. Such analysis is an important aspect of the solution of any real-world problem.

Chapter 8 (Transportation and Assignment Problem): Network theory is introduced through a detailed discussion of the *classical transportation and assignment problems.* A specialized version of the *Simplex Method* for solving such problems is described and illustrated. The important property of *triangularity* of the basis, which simplifies solutions with *integer values*, is illustrated through simple examples.

Chapter 9 (Network Flow Theory): The ideas of the previous chapter are then extended to cover more general network flow problems. Standard network flow concepts such as *trees* (and their properties) are introduced. This is followed by a discussion on solving *maximal flow*, *shortest route*, and *minimum spanning tree* problems. Finally, the *Network Simplex Method* is described for solving the *minimum-cost network-flow* problem. It takes advantage of the tree stucture of the basis to greatly reduce the computations of each iteration.

Appendix A (Linear Algebra): This appendix summarizes all the key linear algebra concepts relevant to solving linear programs.

Appendix B (Linear Equations): This appendix discusses the theory for solving systems of linear equations. The theory for solving linear inequalities makes use of this theory.

SOFTWARE ACCOMPANYING THE BOOK

In modern times, the use of computers has become essential. We have designed the software that accompanies this book to work with the popular operating system Windows 95 as well as with Windows 3.1. The basic software algorithms and their many variations are designed to be powerful and numerically robust.

The software is fully integrated with the exercises. The use of it in conjunction with the text will help students understand key ideas behind the algorithms rather than simply number crunching. A feeling for the path followed by an algorithm to an optimal solution can probably best be gained by scanning the iterates .

To install the software:

- *For Windows 3.1.* Run setup.exe from the WIN31 directory on the enclosed CD-ROM.

- *For Windows 95.* Run setup.exe from the WIN95 directory on the enclosed CD-ROM.

Note: Before running setup please make sure that all programs are **closed** (and not just minimized); otherwise the installation may not execute correctly. If you still have problems with your installation please refer to the file README in the root directory of the CD-ROM.

Data for examples in the book can be found on the disk containing the software. To solve a problem using the software, each exercise in the book specifies which software option to use. For example: the DTZG Primal Simplex Method, Fourier-Motzkin Elimination, and so forth.

Linear Programming 2 and 3.

In a graduate course that we have taught together at Stanford, portions of "Linear Programming 1: Introduction" and "Linear Programming 2: Theory & Implementation" have been used. In addition some of the material in "Linear Programming 3: Structured LPs & Planning Under Uncertainty" has been used in seminars and in large-scale linear optimization.

Professor George B. Dantzig
Department of Operations Research
Stanford University
Stanford, CA 94305

Dr. Mukund N. Thapa
President
Stanford Business Software, Inc.
2680 Bayshore Parkway, Suite 304
Mountain View, CA 94043

DEFINITION OF SYMBOLS

The notation described below will be followed in general. There may be some deviations where appropriate.

- Uppercase letters will be used to represent matrices.
- Lowercase letters will be used to represent vectors.
- All vectors will be column vectors unless otherwise noted.
- Greek letters will typically be used to represent scalars.

$\Re^n$	–	Real space of dimension n.
c	–	Coefficients of the objective function.
A	–	Coefficient matrix of the linear program.
B	–	Basis matrix (nonsingular). Contains basic columns of A.
N	–	Nonbasic columns of A.
x	–	Solution of the linear program (typically the current one).
x_B	–	Basic solution (typically the current one).
x_N	–	Nonbasic solution (typically the current one).
(x, y)	–	The column vector consisting of components of the vector x followed by the components of y. This helps in avoiding notation such as $(x^T, y^T)^T$.
L	–	Lower triangular matrix with 1's on the the diagonal.
U	–	Upper triangular matrix (sometimes R will be used).
R	–	Alternative notation for an upper triangular matrix.
D	–	Diagonal matrix.
$\text{Diag}\,(d)$	–	Diagonal matrix. Sometimes $\text{Diag}\,(d_1, d_2, \ldots, d_n)$.
D_x	–	$\text{Diag}\,(x)$.
I	–	Identity matrix.

e_j	–	jth column of an identity matrix.
e	–	Vector of 1's (dimension will be clear from the context).
E_j	–	Elementary matrix (jth column is different from the identity).
$\|\|v\|\|$	–	The 2-norm of a vector v.
$\det(A)$	–	Determinant of the matrix A.
$A_{\bullet j}$	–	jth column of A.
$A_{i\bullet}$	–	ith row of A.
B^t	–	The matrix B at the start of iteration t.
$B[t]$	–	Alternative form for the matrix B^t.
$\bar{B}$	–	Update from iteration t to iteration $t+1$.
B^{-1}_{ij}	–	Element (i,j) of B^{-1}.
$X \subset Y$	–	X is a proper subset of Y.
$X \subseteq Y$	–	X is a subset of Y.
$X \cup Y$	–	Set union. That is, the set $\{\omega \mid \omega \in X \text{ or } \omega \in Y\}$.
$X \cap Y$	–	The set $\{\omega \mid \omega \in X \text{ and } \omega \in Y\}$.
$X \setminus Y$	–	Set difference. That is, the set $\{\omega \mid \omega \in X, \omega \notin Y\}$
$\emptyset$	–	Empty set.
$\mid$	–	Such that. For example $\{\, x \mid Ax \le b \,\}$ means the set of all x such that $Ax \le b$ holds.
α^n	–	A scalar raised to power n.
$(A)^n$	–	A square matrix raised to power n.
A^T	–	Transpose of the matrix A.
$\approx$	–	Approximately equal to.
$\gg$ $(\ll)$	–	Much greater (less) than.
$\succ$ $(\prec)$	–	Lexicographically greater (less) than.
$\leftarrow$	–	Store in the computer the value of the quantity on the right into the location where the quantity on the left is stored. For example, $x \leftarrow x + \alpha p$.
$O(v)$	–	Implies a number $\le kv$ where k, a fixed constant independent of the value of v, is meant to convey the the notion that k is some small integer value less than 10 (or possibly less than 100) and not something ridiculous like $k = 10^{100}$.
$\text{argmin}_x\, f(x)$	–	is the value of x where $f(x)$ takes on its global minimum value.
$\text{argmin}_i\, \beta_i$	–	is the value of the least index i where β_i takes on its minimum value.
LP	–	Linear program.

CHAPTER 1

THE LINEAR PROGRAMMING PROBLEM

Since the time it was first proposed by one of the authors (George B. Dantzig) in 1947 as a way for planners to set general objectives and arrive at a detailed schedule to meet these goals, linear programming has come into wide use. It has many nonlinear and integer extensions collectively known as the *mathematical programming* field, such as integer programming, nonlinear programming, stochastic programming, combinatorial optimization, and network flow maximization; these are presented in subsequent volumes.

Here then is a formal definition of the field that has become an important branch of study in mathematics, economics, computer science, and decision science (i.e., operations research and management science):

> *Mathematical programming (or optimization theory) is that branch of mathematics dealing with techniques for maximizing or minimizing an objective function subject to linear, nonlinear, and integer constraints on the variables.*

The special case, **linear programming**, has a special relationship to this more general mathematical programming field. It plays a role analogous to that of partial derivatives to a function in calculus—it is the first-order approximation.

> *Linear programming is concerned with the maximization or minimization of a linear objective function in many variables subject to linear equality and inequality constraints.*

For many applications, the solution of the mathematical system can be interpreted as a *program*, namely, a statement of the time and quantity of actions to be performed by the system so that it may move from its given status towards some defined objective.

Linear programming problems vary from small to large: The number of constraints less than 1,000 is considered "small," between 1,000 and 2,000 is considered "medium," and greater than 2,000 is considered "large." Linear programming models can be very large in practice; some have many thousands of constraints and variables. To solve large systems requires special software that has taken years to develop. Other special tools, called *matrix generators*, are often used to help organize the formulation of the model and direct the generation of the coefficients from basic data files. As the size of models that can be solved has grown, so has evolved the art of *model management*. These include, on the input side, model formulation and model updating, and, on the output side, summarizing of the detailed solution output in the form of graphs and other displays (so that the results may be more easily understood and implemented by decision makers).

1.1 SOME SIMPLE EXAMPLES

What follows are four very simple examples of typical linear programming problems; they happen to be similar to the very first applications of the field. The objective of the system in each happens to be the minimization of total costs or maximization of profits measured in monetary units. In other applications, however, the objective could be to minimize direct labor costs or to maximize the number of assembled parts or to maximize the number of trained students having a specified percentage distribution of skills, etc.

With the exception of the "on-the-job training" problem (Example 1.2), each of these examples is so small that the reader should have little difficulty expressing the problem in mathematical terms.

Example 1.1 (A Product Mix Problem) A furniture company manufactures four models of desks. Each desk is first constructed in the carpentry shop and is next sent to the finishing shop, where it is varnished, waxed, and polished. The number of man-hours of labor required in each shop and the number of hours available in each shop are known. Assuming that raw materials and supplies are available in adequate supply and all desks produced can be sold, the desk company wants to determine the optimal product mix, that is, the quantities to make of each type of desk that will maximize profit. This can be represented as a linear programming problem.

Example 1.2 (On-the-Job Training) A manufacturing plant is contracting to make some commodity. Its present work force is too small to produce the amount of the commodity required to meet the specified schedule of orders to be delivered each week for several weeks hence. Additional workers must therefore be hired, trained, and put to work.

The present force can either work and produce at some specified rate of output, or it can train some fixed number of new workers, or it can do both at the same time according to some fixed rate of exchange between output and the number of new workers trained. Even were the crew to spend one entire week training new workers, it would be unable to train the required number. The next week, the old crew and the newly trained workers may either work or train new workers, or may both work and train, and so on.

The commodity being produced is semiperishable so that any amount manufactured before needed will have to be stored at a cost. The problem is to determine the hiring, production, and storage program that will minimize total costs. This is a linear programming problem whose output is a *schedule of activities over time.*

Example 1.3 (The Homemaker's Problem) A family of five lives on the modest salary of the head of the household. A constant problem faced by the homemaker is to plan a weekly menu that reflects the needs and tastes of the family, the limited budget and the prices of foods. The husband must have 3,000 calories per day, the wife is on a 1,500 calorie reducing diet, and the children require 3,000, 2,700, and 2,500 calories per day, respectively.

According to the advice provided by a book on nutrition, these calories must be obtained for each member by foods having no more than a certain amount of fats and carbohydrates and not less than a certain amount of proteins. The diet, in fact, places emphasis on proteins. In addition, each member of the household must satisfy his or her daily vitamin needs. The problem is to assemble a menu each week that will minimize costs based on the current prices for food and subject to these criteria.

This type of linear programming problem, with some additional conditions specified to make the recommended diet more palatable, has been used to plan menus for patients in hospitals. An analogous formulation is used by the agricultural industry to determine the most economical feed mixes for cattle, poultry, and pet foods.

Example 1.4 (A Blending Problem) A type of linear programming problem frequently encountered is one involving blending. Typically a manufacturer wishes to form a mixture of several commodities that he can purchase so that the blend has known characteristics and costs the least. The percent characteristics of the blend are precisely specified.

A manufacturer wishes to produce an alloy (blend) that is 30 percent lead, 30 percent zinc, and 40 percent tin. Suppose there are on the market alloys $j = 1, \ldots, 9$ with the percent composition (of lead, zinc, and tin) and prices as shown in the display below. How much of each type of alloy should be purchased in order to minimize costs per pound of blend?

Alloy	1	2	3	4	5	6	7	8	9	Blend
Lead (%)	20	50	30	30	30	60	40	10	10	30
Zinc (%)	30	40	20	40	30	30	50	30	10	30
Tin (%)	50	10	50	30	40	10	10	60	80	40
Cost ($/lb)	7.3	6.9	7.3	7.5	7.6	6.0	5.8	4.3	4.1	Min

Obviously the manufacturer can purchase alloy 5 alone, but it will cost him $7.60 per pound. On the other hand with $\frac{1}{2}$ pound of alloy 2 and $\frac{1}{4}$ pound each of alloys 8 and 9 he will be able to blend a 30-30-40 mixture at a cost of $5.55 per pound. However, if he buys $\frac{1}{4}$ pound each of alloys, 6, 7, 8, and 9, he will also be able to blend a 30-30-40 mixture at a cost of $5.05. After a few trials of this sort, the manufacturer may well seek a more scientific approach to his problem.

The *quantities* of lead, zinc, and tin in the final blend have not been specified; only their proportions have been given, and it is required to minimize the cost per pound of the output. Often a beginner attempts to formulate the problem without restricting the total amount produced, in which case the material balance equations become difficult to interpret when expressed in terms of percentages instead of amounts.

We shall require that a definite amount of blended metal be produced. It is clear that the most economical purchasing plan for producing one pound of a specified blend can be immediately converted into the most economical purchasing plan for producing n pounds of output simply by multiplying the fractional amounts of each type of alloy by n; and thus we will restrict the *quantity of alloys to those combinations that produce one pound of specified blend of metal.* This stipulation has the further happy result that the percentage requirements of the original statement of the problem now become concrete: the mixture must contain 0.3 pounds of lead, 0.3 pounds of zinc, and 0.4 pounds of tin.

We shall formulate this model by writing down the material balance constraints. The decision variables are

$$x_j \geq 0, \quad j = 1, \ldots, 9,$$

where x_j is the fractional pounds of alloy j to be used in the blend.

There are five items (not four as may have been expected): one for each of the three components (lead, zinc, and tin) of the alloy, the cost of purchasing the alloy, and its weight. As per our discussion above, we shall solve the blending problem for producing exactly one pound of the blend. It is now clear that the problem to be solved is

Minimize the *Objective*

$$7.3x_1 + 6.9x_2 + 7.3x_3 + 7.5x_4 + 7.6x_5 + 6.0x_6 + 5.8x_7 + 4.3x_8 + 4.1x_9 = z$$

subject to

$$\begin{array}{rrrrrrrrrl}
x_1 + & x_2 + & x_3 + & x_4 + & x_5 + & x_6 + & x_7 + & x_8 + & x_9 & = 1 \\
.2x_1 + & .5x_2 + & .3x_3 + & .3x_4 + & .3x_5 + & .6x_6 + & .4x_7 + & .1x_8 + & .1x_9 & = .3 \\
.3x_1 + & .4x_2 + & .2x_3 + & .4x_4 + & .3x_5 + & .3x_6 + & .5x_7 + & .3x_8 + & .1x_9 & = .3 \\
.5x_1 + & .1x_2 + & .5x_3 + & .3x_4 + & .4x_5 + & .1x_6 + & .1x_7 + & .6x_8 + & .8x_9 & = .4
\end{array}$$

and $x_j \geq 0, \; j = 1, \ldots, 9.$

Only minor changes in the model are required in the event the blend specifications are not given precisely but they must lie between certain lower and upper bounds.

▷ **Exercise 1.1** Solve Example 1.4 numerically using the **DTZG Simplex Primal** software option. Find the amount of each type of alloy to purchase and find the minimum cost to produce one pound of the blend.

▷ **Exercise 1.2** Prove that any one of the above equations (excluding the objective) in Example 1.4 can be dropped as redundant.

Example 1.5 (A Transportation Problem) Suppose that a distributor has two canneries labeled 1 and 2, and three warehouses labeled a, b, and c in different geographical locations. The canneries can fill 250 and 450 cases of tins per day, respectively. Each of the warehouses can sell 200 cases per day. The distributor wishes to determine the number of cases to be shipped from the two canneries to the three warehouses so that each warehouse obtains as many cases as it can sell daily at the minimum total transportation cost. The availability of cases at the canneries and the demands which must be met exactly at each warehouse are summarized in the table below:

Cases Available		Cases Demanded	
Cannery	Cases	Warehouse	Cases
1	250	a	200
2	450	b	200
		c	200
Total	700	Total	600

The excess production of 100 cases will be stored without cost. The shipping cost per case from each cannery to each warehouse is given in the Shipping Cost schedule in the display below. The problem is to determine the number of cases that each cannery should ship to each warehouse in order to minimize the total transportation cost.

Shipping Cost ($/case)			
	Warehouses		
Canneries	a	b	c
1	3.4	2.2	2.9
2	3.4	2.4	2.5

We shall formulate this model by writing down the material balance constraints. The decision variables are

$$x_{ij} \geq 0, \quad i = 1, 2, \quad j = a, b, c, \tag{1.1}$$

where x_{ij} is the number of cases to ship from cannery $i = 1, 2$ to warehouse $j = a, b, c$. There are six items: dollars (associated with the cost of shipping), cases available at each of the two canneries, and cases demanded at each of the three warehouses.

The material balance constraints on *availability* are that the number of cases shipped out of each cannery cannot be greater than the number of cases available. Thus,

$$\begin{aligned} x_{1a} + x_{1b} + x_{1c} &\leq 250, \\ x_{2a} + x_{2b} + x_{2c} &\leq 450. \end{aligned} \tag{1.2}$$

The material balance constraints on *demand* are: The amount demanded at each warehouse must be equal to the amount shipped from each cannery to the warehouse. The problem specifies that the demand must be met exactly. Thus,

$$\begin{aligned} x_{1a} + x_{2a} &= 200, \\ x_{1b} + x_{2b} &= 200, \\ x_{1c} + x_{2c} &= 200. \end{aligned} \tag{1.3}$$

Finally, the cost to be minimized is set to an unspecified dollar amount z:

$$3.4x_{1a} + 2.2x_{1b} + 2.9x_{1c} + 3.4x_{2a} + 2.4x_{2b} + 2.5x_{2c} = z.$$

We consolidate below the mathematical constraints of the transportation example.

$$\begin{array}{l} \text{Minimize the } \textit{objective} \\ \quad 3.4x_{1a} + 2.2x_{1b} + 2.9x_{1c} + 3.4x_{2a} + 2.4x_{2b} + 2.5x_{2c} = z \\ \text{subject to} \\ \begin{array}{rrrrrrl} x_{1a} + & x_{1b} + & x_{1c} & & & & \leq 250 \\ & & & x_{2a} + & x_{2b} + & x_{2c} & \leq 450 \\ x_{1a} & & + & x_{2a} & & & = 200 \\ & x_{1b} & & + & x_{2b} & & = 200 \\ & & x_{1c} & & + & x_{2c} & = 200 \end{array} \\ \text{and } x_{1a} \geq 0,\ x_{1b} \geq 0,\ x_{1c} \geq 0,\ x_{2a} \geq 0,\ x_{2b} \geq 0,\ x_{2c} \geq 0. \end{array} \tag{1.4}$$

Properties of the Transportation Problem:

1. *Feasibility Theorem.* If the total availability is not less than the total demand, a solution always exists to (1.1), (1.2), and (1.3).

2. *Infeasibility Theorem.* If the total availability is less than the total demand no solution exists to (1.1), (1.2), and (1.3).

3. *Structure.* The transportation problem has a very special structure. Observe that all the input-output coefficients (excluding those of the objective) are either 1 or 0 with exactly two 1's per column. As a result, the transportation problem can be stored very compactly in a computer since we need to record only the cost coefficients, right-hand sides, and the locations of the coefficients that are 1. This compact storage property will be exploited in the algorithm presented in Chapter 8.

4. *Integer Property.* In a transportation problem, if all the availabilities and demands are positive integers and if the problem has a solution satisfying (1.1), (1.2), and (1.3), then we will show in Chapter 8 that it has at least one optimal solution in which all the variables x_{ij} have integer values.

   Note that the objective function can have only one optimal value; however, there could be many combinations of the variables x_{ij} that generate the same optimal value. If there is exactly one combination of the x_{ij} that generates the optimal value of the objective, the value of each x_{ij} must necessarily turn out to be an integer. If there is more than one combination of x_{ij} values that generate the optimal value of the objective, it can be shown that there are other integer solutions as well as other solutions in which x_{ij} can have noninteger values. All of these properties will also be shown in Chapter 8.

▷ **Exercise 1.3** Solve Example 1.5 numerically using the `DTZG Simplex Primal` software option. Find the optimal amount of shipment from each cannery to warehouse and the minimum cost of the shipments.

▷ **Exercise 1.4** As a way of illustration of the above Infeasibility Theorem, change the number of cases available at Cannery 1 to 100.

▷ **Exercise 1.5** Prove the above Feasibility and Infeasibility Theorems for (1.1), (1.2), and (1.3).

▷ **Exercise 1.6** Generalize the transportation problem to any number of origins (canneries) and any number of destinations (warehouses) and prove the Feasibility and Infeasibility Theorems for this system.

▷ **Exercise 1.7** Prove that if for the transportation problem (1.4) there is more than one optimal integer solution, then noninteger solutions can be found by forming certain weighted linear combinations of two integer solutions.

1.2 MATHEMATICAL STATEMENT

The mathematical definition of a *linear program in* **standard form** is to find values of $x_1 \geq 0$, $x_2 \geq 0, \ldots, x_n \geq 0$ and $\min z$ satisfying

$$\begin{array}{ccccccccc} c_1x_1 & + & c_2x_2 & + & \cdots & + & c_nx_n & = & z \text{ (Min)} \\ a_{11}x_1 & + & a_{12}x_2 & + & \cdots & + & a_{1n}x_n & = & b_1 \\ a_{21}x_1 & + & a_{22}x_2 & + & \cdots & + & a_{2n}x_n & = & b_2 \\ \vdots & & \vdots & & \vdots & & \vdots & & \vdots \\ a_{m1}x_1 & + & a_{m2}x_2 & + & \cdots & + & a_{mn}x_n & = & b_m. \end{array} \tag{1.5}$$

In vector-matrix notation we may restate the above as

$$\begin{array}{lrcll} \text{Minimize} & c^Tx & = & z & \\ \text{subject to} & Ax & = & b, & A: \; m \times n, \\ & x & \geq & 0. & \end{array} \tag{1.6}$$

The definition of a *dual of a linear program in standard form* is to find values of $\pi_1, \pi_2, \ldots, \pi_m$, and $\max v$ satisfying

$$\begin{array}{ccccccccc} b_1\pi_1 & + & b_2\pi_2 & + & \cdots & + & b_m\pi_m & = & v \text{ (Max)} \\ a_{11}\pi_1 & + & a_{21}\pi_2 & + & \cdots & + & a_{m1}\pi_m & \leq & c_1 \\ a_{12}\pi_1 & + & a_{22}\pi_2 & + & \cdots & + & a_{m2}\pi_m & \leq & c_2 \\ \vdots & & \vdots & & \vdots & & \vdots & & \vdots \\ a_{1n}\pi_1 & + & a_{2n}\pi_2 & + & \cdots & + & a_{mn}\pi_m & \leq & c_m. \end{array} \tag{1.7}$$

In vector-matrix notation we may restate the above as

$$\begin{array}{lrcll} \text{Maximize} & b^T\pi & = & v & \\ \text{subject to} & A^T\pi & \leq & c, & A: \; m \times n. \end{array} \tag{1.8}$$

Other definitions of a linear program, all equivalent to each other, are those of linear programs in inequality form, von Neumann symmetric form, and others that will be described later. For many applications it is easy to formulate the model as a system of equations and inequalities with possibly upper and lower bounds on the variables. In many large-scale applications one needs a formal procedure for organizing the basic data of the model and inputting it into the computer.

See Table 1-1 for a standard layout for linear programming data. It is called a *tableau.*

	Activity	(1)	...	(j)	...	(n)		
	Level	$x_1 \geq 0$	...	$x_j \geq 0$	...	$x_n \geq 0$		RHS
	(1)	a_{11}	...	a_{1j}	...	a_{1n}	=	b_1
I	(2)	a_{21}	...	a_{2j}	...	a_{2n}	=	b_2
t	...	...	...	...	...	...	...	...
e	(i)	a_{i1}	...	a_{ij}	...	a_{in}	=	b_i
m	...	...	...	...	...	...	...	...
	(m)	a_{m1}	...	a_{mj}	...	a_{mn}	=	b_m
	Cost	c_1	...	c_j	...	c_n	=	z

Table 1-1: Tableau of Detached Coefficients for a Typical LP

1.3 FORMULATING LINEAR PROGRAMS

Computers are now being applied to almost every aspect of human activity. Every field of science, medicine, engineering, business—you name it—is being computerized in some way. However, before you can put a problem into a computer and efficiently find a solution, you must first abstract it, which means you have to build a mathematical model.

It is the process of abstracting applications from every aspect of life that has given rise to a vast new world of mathematics that has developed for the most part outside mathematics departments. This mathematics, you will see, is just as interesting and exciting as any mathematics that is taught in the standard courses, perhaps more so because it is still new and challenging.

> *The mathematical model of a system is the collection of mathematical relationships which, for the purpose of developing a design or plan, characterize the set of feasible solutions of the system.*

The process of building a mathematical model is often considered to be as important as solving it because this process provides insight about how the system works and helps organize essential information about it. *Models of the real world are not always easy to formulate because of the richness, variety, and ambiguity that exists in the real world or because of our ambiguous understanding of it.* Nevertheless, it is possible to state certain principles that distinguish the separate steps in the model-building process when the system can be modeled as a linear program.

The *linear programming problem* is to determine the values of the variables of the system that (a) are nonnegative or satisfy certain bounds, (b) satisfy a system of linear constraints, and (c) minimize or maximize a linear form in the variables called an objective.

There are two general ways in which we can formulate a problem as a linear program: the column (recipe/activity) approach and the row (material balance) approach. Both ways result in the same final model; the approach you take will depend primarily on how you like to think about and organize the data for the problem.

In certain situations, it is convenient for the modeler to view the system as (i) a collection of activities or processes that may be engaged in rather than (ii) a collection of statements about limitations on the use of scarce resources. As we will see, there are points in common between these two seemingly quite different ways of viewing the system. Indeed, there are benefits to be gained by viewing the system both ways and this is recommended. We shall describe both the approaches and illustrate them through various examples.

1.3.1 THE COLUMN (RECIPE/ACTIVITY) APPROACH

The column approach is to consider a system as decomposable into a number of elementary functions, the *activities*. An activity is thought of as a kind of "black box" into which flow tangible inputs, such as men, material, and equipment, and out of which flow final or intermediate products of manufacture, or trained personnel. An activity is analogous to a recipe in a cookbook. What happens to the inputs inside the "box" is the concern of the engineer in the same way as what chemistry takes place in the cookpot is the concern of a chemist; to the decision maker, only the rates of flow into and out of the activity are of interest. The various kinds of flow are called *items*.

The quantity of each activity is called its *activity level*. To change the activity level it is necessary to change the quantity of each kind of flow into and out of the activity. In linear programming the activity levels are not given but are the decision variables to be determined to meet certain specified requirements.

The steps for formulating a linear program by the column approach are as follows.

Step 1 *Define the Activity Set.* Decompose the entire system under study into all of its elementary functions, the *activities* or *processes* and choose a unit for each type of activity or process in terms of which its quantity or *level* can be measured. For example, manufacturing a desk is an activity. It is defined for the purpose of developing a plan for the *recipe* of items needed to produce one desk. The number of desks manufactured is the level of the activity, which is the decision variable to be determined.

Activity levels are usually denoted by $x_1, x_2, x_3, \ldots$, where x_j is the level of activity j.

Step 2 *Define the Item Set.* Determine the classes of objects, the *items*, that are required as inputs or are produced as outputs by the activities, and choose a unit for measuring each type of item. Obviously the only items that need be considered are those that are potential bottlenecks. Select one item such that the net quantity of it produced by the system as a whole measures the "cost" (or such that its negative measures the "profit" of the entire system). For example, *time in the carpentry shop*, measured in hours, is an item. *Time in the finishing shop*, measured in hours, is a different item, and, *money* is another item, measured in dollars. The negative of the price in dollars at which a desk is sold affects the profit of selling a desk.

In many situations, "costs" are measured in terms of money; however, in other economic situations, they could be measured in terms of labor or any scarce resource whose input is to be conserved or any item whose total output from the system is to be maximized.

The label i is usually used to refer to the type of item consumed or produced by the activities. The role that items play will become clearer in the next step.

Step 3 *Define the Input-Output Coefficients.* Determine the quantity of each item consumed or produced by the operation of each activity at its unit level. These numbers are analogous to the quantities of various ingredients in a cookbook recipe and are called the ***input-output coefficients*** of the activity. They are the factors of proportionality between activity levels and item flows.

The input-output coefficients are usually denoted by a_{ij}, where i refers to the item and j refers to the activity. For example, when manufacturing desks, a_{ij} could be the amount of time in shop i required to manufacture one desk j. If a_{ij} of item i is required by activity j enter it in column j with a plus sign; if it is produced by activity j enter it in column j with a negative sign. Often in economic applications the opposite sign convention for entering is used. The sign convention is arbitrary as long as it is kept consistent. Every item that is either required or produced by an activity j is entered in column j of the tableau.

Step 4 *Specify the Exogenous Flows.* Everything outside the system is called *exogenous.* Specify the exogenous amounts of each item being supplied from the outside to the system as a whole and specify the exogenous amounts required by the outside from the system as a whole. They are usually denoted by b_i for item i and are entered in the rightmost tableau column. Each of these, by our additivity assumption, is equal to the net of the total amounts of each item used by the activities less the total amounts of each item produced by the activities.

These net quantities item by item balance out to the exogenously given right-hand sides of the material balance equations described next.

Step 5 *Set Up the Material Balance Equations.* Assign unknown activity levels $x_1, x_2, x_3, \ldots$, usually nonnegative, to all the activities. Then, for each item, one can easily write the *material balance equation* by referring to the tableau which asserts that the algebraic sum of the flows of that item into each activity (given as the product of the activity levels on the top row by the appropriate input-output coefficients a_{ij}) is equal to the exogenous flow of the item.

There could be a surplus and shortage of items. These should be kept in mind and appropriate surplus and shortage activities should be included. If no costs are associated with the surplus or shortage amount then we could write the constraint as an inequality instead of an equality. However, if the modeler wishes to force the solution not to have any deficit or surplus (or wishes to be sure that all costs, or penalties, associated with a shortage or revenues gained from selling off a surplus are accounted for), then the relation would be written as shown in Table 1-1 as an equation.

The activity approach as defined requires setting up all the activities to be nonnegative and all the constraints (material balances) to be specified as equalities. Hence we will probably not always be successful in completing the model in the first sequence of steps. It frequently happens that certain activities (referred to as slack activities), commonly those related to the disposal of unused resources or the overfulfillment of requirements, are overlooked until the formulation of the material balance equations forces their inclusion. Thus a return from Step 5 to Step 1 will sometimes be necessary before the model is complete.

1.3.2 THE ROW (MATERIAL BALANCE) APPROACH

For many modelers the natural way to set up a linear programming model is to state directly the material balance relations in terms of the decision variables. The steps are as follows.

Step 1 *Define the Decision Variables.* This step is similar to that for the activity approach. Define all the decision variables, that is variables that represent the quantity to produce, buy, etc. For example, the number of desks of type 1 to manufacture is a decision variable. Recall that manufacturing a desk is an activity, and the number of desks manufactured is the level of this activity.

Decision variables are usually denoted by $x_1, x_2, x_3, \ldots,$ where x_j is the number of desks of type j to manufacture.

Step 2 *Define the Item Set.* As in the column approach determine the classes of objects, the *items*, that are considered to be potential bottlenecks and choose a unit for measuring each type of item. See Step 2 of the activity approach for details.

The label i is usually used to refer to a type of item.

Step 3 *Set Up Constraints and the Objective Function.* For each item, write down the constraints associated with the bottleneck by noting how much of each item is used or produced by a unit of each decision variable x_j. This amounts to filling a row of the tableau shown in Table 1-1.

This results in a system of *material balance inequalities* (or material balance equations) depending on whether or not a shortage or surplus of an item is allowed. Next write down the objective function which is formed by multiplying each decision variable by its unit cost (or negative unit profit) and summing.

1.4 EXAMPLES OF MODEL FORMULATION

1.4.1 PRODUCT MIX (COLUMN APPROACH)

We next describe how to formulate the Product Mix Problem described earlier by the Column Approach.

A furniture company manufactures four models of desks. Each desk is first constructed in the carpentry shop and is next sent to the finishing shop, where it is varnished, waxed, and polished. The number of man hours of labor required in each shop is as shown in the display below.

	Desk 1 (hrs)	Desk 2 (hrs)	Desk 3 (hrs)	Desk 4 (hrs)	Available (hrs)
Carpentry Shop	4	9	7	10	6,000
Finishing Shop	1	1	3	40	4,000

Because of limitations in capacity of the plant, no more than 6,000 man hours can be expected in the carpentry shop and 4,000 in the finishing shop in the next six months. The profit (revenue minus labor costs) from the sale of each item is as follows:

	Desk 1	Desk 2	Desk 3	Desk 4
Profit	\$12	\$20	\$18	\$40

Assuming that raw materials and supplies are available in adequate supply and all desks produced can be sold, the desk company wants to determine the optimal product mix, that is, the quantities to make of each type product which will maximize profit.

Step 1 *The Activity Set.* The four manufacturing activities, each of which are measured in desks produced, are

1. Manufacturing Desk 1.
2. Manufacturing Desk 2.
3. Manufacturing Desk 3.
4. Manufacturing Desk 4.

There are other activities, but these will be discussed later.

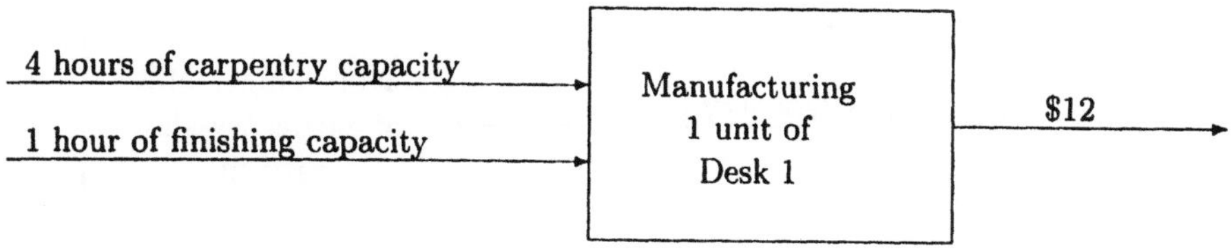

Figure 1-1: Manufacturing Activity 1

Activities	Manufacturing Desks			
Items	1	2	3	4
1. Carpentry capacity (hours)	4	9	7	10
2. Finishing capacity (hours)	1	1	3	40
3. Cost (−Profit) ($)	−12	−20	−18	−40

Table 1-2: Input-Output Coefficients

Step 2 *The Item Set.* The items are

1. Capacity in Carpentry Shop (measured in man hours).
2. Capacity in Finishing Shop (measured in man hours).
3. Costs (measured in dollars).

Step 3 *The Input-Output Coefficients.* Manufacturing activity 1, for example, can be diagramed as shown in Figure 1-1. The table of input-output coefficients for the four manufacturing activities is shown in Table 1-2.

Step 4 *Exogenous flows.* Since capacities in carpentry and finishing are inputs to each of these activities, they must be inputs to the system as a whole. At this point, however, we must face the fact that a feasible program need not use up all of this capacity. The total inputs must not be more than 6,000 carpentry hours and 4,000 finishing hours, but they can be less, and so cannot be specified precisely in material balance equations.

Step 5 *Material balances.* If we went ahead with the formulation anyway, using this data for the exogenous flows, then in order to have a correct mathematical formulation, we would have to write the material balances as inequalities instead of equations. For example, the carpentry capacity limitation is

$$4x_1 + 9x_2 + 7x_3 + 10x_4 \leq 6000,$$

which is not in accordance with our rules for the activity approach.

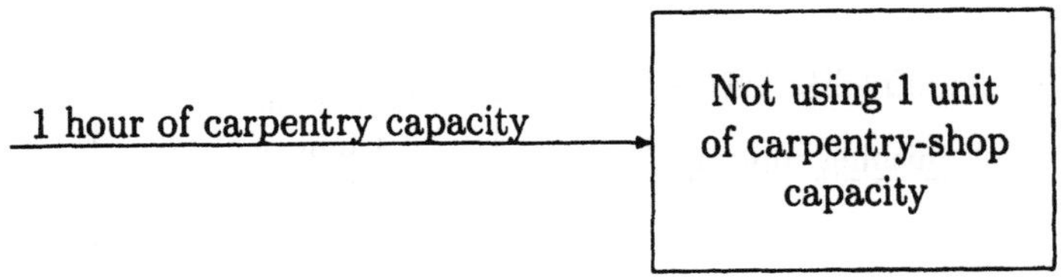

Figure 1-2: Slack Activity 5

Activities	Manufacturing Desks				Slack		Exogenous
Items	x_1	x_2	x_3	x_4	x_5	x_6	
1. Carpentry capacity (hrs)	4	9	7	10	1		6,000
2. Finishing capacity (hrs)	1	1	3	40		1	4,000
3. Cost ($)	−12	−20	−18	−40			z (Min)

Table 1-3: Full Tableau: Product Mix Problem

We see that the model cannot be completed with the lists of activities and items given above, and we have here the situation mentioned in the first section in which a second pass at the initial building of the model is necessary. In this instance all we need to do is add activities to the model that account for the carpentry and finishing capacity not used by the remainder of the program. If we specify "not using capacity" as an activity, we have the two additional activities, called *slack* activities, to add to those listed in Step 1:

5. Not Using Carpentry Shop Capacity (measured in man hours).
6. Not Using Finishing Shop Capacity (measured in man hours).

Activity 5 can be abstracted as diagramed in Figure 1-2. The full tableau of inputs and outputs of the activities and the exogenous availabilities to the system as a whole are shown in Table 1-3.

Thus the linear programming problem is to determine the numbers

$$x_1 \geq 0,\ x_2 \geq 0,\ x_3 \geq 0,\ x_4 \geq 0,\ x_5 \geq 0,\ x_6 \geq 0,$$

and minimum z satisfying

$$\begin{array}{rcl} -12x_1 - 20x_2 - 18x_3 - 40x_4 & = & z \\ 4x_1 + 9x_2 + 7x_3 + 10x_4 + x_5 & = & 6000 \\ x_1 + x_2 + 3x_3 + 40x_4 + x_6 & = & 4000. \end{array}$$

Note that the same values of the x_j's that minimize the cost function will also maximize the profit function p given by

$$12x_1 + 20x_2 + 18x_3 + 40x_4 = p.$$

Thus, a profit maximization problem can be stated as an equivalent to a cost minimization problem. It is obtained by reversing the sign of the coefficients of the objective function of the cost minimization problem.

▷ **Exercise 1.8** Solve the product mix problem numerically using the **DTZG Simplex Primal** software option. Find the optimal amount of each type of desk to manufacture and the maximum profit obtained by manufacturing these amounts.

1.4.2 PRODUCT MIX (ROW APPROACH)

We next describe how to formulate the product mix problem described earlier by the row approach.

Step 1 *Define the Decision Variables.* The decision variables are how many desks to manufacture of each type. Let x_j = the number of desks j to manufacture per month, for $j = 1, 2, 3, 4$. Associated with each of these variables x_j is the activity of manufacturing a desk. With the column approach described in the previous section, only these activities were defined in the first step.

Step 2 *Define the Item Set.* As with the column approach, the items are

1. Capacity in Carpentry Shop (measured in man hours).
2. Capacity in Finishing Shop (measured in man hours).
3. Costs (measured in dollars).

Step 3 *Set Up Constraints and the Objective Function.* The cost item leads to the objective function to be minimized:

$$z = -12x_1 - 20x_2 - 18x_3 - 40x_4.$$

The two capacity items each lead to inequality constraints. Manufacturing one unit of desk 1, one unit of desk 2, one unit of desk 3, and one unit of desk 4 requires 4 hours, 9 hours, 7 hours, and 10 hours respectively of carpentry capacity. The total carpentry capacity cannot exceed 6,000 hours per month. Thus, the material balance inequality for the carpentry item is

$$4x_1 + 9x_2 + 7x_3 + 10x_4 \le 6000.$$

In a similar manner, we can write down the constraint for the finishing shop as

$$1x_1 + 1x_2 + 3x_3 + 40x_4 \le 4000.$$

Thus, the linear programming problem is to determine the numbers

$$x_1 \geq 0, x_2 \geq 0, x_3 \geq 0, x_4 \geq 0,$$

and minimum z satisfying

$$\begin{aligned} -12x_1 - 20x_2 - 18x_3 - 40x_4 &= z \\ 4x_1 + 9x_2 + 7x_3 + 10x_4 &\leq 6000 \\ x_1 + x_2 + 3x_3 + 40x_4 &\leq 4000 . \end{aligned}$$

1.4.3 A SIMPLE WAREHOUSE PROBLEM

Consider the problem of stocking a warehouse with a commodity for sale at a later date. The warehouse can stock only 100 units of the commodity. The storage costs are $1.00 per quarter year for each unit. In each quarter the purchase price equals the selling price. This price varies, however, from quarter to quarter as shown in the display below.

Quarter (t)	*Price ($/Unit)*
1	10
2	12
3	8
4	9

Assuming that the warehouse has an initial stock of 50 units, this suggests that a profit may be realized by selling when the price is high and buying when the price is low. The problem is to determine the optimal selling, storing, and buying plan for a one-year period by quarters.

In each period (quarter) t, we distinguish four types of activities:

Activity	*Quantity*
1. Selling Stock	x_{t1}
2. Storing Stock	x_{t2}
3. Buying Stock	x_{t3}
4. Not Using Capacity (slack)	x_{t4}

and three types of items:

Items
1. Stock
2. Storage Capacity
3. Costs

These activities have the input-output characteristics shown in Figure 1-3 for a typical time period t.

With four quarters, each item and activity appears four times in Table 1-4, the tableau for the warehouse problem, once each quarter with a different time subscript. The problem here is to find the values of $x_{ti} \geq 0$ which satisfy the equations implied by the tableau and which minimize the total cost.

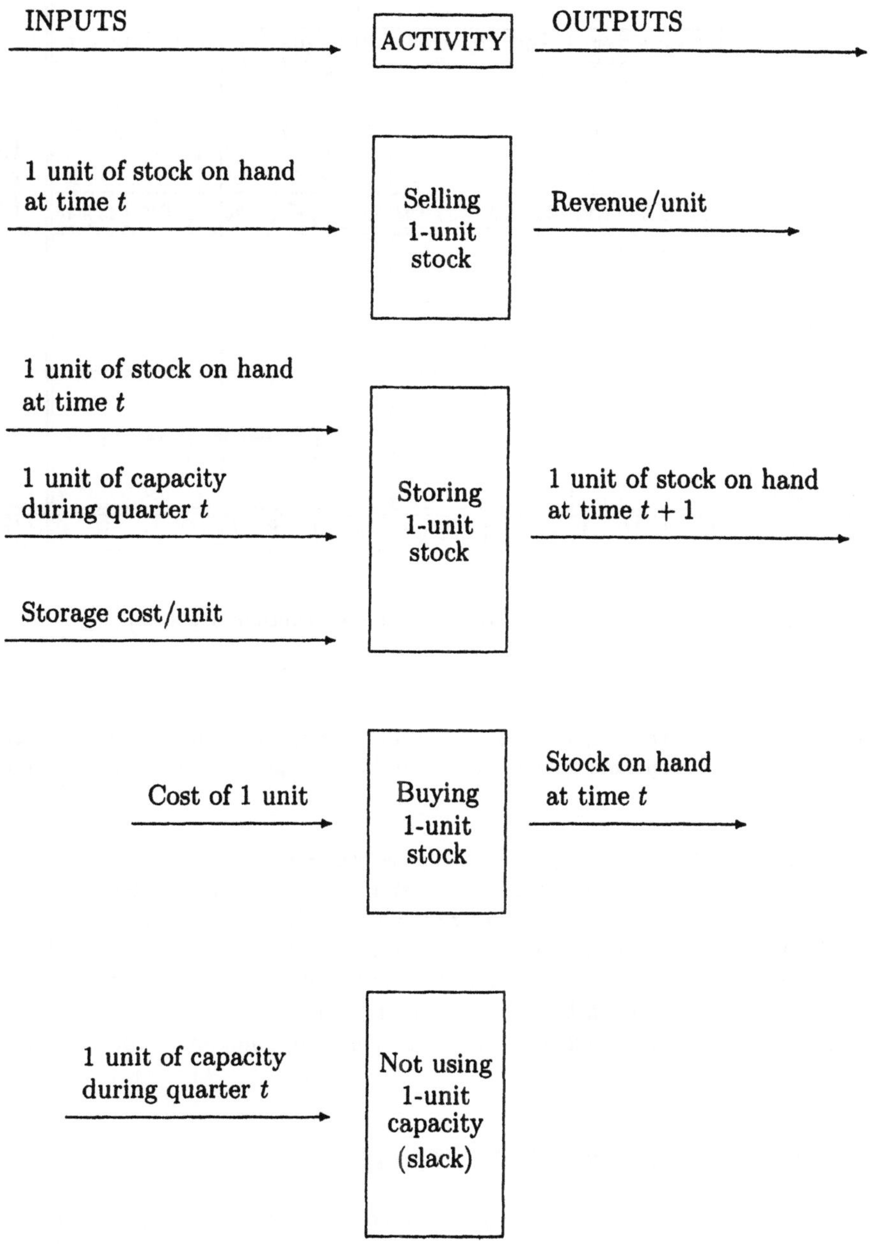

Figure 1-3: Input-Output Characteristics for the Warehouse Problem

	Activities																
	1st Quarter				2nd Quarter				3rd Quarter				4th Quarter				
	Sell	Store	Buy	Slack	Sell	Store	Buy	Slack	Sell	Store	Buy	Slack	Sell	Store	Buy	Slack	Exog-enous
Items	x_{11}	x_{12}	x_{13}	x_{14}	x_{21}	x_{22}	x_{23}	x_{24}	x_{31}	x_{32}	x_{33}	x_{34}	x_{41}	x_{42}	x_{43}	x_{44}	Flows
$t = 0$ Stock	1	1	−1														50
Capac.		1		1													100
$t = 1$ Stock		−1			1	1	−1										0
Capac.						1		1									100
$t = 2$ Stock						−1			1	1	−1						0
Capac.										1		1					100
$t = 3$ Stock										−1			1	1	−1		0
Capac.														1		1	100
Cost	−10	1	10	0	−12	1	12	0	−8	1	8	0	−9	1	9	0	z (Min)

Table 1-4: Tableau Form for the Warehouse Problem

▷ **Exercise 1.9** Solve the simple warehouse problem using the `DTZG Simplex Primal` software option. Find the optimal selling, storing, and buying policy and associated total cost.

▷ **Exercise 1.10** Consider the *cyclic* warehouse problem, where the 4 quarters of each year are followed by four quarters of next year for year after year indefinitely into the future. Assume the levels of corresponding activities in different years in the same season repeat. Further assume that all the data with respect to costs, selling price, and capacity are the same. Instead of having an initial stock of 50 units on hand suppose the problem is to determine the ideal stock level to have on hand at the start of each year so that the net profit per unit is maximized. Formulate the linear programming model to be solved.

1.4.4 ON-THE-JOB TRAINING

The purpose of this example is to illustrate the ability of the linear programming model to cover the many and varied conditions that are so characteristic of practical applications.

A manufacturing plant has a contract to produce 1,500 units of some commodity, C, with the required delivery schedule r_t as shown in the display below.

End of Week	1	2	3	4	5
No. of units	$r_1 = 100$	$r_2 = 200$	$r_3 = 300$	$r_4 = 400$	$r_5 = 500$

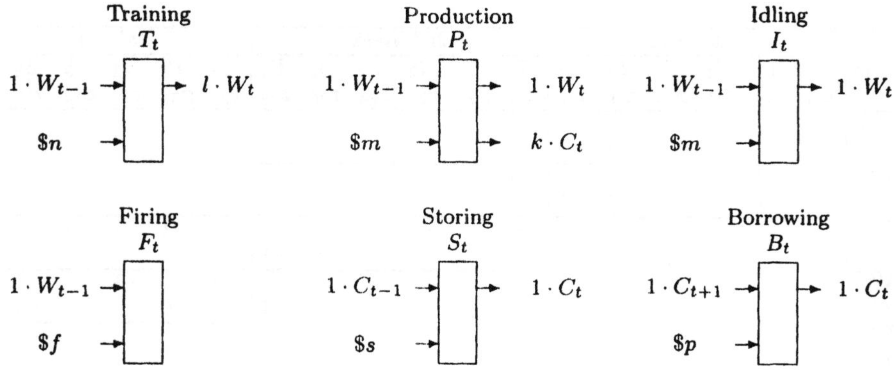

Figure 1-4: Activities for the On-the-Job Training Problem

What hiring, firing, producing, and storing schedule should the manufacturer adopt to minimize the cost of his contract under the following conditions?

1. Each unit of production not delivered on schedule involves a penalty of $p = \$90$ per week until delivery is effective.
2. Any production ahead of schedule requires storage at $s = \$30/\text{unit}/\text{week}$.
3. All required deliveries must be met by the end of the fifth week.
4. Initially there are $g = 20$ workers and $h = 10$ units of C on hand.
5. Each worker used in production during a week can turn out $k = 8$ units of C.
6. Each worker used for training recruits during a week can train $l - 1 = 5$ new workers (that is, produce $l = 6$ trained workers including himself).
7. Wages of a worker are $m = \$300/\text{week}$ when used in production or when idle.
8. Wages of a worker plus $l - 1$ recruits used in training for one week are $n = \$1,800$.
9. The cost to fire one worker is $f = \$300$.

We shall choose for our unit of time a period of one week. At the beginning of each week we shall assign the necessary number of workers and units of C to carry out an activity that takes place during the week. Accordingly, at each of the six times $t = 0, 1, \ldots, 5$, material balance equations for the two items named in the display below will need to be set up:

Type of Item	*Symbol for Item*
Workers	W_t
Commodity	C_t

Item	1st Week						2nd Week						3rd Week					
	T_1	P_1	I_1	F_1	S_1	B_1	T_2	P_2	I_2	F_2	S_2	B_2	T_3	P_3	I_3	F_3	S_3	B_3
	x_{11}	x_{12}	x_{13}	x_{14}	x_{15}	x_{16}	x_{21}	x_{22}	x_{23}	x_{24}	x_{25}	x_{26}	x_{31}	x_{32}	x_{33}	x_{34}	x_{35}	x_{36}
W_0	1	1	1	1														
C_0					1													
W_1	$-l$	-1	-1				1	1	1	1								
C_1		$-k$			-1	-1					1							
W_2							$-l$	-1	-1				1	1	1	1		
C_2						1		$-k$			-1	-1					1	
W_3													$-l$	-1	-1			
C_3												1		$-k$			-1	-1
W_4																		
C_4																		1
W_5																		
C_5																		
Cost	n	m	m	f	s	p	n	m	m	f	s	p	n	m	m	f	s	p

Table 1-5: The Job Training Model (First Three Weeks)

In addition to equations of these types, there will be a cost equation for the *cost item*. In each of five weekly periods, six types of activities named in the display below will need to be set up.

Type of Activity	*Symbol for Activity*
1. Training	T_t
2. Producing	P_t
3. Idling	I_t
4. Firing	F_t
5. Storing	S_t
6. Borrowing	B_t

The input-output characteristics of each of these activities are displayed in Figure 1-4. Except perhaps the borrowing activity, they are straightforward. Each failure to produce enough of commodity C makes it necessary to borrow one unit of commodity C in period t from a competitor and to return one unit to the competitor in the next time period at a penalty cost of p dollars.

These activities are shown in conventional tableau form in Table 1-5. In the fifth week the borrowing activity is omitted because condition (3) on page 19 states that all deliveries must be met by the end of the fifth week. In the sixth week a firing activity F_6 has been introduced to get rid of all workers and to terminate the program.

▷ **Exercise 1.11** Why is it necessary to terminate the program in this manner?

Item	4th Week						5th Week						Exog-enous Flows
	T_4	P_4	I_4	F_4	S_4	B_4	T_5	P_5	I_5	F_5	S_5	F_6	
	x_{41}	x_{42}	x_{43}	x_{44}	x_{45}	x_{46}	x_{51}	x_{52}	x_{53}	x_{54}	x_{55}	x_{64}	
W_0													g
C_0													h
W_1													0
C_1													$-r_1$
W_2													0
C_2													$-r_2$
W_3	1	1	1	1									0
C_3					1								$-r_3$
W_4	$-l$	-1	-1				1	1	1	1			0
C_4		$-k$			-1	-1					1		$-r_4$
W_5							$-l$	-1	-1			1	0
C_5						1		$-k$			-1		$-r_5$
Cost	n	m	m	f	s	p	n	m	m	f	s	f	z (Min)

Table 1-6: The Job Training Model (Continued)

▷ **Exercise 1.12** Assuming that firing is the opposite of hiring, give reasons why it is better to treat these as two nonnegative activities rather than as a single activity with positive and negative activity levels.

▷ **Exercise 1.13** Solve the simple job training model numerically using the **`DTZG Simplex Primal`** software option. Find the optimal hiring, firing, and storing schedule that the manufacturer should adopt.

1.5 BOUNDS

In a linear program in standard form the levels of the activities are nonnegative. In many real-world problems the levels of the activities are between bounds.

NONNEGATIVITY

Typically, in linear programming models, the levels of activities are nonnegative. For example, it is not possible to train a negative number of workers or to combine negative quantities of food items to determine the optimal diet. A subtle example of nonnegativity occurs in a well-known classic: the Mad Hatter, you may recall, in *Alice's Adventures in Wonderland*, was urging Alice to have some more tea, and Alice was objecting that she couldn't see how she could take more when she hadn't had any. The hatter replied: "You mean, you don't see how you can take *less* tea. It is very easy to take more than nothing."

Lewis Carroll, the author, was a mathematician, and his point was probably lost on his pre-linear-programming audience, for why should one emphasize the obvious fact that the activity of "taking tea" cannot be done in negative quantity? Perhaps it was Carroll's way of saying that mathematicians had been so busy for centuries extending the number system, from integers to fractions to negative to imaginary numbers, that they had forgotten the art of keeping the variables of their problems in their original nonnegative range. This characteristic of the variables of the linear programming model in most situations is known as the *nonnegativity assumption*. In linear programs, nonnegativity restrictions on variables are denoted by $x_j \geq 0$.

UPPER AND LOWER BOUNDS

Sometimes an activity level is required to be not less than some quantity called a *lower bound*. This bound may be positive or negative. There may be other restrictions on the variables as well, such that they cannot exceed a certain quantity called an *upper bound*. For activity j, this can be represented by $l_j \leq x_j \leq u_j$.

In some of applications of linear programs, variables may be allowed to have negative values. For example, in financial applications, there may be no restriction on the sign of the level of an activity measuring cash flow. In certain situations it may even be advantageous for computational reasons to restrict certain variables to always be nonpositive or to allow certain variables to be temporarily negative.

1.6 AXIOMS

A linear programming model satisfies certain assumptions (or axioms), namely proportionality, additivity, and continuity. Other types of mathematical programs do not satisfy these, for example, integer program models do not satisfy the axiom of continuity.

PROPORTIONALITY

For example, suppose 1 slice of bread provides 77.5 calories; if the number of slices is doubled it provides 155 calories. That is, in the linear programming model the quantities of flow of various items into and out of the activity are always proportional to the activity level. The ingredients to make two loaves of bread are double those for one loaf. If we wish to double the activity level, we simply double all the corresponding flows for the unit activity level.

In general, the *proportionality assumption* implies that if a_{ij} units of the ith item are required by 1 unit level of the jth activity, then x_j units of the jth activity require $a_{ij}x_j$ units of item i. The proportionality assumptions also implies that if it costs c_j to buy 1 unit level of the jth activity then it costs c_jx_j to buy x_j units of the jth activity.

ADDITIVITY

For example, if 2 slices of bread provide 155 calories and a boiled egg provides 80 calories, then 235 calories are provided by eating 2 slices of bread and 1 boiled egg.

In general, the *additivity assumption* implies that if a_{ij} units of the ith item are provided by 1 unit of the jth activity and a_{ik} units of the ith item are provided by 1 unit of the kth activity then $a_{ij}x_j + a_{ik}x_k$ units of the ith item are provided by x_j units of the jth activity and x_k units of the kth activity. The additivity assumption also implies that if it costs c_j to buy 1 unit of the jth activity and costs c_k to buy 1 unit of the kth activity, then it costs $c_jx_j + c_kx_k$ to buy x_j units of the jth activity and x_k units of the kth activity. That is, the additivity assumption implies that the objective function is additively *separable* in the variables; there are no mixed variable terms like $c_{kj}x_kx_j$.

CONTINUITY

The activity levels, or variables, can take on any real values within their allowable range. Thus, if a problem requires that some activity level must take on one of a finite set of values (such as a discrete number of real or integer values), the problem cannot be represented as a linear program. Such problems can be reformulated as integer programs, which, in general, belong to a class of problems that have been shown to be much harder to solve than linear programming problems.

1.7 NOTES & SELECTED BIBLIOGRAPHY

From the time that the Simplex Method was first proposed by George Dantzig in 1947, applications and new theories have grown at an astounding rate. They have grown so rapidly that it is not possible to treat every aspect of linear programming and extensions here. In fact, the early funding of the development of computers was done to make it possible to solve linear programs (see the Foreword)! For another brief history of linear programming see Orden [1993]. A classic book on linear programming is due to Dantzig [1963]. For a history of mathematical programming, see Lenstra, Rinnooy Kan, & Schrijver [1991].

Since the early 1950s many areas that we collectively call mathematical programming began to emerge. These subfields have all grown rapidly, with linear programming playing a fundamental role in their development. They are briefly described in the foreword and are: nonlinear programming, commercial applications, software, network flow theory, large scale methods, stochastic programming, integer programming, complementary pivot theory, computational complexity, and polynomial time algorithms.

One of the first known applications of the Simplex Algorithm was the determination of an adequate diet that was of least cost. J. Cornfield, of the U.S. government, formulated such a mathematical model in 1940. Later, in the fall of 1947, J. Laderman, of the Mathematical Tables Project of the National Bureau of Standards, undertook as a test of the newly proposed Simplex Method, what was the first large scale computation in this field. It was a system with 9 equations and 77 variables. This problem took approximately 120 man-days to solve using hand-operated desk calculators. Today such a problem is considered tiny and can be solved in a matter of seconds on a personal computer! This particular problem was one that had been studied by G.J. Stigler [1945], who had determined a nonoptimal solution by selecting a handful of food combinations to be examined in an attempt to reduce the annual cost.

Among other early applications of linear programming were these: scheduling job shop production (Dantzig [1957a], Jackson [1957], and Salveson [1953]); applications to the oil industry (for example, Manne [1956], Charnes, Cooper, & Mellon [1952], Garvin, Crandall, John, & Spellman [1957]); food processing industry (Henderson & Schlaifer [1954] and Fisher & Schruben [1953]); the iron & steel industry (Fabian [1954, 1955, 1958]); metalworking industries (Lewis [1955], Maynard [1955], and Morin [1955]); paper mills (Doig & Belz [1956], Land & Doig [1960], Eisemann [1957], and Paull & Walter [1955]); optimal routing of messages in a communications network (Kalaba & Juncosa [1956]); contract award problems (Gainen [1956], Goldstein [1952]); routing of aircraft and ships (Dantzig & Fulkerson [1954]; Ferguson & Dantzig [1955, 1956]); investment in electric power (Massé & Gibrat [1957]); among others.

Since the early days, the number of applications has exploded, and it is impossible to even attempt to list them. An example of a commercially successful application of network analysis is the award-winning study by Klingman, Philips, Steiger, & Young [1987] and Klingman, Philips, Steiger, Wirth & Young [1986] at Citgo Petroleum Corporation. It was developed with full top management and support, and is estimated to have saved Citgo approximately $2.4 million as a result of better pricing, transportation, and coordination. Many successful applications have been those for the oil industry (Rigby, Lasdon, & Waren [1995], and Thapa [1991,1992]).

Applications that result in savings to management are published in *Interfaces*, *Management Science*, etc. For example, valuation and planning of New Zealand Plantation forests (Manley & Threadgill [1991]); forest management (Vertinsky, Brown, Schreier, Thompson, van Kooten [1994] mortgage valuation models (Ben-Dov, Hayre, & Pica [1992]); telephone network planning (Jack, Kai & Shulman [1992]); managing consumer credit delinquency (Makuch, Dodge, Ecker, Granfors, & Hahn [1992]); freight routing using network optimization (Roy & Crainic [1992]); plant closure (Clements & Reid [1994]); optimal leveraged lease analysis through linear programming (Litty [1994], and Thapa [1984a]); portfolio optimization (Feinstein & Thapa [1993]); strategic and Operational Management in the Steel Industry (Sinha, Chandrasekaran, Mitter, Dutta, Singh, Choudhry, & Roy [1995]); supply chain management (Arntzen, Brown, Harrison, & Trafton [1995]); a new linear programming benchmarking technique (see Sherman & Ladino [1995]). Recent advances in stochastic linear programming have made it possible to build stochastic linear programs for a variety of problems, for example, portfolio optimization (Dantzig & Infanger [1993]), asset/liability management (see Cariño, Kent, Myers, Stacy, Sylvanus, Turner, Watanabe, & Ziemba [1994]); and animal feed formulation (Roush, Stock, Cravener, & D'Alfonso [1994]).

Extensions of linear programming have been applied to numerous areas. To give you an idea, a very small set includes mixed integer linear programming formulations in bulk sugar deliveries (Katz, Sadrian, & Patrick T. [1994] and Vliet, Boender, Rinnooy Kan [1992]); balancing workloads (Grandzol & Traaen [1995]); telecommunications (Cox, Kuehner, Parrish, & Qiu [1993]).

Various other linear programming applications can be found in, for example, Bradley, Hax, & Magnanti [1977], Hillier & Lieberman [1990]. For additional reading on modeling, see, for example, Ackoff & Rivett [1963], Gass [1991], Morris [1967], Starfield, Smith, & Bleloch [1990], and Williams [1985].

The product mix problem, cannery example, on-the-job training, homemaker's problem, and the warehouse example are based on examples in Dantzig [1963].

1.8 PROBLEMS

1.1 *Dantzig [1963].* If an activity such as steel production needs capital such as bricks and cement to build blast furnaces, what would the negative of these activities imply if they were used as admissible activities?

1.2 *A Machine Problem (Kantorovich [1939]).* Formulate the following problem. An assembled item consists of two different metal parts. The milling work can be done on different machines: milling machines, turret lathes, or on automatic turret lathes. The basic data are available in the following table:

Productivity of the Machines for Two Parts			
Type of Machine	Number of Machines	Maximum Output per Machine per Hour	
		First Part	Second Part
Milling Machines	3	10	20
Turret Lathes	3	20	30
Automatic Turret Lathes	1	30	80

(a) Divide the work time of each machine to obtain the maximum number of completed items per hour.

(b) Prove that an optimal solution has the property that there will be no slack time on any of the machines and that equal numbers of each part will be made.

(c) State the dual of the primal problem.

1.3 Generalize Problem 1.2 to n machines and m parts, where the objective is to produce the largest number of completed assemblies.

(a) Show, in general, that if each machine is capable of making each part, and there is no value to the excess capacity of the machines or unmatched parts, any optimal solution will have only matched parts and will use all the machine capacity. What can happen if some machines are incapable of producing certain parts?

(b) State the dual of the primal problem.

(c) Suppose there are two types of assemblies instead of one and a "value" can be attached to each. Maximize the weighted output.

1.4 *The Chicken and Egg Problem (Kemeny in Dantzig [1963]).* Suppose it takes a hen two weeks to lay 12 eggs for sale or to hatch 4. What is the best laying and hatching program if at the end of the fourth period all hens and chicks accumulated during the period are sold at 60 cents apiece and eggs at 10 cents a piece. Formulate the problem assuming, in turn

(a) An initial inventory of 100 hens and 100 eggs,

(b) 100 hens and zero eggs,

(c) 100 hens and zero eggs and also a final inventory of 100 hens and zero eggs.

1.5 A small refinery blends five raw gasoline types to produce two grades of motor fuel: regular and premium. The number of barrels per day of each raw gasoline type available, the performance rating, and cost per barrel are given in the following table:

Raw Gasoline Type	Performance Rating	Barrels/day	Cost/barrel ($)
1	70	2000	0.80
2	80	4000	0.90
3	85	4000	0.95
4	90	5000	1.15
5	99	5000	2.00

Regular motor fuel must have a performance rating of at least 85 and premium at least 95. The refinery's contract requires that at least 8,000 barrels/day of premium be produced; at the same time, the refinery can sell its entire output of both premium and regular for $3.75/barrel and $2.85/barrel, respectively. Assume the performance rating of a blend is proportional, i.e., a 50–50 mixture of raw gasoline types 1 and 2 has a performance of 75.

Formulate a linear program to maximize the refinery's profit. Be sure to define all of your variables.

1.6 *Optimum Blending Of Residual Fuel Oil In a Refinery (Soares, Private Communication in 1986).* Residual fuel oil is the major by-product of fuel refineries. The main uses for residual fuel are in the industrial and electric utility sectors, as well as for space heating and as marine Bunker C fuel. Rigid product specifications, combined with continually changing crudes, refinery operating conditions, and market economics, creates a need for a quick and easy-to-use technique for developing an optimum blend recipe for residual fuel. The reason for this is that by the time crude oil has been refined, the optimal blend of residual fuel oil that is obtained from a possibly large refinery linear programming model may no longer be valid. Thus, it is important to be able to quickly determine a new optimal blend recipe.

Critical properties of residual fuel oil include gravity, sulfur content, viscosity, and flash point. These properties are described next.

- *Gravity.* API gravity is used widely in the petroleum industry and is defined by the American Petroleum Institute as follows:

$$\text{API} = (141.5/\text{specific gravity}) - 131.5,$$

  where specific gravity is the ratio of the density of the material to the density of water, and density is defined to be the ratio of mass (weight) to volume.

  Water, with a specific gravity of 1.0, has an API gravity of 10.0, and fuels heavier than water will have an API gravity below 10.0. Low API gravity fuels, being heavier, have slightly higher heating values; however, at gravities below 10 API, water and entrained sediment will not settle out of the fuel.

  API gravity does not blend linearly; however, specific gravity does blend linearly. Thus, you will need to convert the API gravity specifications to specific gravity.

- *Sulfur.* High sulfur fuels require higher cold-end temperatures in the air-preheaters and economizers of boilers so as to protect against corrosion and resulting fouling of the boiler tubes. In addition, atmospheric pollution regulations limit the maximum sulfur content of residual fuels.

Typically sulfur concentrations are specified as a percentage by weight, thus, you will need to be careful that you do not apply this percentage to a variable that has volume units. Specific gravity can be used to convert between weight and volume. For example, weight = density × volume.

- *Viscosity.* This is a measure of the ability of the fuel to flow. Viscosity is the single most important property because of the difficulties involved in the handling and atomizing of such fuels at the burner tips. Viscosity is measured in units of centistokes (cs) at 122 degrees Fahrenheit temperature.

  Although viscosity is highly nonlinear, when converted to a Viscosity Blend Index (VBI) linear blending is possible. The conversion to VBI is a table look up and has already been done for you in this case. VBI can be applied to variables that have volume units.

- *Flash Point.* This is the temperature at which the vapor above the fuel will momentarily flash or explode when in the presence of a flame. Flash point is an indicator of the temperature at which the fuel can be handled without danger of a fire. A low flash point is extremely difficult to blend off; consequently, it is most desirable to start off with components that all meet flash point specification. Assume that all the components meet flash point specification.

Frequent changes in the quality of crude oil run (high or low sulfur), type of asphalt produced (heavy or light), and economics of the finished products market create a need to develop a quick and easy-to-use method for determining an optimum blend recipe for finished residual fuel. Assume that a larger refinery model has been run and it has been determined that the best strategy is to blend to produce an optimum finished residual fuel oil.

Use the information in Table 1-7 to develop a linear programming model to provide an optimal blend recipe. For simplicity, consider only three refinery produced streams for use in blending residual fuel oil: asphalt flux, clarified oil, and kerosene distillate. Market conditions are such that residual fuel can be sold at 60.0 cents/gallon and the best possible alternate disposition of the constituent streams are as shown in Table 1-8. Finally, assume that the cost of blending the constituent streams to form residual fuel oil is negligible.

Model Formulation and Analysis.

Formulate and solve the resulting linear program by using the **DTZG Simplex Primal** software option on it to determine the optimum residual fuel mix (by fractional volume). Perform any suitable sensitivity analysis that you can think of. *Make sure you justify whatever you choose to do and choose not to do.* Write a report that is organized so that it is easy for management to read and take prompt action. The following is how you should organize your report.

- First report a complete summary of your LP run(s) indicating clearly what must be done and why.
- Next report details of your LP run(s) and any sensitivity runs/analysis that you may have performed. Justify whatever you do.

	Properties			
Type of Stream	API Gravity	Sulfur Weight (%)	Viscosity % cs at 122°F	Viscosity Blend Index
Asphalt Flux	7.5	2.39	1.5	0.966
Clarified Oil	-3.0	2.20	96.5	0.740
Kerosene	38.5	0.20	1.3	0.347
Product Specifications				
Residual fuel (Max)	18.0	2.00	640.0	0.808
Residual fuel (Min)	10.0	None	92.0	0.738

Table 1-7: Stream Properties and Product Specifications

Constituent Stream	*Price*
Asphalt flux	61.7 cents/gallon
Clarified Oil	40.0 cents/gallon
Kerosene	76.0 cents/gallon

Table 1-8: Selling Prices for the Constituent Streams

- **Describe your model formulation in an appendix.**
- **In a second appendix indicate if you ran into any numerical problems. Justify your observations.**

1.7 ***Adapted from a model developed by Thapa [1991] at Stanford Business Software, Inc., and by G. Soares.*** **A small refinery would like to optimally distribute gasoline through the use of various exchanges. The following describes the problem they face.**

- **A fixed amount of gasoline is manufactured every month. G grades (for example, unleaded, premium, super-unleaded, etc.) of gasoline are manufactured at the refinery. Exact manufacturing costs are difficult to get a handle on. Thus, the company assumes that a base amount of one grade, unleaded, is manufactured at 0 cost and the other volumes are generated from it at given manufacturing differentials (different for each grade) of a few cents per gallon.**
- **The refinery has exchange contracts with exchange partners. There is a total of P exchange partners, and the partners lift gasoline from the refinery up to a maximum prespecified amount by grade.**

  Then gasoline is taken back by the refinery at various terminals owned by the partners. A location differential (of the order of a few cents or fraction of cents) independent of the grades is incurred by the refinery.

  The gasoline can also be taken back from the partners at a supply source where exchanges take place. Here too a location differential (of the order of a few cents or fraction of cents) independent of the grades is incurred by the refinery. There are a total of E supply sources; currently there is only 1 supply source but there may be more in the future. Usually, the refinery gets a credit for gasoline lifted back at the supply source.

- From the supply source the gasoline can be shipped via pipeline (at a cost of a few cents per gallon) to various terminals owned by the partners.
- It is possible to also obtain a different mix of grades from the partners than was given to the partners. A regrade differential is incurred in the event that a different mix of grades is lifted. For example, suppose that a partner takes 500,000 gallons of unleaded and 500,000 gallons of premium at the refinery. Then the refinery takes back (at the supply source or terminal) 600,000 gallons of unleaded and 400,000 gallons of premium. Then the partner owes the refinery money for having taken more of premium, a higher valued grade. Note that these costs are computed by assuming that, one of the grades is the base grade, for example unleaded in the above example.
- The refinery supplies a total of S stations with these grades of gasoline. Some stations are supplied directly and others are supplied through exchanges and other terminals. Each station has a prespecified demand for each grade. The demand must always be met.
- Due to physical constraints some grades must be supplied together, that is, *split-loading* is not allowed for certain groups of grades. Typically, super-unleaded is supplied only from the refinery, and the other grades must all come either from the refinery or from a terminal. That is, unleaded cannot come from one terminal and premium from another terminal. If the economics dictate, it is possible, however, for a station to be supplied a portion of a group of grades from one terminal and the balance from another terminal.
- There is a freight cost of the order of a few cents per gallon (independent of the grade) for shipping gasoline from a terminal or from the refinery to the stations.
- It is known in advance which terminals are clearly uneconomical; this reduces the model size since only a few terminals can supply each station. Typically, between 5 and 15 terminals supply each station.

Do the following:

(a) Formulate the retail distribution model described above.

(b) Suppose that the refinery would like to analyze buy-and-sell options at the refinery, terminals, and supply sources. Incorporate this into your model. How would you model incremental sales (assuming one customer only); that is, for example, the first 100,000 gallons of unleaded are sold at 57 cents/gallon and the next 200,000 gallons of unleaded are sold at 55 cents/gallon.

(c) It may be advantageous to blend different grades of gasoline in the truck once it is picked up at a terminal. For example, 86 octane gasoline can be obtained by blending 25% of 80 octane and 75% of 88 octane. In your formulation, incorporate blending of gasoline at terminals only. Assume a small cost for the blending.

(d) In some instances it is possible to enhance a grade of gasoline by adding an octane enhancer in the truck. For example, 1 gallon of 86 octane gasoline can be obtained by enhancing 1 gallon of 84 octane gasoline. In your formulation, incorporate grade conversion at terminals only. Assume a small cost for the grade conversion.

(e) How would you modify your formulation to allow the user to specify for each station any combination of products to be supplied as a group; i.e., different split-loading restrictions for each station.

Comment: The making and distribution of asphalt can be formulated as a very similar model. In this case, however, because asphalt is a seasonal product and inventories need to be maintained, a multi-time period is needed (see Thapa [1992]).

1.8 *Adapted from a model developed by Hogan and enhanced by Thapa at Stanford Business Software, Inc., in 1990.* A gas and electric company has been ordered by the Public Utilities Commission (PUC) to allow customers to bid for the extra gas transportation capacity on their pipelines. The pipeline has many links but for simplicity assume that we are concerned only with one link.

- The gas company cannot make a profit on the selling of this capacity but must be fair in assigning capacity.
- The actual capacity available in the pipeline depends on usage by core customers. Once core usage is satisfied, the other capacity is available to the noncore customers. Assume that there are N noncore customers.
- Assume that the available capacity is prioritized into blocks $1, \ldots, K$, where block K has the lowest probability of being available. That is, the capacity in each block is known with a given probability.
- Noncore customers bid for the maximum capacity they would like in each priority block. They also assign a price to their bid in each priority block without knowledge of the prices assigned by other noncore customers. Furthermore, they also indicate the maximum capacity they would like to receive over all blocks.
- Assume that if a customer bids for block k, then this bid is also available for all higher priority blocks, i.e., for blocks $1, \ldots, k$.

The process is best illustrated by an example. Suppose that there are 2 customers and that there are two priority blocks. The bids and prices are illustrated in the table below:

	Bid Price for Block		Max Bid
	2	1	
Customer 1	15	17	250
Customer 2	30	35	250
Max Capacity	200	200	

The optimal award is as follows:

- Customer 1 gets 150 in block 2.

- Customer 2 gets the entire bid amount, that is, 200 in block 1 and 50 in block 2.

The willingness to pay is then 10,750= $150 \times 15 + 50 \times 30 + 200 \times 35$. The revenue to the company is somewhat lower because the market clearing prices are lower. To understand this, observe that Customer 1 is the marginal bidder since he/she gets the last available unit in block 2. Thus, the resulting marginal price is 15 for that block which is what both customers actually pay. It appears that Customer 2 is indifferent between block 1 and block 2, which is cheaper by 5. Thus, the market clearing price for block 1 is $20 = 15 + 5$, which is what Customer 2 actually pays. This results in a total revenue to the company of 7,000= $150 \times 15 + 50 \times 15 + 200 \times 20$. With this in mind do the following:

(a) Formulate the linear program to maximize the benefits as measured by the willingness to pay.

(b) Once the bids are assigned, the customer pays the market clearing price for the block. Write down the dual of the problem and show that the dual prices are the market clearing prices. This implies that the revenue earned is actually not necessarily the value of the objective function! Why?

(c) It turns out that linear programs often have multiple solutions in practice. This problem is not an exception; the implication here is that it is possible for only one of two identical bids to be awarded. How would you modify the model to distribute the bids fairly. Hint: One approach is to use a quadratic function; in this case, it is no longer a linear program.

(d) Some customers have a minimum acceptable quantity that they would accept. Modify your formulation to reflect this. Is the resulting formulation a linear program? Why?

1.9 *Adapted from a model developed by Thapa [1984] at Stanford Business Software, Inc.* This is a simplified version of a leveraged lease model. A leasing company (lessor) obtains a loan on a piece of equipment and leases it out to a customer (lessee) who pays rent every month for 30 years. Formulate the leveraged leasing model as a linear programming model assuming

- The goal is to minimize the present value of the rents received while obtaining the desired yield. The present value of cash in time period t is the cash divided by $(1 + r)^t$, where r is the monthly discount rate (interest rate) for the lessee.

- The IRS requires that the sum of the rents received in each year must be within the range $[0.9 * \text{AVG}, 1.1 * \text{AVG}]$, where AVG is the average yearly rent computed over the rents received for the entire lease period of 360 months.

- The lessor's loan principal is equal to the asset price plus the fee minus the equity paid by the lessor.

- The rent received in each time period must be greater than or equal to the debt service. The debt service in a time period is defined to be the sum of the interest payment and loan principal repayment in that time period. The interest in any period is defined to be the product of the bank interest rate charged to the lessor times the principal balance in that time period.

- The present value of the cash flow over all time periods for a prespecified yield (interest rate) must be equal to the investment amount (total loan principal) for the lessor. The cash flow for the lessor is the difference between the rents and debt service in each time period.

1.10 *Problem Under Uncertainty.* Suppose that there are three canneries that ship cases of cans to five warehouses. The number of cases available during each season at each of the canneries is known in advance and is shown in the table below together with the cost to ship per case to each of the warehouses.

	Availability	Shipping Cost ($/case)				
		Warehouses				
Canneries	Cases	a	b	c	d	e
1	50,000	0.9	2.0	1.8	1.7	2.5
2	75,000	0.6	1.6	1.4	1.8	2.5
3	25,000	2.7	1.8	1.5	1.0	0.9

The seasonal demand at each of the warehouses is uncertain and is shown in the table below:

	Demand at Probability		
Warehouse	15%	55%	30%
a	15,000	20,000	30,000
b	16,000	20,000	28,000
c	17,000	20,000	26,000
d	18,000	20,000	24,000
e	19,000	20,000	22,000

Assume all cases left over at the end of the season must be disposed of at a loss of $1 per case (they cannot be stored any longer because the food in the cans will spoil). Failure to supply demand during a season is penalized at $0.25 per case as the discounted estimated loss of all future sales (turning a customer away runs the risk that the customer will not return by becoming the customer of another supplier). Use the `DTZG Simplex Primal` software option to determine what shipping schedule will optimize the total shipping cost and expected net revenues?

1.11 *Ph.D. Comprehensive Exam, September 25, 1976, at Stanford.* You have been called to appear as an expert witness before the congressional committee that is reviewing the new budget of the Department of Energy. In the past, this department and its predecessor agencies have provided a substantial amount of financial support for the development of mathematical programming, complementarity, and fixed-point algorithms. Congressman Blank, a member of this committee, is hostile to this type of research. He has just made newspaper headlines by reading out the titles of some of the more esoteric publications in this area.

You are asked to prepare a non-technical statement (not to exceed 500 words in length) explaining the relevance of such research to the area of energy policy. Recall that most congressmen have been trained as lawyers, that they have not had college-level courses in mathematics, and that they are skeptical about mathematical reasoning.

1.12 *Ph.D. Comprehensive Exam, September, 1982, at Stanford.* It is often said that there is a similarity between market mechanisms and mathematical programming models. For what types of applications does this seem valid? Give an example in which the analogy breaks down, and explain why.

CHAPTER 2

SOLVING SIMPLE LINEAR PROGRAMS

Linear programs, except possibly very tiny ones or very special cases, require a computer for solution. When linear problems have exactly two variables subject to many inequality constraints or exactly two equations in many nonnegative variables, it is possible to solve them graphically. In Section 2.1 we illustrate how to solve the first class graphically. In Section 2.2 we illustrate the second class and also introduce the concept of *duality* and the role that it plays in the solution. Finally in Section 2.3 we show how to solve simple linear programs algebraically using the Fourier-Motzkin Elimination Method.

2.1 TWO-VARIABLE PROBLEM

Consider the following two-variable case:

$$\begin{array}{lrcrcl}
\text{Minimize} & -2x_1 & - & x_2 & = & z \\
\text{subject to} & x_1 & + & x_2 & \leq & 5 \\
 & 2x_1 & + & 3x_2 & \leq & 12 \\
 & x_1 & & & \leq & 4 \\
\multicolumn{6}{l}{\text{and } x_1 \geq 0,\ x_2 \geq 0.}
\end{array}$$

To solve this problem graphically we first shade the region in the graph in which all the feasible solutions must lie and then shift the position of the objective function line $-2x_1 - x_2 = z$ by changing the value of its right hand side z until the objective function cuts the feasible region with the lowest possible value for the objective.

The feasible region is the set of points with coordinates (x_1, x_2) that satisfy all the constraints. It is shown in Figure 2-1 as the shaded region. It is determined

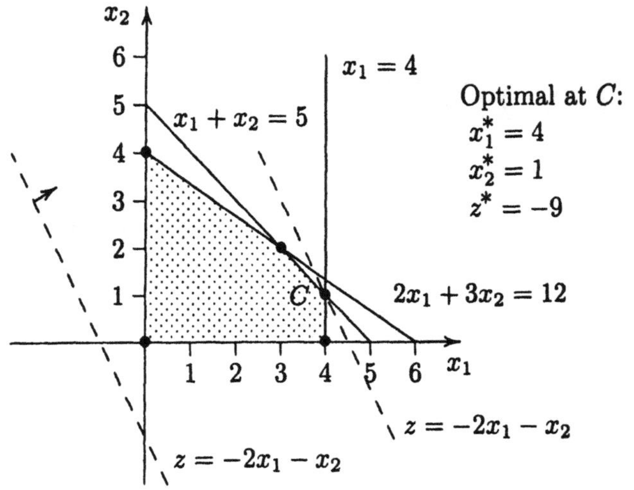

Figure 2-1: Graphical Solution of a Two-Variable LP

as follows: The nonnegativity constraints $x_1 \geq 0$ and $x_2 \geq 0$ clearly restrict all the feasible solutions to lie in the first (or northeast) quadrant. The first constraint, $x_1 + x_2 \leq 5$, implies that all the feasible x_1 and x_2 must lie to one side of its boundary, the line $x_1 + x_2 = 5$. The side on which the feasible x_1 and x_2 must lie is easily determined by first checking whether the origin lies on the feasible side of the line; in this case it is easy to see that the feasible side is on the same side as the origin since $(0, 0)$ obviously satisfies $x_1 + x_2 \leq 5$. In a similar manner we can check which side of the boundary of the other two constraints is the feasible side.

The objective is to minimize the linear function $z = -2x_1 - x_2$. If we fix for the moment the value of z to be zero we see that the objective function can be represented as a line of slope -2 that passes through the origin. Translating this objective line (i.e., moving it without changing its slope) to a different position is equivalent to choosing a different value for z. Clearly, translating the line in a Southwest direction away from the feasible region is pointless. The origin is an *extreme point* (corner) of the feasible region but is not an optimal solution point since translating the objective line into the feasible region results in a smaller value for the objective (for example, draw the objective line with $z = -3$). Thus translating the objective function in a northeast direction is desirable since it results in a smaller and smaller objective function value. However, moving the objective function line past the extreme point marked $C = (4, 1)$ in Figure 2-1 causes the line to no longer intersect the feasible region. Thus, the extreme point C, which is the intersection of the boundary of constraints 1 and 3, must be the optimal point for this two-dimensional LP. At the optimal solution point $(x_1, x_2) = (4, 1)$, the minimum (optimal) value of the objective function is -9. We will prove later that bounded linear programs that have feasible solutions always have optimal solutions that are extreme points. If they have more than one optimal extreme point, then any weighted average of these extreme points is also an optimal solution.

▷ **Exercise 2.1** Use the `DTZG Simplex Primal` software option to verify that the above solution is correct.

▷ **Exercise 2.2** Prove that the two-variable problem can have at most two optimal extreme points.

▷ **Exercise 2.3** Construct a graphical example in three dimensions to show that a three-variable problem can have more than three optimal extreme points.

The following cases can arise for a minimization problem (analogous results hold if one is maximizing):

1. If the constraints are such that there is no feasible region, then no solution exists.

2. If the objective function line can be moved indefinitely away from a feasible point in a direction that decreases z and still intersects the feasible region, then the feasible region is unbounded and there is a sequence of feasible points (x_1, x_2) for which the corresponding values of z approach $-\infty$.

3. If the objective function line can be moved only a finite amount by decreasing the value of z while still intersecting the feasible region, then the last feasible point touched by the objective function line, if unique, yields the unique optimal solution, and the corresponding value of z is the minimum value for the objective. If not unique, then any point on the segment of the boundary last touched yields an optimal solution and the minimum value for the objective.

▷ **Exercise 2.4** Draw a graph of a two-variable linear program to illustrate each of the above three cases.

▷ **Exercise 2.5** Construct an example where the set of points (x_1, x_2) where z is minimized is (a) a line segment; (b) an infinite line segment that is bounded at one end;(c) an infinite line segment not bounded on either end.

2.2 TWO-EQUATION PROBLEM

In order to illustrate how to solve a two-equation problem graphically, we shall make use of the product mix problem described in Section 1.4.1. The problem, repeated here for convenience, is to minimize z subject to $x_j \geq 0$ and

$$\begin{array}{rcrcrcrcrcrcl} 4x_1 &+& 9x_2 &+& 7x_3 &+& 10x_4 &+& x_5 & & &=& 6000 \\ x_1 &+& x_2 &+& 3x_3 &+& 40x_4 & & &+& x_6 &=& 4000 \\ -12x_1 &-& 20x_2 &-& 18x_3 &-& 40x_4 & & & & &=& z. \end{array} \tag{2.1}$$

2.2.1 GRAPHICAL SOLUTION

Clearly the techniques of the last section cannot be applied directly since it is not easy to visualize the equations as objects in the six-dimensional space of points whose coordinates are $(x_1, x_2, x_3, x_4, x_5, x_6)$. Fortunately, this problem can be converted to one that involves finding a way to average a set of points in a two-dimensional space to attain a specified average value while simultaneously minimizing the average cost associated with these points.

To convert the product mix problem (2.1) to one that can be solved graphically, it is first necessary to modify the units used to measure the quantity of items and activity levels and also to redefine the activity levels so that the activity levels sum to unity. Algebraically, this is done by first adding the two equations to form a new equation. This allows us to drop one of the original equations as now redundant. We next change the units for measuring the x_j's so that they sum to unity. Operationally we can do this by introducing as a new item *total capacity*, which is the sum of the carpentry capacity and the finishing capacity.

$$\begin{array}{rcrcrcrcrcrcl} 5x_1 &+& 10x_2 &+& 10x_3 &+& 50x_4 &+& x_5 &+& x_6 &=& 10000 \\ 4x_1 &+& 9x_2 &+& 7x_3 &+& 10x_4 &+& x_5 & & &=& 6000 \\ x_1 &+& x_2 &+& 3x_3 &+& 40x_4 & & &+& x_6 &=& 4000 \\ -12x_1 &-& 20x_2 &-& 18x_3 &-& 40x_4 & & & & &=& z. \end{array} \tag{2.2}$$

We then drop, for example, the finishing capacity equation, which is now redundant. Next we *change the column units* that are used for measuring activity levels so that 1 new unit of each activity requires the full 6,000 + 4,000 = 10,000 hours of total capacity.

To change units in (2.1) note that one unit of the first activity requires $4+1=5$ hours of total capacity; thus, 2,000 units of the first activity would require 10,000 hours of capacity and is equivalent to *one new unit* of the first activity. In general, if y_1 is the number of new units, then $2000y_1 = x_1$ old units of the first activity. The relationship for each activity between the old and new activity levels after such a change in units is

$$\begin{array}{lll} 2000y_1 = x_1, & 1000y_2 = x_2, & 1000y_3 = x_3, \\ 200y_4 = x_4, & 10000y_5 = x_5, & 10000y_6 = x_6. \end{array} \tag{2.3}$$

It is also convenient to *change the row units* that are used to measure capacity and costs. Let 10,000 hours = 1 new capacity unit; \$10,000 = 1 new cost unit, i.e., $10000\bar{z} = z$. Then it is easy to see that the product mix model in Table 1-3 will become, after the changes in the units for activities and items given above, Table 2-1.

▷ **Exercise 2.6** Show that the product mix model as described in Table 1-3 becomes, after the changes in the units for activities and items given by (2.3), Table 2-1.

Items	Activities: Type: Levels:	Manufacturing Desks (1) y_1	(2) y_2	(3) y_3	(4) y_4	Slacks Carp. y_5	Fin. y_6	Exogenous
(0): Total Capacity (10,000 hrs)		1.0	1.0	1.0	1.0	1.0	1.0	1.0
(1): Carpentry Capacity (10,000 hrs)		.8	.9	.7	.2	1.0		0.6
(3): Cost ($10,000)		−2.4	−2.0	−1.8	−.8			$\bar{z}$ (Min)

Table 2-1: The Converted Product Mix Model

Note that the coefficients in the top row of Table 2-1 (Item 0) are now all 1s. If we set this fact aside for the moment, then for the purpose of graphing the data in a column we plot the two remaining coefficients in each column j as the coordinates of a point A_j in two-dimensional space. That is,

$$A_1 = \begin{pmatrix} .8 \\ -2.4 \end{pmatrix}; \; A_2 = \begin{pmatrix} .9 \\ -2.0 \end{pmatrix}; \; A_3 = \begin{pmatrix} .7 \\ -1.8 \end{pmatrix};$$
$$A_4 = \begin{pmatrix} .2 \\ -0.8 \end{pmatrix}; \; A_5 = \begin{pmatrix} 1.0 \\ 0.0 \end{pmatrix}; \; A_6 = \begin{pmatrix} 0.0 \\ 0.0 \end{pmatrix}.$$

The right hand side is a point whose coordinates are

$$R = \begin{pmatrix} .6 \\ \bar{z} \end{pmatrix}.$$

Thus, the coordinates of each point A_j are plotted as a point labeled A_j in Figure 2-2. Its first coordinate is the coefficient for the *carpentry* capacity and the second coordinate is the cost coefficient of activity j. The right hand side R is plotted as a "requirements" line rather than a point since its v coordinate $\bar{z}$ is a variable quantity to be determined.

In physics, if one is given a set of points $A_1, A_2, \ldots, A_n$ with given relative weights ($y_1 \geq 0, y_2 \geq 0, \ldots, y_n \geq 0$), where $\sum_{i=1}^{n} y_i = 1$, then the *center of gravity* G of the set of points $A_1, A_2, \ldots, A_n$ is found by the formula

$$G = A_1 y_1 + A_2 y_2 + \cdots + A_n y_n, \tag{2.4}$$

where the weights sum to unity:

$$y_1 + y_2 + \cdots + y_n = 1. \tag{2.5}$$

Relation (2.4) is in vector notation; it means that the relation holds if the first coordinate of G and A_j for $j = 1, \ldots, n$ are substituted for G and A_j, and the relation is also true if the second coordinate is substituted.

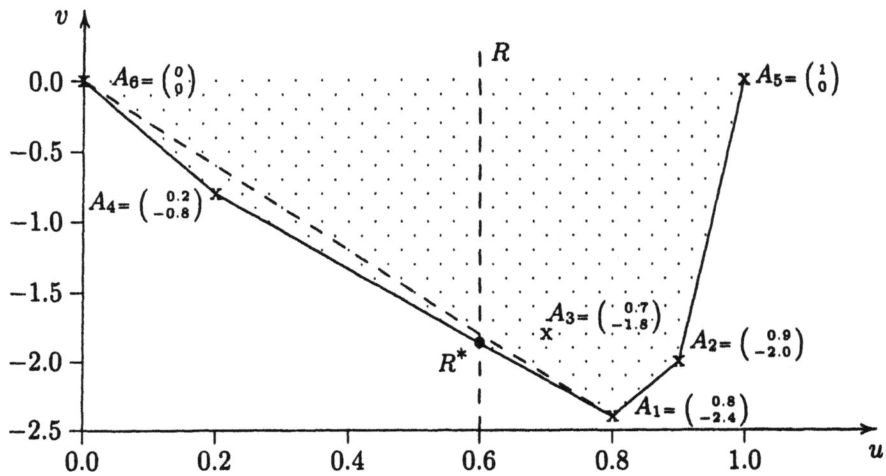

Figure 2-2: Graphical Solution of the Product Mix Problem

In our application, the center of gravity G is specified as the lowest point on the requirement line that can be generated by assigning nonnegative weights to the points. The problem becomes one of determining the weights as unknowns so as to achieve this lowest point. Because the unknowns $y_j \geq 0$ sum to unity, the problem is therefore one of *assigning* nonnegative weights to points $A_1, A_2, \ldots, A_6$ such that their center of gravity lies on the requirement line given by R at a point where the cost coordinate $\bar{z}$ is minimized.

The optimum solution to the product mix problem is easily found by inspection of the graph in Figure 2-2. Clearly, the point R^* has the minimum cost coordinate, which is found by assigning zero weights y_j to all points, except A_1 and A_4, and appropriately weighting the latter so that the *center of gravity* of A_1 and A_4 has abscissa 0.6. To determine the two remaining weights y_1, y_4, set $y_2 = 0$, $y_3 = 0$, $y_5 = 0$, and $y_6 = 0$ in Table 2-1. Recalling from (2.5) that the sum of the weights must equal unity, this results in

$$\begin{aligned} .8y_1 + .2y_4 &= 0.6 \\ y_1 + y_4 &= 1.0, \end{aligned}$$

whence solving the two equations in two unknowns for y_1 and y_4,

$$y_1 = \frac{2}{3},\ y_4 = \frac{1}{3}.$$

The corresponding cost $\bar{z}$ is given by

$$\bar{z} = -2.4y_1 - 0.8y_4 = -\tfrac{5.6}{3}.$$

Thus the optimal solution is to manufacture $x_1 = 2000y_1 = \frac{2}{3} \times 2000$ desks of Type 1, $x_4 = 200y_4 = \frac{1}{3} \times 200$ desks of Type 4, and none of the other types of desks. This will use the full capacity of the plant since the slack variables y_5 and y_6 are zero. The minimum cost $z = 10000\bar{z} = 10000 \times (-5.6/3)$, a profit of \$18,666.67.

Despite the fact that the material balance equation for finishing capacity was omitted in the above calculation, the limitation of 4,000 hours on the use of this capacity is completely accounted for by this solution. As noted earlier, this is because the adding of the *total capacity* equation to the system and dropping one of the remaining redundant equations yields an equivalent model that properly takes into account the limitation on the amount of each type of capacity available.

▷ **Exercise 2.7** Use the `DTZG Simplex Primal` software option to verify the correctness of the above solution. Change the profit on desk 1 to be \$8 instead of \$12 and rerun the software. How does the solution change?

ALGEBRAIC CHECK OF OPTIMALITY

We can check algebraically whether our choice of A_1, A_4 in Figure 2-2 is correct by first determining that the calculated values for y_1 and y_4 satisfy nonnegativity and then testing to see whether the estimate of every point in the shaded region has value v greater than or equal to that of the point on the line joining A_1 to A_4 with the same abscissa u. If the latter is true we say the shaded region lies on or above the extended line joining A_1 to A_4. The extended line joining A_1 to A_4 is called the *solution line.* Clearly points A_2, A_3, A_5, and A_6 lie above the solution line in Figure 2-3, and therefore it is intuitively clear (and can be shown rigorously) that the feasible solution $y_1 = 2/3$, $y_4 = 1/3$ is optimal.

2.2.2 THE DUAL LINEAR PROGRAM

Another way to view the linear program, called the *dual* view, is to consider the set $\mathcal{L}$ of lines L in the (u, v) plane such that all points $A_1, A_2, \ldots, A_n$ lie on or above each line of $\mathcal{L}$ (See Figure 2-3). The line L in $\mathcal{L}$ that we are most interested in is the *solution line,* which is the line L in $\mathcal{L}$ that intersects the requirements line R at the highest point R^*.

We can state this dual problem algebraically. The equation of a *general line L* in the (u, v) plane is

$$v = \pi_1 + \pi_2 u$$

where π_1 is the intercept and π_2 the slope. In order that the shaded region lies on or above this line, each of the points $A_1, A_2, \ldots, A_6$ in the shaded region must lie on or above the line. In order to test whether or not the point $A_2 = (.9, -2.0)$, for example, lies on or above L substitute the coordinate $u = .9$ of A_2 into its equation; if the v coordinate of A_2 is greater than or equal to the value of the ordinate $v = \pi_1 + \pi_2(.9)$ of L, then A_2 lies on or above L. Thus, our test for A_2 is $\pi_1 + \pi_2(.9) \leq -2.0$, and the test for the entire set of points $A_1, A_2, \ldots, A_6$ lying on

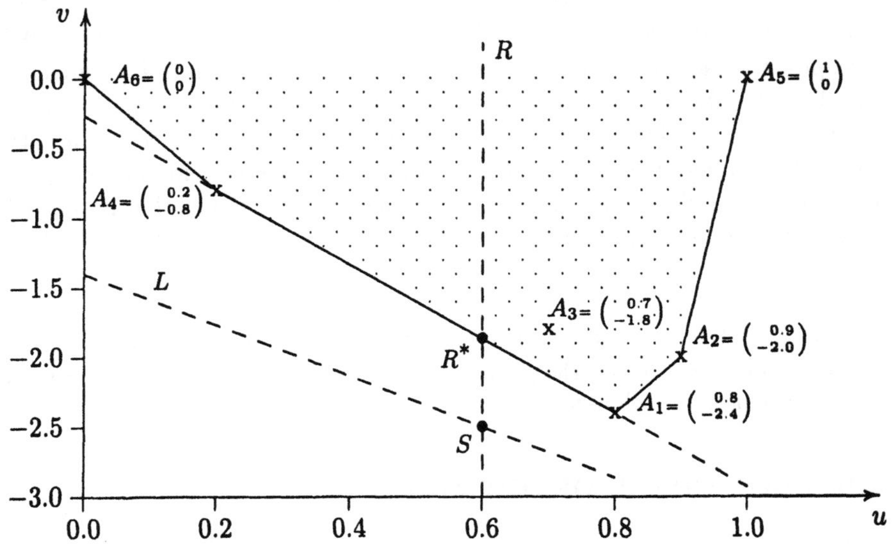

Figure 2-3: Optimality Check—The Product Mix Problem

or above the line L is:

$$\begin{aligned} \pi_1 + \pi_2(0.8) &\leq -2.4, \\ \pi_1 + \pi_2(0.9) &\leq -2.0, \\ \pi_1 + \pi_2(0.7) &\leq -1.8, \\ \pi_1 + \pi_2(0.2) &\leq -0.8, \\ \pi_1 + \pi_2(1.0) &\leq \ \ 0.0, \\ \pi_1 + \pi_2(0.0) &\leq \ \ 0.0. \end{aligned} \tag{2.6}$$

Let $S = (.6, \bar{v})$ for some $\bar{v}$ be the intersection of the vertical line $u = .6$ with $v = \pi_1 + \pi_2 u$. In order for S to lie on the line L, we must have $\bar{v} = \pi_1 + \pi_2(.6)$. The line L below the shaded region whose $v = \bar{v}$ coordinate of S is *maximum*, is the line L with slope π_2 and intercept π_1 that satisfies (2.6) and maximizes $\bar{v}$ given by (2.7):

$$\pi_1 + \pi_2(.6) = \bar{v} \text{ (Max)}. \tag{2.7}$$

It is clear from the figure that the line in the figure on or below the convex feasible region (shaded area) with maximum intercept with the requirement line is the same as the optimal solution line for the original (primal) problem, namely the line passing through the optimal pair of points A_1 and A_4.

The problem of finding π_1, π_2, and max $\bar{v}$ satisfying (2.6) and (2.7) is called the *dual* of our original (*primal*) problem (2.1). The obvious observation that the ordinate of R^* for these two problems satisfies

$$\max\ \bar{v} = \min\ z \tag{2.8}$$

is a particular instance of the famous von Neumann *Duality Theorem 5.3* for linear programs.

In summary, if we conjecture that some pair like A_1, A_4 (obtained by visual inspection of the graph or otherwise) is an optimal choice, it is an easy matter to verify our choice by checking whether (i) the intersection with the requirement line lies *between* the selected two points and (ii) all points $A_1, A_2, \ldots, A_6$ lie *on or above* the extended line joining the selected two points. To check the first, we solve (as we did earlier)

$$\begin{aligned} .8y_1 + .2y_4 &= 0.6 \\ y_1 + y_4 &= 1.0 \end{aligned} \tag{2.9}$$

to see whether y_1 and y_4 are nonnegative. We obtain $y_1 = 1/3$, $y_4 = 2/3$, which are positive, so S lies *between* A_1 and A_4. These values, with remaining $y_j = 0$, satisfy the primal system (2.1). To check the second set of conditions, we must first determine the coefficients π_1 and π_2 of the equation of the line $v = \pi_1 + \pi_2 u$ which passes through A_1 and A_4. Thus π_1 and π_2 are found by substituting the coordinates of A_1 and A_4 into the equation of the line:

$$\begin{aligned} \pi_1 + \pi_2(.8) &= -2.4 \\ \pi_1 + \pi_2(.2) &= -0.8. \end{aligned} \tag{2.10}$$

Solving the two equations in two unknowns yields $\pi_1 = 5.6/3$, $\pi_2 = -8/3$. We then substitute these values of π_1 and π_2 into (2.6) to see whether they satisfy the dual system of relations. Since (2.6) is satisfied, the test is complete, and the solution $y_1 = 1/3$, $y_4 = 2/3$, and all other $y_j = 0$ is optimal.

2.3 FOURIER-MOTZKIN ELIMINATION

In this section we will show how to solve simple linear programs algebraically. For small linear programming problems, the Fourier-Motzkin Elimination (FME) Method can be used. This method reduces the number of variables in a system of linear inequalities by one in each iteration of the algorithm. Its drawback is that it can greatly increase the number of inequalities in the remaining variables.

Before proceeding we state the following definitions:

Definition (Consistent, Solvable, or Feasible): A mixed system of equations or inequalities having one or more solutions is called *consistent* or *solvable* or *feasible*.

Definition (Inconsistent, Unsolvable, or Infeasible): If the solution set is empty, the system is called *inconsistent* or *unsolvable* or *infeasible*.

Definition (Equivalent Systems): Mixed systems of linear equations or inequalities are said to be *equivalent* if their solution sets are the same.

2.3.1 ILLUSTRATION OF THE FME PROCESS

Note that is legitimate to generate a new inequality by positive linear combinations of existing inequalities pointing in the same direction. Moreover, if existing inequalities have coefficients, of say x_j, with opposite signs then we can find a positive linear combination that generates an inequality with coefficients of x_j equal to 0. We will apply these two ideas to solve a system of linear inequalities by the Fourier-Motzkin Elimination process.

Example 2.1 (Fourier-Motzkin Elimination) Find a feasible solution to the system of inequalities in the variables x_1, x_2, z by eliminating x_1, x_2, and z in turn.

$$\begin{aligned} x_1 \qquad\quad &\geq 0 \\ x_1 + 2x_2 &\leq 6 \\ x_1 + x_2 &\geq 2 \\ x_1 - x_2 &\geq 3 \\ x_2 &\geq 0 \\ -2x_1 - x_2 &\leq z. \end{aligned}$$

Note that the inequalities do not all point in the same direction. Rewrite the above inequalities by reversing signs as necessary so that they point in the same direction. Also, positively rescale if necessary so that all the nonzero coefficients of x_1 are $+1$ or -1.

$$\begin{array}{rcl} x_1 & \geq & 0 \\ -x_1 - 2x_2 & \geq & -6 \\ x_1 + x_2 & \geq & 2 \\ x_1 - x_2 & \geq & 3 \\ 2x_1 + x_2 + z & \geq & 0 \\ x_2 & \geq & 0 \end{array} \implies \begin{array}{rcl} x_1 & \geq & 0 \\ -x_1 - 2x_2 & \geq & -6 \\ x_1 + x_2 & \geq & 2 \\ x_1 - x_2 & \geq & 3 \\ x_1 + \frac{1}{2}x_2 + \frac{1}{2}z & \geq & 0 \\ x_2 & \geq & 0. \end{array} \tag{2.11}$$

From the representation (2.11) it is clear that there are three classes of inequalities with respect to x_1. The first class consists of inequalities in which the coefficient of x_1 is $+1$; the second class consists of inequalities in which the coefficient of x_1 is -1; and the third class of inequalities are those in which the coefficient of x_1 is 0. We eliminate the variable x_1 by adding each inequality in the first class to each inequality in the second class. This results in the following system that appears on the left in (2.12) which, in turn, can be rewritten as the system that appears on the right of (2.12).

$$\begin{array}{rcl} -2x_2 & \geq & -6 \\ -x_2 & \geq & -4 \\ -3x_2 & \geq & -3 \\ -\frac{3}{2}x_2 + \frac{1}{2}z & \geq & -6 \\ x_2 & \geq & 0 \end{array} \implies \begin{array}{rcl} -x_2 & \geq & -3 \\ -x_2 & \geq & -4 \\ -x_2 & \geq & -1 \\ -x_2 + \frac{1}{3}z & \geq & -4 \\ x_2 & \geq & 0. \end{array} \tag{2.12}$$

The next step is to repeat the process with (2.12) by eliminating x_2:

$$\begin{array}{rcl} 0 & \geq & -3 \\ 0 & \geq & -4 \\ 0 & \geq & -1 \\ \frac{1}{3}z & \geq & -4 \end{array} \implies \begin{array}{rcl} 0 & \geq & -3 \\ 0 & \geq & -4 \\ 0 & \geq & -1 \\ z & \geq & -12. \end{array}$$

Obviously 0 is greater than or equal to -3, -4, or -1, so that the only inequality of interest left is

$$z \geq -12. \tag{2.13}$$

It is not possible to eliminate z and therefore the process stops with (2.13) and we are ready for the back-substitution steps. Choose any value for z that satisfies (2.13), for example, $z = -12$, which is the smallest value that z can take which keeps the original system feasible. Setting $z = -12$ in (2.12), it turns out that any value of x_2 in the range $0 \leq x_2 \leq 0$ may be chosen, implying $x_2 = 0$. Substituting $z = -12$ and $x_2 = 0$ in (2.11) yields $6 \leq x_1 \leq 6$ implying $x_1 = 6$.

Thus, the solution that minimizes z, for this relatively small problem, was found easily, is unique. It is $x_1 = 6$, $x_2 = 0$, and $z = -12$.

▷ **Exercise 2.8** Show that other feasible solutions can be obtained that satisfy $z \geq -12$.

▷ **Exercise 2.9** Prove that the variable z can never be eliminated during the FME process.

It is straightforward to see that by choosing the smallest z we actually solved the following linear program.

$$\begin{array}{lrcrcl}
\text{Minimize} & -2x_1 & - & x_2 & = & z \\
\text{subject to} & x_1 & + & 2x_2 & \leq & 6 \\
& x_1 & + & x_2 & \geq & 2 \\
& x_1 & - & x_2 & \geq & 3 \\
& x_1 & & & \geq & 0 \\
& & & x_2 & \geq & 0.
\end{array}$$

▷ **Exercise 2.10** Apply the FME process to the linear program

$$\begin{array}{lrcrcl}
\text{Minimize} & -2x_1 & - & x_2 & = & z \\
\text{subject to} & x_1 & + & 2x_2 & \geq & 6 \\
& x_1 & + & x_2 & \geq & 2 \\
& x_1 & - & x_2 & \geq & 3 \\
& x_1 & & & \geq & 0 \\
& & & x_2 & \geq & 0
\end{array}$$

to show that it has a class of feasible solutions in which z tends to $-\infty$. Use the `Fourier-Motzkin Elimination` software option to verify this.

▷ **Exercise 2.11** Apply the FME process to the linear program

$$\begin{array}{lrcrcl}
\text{Minimize} & -2x_1 & - & x_2 & = & z \\
\text{subject to} & x_1 & + & 2x_2 & \leq & 6 \\
& x_1 & + & x_2 & \geq & 2 \\
& x_1 & + & x_2 & \leq & 1 \\
& x_1 & - & x_2 & \geq & 3 \\
& x_1 & & & \geq & 0 \\
& & & x_2 & \geq & 0
\end{array}$$

to show that it is infeasible because the process generates an *infeasible inequality* of the form

$$0x_1 + 0x_2 \geq d, \qquad \text{where } d > 0.$$

Use the `Fourier-Motzkin Elimination` software option to verify this.

LEMMA 2.1 (FME Applied to a General Linear Program) *The FME process can be applied to the following general linear program:*

$$\begin{array}{rrcl} \text{Minimize} & \displaystyle\sum_{j=1}^{n} c_j x_j & = & z \\ \text{subject to} & \displaystyle\sum_{j=1}^{n} a_{ij} x_j & \geq & b_i \text{ for } i = 1, \ldots, m \\ & x_j & \geq & 0 \text{ for } j = 1, \ldots, n, \end{array} \tag{2.14}$$

to obtain a feasible optimal solution if it exists or to determine that it is infeasible because it generates an infeasible inequality

$$\sum_{j=1}^{n} 0x_j \geq d, \qquad \textit{where } d > 0,$$

or to obtain a class of feasible solutions in which z tends to $-\infty$.

Generating an infeasible inequality to show that a system of linear inequalities is infeasible is formalized through the Infeasibility Theorem, which we state and prove in Section 2.4.

2.3.2 THE FOURIER-MOTZKIN ELIMINATION ALGORITHM

Algorithm 2.1 (FME) This algorithm is only practical for a very small system of m inequalities in n variables. Given a system of inequalities labeled as $\mathcal{A}$, initiate the process by setting $\mathcal{R}_1 = \mathcal{A}$ and $k = 0$.

1. Set $k \leftarrow k + 1$. If $k = n + 1$ go to Step 7.
2. Rewrite the inequalities $\mathcal{R}_k$ so that the variable x_k appears by itself with a coefficient of -1, 1, or 0 on one side of each inequality written as a $\geq$ inequality. Consider all those constraints with zero coefficients for x_k as a part of the reduced system.
3. *All the coefficients of x_k are zero.* Mark the value of x_k as "arbitrary," set

   $$\mathcal{R}_{k+1} \leftarrow \mathcal{R}_k$$

   and go to Step 1.
4. *The coefficients of x_k are all $+1$ (or all -1).* If $k < n$, assign arbitrary values $x_{k+1}, \ldots, x_n$. Go to Step 9.
5. *All coefficients of x_k are a mix of* 0 *and* $+1$ *(or all coefficients are a mix of* 0 *and* -1*).* Call the constraints with zero coefficients for x_k the reduced system R_{k+1} and go to Step 1.

6. *This is the case where there is at least one pair of inequalities with a* $+1$ *and a* -1 *coefficient for* x_k. For each such pair augment the reduced system by their sum. Set $\mathcal{R}_{k+1}$ to be the reduced system and go to Step 1.
7. *Feasibility Check.* Check the right-hand sides of $\mathcal{R}_{n+1}$. If any of them is positive report the original system as being infeasible and stop; else, set $k \leftarrow n$.
8. *Determine* x_n. If $k = n$, determine a feasible value for x_n from $\mathcal{R}_n$. Set $k \leftarrow k - 1$.
9. *Back Substitution.* Start with $j = k$. While $j \geq 1$ do the following.
   (a) If x_j has been marked as arbitrary, assign it an arbitrary value (for example 0). If x_j is not marked as arbitrary and $j < n$, substitute the values for $x_{j+1}, \ldots, x_n$ in $\mathcal{R}_j$ and determine a feasible value for x_j.
   (b) Set $k \leftarrow k - 1$.

Except for very small problems, a more efficient algorithm is needed because in general, if half of the m coefficients of x_1 appear with opposite sign, then the elimination would cause the number of inequalities to grow on the next step from m to $(m/2)^2$. Thus, it is possible after a few eliminations that the number of remaining inequalities could become too numerous. However, from a theoretical point of view, the method is a powerful one since it can be used to prove such fundamental theorems as the infeasibility theorem and the duality theorem.

▷ **Exercise 2.12** Show that if the worst case described above could occur at each iteration, then at the end of n iterations the number of inequalities could grow to

$$\frac{1}{2^{2^n-2}}\left(\frac{m}{2}\right)^{2^n},$$

where m is the number of inequalities at the start of the FME algorithm and n is the number of variables.

2.3.3 FOURIER-MOTZKIN ELIMINATION THEORY

In this section we provide the theoretical background for the Fourier-Motzkin Elimination process.

WHY IT WORKS

Suppose that we wish to find solutions to the following system:

$$\sum_{j=1}^{n} a_{ij}x_j \geq b_i, \qquad \text{for } i = 1, \ldots, m, \tag{2.15}$$

where all inequalities are written as in (2.15) with variables on the left and constants on the right of the $\geq$ symbol. Since this problem is trivial if $m = 1$ or $n = 1$, we assume, to simplify the discussion, that $m > 1$ and $n > 1$.

In outline, the Fourier-Motzkin elimination process begins by eliminating x_1 by adding every pair of inequalities in which x_1 appears with a $+1$ in one and -1 in the other. This generates a new system of inequalities, called the *reduced system*, in which x_1 appears with zero coefficient in all its inequalities. The process is repeated with the new system except now x_2 is eliminated followed by $x_3, x_4, \dots, x_n$ in turn. The iterative process stops either when

1. it is not possible to carry out the elimination procedure on the next variable to be eliminated because all the next variable coefficients are $+1$ (or all -1), or

2. all variables have already been eliminated.

For the first of these possibilities, it is easy to find a feasible solution; for the second of these possibilities it is easy to find a feasible solution of the form $\sum_j 0x_j \geq \gamma$, or show that none exists because $\gamma > 0$. If the final solution is feasible, then a feasible solution for the original system can be found by a sequence of back-substitution steps.

We can reformulate (2.15) by partitioning its constraints into 3 groups, h, k, and l, depending on whether a particular constraint has its x_1 coefficient "> 0," "< 0," or "$= 0$." After dividing by the absolute value of the coefficient of x_1 when nonzero and rearranging the terms and the order of the inequalities, we can write these as

$$x_1 + \sum_{j=2}^{n} D_{hj}x_j \geq d_h, \quad h = 1, \dots, H, \tag{2.16}$$

$$-x_1 + \sum_{j=2}^{n} E_{kj}x_j \geq e_k, \quad k = 1, \dots, K, \tag{2.17}$$

and the remainder

$$\sum_{j=2}^{n} F_{lj}x_j \geq f_l, \quad l = 1, \dots, L. \tag{2.18}$$

Note that $H + K + L = m$, where we assume for the moment that $H \geq 1$ and $K \geq 1$.

Clearly this is equivalent to system (2.15). That is, any solution (if one exists) to (2.16), (2.17), and (2.18) is a solution to (2.15) and vice versa. We refer to (2.16), (2.17), and (2.18) as the original system. Assume that $(x_1, x_2, \dots x_n) = (x_1^o, x_2^o, \dots x_n^o)$ is a feasible solution. Setting aside (2.18) for the moment, the "elimination" of x_1 is done by adding the hth constraint of (2.16) to the kth constraint of (2.17), thus obtaining

$$\sum_{j=2}^{n} E_{kj}x_j + \sum_{j=2}^{n} D_{hj}x_j \geq e_k + d_h. \tag{2.19}$$

When we say x_1 has been "eliminated" from (2.19), we mean that the coefficient of x_1 is zero. We do this for every combination of $h = 1, \dots, H$, and $k = 1, \dots, K$. The new system of inequalities obtained by eliminating x_1 consists of the L inequalities (2.18) and the $H \times K$ inequalities (2.19). Since the $L + H \times K$ inequalities are implied by (2.16), (2.17), and (2.18), it follows that $(x_1, x_2, \ldots x_n) = (x_1^o, x_2^o, \ldots x_n^o)$ is a feasible solution to the "reduced" system (2.18) and (2.19). When we say that the system (2.18) and (2.19) is "reduced," we mean it has at least one less variable than the original system.

▷ **Exercise 2.13** When is it legal to take a linear combination of linear inequalities. Illustrate by way of examples, cases when it is legal and when it is illegal to do so.

LEMMA 2.2 (Equivalence After Elimination) *If a feasible solution exists for the original system* (2.16), (2.17), *and* (2.18), *then one exists for the reduced system* (2.18) *and* (2.19) *and vice versa.*

Proof. We have just shown that if (2.16), (2.17), and (2.18) hold for some choice of $(x_1, x_2, \ldots, x_n) = (x_1^o, x_2^o, \ldots, x_n^o)$, then (2.18) and (2.19) hold for the same values of $(x_1, x_2, \ldots, x_n) = (x_1^o, x_2^o, \ldots, x_n^o)$. Conversely, given a feasible solution $(x_2, x_3, \ldots, x_n) = (x_2^1, x_3^1, \ldots, x_n^1)$ to the eliminated system (2.18) and (2.19), we wish to show that we can find an $x_1 = x_1^1$ together with $(x_2, x_3, \ldots x_n) = (x_2^1, x_3^1, \ldots x_n^1)$ such that (2.16), (2.17), and (2.18) hold. This is easily done by choosing any $x_1 = x_1^1$ satisfying (2.20):

$$\min_{1 \le k \le K} \left(\sum_{j=2}^{n} E_{kj} x_j^1 - e_k \right) \ge x_1^1 \ge \max_{1 \le h \le H} \left(- \sum_{j=2}^{n} D_{hj} x_j^1 + d_h \right). \tag{2.20}$$

That such an $x_1 = x_1^1$ exists follows because we can rewrite (2.19) as

$$\sum_{j=2}^{n} E_{kj} x_j^1 - e_k \ge - \sum_{j=2}^{n} D_{hj} x_j^1 + d_h \tag{2.21}$$

for every pair (h, k) and hence for the h that maximizes the left hand side and the k that minimizes the right-hand side of (2.21); this implies that (2.20) holds. Therefore, for every h and k combination

$$\sum_{j=2}^{n} E_{kj} x_j^1 - e_k \ge x_1^1 \ge - \sum_{j=2}^{n} D_{hj} x_j^1 + d_h$$

and

$$\sum_{j=2}^{n} F_{lj} x_j^1 \ge f_l, \quad l = 1, \ldots, L.$$

∎

▷ **Exercise 2.14** Prove that Lemma 2.2 implies that if the original system is infeasible then so is the reduced system and vice versa.

Thus, the final reduced system of inequalities consists of (2.21), and the set of inequalities (2.18), which we set aside, i.e.,

$$\sum_{j=2}^{n} E_{kj}x_j - e_k \geq -\sum_{j=2}^{n} D_{hj}x_j + d_h, \ h = 1,\dots,H, \ \ k = 1,\dots,K \tag{2.22}$$

and

$$\sum_{j=2}^{n} F_{lj}x_j \geq f_l, \ l = 1,\dots,L. \tag{2.23}$$

If in fact $H = 0$ or $K = 0$, then (2.22) is vacuous and the reduced system consists of (2.23) only. If $H = 0$ and $L = 0$ (or $K = 0$ and $L = 0$), the reduced system is vacuous and we terminate the elimination procedure.

The process of moving from (2.15) to (2.22) and (2.23) is called eliminating x_1 by the *Fourier-Motzkin Elimination (FME)* process. Of course, there is nothing forcing us to stop here. If we wish, we could proceed to eliminate x_2 from the reduced system provided the reduced system is not vacuous. We keep on eliminating, in turn, $x_1, x_2, \dots$ until at some step $k \leq n$, the reduced system is vacuous or all the variables are eliminated.

However, we pause to observe that there are two cases that could cause our elimination of x_1 (or, at a future step, x_k) to be impossible to execute.

Case 1. **All coefficients of x_1 in (2.15) are equal to 0.** If x_1 is the last variable to be eliminated (no more x_j remain to be eliminated) then *terminate.* In the latter situation terminate infeasible if $b_i > 0$ for some i, otherwise x_1 may be chosen arbitrarily. If x_1 is not the last variable to be eliminated, then the "elimination" results in just system (2.18) which we had set aside. In the latter situation, we declare x_1 "eliminated" and proceed to solve (2.18). If feasible then any solution $x_2, x_3, \dots, x_n$ to (2.18) with x_1 arbitrary is a solution to (2.15).

Case 2. **The original system (2.15) consists of just (2.16) or (2.17) (but not both) and (2.18).** If there are no relations (2.18), then *terminate.* In the latter situation choose $x_2, x_3, \dots, x_n$ (if any remain) arbitrarily; x_1 can then be chosen sufficiently positive to satisfy just (2.16) or sufficiently negative to satisfy (2.17). If relations (2.18) are not vacuous then we declare x_1 "eliminated" and (2.18) as the reduced system. Any solution to (2.18) if it exists can be substituted into (2.16) or (2.17), and a value of x_1 can be found sufficiently positive or negative as above.

INFEASIBILITY MULTIPLIERS

Definition: An inequality is said to be a *nonnegative linear combination* of inequalities (2.15) if it is formed by multiplying each inequality i by some $y_i \geq 0$ for $i = 1, \dots, m$ and summing.

▷ **Exercise 2.15** Prove that each inequality of (2.22) and (2.23) is formed as either a positive linear combination of inequalities of the original system or is one of the original inequalities.

It is evident from Exercise 2.15 that each inequality generated by the elimination of x_1 can also be formed by nonnegative linear combinations (not all zero) of the first system of inequalities. The third system of inequalities that are each formed by the elimination of x_2 can also be formed in the same way by nonnegative linear combinations (not all zero) of the second system of inequalities, which in turn were formed by nonnegative linear combinations of the first system. Hence the third system of inequalities can be formed by nonnegative linear combinations of the first system of inequalities. Eventually the FME process terminates with either a vacuous set of inequalities and a feasible solution or a final system of inequalities consisting of some set of P inequalities of the form

$$0x_1 + 0x_2 + \cdots + 0x_n \geq \Gamma_i, \qquad i = 1, \ldots, P,$$

where Γ_i is some constant. The original system is feasible depending on whether or not all $\Gamma_i \leq 0$. Each such inequality could have been generated directly by a non-negative linear combination of the relations of the original system using multipliers $y_1 \geq 0, y_2 \geq 0, \ldots\ y_m \geq 0$ not all zero. Applying such a set of nonnegative weights $(y_1, y_2, \ldots, y_m)$ to the system (2.15) and adding we have

$$\sum_{k=1}^{m} y_k \left(\sum_{j=1}^{n} a_{kj} x_j \right) \geq \sum_{k=1}^{m} y_k b_k$$

which we can rewrite as

$$\sum_{j=1}^{n} \left(\sum_{k=1}^{m} y_k a_{kj} \right) x_j \geq \sum_{k=1}^{m} y_k b_k. \tag{2.24}$$

Since all variables have been eliminated, it must be that for each such set of weights $(y_1, y_2, \ldots, y_m)$,

$$\sum_{k=1}^{m} y_k a_{kj} = 0, \qquad \text{for } j = 1, \ldots, n.$$

Thus, we note the following two cases:

Case Feasible: If $\sum_{k=1}^{m} y_k b_k \leq 0$ for each such set of m nonnegative weights $(y_1, y_2, \ldots, y_m)$, then (2.24), and hence (2.15), is feasible.

Case Infeasible: If $\sum_{k=1}^{m} y_k b_k > 0$ for one or more such set of m nonnegative weights $(y_1, y_2, \ldots, y_m)$, then (2.24), and hence (2.15), is infeasible.

2.4 INFEASIBILITY THEOREM

A system of inequalities can be written with the inequalities all pointing in the same direction by multiplying those that do not through by -1. Assuming this is done, a system of inequalities is clearly infeasible if we can exhibit a nonnegative linear combination of the inequalities that is an ***infeasible inequality***, that is, an inequality of the form

$$0x_1 + 0x_2 + \cdots + 0x_n \geq d \text{ with } d > 0. \tag{2.25}$$

The Infeasibility Theorem (Theorem 2.3 below) states that if a system of linear inequalities is infeasible, then we can always find a nonnegative linear combination that results in (2.25).

▷ **Exercise 2.16** Prove that (2.25) is the only type of single linear inequality that is infeasible.

▷ **Exercise 2.17** Typically, linear programs are stated in the form of equations in nonnegative variables. Prove that the only single infeasible equation in nonnegative variables x_j is of the form

$$\alpha_1 x_1 + \alpha_2 x_2 + \cdots + \alpha_n x_n = d, \tag{2.26}$$

with $\alpha_j \geq 0$ for all j and $d < 0$ (or $\alpha_j \leq 0$ for all j and $d > 0$).

▷ **Exercise 2.18 (Converting Equalities to Inequalities)** Show how to convert the following system in m linear equations in n nonnegative variables

$$\begin{aligned} \sum_{j=1}^{n} a_{ij}x_j &= b_i, \quad \text{for } i = 1 \text{ to } m, \\ x_j &\geq 0, \quad \text{for } j = 1 \text{ to } n, \end{aligned}$$

to a system of linear inequalities in nonnegative variables by two different methods, one that replaces the equations by $2m$ inequalities and one that replaces them by only $m + 1$ inequalities.

▷ **Exercise 2.19** Most software to solve linear programs internally converts a system of inequalities to a system of equalities in bounded variables, where some bounds may be $+\infty$ or $-\infty$. Show how this can be done.

THEOREM 2.3 (Infeasibility Theorem) *The system of linear inequalities*

$$\sum_{j=1}^{n} a_{ij}x_j \geq b_i, \quad \textit{for } i = 1, \ldots, m \tag{2.27}$$

is infeasible if and only if there exists a nonnegative linear combination of the inequalities that is an infeasible inequality.

Comment: In matrix notation, the system $Ax \geq b$ is infeasible if and only if there exists a vector $y \geq 0$ such that $y^T Ax \geq y^T b$ is an infeasible inequality, namely one where $y^T A = 0$ and $y^T b > 0$.

Proof. The theorem states that the system (2.27) is infeasible if and if only there exist $y_k \geq 0$, for $k = 1, \ldots, m$, such that

$$\sum_{k=1}^{m} y_k a_{kj} = 0,\ j = 1, \ldots, n, \text{ and } \sum_{k=1}^{m} y_k b_k > 0. \tag{2.28}$$

If (2.27) is infeasible then the y_k obtained using the Fourier-Motzkin Elimination (FME) for Case Infeasible in Section 2.3.3 can be used to obtain an infeasible inequality and hence $(y_1, y_2, \ldots, y_m)$ satisfying (2.28). Thus (2.27) is infeasible implies that there exists $y_k \geq 0$, $k = 1, \ldots, m$ such that (2.28) holds. On the other hand, if (2.28) holds, then it is obvious that system (2.27) is infeasible because multiplying (2.27) by $y_1 \geq 0, y_2 \geq 0, \ldots, y_m \geq 0$ and summing results in an infeasible inequality of the form (2.25). ∎

▷ **Exercise 2.20** Consider the system of linear equations

$$\begin{array}{rcrcrcl} x_1 & + & x_2 & & & = & 1 \\ & - & x_2 & + & x_3 & = & 1 \\ x_1 & & & + & x_3 & = & 1. \end{array}$$

Note that the sum of the first two equations results in $x_1 + x_3 = 2$, which contradicts the third equation. Show that eliminating x_1 and x_2 in turn results in an equation, called an *infeasible equation*,

$$0x_1 + 0x_2 + 0x_3 = d$$

where $d \neq 0$. What multipliers applied to the original system of equations results in the infeasible equation above? Show how the process of elimination can be used to find these multipliers.

COROLLARY 2.4 (Infeasible Equation) *If a system of linear equations in nonnegative variables is infeasible, there exists a linear combination of the equations that is an infeasible equation in nonnegative variables.*

▷ **Exercise 2.21** Prove Corollary 2.4 by converting the system of equations into a system of inequalities and applying Theorem 2.3.

2.5 NOTES & SELECTED BIBLIOGRAPHY

Fourier [1826] first proposed the Fourier-Motzkin Elimination (FME) Method. The paper in its original form is accessible in Fourier [1890]; an English translation was done by Kohler [1973]. Fourier's method was later reintroduced by T. Motzkin [1936] and is sometimes referred to as the Motzkin Elimination Method. A good discussion of the method can be found in Kuhn [1956].

As originally mentioned by Fourier, the efficiency of the FME process can be greatly improved by detecting and removing redundant inequalities, where an inequality is said to be redundant if its removal does not affect the feasible set of solutions. In general the detection of redundant inequalities is very difficult. (For a discussion on how to identify redundant constraints so as to be able to obtain the set of feasible solutions with the least number of constraints see, for example, Adler [1976], Luenberger [1973], and Shefi [1969].) However, it is possible to *detect redundancies* in a computationally efficient manner when they occur as a result of combining inequalities during the iterations of the FME process, see Duffin [1974]. Unfortunately, even this is not enough to make the method competitive with the Simplex Method for solving linear programs.

2.6 PROBLEMS

2.1 The Soft Suds Brewing and Bottling Company, because of faulty planning, was not prepared for the Operations Research Department. There was to be a big party at Stanford University, and Gus Guzzler, the manager, knew that Soft Suds would be called upon to supply the refreshments. However, the raw materials required had not been ordered and could not be obtained before the party. Gus took an inventory of the available supplies and found the following:

Malt	75 units,
Hops	60 units,
Yeast	50 units.

Soft Suds produces two types of pick-me-ups: light beer and dark beer, with the following specifications:

	Requirement per gallon		
	Malt	Hops	Yeast
Light beer	2	3	2
Dark beer	3	1	5/3

The light beer brings \$2.00/gallon profit, the dark beer \$1.00/gallon profit. Knowing the O.R. Department will buy whatever is made available, formulate the linear program Gus must solve to maximize his profits, and solve it graphically. Be sure to define all of your variables.

2.2 For the linear program

$$\begin{array}{lrcl} \text{Maximize} & x_1 + x_2 & = & z \\ \text{subject to} & -x_1 + x_2 & \leq & 2 \\ & x_1 + 2x_2 & \geq & 2 \end{array}$$

plot the feasible region graphically and show that the linear program is unbounded.

2.3 Consider the following linear program:

$$
\begin{array}{lrcrcl}
\text{Maximize} & 3x_1 & + & x_2 & & \\
\text{subject to} & -x_1 & + & x_2 & \le & -1 \\
 & -3x_1 & - & x_2 & \le & -1 \\
 & 4x_1 & + & 2x_2 & \le & 1 \\
 & & & 2x_2 & \le & -1 \\
 & \multicolumn{5}{l}{x_1 \ge 0,\ x_2 \ge 0.}
\end{array}
$$

(a) Plot it graphically and identify all the corner point solutions.
(b) Solve it graphically.
(c) Solve it with the **DTZG Simplex Primal** software option.
(d) Solve it by hand by the FME algorithm.
(e) Solve it by the **Fourier-Motzkin Elimination** software option.

2.4 Consider the following two-equation linear program:

$$
\begin{array}{lrcrcrcrcrcl}
\text{Minimize} & 2x_1 & + & 3x_2 & + & x_3 & + & 5x_4 & + & x_5 & = & z \\
\text{subject to} & 4x_1 & + & 2x_2 & + & 3x_3 & + & x_4 & + & 4x_5 & \le & 50 \\
 & 3x_1 & + & 7x_2 & + & x_3 & + & 3x_4 & + & 2x_5 & \le & 100 \\
 & \multicolumn{11}{l}{x_j \ge 0,\ j = 1, \ldots, 5.}
\end{array}
$$

(a) Solve it using the **DTZG Simplex Primal** software option.
(b) Solve it graphically.
(c) Use the graphical representation to write down its dual linear program. Use it to verify optimality.

2.5 Consider the following two-equation linear program:

$$
\begin{array}{lrcrcrcrcrcl}
\text{Minimize} & -5x_1 & + & 3x_2 & - & 2x_2 & + & x_4 & - & 2x_5 & = & z \\
\text{subject to} & x_1 & + & x_2 & + & 3x_3 & + & 2x_4 & + & x_5 & \le & 1000 \\
 & 5x_1 & + & 3x_2 & + & x_3 & + & 5x_4 & + & 2x_5 & \le & 2000 \\
 & \multicolumn{11}{l}{x_j \ge 0,\ j = 1, \ldots, 5.}
\end{array}
$$

(a) Solve it using the **DTZG Simplex Primal** software option.
(b) Solve it graphically.
(c) Use the graphical representation to write down its dual linear program. Use it to verify optimality.

2.6 Consider the data for Example 1.4 on page 3. Suppose now that the manufacturer wishes to produce an alloy (blend) that is 35 percent lead, 30 percent zinc, and 35 percent tin.

(a) Formulate this problem and solve it graphically.
(b) Solve it using the **DTZG Simplex Primal** software option.

2.7 Consider the following two-variable linear program:

$$
\begin{array}{lrcrcl}
\text{Minimize} & x_1 & + & x_2 & = & z \\
\text{subject to} & x_1 & + & 2x_2 & \ge & 2 \\
 & 3x_1 & + & 2x_2 & \le & 1 \\
 & x_1 & + & x_2 & \ge & 1 \\
\multicolumn{6}{l}{x_1 \ge 0,\ x_2 \ge 0.}
\end{array}
$$

(a) Plot the region graphically and show that it is empty.
(b) Solve it with the **DTZG Simplex Primal** software option.
(c) Solve it by hand using the FME process.
(d) Solve it using the **Fourier-Motzkin Elimination** software option.

2.8 Consider the following two-variable linear program:

$$\begin{array}{lrcrcl} \text{Minimize} & 9x_1 & + & 8x_2 & = & z \\ \text{subject to} & x_1 & - & 2x_2 & \le & 3 \\ & 3x_1 & - & 4x_2 & \ge & 5 \\ & 6x_1 & - & 7x_2 & = & 8 \\ & \multicolumn{5}{l}{x_1 \ge 0,\ x_2 \ge 0.} \end{array}$$

(a) Plot it graphically and identify all the corner point solutions.
(b) Solve it graphically.
(c) Solve it with the **DTZG Simplex Primal** software option.
(d) Solve it by hand by the FME algorithm.
(e) Solve it by the **Fourier-Motzkin Elimination** software option.

2.9 Consider the following two-variable linear program:

$$\begin{array}{lrcrcl} \text{Minimize} & -x_1 & + & 4x_2 & = & z \\ \text{subject to} & -3x_1 & + & x_2 & \le & 6 \\ & x_1 & + & 2x_2 & \le & 4 \\ & x_1 & & & \ge & -1 \\ & & & x_2 & \ge & -3. \end{array}$$

(a) Plot the region graphically and solve the problem.
(b) Reformulate the problem so that the lower bounds on all the variables are 0.
(c) Plot the reformulated problem graphically and re-solve the problem. Derive the solution to the original problem from this solution.

2.10 Consider the following two-variable system of inequalities.

$$\begin{array}{rcrcl} x_1 & + & 2x_2 & \ge & 2 \\ 3x_1 & + & 2x_2 & \le & 1 \\ x_1 & & & \ge & 0 \\ & & x_2 & \ge & 0 \end{array}$$

(a) Solve the problem by the FME process.
(b) Plot graphically and show that it is infeasible by showing that the set of feasible points is empty.

2.11 Graphically show that the two-variable linear program

$$\begin{array}{lrcrcl} \text{Minimize} & -x_1 & - & 2x_2 & = & z \\ \text{subject to} & -x_1 & + & x_2 & \le & -2 \\ & 4x_1 & + & x_2 & \le & 4 \\ & \multicolumn{5}{l}{x_1 \ge 0,\ x_2 \ge 0} \end{array}$$

has no feasible solution.

2.12 Use the **DTZG Simplex Primal** software option to solve

$$\begin{array}{lrcl} \text{Maximize} & 3x_1 + 2x_2 + x_3 & = & z \\ \text{subject to} & 5x_1 - 2x_2 + x_3 & \leq & 6 \\ & 2x_1 - x_2 - x_3 & \leq & 4 \\ & 9x_1 - 4x_2 - x_3 & \geq & 15 \\ & x_1 \geq 0,\ x_2 \geq 0,\ x_3 \geq 0. & & \end{array}$$

2.13 Consider the following linear program

$$\begin{array}{lrcl} \text{Minimize} & 2x_1 - x_2 + 3x_3 + 7x_4 - 5x_5 & = & z \\ \text{subject to} & x_1 + 2x_2 + x_3 + x_4 + 6x_5 & \leq & 10 \\ & 2x_1 + 3x_2 + 4x_3 + x_4 + 2x_5 & \geq & 4 \\ & 3x_1 + 2x_2 \qquad\qquad + 3x_4 + x_5 & \leq & 8 \\ & x_1 \geq 0,\ x_2 \geq 0. & & \end{array}$$

(a) Solve it with the **DTZG Simplex Primal** software option.
(b) Solve it by hand by the FME algorithm.
(c) Solve it by the **Fourier-Motzkin Elimination** software option.

2.14 Consider the linear program

$$\begin{array}{lrcl} \text{Maximize} & x_1 + 3x_2 + 2x_3 & = & z \\ \text{subject to} & x_1 + x_2 + x_3 & = & 1 \\ & 7x_1 + 2x_2 + 3x_3 & \leq & 20 \\ & x_1 + 5x_2 + 4x_3 & \leq & 30 \\ & x_1 \geq 0,\ x_2 \geq 0,\ x_3 \geq 0. & & \end{array}$$

(a) Solve it by the **DTZG Simplex Primal** software option.
(b) Solve it by hand by the FME algorithm.
(c) Solve it by the **Fourier-Motzkin Elimination** software option.

2.15 *Degeneracy.* Look at the feasible region defined by

$$\begin{array}{rcl} x_1 + 2x_2 & \leq & 8 \\ x_1 + x_2 & \leq & 6 \\ x_1,\ x_2 & \geq & 0. \end{array} \tag{2.29}$$

(a) Draw the feasible region in (x_1, x_2)-space and label the constraints.
(b) Notice that including nonnegativity, we have four constraints. What is the solution corresponding to each extreme point of the feasible region?
(c) Suppose we add the constraint

$$x_2 \leq 4. \tag{2.30}$$

In your diagram, the extreme point $(4, 0)$ of the feasible region is now the intersection of three constraints, and any two of them will uniquely specify that extreme point. Show that there are three ways to do this.

(d) When there is more than one way to specify an extreme point, the extreme point is said to be degenerate. In part (c) we created an example of degeneracy by using a redundant system of inequalities. The redundancy can be seen in the diagram in that we could remove one of the constraints without changing the feasible region. Give an example of degeneracy with a *nonredundant* system of inequalities. Draw a picture to demonstrate this.

2.16 *Sensitivity to Changes in One Objective Coefficient.* Consider

$$\begin{array}{ll} \text{Minimize} & 2x_1 + c_2x_2 = z \\ \text{subject to} & 3x_1 + 2x_2 \leq 4 \\ & 2x_1 - 3x_2 \geq 6 \\ & x_1 \geq 0,\ x_2 \geq 0. \end{array}$$

(a) Solve the linear program graphically for $c_2 = 0$.
(b) By adjusting c_2, determine graphically the range of c_2 for which the solution stays optimal.

2.17 *Sensitivity to Changes in One Right-Hand Side Value.* Consider

$$\begin{array}{ll} \text{Minimize} & x_1 + 3x_2 = z \\ \text{subject to} & 3x_1 - 2x_2 \leq b_1 \\ & x_1 + 3x_2 = 2 \\ & x_1 \geq 0,\ x_2 \geq 0. \end{array}$$

(a) Solve the linear program graphically for $b_1 = 2$.
(b) By adjusting b_1, determine graphically the range of b_1 for which the solution stays optimal.
(c) What is the range of the objective value for the range of b_1 in Part(b)?
(d) Deduce the relationship between the change in b_1 to the change in objective value?
(e) By adjusting b_1, determine graphically the range of b_1 for which the solution stays feasible.

2.18 *Sensitivity to Changes in a Matrix Coefficient.* Consider

$$\begin{array}{ll} \text{Minimize} & 3x_1 + 2x_2 = z \\ \text{subject to} & x_1 + x_2 \geq 1 \\ & 3x_1 - a_{22}x_2 \leq 6 \\ & x_1 - 2x_2 \leq 4 \\ & x_1 \geq 0,\ x_2 \geq 0. \end{array}$$

(a) Solve the linear program graphically for $a_{22} = 2$.
(b) By adjusting a_{22}, determine graphically the range of a_{22} for which the solution stays optimal.
(c) What is the range of the objective value for the range of a_{22} in Part 2?
(d) By adjusting a_{22}, determine graphically the range of a_{22} for which the solution stays feasible.

2.19 *Shadow Price.*

$$\begin{array}{ll} \text{Maximize} & 2x_1 + x_2 = z \\ \text{subject to} & x_1 + 2x_2 \leq 14 \\ & 3x_1 - x_2 \geq 2 \\ & x_1 + 4x_2 \leq 18 \\ & x_1 \geq 0,\ x_2 \geq 0. \end{array}$$

(a) Solve the problem graphically.

(b) The *shadow price* of an item i is the change in objective value as a result of unit change in b_i. Find the shadow prices on each constraint graphically.

(c) How much would resource 1 have to increase to get an objective value increase of 8?

(d) How much would resource 1 have to increase to get an objective value increase of 12? Is this increase possible?

2.20 Consider the linear program:

$$\begin{array}{lrcl} \text{Maximize} & 2x_1 + x_2 & = & z \\ \text{subject to} & 2x_1 + x_2 & \leq & 10 \\ & 3x_1 - x_2 & \geq & 2 \\ & -x_1 + x_2 & \leq & 4 \\ & x_1 \quad & \leq & 5 \\ & x_1 \geq 0, & x_2 \geq & 0. \end{array}$$

(a) Plot the feasible region graphically.

(b) Show graphically that multiple optimal solutions exist.

(c) Write down the set of feasible optimal solutions (not necessarily corner points).

2.21 Consider the linear program:

$$\begin{array}{lrcl} \text{Minimize} & 2x_1 + 5x_2 & = & z \\ \text{subject to} & x_1 + x_2 & \leq & 1 \\ & 2x_1 + 2x_2 & \leq & 5 \\ & x_1 \quad & \leq & 3 \\ & x_2 & \leq & 5 \\ & x_1 \geq 0, & x_2 \geq & 0. \end{array}$$

(a) Plot the feasible region graphically.

(b) Identify the redundant inequalities.

(c) Solve the problem graphically.

2.22 Consider the linear program:

$$\begin{array}{rcl} x_1 + x_2 & \geq & 1 \\ x_1 - x_2 & \leq & 2 \\ x_1 + x_2 + x_3 & \leq & 5. \end{array}$$

(a) Plot graphically.

(b) Solve by hand, using the FME process.

(c) Show how to generate the class of all solutions for the problem.

2.23 Show by a graphical representation whether there is no solution, one solution, or multiple solutions to the following systems of inequalities:

(a)

$$\begin{aligned} x_1 + x_2 &\geq 1 \\ x_1 &\geq 0 \\ x_2 &\geq 0. \end{aligned}$$

(b)

$$\begin{aligned} x_1 + x_2 &\geq 1 \\ x_1 + 2x_2 &\leq 4 \\ -x_1 + 4x_2 &\geq 0 \\ x_1 &\geq 0 \\ x_2 &\geq 0. \end{aligned}$$

(c)

$$\begin{aligned} x_1 + x_2 &\geq 1 \\ x_1 + 2x_2 &\leq 4 \\ -x_1 + 4x_2 &\geq 0 \\ -x_1 + x_2 &\geq 1 \\ x_1 &\geq 0 \\ x_2 &\geq 0. \end{aligned}$$

2.24 Consider the following set of inequalities:

$$\begin{aligned} -x_1 + 2x_2 + x_3 &\leq 1 \\ x_1 - x_2 - x_3 &\leq 0 \\ x_1 - x_2 - x_3 &\leq -1 \\ - x_2 &\leq 0. \end{aligned}$$

(a) Apply the FME process by hand. Stop the algorithm as soon as you encounter the inequality $0 \leq 0$. Note that this is possible even if more variables remain to be eliminated.

(b) Find the nonnegative multipliers $\pi_1 \geq 0$, $\pi_2 \geq 0$, $\pi_3 \geq 0$, and $\pi_4 \geq 0$ of the original system that gives the trivial inequality $0 \leq 0$.

(c) Show that for all $i = 1, \ldots, 4$ for which $\pi_i > 0$, we can replace the ith inequality by an equality without changing the set of feasible solutions.

(d) Prove a generalization of the above, i.e., if a nonnegative linear combination of a given system of linear inequalities is the trivial inequality $0 \leq 0$, then that system is equivalent to the system in which all the inequalities corresponding to the positive multipliers are replaced by equalities.

2.25 Consider the following linear program:

$$\begin{array}{lrcl} \text{Maximize} & x + y & & \\ \text{subject to} & 8x + 3y & \leq & 24 \\ & 5x + 7y & \leq & 35 \\ & -x + y & \leq & 4 \\ & y & \geq & -2. \end{array} \tag{2.31}$$

(a) Solve (2.31) using the Fourier-Motzkin Elimination process. Find the optimal values of x and y as well as the optimal objective value. Hint: it is convenient to introduce another variable z to keep track of the objective value. That is, one possibility is $z - x - y \le 0$.

(b) If we change the right-hand side of the third inequality in (2.31) from 4 to -7 then the system becomes infeasible. Use the Fourier-Motzkin Elimination process to find the infeasibility multipliers; that is, the multipliers $y_1, y_2, \ldots, y_m$ that resulted in the final inequality being infeasible. Note: It is not necessary to start from scratch. The elimination you have already done should help.

2.26 The infeasibility theorem for inequality systems is called a *theorem of the alternative* when stated as

Theorem of the Alternative. *Either* there exists an x such that $Ax \ge b$ *or* there exists a $y \ge 0$ such that $A^T y = 0$ and $y^T b > 0$ *but not both.*

Prove the following theorems of the alternative by using the above theorem

(a) *Either* there exists an $x \ge 0$ such that $Ax \le b$ *or* there exists a $y \ge 0$ such that $A^T y \ge 0$ and $y^T b < 0$ *but not both.*

(b) *Either* there exists an $x > 0$ (i.e., $x_i > 0$ for all i) such that $Ax = 0$ *or* there exists a π such that $0 \ne A^T \pi \ge 0$ *but not both.*

2.27 Use calculus to solve

$$\begin{array}{lrcl} \text{Minimize} & x_1 + 2x_2 + 3x_3 & = & z \\ \text{subject to} & x_1 + x_2 + x_3 & = & 1 \\ \multicolumn{4}{l}{\text{and } x_1 \ge 0,\ x_2 \ge 0,\ x_3 \ge 0} \end{array}$$

as follows. The nonnegativity of x_1, x_2, x_3 may be circumvented by setting $x_1 = u_1^2$, $x_2 = u_2^2$, $x_3 = u_3^2$:

$$\begin{array}{lrcl} \text{Minimize} & u_1^2 + 2u_2^2 + 3u_3^2 & = & z \\ \text{subject to} & u_1^2 + u_2^2 + u_3^2 & = & 1. \end{array}$$

Form the Lagrangian

$$L(u_1, u_2, u_3) = u_1^2 + 2u_2^2 + 3u_3^2 - \lambda(u_1^2 + u_2^2 + u_3^2 - 1).$$

Setting $\partial L/\partial u_1 = 0$, $\partial L/\partial u_2 = 0$, $\partial L/\partial u_3 = 0$ results in

$$u_1(1 - \lambda) = 0, \quad u_2(2 - \lambda) = 0, \quad u_3(3 - \lambda) = 0.$$

(a) Try to complete the solution by analyzing which member of each pair is zero.

(b) Consider the general linear program

$$\begin{array}{lll} \text{Minimize} & c^T x & \\ \text{subject to} & Ax = b, & A : m \times n, \\ & x \ge 0. & \end{array}$$

Substitute for each x_j, $u_j^2 = x_j$; form the Lagrangian; and set $\partial L/\partial u_j = 0$ for all j. Try to discover why the classical approach is not a practical one.

CHAPTER 3

THE SIMPLEX METHOD

In this chapter we develop the *Dantzig Simplex Method* for solving linear programming problems. For the origins of this method see the Foreword. The Simplex Method solves linear programs by moving along the boundaries from one vertex (extreme point) to the next. Later on, in Chapter 4, we will discuss techniques for solving linear programs that have become very popular since the 1980s that move instead through the interior of the feasible set of solutions.

The Simplex Method is a very efficient procedure for solving large practical linear programs on the computer. Classes of examples, however, have been constructed where the number of pivot steps required by the method can grow by an exponential function of the dimensions of the problem. Never to our knowledge has anything like this *worst-case* performance been observed in real world problems. Nevertheless, such unrealistic examples have stimulated the development of a theory of alternative methods for each of which the number of steps are guaranteed not to grow exponentially with problem size whatever the structure of the matrix of coefficients.

Given a *fixed* problem, say one with $m = 1000$ rows with a highly specialized structure the goal is find the best algorithm for solving it. This theory so far has provided us with little or no guide as to which algorithm is likely to be best.

Most of the discussion in this chapter and other chapters will refer to a linear program in *standard form*, that is,

$$\begin{array}{lllllllll} c_1x_1 & + & c_2x_2 & + & \cdots & + & c_nx_n & = & z \text{ (Min)} \\ a_{11}x_1 & + & a_{12}x_2 & + & \cdots & + & a_{1n}x_n & = & b_1 \\ a_{21}x_1 & + & a_{22}x_2 & + & \cdots & + & a_{2n}x_n & = & b_2 \\ \vdots & & \vdots & & \vdots & & \vdots & & \vdots \\ a_{m1}x_1 & + & a_{m2}x_2 & + & \cdots & + & a_{mn}x_n & = & b_m, \\ \multicolumn{9}{l}{\text{and } x_1 \geq 0,\ x_2 \geq 0,\ \ldots,\ x_n \geq 0.} \end{array} \tag{3.1}$$

It is assumed that the reader is familiar with simple matrix notation and operations. See Appendix A for definitions of these basic concepts. In matrix notation (3.1) can be rewritten as

$$\begin{array}{lll} \text{Minimize} & c^T x & \\ \text{subject to} & Ax = b, & A:\ m \times n, \\ & x \geq 0, & \end{array} \tag{3.2}$$

where A is a rectangular matrix of dimension $m \times n$, b is a column vector of dimension m, c is a column vector of dimension n, x is a column vector of dimension n, and the superscript T stands for transpose.

We start by illustrating the method graphically in Section 3.1. In Section 3.2 the Simplex Algorithm will be described; its use, as part of the Simplex Method, will be developed in Section 3.3. Next we will examine linear programs in *standard form with bounds*; these are systems whose nonnegativity constraints have been replaced by upper and lower bounds on each variable x_j as shown below:

$$\begin{array}{lll} \text{Minimize} & c^T x & \\ \text{subject to} & Ax = b, & A:\ m \times n, \\ & l \leq x \leq u. & \end{array} \tag{3.3}$$

3.1 GRAPHICAL ILLUSTRATION

To visualize graphically how the Simplex Algorithm solves a linear program in standard form, consider the example in Figure 3-1, which is the two-variable problem discussed earlier in Section 2.1. The points labeled O, A, B, C, D are the vertices (or extreme points), where C is the optimal vertex. The segments OA, AB, BC, CD, DO are the edges (or the boundaries) of the feasible region. Starting at O, say, the Simplex Algorithm either moves from O to A to B to C, or moves from O to D to C, depending on the criteria used to decide whether to move from O to A or O to D.

3.2 THE SIMPLEX ALGORITHM

The Simplex Algorithm described in this section assumes that an initial feasible solution (in fact that an initial *basic feasible* solution) is given to us. If an initial feasible solution is not given, finding a feasible solution to a linear program can be done by solving a different linear program, one with the property that an obvious starting feasible solution is available.

3.2.1 CANONICAL FORM AND BASIC VARIABLES

The Simplex Method finds an optimal solution (or determines it does not exist) by a sequence of *pivot steps* on the original system of equations (3.1). For example,

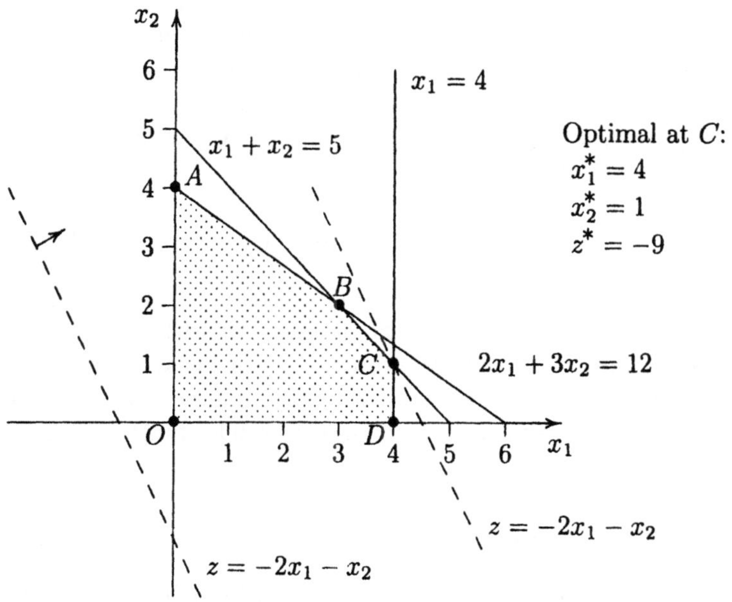

Figure 3-1: Graphical Solution of a Two-Variable LP

consider the problem of minimizing z for $x_j \geq 0$ where

$$\begin{array}{rcrcrcrcrcl} 2x_1 &+& 2x_2 &+& 2x_3 &+& x_4 &+& 4x_5 &=& z \\ 4x_1 &+& 2x_2 &+& 13x_3 &+& \mathbf{3x_4} &+& x_5 &=& 17 \\ x_1 &+& x_2 &+& 5x_3 &+& x_4 &+& x_5 &=& 7. \end{array} \tag{3.4}$$

A pivot consists in choosing some nonzero element (called the *pivot*) in the array such as $\mathbf{3x_4}$ and using it to eliminate x_4 from the remaining equations by first dividing its equation by 3, obtaining

$$\begin{array}{rcrcrcrcrcl} 2x_1/3 &+& 4x_2/3 &-& 7x_3/3 & & &+& 11x_5/3 &=& z - 17/3 \\ 4x_1/3 &+& 2x_2/3 &+& 13x_3/3 &+& x_4 &+& x_5/3 &=& 17/3 \\ -1x_1/3 &+& \mathbf{1x_2/3} &+& 2x_3/3 & & &+& 2x_5/3 &=& 4/3. \end{array} \tag{3.5}$$

If we pivot again by choosing say $\mathbf{x_2/3}$ as the pivot we obtain

$$\begin{array}{rcrcrcrcrcl} 2x_1 & & & - & 5x_3 & & & + & 1x_5 &=& z - 11 \\ 2x_1 & & & & 3x_3 &+& 1x_4 &-& x_5 &=& 3 \\ -x_1 &+& 1x_2 &+& 2x_3 & & &+& 2x_5 &=& 4. \end{array} \tag{3.6}$$

We say that the original system (3.4) is *equivalent* to (3.6) because it has the same solution set. Rewriting (3.6) we obtain

$$\begin{array}{rcrcrcrcrcrcl} (-z) &+& 2x_1 & & & - & 5x_3 & & & + & 1x_5 &=& 11 \\ & & 2x_1 & & & & 3x_3 &+& 1x_4 &-& x_5 &=& 3 \\ &-& x_1 &+& 1x_2 &+& 2x_3 & & &+& 2x_5 &=& 4. \end{array} \tag{3.7}$$

We say that (3.7) is in *canonical form* with respect to variables $(-z)$, x_4, x_2, which are called the dependent variables, or *basic variables*, because these values have been expressed in terms of the independent, or *nonbasic variables*. In practice, z is referred to as the *objective variable* and the other dependents as basic.

The *basic feasible* solution is found by setting the values of the nonbasics to zero. In (3.7) it can be read off by inspection:

$$z = 11,\ x_B = (x_4, x_2) = (3, 4),\ x_N = (x_1, x_3, x_5) = (0, 0, 0). \tag{3.8}$$

Note that in this example the basic solution turned out to be nonnegative. This is a necessary requirement for applying the Simplex Algorithm.

Note that choosing $(-z)$ and any arbitrary set of variables as basic variables to create the canonical form will not necessarily yield a basic feasible solution to (3.4). For example, had the variables x_1 and x_4 been chosen for pivoting, the basic solution would have been

$$z = 3,\ x_1 = -4,\ x_4 = 11,\ x_2 = x_3 = x_5 = 0,$$

which is *not feasible* because x_1 is negative. We now formalize the concepts discussed so far.

Pivoting forms the basis for the operations to reduce a system of equations to a canonical form, and as we shall see later, to maintain it in such form. The detailed steps for pivoting on a term $a_{rs}x_s$, called the *pivot term*, where $a_{rs} \neq 0$, are as follows:

1. Replace the rth equation by the rth equation multiplied by $(1/a_{rs})$.
2. For each $i = 1, \ldots, m$ except $i = r$, replace the ith equation by the sum of the ith equation and the replaced rth equation multiplied by $(-a_{is})$.

Since pivoting is a process that inserts and deletes redundant equations, it does not alter the solution set, and the resulting system is equivalent to the original system.

> *Definition (Canonical Form)*: A system of m equations in n variables x_j is said to be in *canonical form* with respect to an *ordered set* of variables $(x_{j_1}, x_{j_2}, \ldots, x_{j_m})$ if and only if x_{j_i} has a unit coefficient in equation i and a zero coefficient in all other equations.

System (3.9) below, with $x_B = (x_1, x_2, \ldots, x_m)^T$ and $x_N = (x_{m+1}, \ldots, x_n)^T$ is canonical because for each i, the variable x_i has a unit coefficient in the ith equation and zero elsewhere:

$$Ix_B + \bar{A}x_N = \bar{b}. \tag{3.9}$$

> *Definition (Basic Solution)*: The special solution obtained by setting the independent variables equal to zero and solving for the dependent variables is called a *basic solution*.

Thus, if (3.9) is the canonical system of (3.1) with basic variables $x_1, x_2, \ldots, x_m$, the corresponding basic solution is $x_B = \bar{b}$ and $x_N = 0$, i.e.,

$$x_1 = \bar{b}_1, x_2 = \bar{b}_2, \ldots, x_m = \bar{b}_m; \quad x_{m+1} = x_{m+2} = \cdots = x_n = 0. \tag{3.10}$$

Definition (Degeneracy): A basic solution is *degenerate* if the value of one or more of the dependent (basic) variables is zero. In particular, the basic solution (3.10) is degenerate if $\bar{b}_i = 0$ for at least one i.

A set of columns (of a system of equations in detached coefficient form) is said to be a *basis* if they are linearly independent and all other columns can be generated from them by linear combinations. To simplify the discussion we shall assume that the original system of equations (3.1) is of full rank.

Definition (Basis): In accordance with the special usage in linear programming, the term *basis* refers to the columns of the original system (in detached coefficient form), *assumed to be full rank*, corresponding to the ordered set of basic variables where the order of a basic variable is i if its coefficient is $+1$ in row i of the canonical equivalent.

Definition (Basic Columns/Activities): The columns of the basis are called *basic columns* (or *basic activities*).

The Simplex Algorithm is always initiated with a system of equations in canonical form with respect to some ordered set of basic variables. For example, let us suppose we have the canonical system (3.11) below with basic variables $(-z)$, $x_1, x_2, \ldots, x_m$. The $(m+1)$-equation $(n+1)$-variable canonical system (3.11) is equivalent to the standard form (3.1).

Our problem is to find values of $x_1 \geq 0$, $x_2 \geq 0$, ..., $x_n \geq 0$, and min z satisfying

$$\begin{array}{rcl}
-z \quad\quad\quad\quad + \bar{c}_{m+1}x_{m+1} + \cdots + \bar{c}_j x_j + \cdots + \bar{c}_n x_n & = & -\bar{z}_0 \\
x_1 \quad\quad\quad + \bar{a}_{1,m+1}x_{m+1} + \cdots + \bar{a}_{1j}x_j + \cdots + \bar{a}_{1n}x_n & = & \bar{b}_1 \\
x_2 \quad\quad + \bar{a}_{2,m+1}x_{m+1} + \cdots + \bar{a}_{2j}x_j + \cdots + \bar{a}_{2n}x_n & = & \bar{b}_2 \\
\ddots \quad\quad\quad \vdots \quad\quad\quad\quad \vdots \quad\quad\quad\quad \vdots & & \vdots \\
x_m + \bar{a}_{m,m+1}x_{m+1} + \cdots + \bar{a}_{mj}x_j + \cdots + \bar{a}_{mn}x_n & = & \bar{b}_m,
\end{array} \tag{3.11}$$

where $\bar{a}_{ij}$, $\bar{c}_j$, $\bar{b}_i$, and $\bar{z}_0$ are constants. In matrix notation, the same system can be written compactly as:

$$\begin{pmatrix} 1 & 0 & \bar{c} \\ 0 & I & \bar{A} \end{pmatrix} \begin{pmatrix} -z \\ x_B \\ x_N \end{pmatrix} = \begin{pmatrix} -\bar{z}_0 \\ \bar{b} \end{pmatrix}, \tag{3.12}$$

where $x_B = (x_1, x_2, \ldots, x_m)^T$ and $x_N = (x_{m+1}, x_{m+2}, \ldots, x_n)^T$. In this canonical form, the basic solution is

$$z = \bar{z}_0, \quad x_B = \bar{b}, \quad x_N = 0. \tag{3.13}$$

In the Simplex Algorithm it is *required* that this initial basic solution be *feasible*, by which we mean that

$$\bar{x} = \bar{b} \geq 0. \tag{3.14}$$

If such a solution is not readily available, we describe a Phase I procedure in Section 3.3 for finding such a feasible solution if it exists.

> *Definition (Feasible Canonical Form)*: If (3.14) holds, the linear program is said to be in *feasible canonical form.*

▷ **Exercise 3.1** Why is (3.14) sufficient for the basic solution (3.13) to be feasible for (3.11).

3.2.2 IMPROVING A NONOPTIMAL BASIC FEASIBLE SOLUTION

Given the canonical form, it is easy to read off the associated basic solution. It is also easy to determine, by inspecting $\bar{b}$, whether or not the basic solution (3.13) is feasible; and if it is feasible, it is easy (provided the basic solution is nondegenerate) to determine by inspecting the "modified" objective equation in (3.11) whether or not (3.13) is optimal.

> *Definition (Reduced Costs or Relative Cost Factors)*: The coefficients $\bar{c}_j$ in the cost or objective form of the canonical system (3.11) are called *relative cost factors*—"relative" because their values depend on the choice of the basic set of variables. These relative cost factors are also called the *reduced costs* associated with a basic set of variables.

Continuing with our example from Section 3.2, we redisplay (3.7) with $\mathbf{3x_3}$ boldfaced.

$$\begin{array}{rrrrrrl} (-z) & +\ 2x_1 & & -\ 5x_3 & & +\ \ x_5 & = -11 \\ & +\ 2x_1 & & +\ \mathbf{3x_3} & +\ x_4 & -\ \ x_5 & = \ \ \ 3 \\ & -\ \ x_1 & +\ x_2 & +\ 2x_3 & & +\ 2x_5 & = \ \ \ 4. \end{array} \tag{3.15}$$

The boldfaced term will be used later to improve the solution. The basic feasible solution to (3.15) can be read off by inspection:

$$z = 11,\ x_B = (x_4, x_2) = (3, 4),\ x_N = (x_1, x_3, x_5) = (0, 0, 0). \tag{3.16}$$

One relative cost factor in the canonical form (3.15) is negative, namely $\bar{c}_3 = -5$, which is the coefficient of x_3. If x_3 is increased to any positive value while holding the values of the other nonbasic at zero and adjusting the basic variables so that the equations remain satisfied, it is evident that the value of z would be reduced, because the corresponding value of z is given by

$$z = 11 - 5x_3. \tag{3.17}$$

It seems reasonable, therefore, to try to make x_3 as large as possible, since the larger the value of x_3, the smaller will be the value of z. However, in this case, the value of x_3 cannot be increased indefinitely while the other nonbasic variables remain zero because the corresponding values of the basic variables satisfying (3.15) are

$$\begin{aligned} x_4 &= 3 - 3x_3 \\ x_2 &= 4 - 2x_3. \end{aligned} \tag{3.18}$$

We see that if x_3 increases beyond $3 \div 3$, then x_4 becomes negative, and that if x_3 increases beyond $4 \div 2$ then x_2 also becomes negative. Obviously, the largest permissible value of x_3 is the smaller of these, namely $x_3 = 1$, which yields upon substitution in (3.17) and (3.18) a new feasible solution (in fact a basic feasible solution) with lower cost:

$$z = 6, \ x_3 = 1, \ x_2 = 2, \ x_1 = x_4 = x_5 = 0. \tag{3.19}$$

This solution reduces z from 11 to 6.

Our immediate objective is to discover whether or not this new solution is minimal. This time a short cut is possible since the new canonical form changes with respect to only one basic variable, i.e., by making x_4 nonbasic since its value has dropped to zero and making x_3 basic because its value is now positive. A new canonical form with new basic variables, x_3 and x_2, can be obtained directly by one pivot from the old canonical form, which has x_4 and x_2 basic. Choose as *pivot term* that x_3 term in the equation that limited the maximum amount by which the basic variables, x_2 and x_4, could be adjusted without becoming negative, namely the term $3x_3$, which we boldfaced $\mathbf{3x_3}$. Pivoting on $3x_3$, the new canonical form relative to $(-z)$, x_3, and x_2 becomes

$$\begin{array}{rrrrrrl} (-z) & + \frac{16}{3}x_1 & & & + \frac{5}{3}x_4 & - \frac{2}{3}x_5 & = -6 \\ & + \frac{2}{3}x_1 & & + x_3 & + \frac{1}{3}x_4 & - \frac{1}{3}x_5 & = 1 \\ & - \frac{7}{3}x_1 & + x_2 & & - \frac{2}{3}x_4 & + \mathbf{\frac{8}{3}x_5} & = 2. \end{array} \tag{3.20}$$

Note that the basic solution,

$$z = 6, \ x_B = (x_3, x_2) = (1, 2), \ x_N = (x_1, x_4, x_5) = (0, 0, 0),$$

is the same as that obtained by setting $x_1 = 0$, $x_5 = 0$, and increasing x_3 to the point where $x_4 = 0$. Since the solution set of the canonical forms before and after pivoting are the same, the values of x_2 and x_3 are uniquely determined when $(x_1, x_4, x_5) = 0$ whether obtained via (3.18) or by inspecting the right-hand side of (3.20).

This gives a new basic feasible solution with $z = 6$. Although the value of z has been reduced, it can clearly still be improved upon since $\bar{c}_5 = -2/3$. Furthermore, as before, the coefficient $\bar{c}_5 = -2/3$ together with the fact that the basic solution is nondegenerate indicates that the solution still is not minimal and that a better solution can be obtained by keeping the other nonbasic variables, $x_1 = x_4 = 0$, and solving for new values for x_2, x_3, and z in terms of x_5:

$$\begin{aligned} -z &= -6 + \tfrac{2}{3}x_5 \\ x_3 &= 1 + \tfrac{1}{3}x_5 \\ x_2 &= 2 - \tfrac{8}{3}x_5. \end{aligned} \tag{3.21}$$

Therefore we increase x_5 to the maximum possible while keeping x_3 and x_2 nonnegative. Note that the second relation in (3.21) places no bound on the increase of x_5, but that the third relation restricts x_5 to a maximum of $(\frac{8}{3} \div 2)$ at which value x_2 is reduced to zero. Therefore, the *pivot term*, $\mathbf{\frac{8}{3}x_5}$ in the third equation of (3.20) is used for the next elimination. Since the value of x_2 has dropped to zero and x_5 has become positive, the new set of basic variables is x_3 and x_5. Reducing system (3.20) to canonical form relative to $x_3, x_5, (-z)$ gives

$$\begin{array}{rrrrrrl} (-z) & + \frac{19}{4}x_1 & + \frac{1}{4}x_2 & & + \frac{3}{2}x_4 & & = -\frac{11}{2} \\ & + \frac{3}{8}x_1 & + \frac{1}{8}x_2 & + x_3 & + \frac{1}{4}x_4 & & = \frac{5}{4} \\ & - \frac{7}{8}x_1 & + \frac{3}{8}x_2 & & - \frac{1}{4}x_4 & + x_5 & = \frac{3}{4} \end{array} \tag{3.22}$$

and the basic feasible solution

$$z = \frac{11}{2},\; x_3 = \frac{5}{4},\; x_5 = \frac{3}{4},\; x_1 = x_2 = x_4 = 0. \tag{3.23}$$

Since all the relative cost factors for the nonbasic variables are now positive, this solution is minimal. In fact it is the unique minimal solution because all the relative cost factors are strictly positive; if any of the relative cost factors for the nonbasic variables, say x_j, were zero, we could exchange this x_j with one of the basic variables without changing the value of the objective function. Note that it took two pivot iterations on our initial canonical system (3.15) to find this optimal solution.

Key Components of the Simplex Algorithm. The example illustrates the following two key components of the Simplex Algorithm:

1. *Optimality Test.* If all the relative cost factors are nonnegative, the basic feasible solution is optimal.

2. *Introducing a Nonbasic Variable into the Basis.* When bringing a nonbasic variable into the basis, the amount by which we can increase it is constrained by requiring that the adjusted values of the basic variables remain nonnegative.

▷ **Exercise 3.2 (Infeasible Problem)** It is obvious that the linear program shown below is infeasible. Show algebraically by generating an infeasible inequality that this is indeed the case.

$$\begin{array}{lrcrcr} \text{Minimize} & x_1 & + & x_2 & & \\ \text{subject to} & x_1 & + & x_2 & = & -2 \\ & x_1 & & & \geq & 0 \\ & & & x_2 & \geq & 0. \end{array}$$

▷ **Exercise 3.3 (Unique Minimum)** Determine by inspection the basic solution to

$$\begin{array}{lrcl} \text{Minimize} & \frac{7}{2}x_1 & & = z + 15 \\ \text{subject to} & x_1 + x_2 & & = 3 \\ & \frac{3}{2}x_1 & + x_3 & = 4 \\ \multicolumn{4}{l}{\text{and } x_1 \geq 0,\ x_2 \geq 0,\ x_3 \geq 0.} \end{array}$$

Why is it feasible, optimal, and unique?

▷ **Exercise 3.4 (Multiple Minima)** Prove that the basic solution $z = -15$, $x_1 = 8/3$, $x_2 = 1/3$ is a feasible optimal solution to

$$\begin{array}{lrrcl} \text{Minimize} & & 0x_3 & = & z + 15 \\ \text{subject to} & x_2 & - \frac{2}{3}x_3 & = & \frac{1}{3} \\ & x_1 & + \frac{2}{3}x_3 & = & \frac{8}{3} \\ \multicolumn{5}{l}{\text{and } x_1 \geq 0,\ x_2 \geq 0,\ x_3 \geq 0,} \end{array}$$

but that it is not unique. Can you find another optimal basic feasible solution? Are there any nonbasic feasible solutions that are also optimal?

▷ **Exercise 3.5 (Unbounded Class of Solutions)** Reduce

$$\begin{array}{lrrl} \text{Minimize} & -x_1 & - x_2 & = z \\ \text{subject to} & x_1 & - x_2 & = 1 \\ & x_1 & & \geq 0 \\ & & x_2 & \geq 0 \end{array}$$

to feasible canonical form and generate a class of solutions that in the limit cause the objective function to go to $-\infty$.

3.2.3 THE SIMPLEX ALGORITHM

Algorithm 3.1 (Simplex Algorithm) Assume that a linear program in standard form has been converted to a feasible canonical form

$$\begin{array}{rcl} (-z) + 0x_B + \bar{c}^T x_N & = & -\bar{z}_0 \\ Ix_B + \bar{A}x_N & = & \bar{b}, \end{array} \tag{3.24}$$

which was shown earlier in Equations (3.11). Then the initial basic feasible solution is

$$x_B = \bar{b} \geq 0, \quad x_N = 0, \quad z = \bar{z}_0.$$

The algorithmic steps are as follows:

1. *Smallest Reduced Cost.* Find

$$s = \operatorname*{argmin}_j \bar{c}_j, \tag{3.25}$$

where s is the index j (argument) where $\bar{c}_j$ attains a minimum, that is,

$$\bar{c}_s = \min_j \bar{c}_j. \tag{3.26}$$

2. *Test for Optimality.* If $\bar{c}_s \geq 0$, report the basic feasible solution as optimal and **stop**.
3. *Incoming Variable.* If $\bar{c}_s < 0$, then s is the index of the incoming basic variable.
4. *Test for unbounded z.* If $\bar{A}_{\bullet s} \leq 0$ report the class of feasible solutions $x_B = \bar{b} - A_{\bullet s} x_s$, $x_j = 0$, j nonbasic and $j \neq s$, and $z = \bar{z}_0 + \bar{c}_s x_s$ such that $z \to -\infty$ as $x_s \to \infty$, and **stop**. This requires reporting the basic feasible solution, the incoming column index s, and the column $\bar{A}_{\bullet s}$.
5. *Outgoing Variable.* Choose the outgoing basic variable x_{j_r} and the value of $\bar{x}_s$, the incoming basic variable, as

$$\bar{x}_s = \frac{\bar{b}_r}{\bar{a}_{rs}} = \min_{\{i \mid \bar{a}_{is} > 0\}} \frac{\bar{b}_i}{\bar{a}_{is}} \geq 0, \quad (\bar{a}_{rs} > 0). \tag{3.27}$$

   In the case of ties, let $\mathcal{R}$ be the set of rows k tied:

$$\mathcal{R} = \left\{ k \;\middle|\; \frac{\bar{b}_k}{\bar{a}_{ks}} \leq \frac{\bar{b}_i}{\bar{a}_{is}},\ \bar{b}_i \geq 0,\ \bar{a}_{is} > 0,\ \bar{a}_{ks} > 0,\ i = 1, \ldots, m \right\}. \tag{3.28}$$

   NONDEGENERATE CASE: If $\bar{b}_k > 0$ for all $k \in \mathcal{R}$, choice of k among the ties is arbitrary.

   DEGENERATE CASE: If $\bar{b}_k = 0$ for more than one $k \in \mathcal{R}$, the *Random Choice Rule* can be used; that is, choose r at random (with equal probability).
6. Pivot on $\bar{a}_{rs}$ to determine a new basic feasible solution, set $j_r = s$ and return to Step 1. Note that the pivot step is made regardless of whether or not the value of z decreases.

The *DEGENERATE CASE* where $\bar{b}_k = 0$ for more than one $k \in \mathcal{R}$ is often ignored in practice, that is, $r \in \mathcal{R}$ is chosen arbitrarily or from among those i with max $\bar{a}_{is}$. See Problems 3.14 and 3.15 for examples where the rule used results in a sequence of pivots that repeats, called *cycling in the Simplex Algorithm.* Several techniques besides the *Random Choice Rule* exist for avoiding cycling in the Simplex Algorithm. A very simple and elegant (but not necessarily efficient) rule due to R. Bland is as follows.

BLAND'S RULE

Whenever the regular choice for selecting the pivot in the Simplex Method would result in a 0 change of the objective value of the basic feasible solution, then instead of the regular choice, do the following:

1. *Incoming Column.* Choose the pivot column $j = s$ with relative cost $\bar{c}_j < 0$ having the smallest index j.

2. *Outgoing Column.* Choose the outgoing basic column j_r among those eligible for dropping with smallest index j_i.

3.2.4 THEORY BEHIND THE SIMPLEX ALGORITHM

In this section we discuss the technical details behind the Simplex Algorithm as described so far.

THEOREM 3.1 (Optimality Test) *A basic feasible solution is a minimal feasible solution with total cost* $\bar{z}_0$ *if all relative cost factors are nonnegative:*

$$\bar{c}_j \geq 0 \quad \textit{for} \quad j = 1, \ldots, n. \tag{3.29}$$

Proof. Referring to the canonical form (3.11), it is obvious that if the coefficients of the modified cost form are all positive or zero, the smallest value of $\sum \bar{c}_j x_j$ is greater or equal to zero whatever be the choice of nonnegative x_j. Thus, $z \geq \bar{z}_0$ for all feasible choices of x. In the particular case of the basic feasible solution (3.13) however, we have $z = \bar{z}_0$; hence $\min z = \bar{z}_0$ and the solution is optimal. ∎

This proof shows that for all solutions $x_j \geq 0$ that satisfy the canonical form (3.11), the basic solution has the smallest value of z. This proof also shows that for all solutions that satisfy the original system (3.1), the basic solution is optimal because the original system (3.1) and (3.11) are equivalent, i.e., have the same feasible solution set. It turns out, see Exercise 3.6, that the converse of Theorem 3.1 is true only if the linear program is nondegenerate.

▷ **Exercise 3.6** Consider the two systems (3.30a) and (3.30b):

$$\text{(a)}\quad \begin{array}{rcrcrcl} (-z) & & & & - x_3 & = & 0 \\ & & x_1 & & + x_3 & = & 0 \\ & & & x_2 & + x_3 & = & 1 \end{array} \qquad \text{(b)}\quad \begin{array}{rcrcrcl} (-z) & + & x_1 & & & = & 0 \\ & & x_1 & & + x_3 & = & 0 \\ & - & x_1 & + x_2 & & = & 1 \end{array} \tag{3.30}$$

Prove that the two systems (*a*) and (*b*) shown in (3.30) are equivalent. Show that the basic solutions relative to the two different sets of basic variables are the same and both are optimal but that we cannot tell that the basic feasible solution associated with Equation (3.30a) is optimal just by inspecting its relative cost factors, i.e., the coefficients in the z equation of the canonical form.

THEOREM 3.2 (Multiple Minima) *Given a minimal basic feasible solution* (x^*, z^*) *with relative cost factors* $\bar{c}_j \geq 0$*, then any other feasible solution* (x, z)*, not necessarily basic, with the property that* $x_j = 0$ *for all* $\bar{c}_j > 0$ *is also a minimal solution; moreover, any other feasible solution* (x, z) *with the property that* $x_j > 0$ *and* $\bar{c}_j > 0$ *for some* j *cannot be a minimal solution.*

COROLLARY 3.3 (Unique Optimum) *A basic feasible solution is the unique minimal feasible solution if* $\bar{c}_j > 0$ *for all nonbasic variables.*

▷ **Exercise 3.7** Prove Theorem 3.2 and Corollary 3.3.

As we have seen in the numerical example, the canonical form provides an easy criterion for testing the optimality of a basic feasible solution. Furthermore, if the criterion is not satisfied, another solution is generated by pivoting that reduces the value of the objective function (except for certain degenerate cases).

We now formalize this procedure of improving a nonoptimal basic feasible solution. *In general*, if at least one relative cost factor $\bar{c}_j$ in the canonical form (3.11) is negative, it is possible, assuming nondegeneracy (i.e., all $\bar{b}_i > 0$), to construct a new basic feasible solution with an objective value lower than $z = \bar{z}_0$. The lower objective value solution is obtained by increasing the value of one of the nonbasic variables x_s that has $\bar{c}_s < 0$ and adjusting the values of the basic variables accordingly. Which x_s to choose when there are several $\bar{c}_j < 0$ has been the subject of much study. One commonly used rule is to choose the $j = s$ that gives the maximum decrease of the objective z per unit increase of a nonbasic variable x_j:

$$s = \operatorname*{argmin}_j \bar{c}_j < 0. \tag{3.31}$$

▷ **Exercise 3.8** Show that there exists a positive constant λ_k such that rescaling the units for measuring x_k by λ_k causes the adjusted $\bar{c}_k = \min_j \bar{c}_j$ for any arbitrary x_k, where $\bar{c}_k < 0$, to be chosen as the new incoming basic variable.

Criterion (3.31) is commonly used in practice because it typically leads to significantly fewer iterations than just using any arbitrary $j = s$ such that $\bar{c}_j < 0$. One reason why this choice may be better than an arbitrary one is that the columns of practical models are not likely to be arbitrarily scaled relative to one another. There are other more complex rules, including those that are scale invariant, that are more computationally efficient for large problems.

Using the canonical form (3.11), we construct a solution in which x_s takes on some positive value; the values of all other nonbasic variables, x_j, $j \neq s$, are frozen temporarily at zero; and the values of z and the basic variables x_B, whose indices (denoted by the subscript B) are $j_1, \ldots, j_m$, are adjusted to take care of the increase in x_s:

$$\begin{aligned} z &= \bar{z}_0 + \bar{c}_s x_s \\ x_B &= \bar{b} - \bar{A}_{\bullet s} x_s, \end{aligned} \tag{3.32}$$

where $\bar{c}_s < 0$. Since $\bar{c}_s$ is negative, it is clear that we can make z as small as possible by making x_s as large as possible. However, we have to retain feasibility, and thus the only thing that prevents us from setting x_s infinitely large is the possibility that the value of one of the basic variables in (3.32) will become negative. It is straightforward to see that if all the components of $\bar{A}_{\bullet s}$ are nonpositive then x_s can be made arbitrarily large without violating feasibility. This establishes

THEOREM 3.4 (Unbounded Linear Program) *If in the canonical system, for some s, all coefficients $\bar{a}_{is}$ are nonpositive and $\bar{c}_s$ is negative, then a class of feasible solutions can be constructed where the set of z values has no lower bound: namely, $z = \bar{z}_0 + \bar{c}_s x_s$ and $x_B = \bar{b} - \bar{A}_{\bullet s} x_s \geq 0$ where $x_s \to \infty$, $x_j = 0$ for all nonbasic $j \neq s$.*

COROLLARY 3.5 (Representation of an Unbounded Solution) *The infinite class of feasible solutions generated in Theorem 3.4 is the sum of a basic feasible solution and a nonnegative scalar multiple of a nonnegative (nontrivial) homogeneous solution.*

▷ **Exercise 3.9** Prove Corollary 3.5 by showing that if the linear program is unbounded, the homogeneous solution is $(x_B, x_s, 0) = (-\bar{A}_{\bullet s}, 1, 0) \geq 0$.

On the other hand, if at least one $\bar{a}_{is}$ is positive, it will not be possible to increase the value of x_s indefinitely, because from (3.32) whenever for this i, $x_s > \bar{b}_i/\bar{a}_{is}$, the value of x_{j_i} will be negative. To maintain feasibility, we can only increase x_s to the smallest ratio of $\bar{b}_i/\bar{a}_{is}$ over all positive $\bar{a}_{is}$. That is,

$$\bar{x}_s = \frac{\bar{b}_r}{\bar{a}_{rs}} = \min_{\{\, i \mid \bar{a}_{is} > 0 \,\}} \frac{\bar{b}_i}{\bar{a}_{is}} \geq 0, \quad (\bar{a}_{rs} > 0), \tag{3.33}$$

where it should be particularly noted that *only* those i and r are considered for which $\bar{a}_{is} > 0$ and $\bar{a}_{rs} > 0$.

If more than one $\bar{b}_i/\bar{a}_{is} = \bar{x}_s$ tie for a minimum in (3.33) and $\bar{x}_s > 0$, then arbitrarily choose any such i for r. Typically, in this case, the i chosen is the one for which $\bar{a}_{is}$ is the smallest.

▷ **Exercise 3.10** Show that if more than one $\bar{b}_i/\bar{a}_{is} = \bar{x}_s$ tie for a minimum in (3.33) then the next iteration results in a degenerate solution.

In general, if $\bar{x}_s = 0$ in (3.33), one or more $\bar{b}_i$ are zero (i.e., the basic feasible solution is degenerate); moreover, from (3.32), the value z will not decrease during such an iteration. If there is no decrease, there is the possibility of cycling. To avoid this possibility of cycling, one could choose r at random from those i such that $\bar{a}_{is} > 0$ and $\bar{b}_i = 0$. For example, if $\bar{a}_{1s} > 0$ and $\bar{a}_{2s} > 0$ but $\bar{b}_1 = \bar{b}_2 = 0$, flip a coin to decide whether $r = 1$ or $r = 2$. Under this random choice rule it can be proved that the algorithm will *almost surely* (with probability one) terminate in a finite number of iterations.

If the basic solution is *nondegenerate*, we have:

THEOREM 3.6 (Decrease Under Nondegeneracy) *If in the canonical system, the basic solution is nondegenerate and a relative cost factor $\bar{c}_s$ is negative for some s and for this s at least one coefficient $\bar{a}_{is}$ is positive, then the nondegenerate basic feasible solution can be modified into a new basic feasible solution with a lower total cost z.*

▷ **Exercise 3.11** Prove Theorem 3.6.

Specifically, we shall now show that the replacing of x_{j_r} by x_s in the set of basic variables $x_{j_1}, x_{j_2}, \ldots, x_{j_m}$ results in a new set that is basic and a corresponding basic solution that is feasible. Assuming nondegeneracy, $\bar{b}_r > 0$. Since $\bar{a}_{rs} > 0$, we have $\bar{x}_s > 0$ by (3.33) and $z < \bar{z}_0$ by (3.32). By construction of $\bar{x}_s$ through Equation (3.33), $x_{j_r} = 0$ and the remaining variables $x_B \geq 0$, which implies that the new solution is feasible. In order to show that the new solution is basic observe that since $\bar{a}_{rs} > 0$, we may use the rth equation of the canonical form (3.11) and $\bar{a}_{rs}$ as pivot element to eliminate the variable x_s from the other equations and minimizing form. Only this one pivot operation is needed to reduce the system to canonical form relative to the new set of variables. This fact and the way s is selected constitutes the key to the computational efficiency of the Simplex Method. The basic solution associated with the new set of basic variables is unique; see Theorem B.6 of the linear equation review in Appendix B on page 352.

THEOREM 3.7 (Finite Termination under Nondegeneracy) *Assuming nondegeneracy at each iteration, the Simplex Algorithm will terminate in a finite number of iterations.*

Proof. There is only a finite number of ways to choose a set of m basic variables out of n variables. If the algorithm were to continue indefinitely, it could only do so by repeating the same set of basic variables as that obtained on an earlier iteration—hence, the same canonical system and the same value of z. (See the Uniqueness Theorem B.6 on page 352.) This repetition cannot occur since the value of z strictly decreases with each iteration under nondegeneracy. ∎

However, when degenerate solutions occur, we can no longer argue that the procedure will necessarily terminate in a finite number of iterations, because under degeneracy it is possible for $\bar{b}_r = 0$ in (3.33), in which case the value of z does not decrease. The procedure will not terminate if this were to happen an infinite number of iterations in a row; however, this can only happen if the same set of basic variables recur. If one were to continue, with the same selection of s and r for each iteration as before, the same basic set would recur after, say, k iterations, and again after $2k$ iterations, etc., indefinitely. There is therefore the possibility of *cycling* in the Simplex Algorithm. In fact, examples have been constructed by E.M.L. Beale, A.J. Hoffman, H. Kuhn, and others to show that this can happen. Although in practice almost no problems have been encountered that would cycle even if no special rules were used to prevent cycling, still, such rules are useful in reducing the number of iterations in cases of near degeneracy.

3.3 SIMPLEX METHOD

The Simplex Method is applied to a linear program in standard form (3.1). It employs the Simplex Algorithm presented in Section 3.2.3 in two phases. In Phase I a starting basic feasible solution is sought to initiate Phase II or to determine that no feasible solution exists. If found, then in Phase II an optimal basic feasible solution or a class of feasible solutions with $z \to -\infty$ is sought.

Many problems encountered in practice often have a starting feasible canonical form readily at hand. The Phase I procedure is, of course, not necessary if one is available. For example, one can immediately construct a great variety of starting basic feasible solutions for the important class called "transportation" problems (see Chapter 8). Other models, such as economic models, often contain storage and slack activities, permitting an obvious starting solution that uses these storage and slack activities. Such a solution, if basic, may be far away from the optimum solution, but it provides an easy start. Even if not basic, it can be modified into a basic feasible solution in no more than k pivot steps, where $k \leq n - m$.

However, there are many problems encountered in practice where no obvious starting feasible canonical form is available and a Phase I approach is required. Initially nothing may be known (mathematically speaking) about the problem. It is up to the algorithm to determine whether or not there are

1. *Redundancies:* This could occur, for example, if an equation balancing money flow had been obtained from the equations balancing material flows by multiplying price by quantity and summing. The classic transportation problem (Example 1.5 on page 4) provides a second example; and, the blending problem (Example 1.4 on page 3) provides a third example.

2. *Inconsistencies:* This could be caused by input errors, the use of inconsistent data, or by the specification of requirements that cannot be filled from the available resources. For example, one may pose a problem in which resources are in short supply, and the main question is whether or not a feasible solution exists.

The Phase I procedure, which uses the Simplex Algorithm itself to provide a starting feasible canonical form (if it exists) for Phase II, has several important features:

1. No assumptions are made regarding the original system; it may be redundant, inconsistent, or not solvable in nonnegative numbers.

2. No eliminations are required to obtain an initial solution in canonical form for Phase I.

3. The end product of Phase I is a basic feasible solution (if it exists) in canonical form ready to initiate Phase II.

3.3.1 THE METHOD

The first step of the Simplex Method is the introduction into the standard form of additional terms in additional nonnegative variables in such a way that the resulting augmented problem is in canonical form. Historically these have been called by many authors *artificial*, or *error*, or *logical variables.* We prefer the term artificial variables as a reminder that we need to drive them out of the initial feasible basis of Phase I.

The objective is replaced by a new objective w (instead of z), which is the sum of the artificial variables. Now at this point the Simplex Algorithm is employed. It consists of a sequence of pivot operations, referred to as Phase I, that produces a succession of different canonical forms with the property that the sum of the artificial variables decreases with each iteration. The objective is to drive this sum to zero. If we succeed in doing so we have found a basic feasible solution to the original system with which to initiate Phase II.

Example 3.1 (Illustration of the Simplex Method) We now illustrate the Simplex Method by carrying out the steps on the following problem: Find min z, $x \geq 0$, such that

$$\begin{array}{rcrcrcrcrcl} 2x_1 & + & 1x_2 & + & 2x_3 & + & x_4 & + & 4x_5 & = & z \\ 4x_1 & + & 2x_2 & + & 13x_3 & + & 3x_4 & + & x_5 & = & 17 \\ x_1 & + & x_2 & + & 5x_3 & + & x_4 & + & x_5 & = & 7. \end{array} \tag{3.34}$$

If any of the constant terms are negative we change the signs of the corresponding equations. We can now initiate Phase I by adding the artificial variables $x_6 \geq 0$ and $x_7 \geq 0$ and the artificial objective w as shown below.

$$\begin{array}{rcrcrcrcrcrcrcl} & & & & & & & & & & x_6 & + & x_7 & = & w \\ 4x_1 & + & 2x_2 & + & 13x_3 & + & 3x_4 & + & x_5 & + & x_6 & & & = & 17 \\ x_1 & + & x_2 & + & 5x_3 & + & x_4 & + & x_5 & & & + & x_7 & = & 7. \end{array} \tag{3.35}$$

The Simplex Algorithm requires the equations to be in *feasible canonical form* at the start, which can be easily done by subtracting the second and third equations from the w equation to get the starting tableau shown in Table 3-1. The steps for the minimization of w in Phase I are similar to those for minimizing z. On the first iteration (see Table 3-1) the value of w is reduced from 24 to 6/13; on the second iteration (see Table 3-1) w is reduced to zero. The basic feasible solution has basic variables $x_3 = 5/4$, $x_5 = 3/4$. Substituting them into the z equation with the nonbasics set at zero, we get $z = 11/2$. Variables x_6 and x_7 are nonbasic artificial and hence are made ineligible for pivoting in Phase II in order to prevent the reintroduction of artificials x_6 and x_7 back into the solution. (In addition, the general rule is that we make ineligible for pivoting in Phase II all variables, artificial or not, that have positive relative cost factors $\bar{d}_j$ in the w equation.) Next, the z equation is reintroduced and basic variables x_3 and x_5 are eliminated as shown in Table 3-1. On the third iteration (see Table 3-1) the value of z dropped from $z = \frac{11}{2}$ (iteration 2) to $z = 4$, which turns out to be minimum because all the relative cost factors (top row) are greater than or equal to zero. The optimal solution is $z = 4$, $x_2 = 2$, $x_3 = 1$, and all other $x_j = 0$.

3.3.2 PHASE I/PHASE II ALGORITHM

An outline of the detailed steps involved is shown below.

Algorithm 3.2 (The Simplex Method)

1. *Make each b_i nonnegative.* Modify the original system of equations (3.1) so that all the constant terms b_i are nonnegative by multiplying an equation whose b_i is less than zero by -1.

Iteration 0 (Phase I)

Basic Variables	OBJ	Original Variables					Artificial Variables		RHS
	$-w$	x_1	x_2	x_3	x_4	x_5	x_6	x_7	
$-w$	1	-5	-3	-18	-4	-2			-24
x_6		4	2	**13**	3	1	1		17
x_7		1	1	5	1	1		1	7

Iteration 1 (Phase I)

$-w$	1	$\frac{7}{13}$	$-\frac{3}{13}$		$\frac{2}{13}$	$-\frac{8}{13}$	$\frac{18}{13}$		$-\frac{6}{13}$
x_3		$\frac{4}{13}$	$\frac{2}{13}$	1	$\frac{3}{13}$	$\frac{1}{13}$	$\frac{1}{13}$		$\frac{17}{13}$
x_7		$-\frac{7}{13}$	$\frac{3}{13}$		$-\frac{2}{13}$	$\mathbf{\frac{8}{13}}$	$-\frac{5}{13}$	1	$\frac{6}{13}$

Iteration 2 (Phase I Optimal)

$-w$	1						1	1	0
x_3		$\frac{3}{8}$	$\frac{1}{8}$	1	$\frac{1}{4}$		$\frac{1}{8}$	$-\frac{1}{8}$	$\frac{5}{4}$
x_5		$-\frac{7}{8}$	$\frac{3}{8}$		$-\frac{1}{4}$	1	$-\frac{5}{8}$	$\frac{13}{8}$	$\frac{3}{4}$

Iteration 2 (Phase II Start: Introduce z)

Basic Variables	OBJ	Original Variables					Artificial Variables		RHS
	$-z$	x_1	x_2	x_3	x_4	x_5	x_6	x_7	
$-z$	1	2	1	2	1	4	0	0	0
x_3		$\frac{3}{8}$	$\frac{1}{8}$	1	$\frac{1}{4}$		$\frac{1}{8}$	$-\frac{1}{8}$	$\frac{5}{4}$
x_5		$-\frac{7}{8}$	$\frac{3}{8}$		$-\frac{1}{4}$	1	$-\frac{5}{8}$	$\frac{13}{8}$	$\frac{3}{4}$

Iteration 2 (Phase II: Updated z)

$-z$	1	$\frac{19}{4}$	$-\frac{3}{4}$		$\frac{3}{2}$		$\frac{9}{4}$	$-\frac{25}{4}$	$-\frac{11}{2}$
x_3		$\frac{3}{8}$	$\frac{1}{8}$	1	$\frac{1}{4}$		$\frac{1}{8}$	$-\frac{1}{8}$	$\frac{5}{4}$
x_5		$-\frac{7}{8}$	$\mathbf{\frac{3}{8}}$		$-\frac{1}{4}$	1	$-\frac{5}{8}$	$\frac{13}{8}$	$\frac{3}{4}$

Iteration 3 (Phase II Optimal)

$-z$	1	3			1	2	1	$\frac{19}{2}$	-4
x_3		$\frac{2}{3}$		1	$\frac{1}{3}$	$-\frac{1}{3}$	$\frac{1}{3}$	$-\frac{2}{3}$	1
x_2		$-\frac{7}{3}$	1		$-\frac{2}{3}$	$\frac{8}{3}$	$-\frac{5}{3}$	$\frac{13}{3}$	2

Table 3-1: Simplex Method: Tableau Form Example

2. *Add artificial variables.* In order to set up an initial feasible solution for Phase I, augment the system of equations to include a basic set,

$$x_a = (x_{n+1}, x_{n+2}, \dots, x_{n+m}) \geq 0,$$

of *artificial variables* so that the system becomes

$$\begin{array}{rcrcl} -w & & + & e^T x_a & = 0 \\ & Ax & + & I x_a & = b \\ & x & & & \geq 0 \\ & & & x_a & \geq 0, \end{array} \tag{3.36}$$

where $e = (1, 1, \dots, 1)^T$.

3. *Do Phase I.* Use the Simplex Algorithm to find a solution to (3.36) that minimizes the sum of the artificial variables w:

$$w = \sum_{j=n+1}^{n+m} x_j. \tag{3.37}$$

Equation (3.37) is called the *infeasibility form.* The initial feasible canonical system for Phase I is obtained by selecting as basic variables $x_a, (-w)$, and eliminating x_a from the infeasibility form by subtracting the sum of the last m equations of (3.36) from the first equation, yielding

$$\begin{array}{rcrcl} -w & + & d^T x & & = -\bar{w}_0 \\ & & Ax & + \; I x_a & = b \\ & & x & & \geq 0 \\ & & & x_a & \geq 0, \end{array} \tag{3.38}$$

where $b_i \geq 0$ and

$$\begin{aligned} d &= -A^T e \\ -\bar{w}_0 &= -e^T b. \end{aligned} \tag{3.39}$$

Writing (3.38) in detached coefficient form constitutes the *initial tableau* for Phase I (see Table 3-2).

$-w$	x	x_a	RHS
1	d		$-\bar{w}_0$
	A	I	b

Table 3-2: Initial Tableau for Phase I

4. *Terminate if* $\min w > 0$ *at the end of Phase I.* No feasible solution exists to the original problem.

5. *Set up Phase II if* $\min w = 0$. Start Phase II of the Simplex Method by

   (a) dropping from further consideration all nonbasic nonartificial variables x_j whose corresponding coefficients $\bar{d}_j$ are positive (not zero) in the final updated w-equation;

(b) dropping from further consideration all nonbasic artificial variables;

(c) dropping the linear form w (as modified by various eliminations);

(d) Introducing the linear form z after first eliminating all the basic nonartificial variables and then augmenting the resulting z form by basic artificial terms with zero coefficients.

6. *Do Phase II.* Apply the Simplex Algorithm to the feasible canonical form just obtained and iterate to find a solution that minimizes the value of z or generates a class of solutions such that $z \to -\infty$.

▷ **Exercise 3.12** Why is Equation (3.37) called the infeasibility form?

▷ **Exercise 3.13** Referring to Step 5d, show how the optimal canonical form for Phase I can be used to eliminate the term in the z equation in order to initiate Phase II.

▷ **Exercise 3.14** Show that if all the artificials are out of the basis at the end of Phase I, then $w = \sum_{j=n+1}^{n+m} x_j$.

3.3.3 THEORY BEHIND PHASE I

The above procedure for Phase I deserves a little more discussion. It is clear that if there exists a feasible solution to the original system (3.1) then this same solution also satisfies (3.36) with the artificial variables set equal to zero; that is, $w = 0$ in this case. From (3.37), the smallest possible value for w is zero since w is the sum of nonnegative variables. Hence, if feasible solutions exist, the minimum value of w will be $w = 0$. Conversely, if a solution is obtained for (3.36) with $w = 0$, it is clear that all $x_{n+i} = 0$ and the values of x_j for $j \le n$ constitute a feasible solution to (3.1). It also follows that if $\min w > 0$, then no feasible solutions to (3.1) exist. Note that *the Phase I procedure cannot result in an unbounded problem* since this would imply falsely that w defined as a sum of nonnegative variables has no lower bound.

Whenever the original system contains redundancies and often when degenerate solutions occur, artificial variables will remain as part of the basic set of variables at the end of Phase I. Thus, it is necessary that their values during Phase II never exceed zero. We will now give three different ways (including Step 5 of Algorithm 3.2) in which this can be accomplished.

1. One way was given in Step 5 of Algorithm 3.2 above by dropping all non-artificial variables whose relative cost factors for w were positive and dropping all nonbasic artificial variables. To see this we note that the w equation at the end of Phase I satisfies

$$\sum_{j=1}^{n+m} \bar{d}_j x_j = w - \bar{w}_0, \tag{3.40}$$

where $\bar{d}_j \geq 0$ and $\bar{w}_0 = 0$, since feasible solutions exist. For feasibility, w must remain zero in Phase II, which means that every x_j corresponding to $\bar{d}_j > 0$ must be zero; hence, all such x_j can be set equal to zero and dropped from further consideration in Phase II. We can also drop any nonbasic artificials as no longer of any interest. Our attention is now confined only to variables whose corresponding $\bar{d}_j = 0$. All feasible solutions involving only these remaining variables now have $w = 0$ by (3.40), and therefore the remaining artificial variables that sum to w are also zero and will remain zero as we subsequently pivot. Consequently, the feasible solution to the modified augmented problem is also a feasible solution for the original problem.

A variant of this method is to treat the z-equation as just another constraint with z unrestricted in sign during Phase I. This automatically eliminates the basic variables x_B from the z-equation on each iteration. It does not require any manipulation of data structures; but it can involve more computations.

2. A second way to maintain the basic artificial variables at zero values during Phase II is to try to eliminate (if possible) all artificial variables still in the basic set. This can be done by choosing a pivot in a row r corresponding to such an artificial variable and in any column $s \leq n$ such that $\bar{a}_{rs} \neq 0$. If all coefficients in such a row for $j = 1, \ldots, n$ are zero, the row may be deleted because the corresponding equation in the original system is redundant, or the row may be left in if that is more convenient.

3. A third way is to keep the w equation during Phase II and treat the $(-w)$ variable as just another variable that is restricted to nonnegative values. The system is then augmented by introducing the z-equation after eliminating the basic variables from it. Since $w \geq 0$ is always true, the added condition $(-w) \geq 0$ implies $w = 0$ during Phase II.

▷ **Exercise 3.15** Solve the following problem by all three ways of transitioning from Phase I to Phase II. Find max z, $x \geq 0$ such that

$$\begin{array}{rcrcrcrcl} 10x_1 & + & 10x_2 & + & 20x_3 & + & 30x_4 & = & z \\ x_1 & & & + & x_3 & + & x_4 & = & 1 \\ & & x_2 & + & x_3 & + & x_4 & = & 2 \\ 3x_1 & + & 2x_2 & + & 2x_3 & + & x_4 & = & 7. \end{array} \tag{3.41}$$

Replace the last constraint by $3x_1 + 2x_2 + 5x_3 + 5x_4 = 7$ and solve the problem again using all three ways of transitioning from Phase I to Phase II. Explain what properties these two examples have that make them interesting.

THEOREM 3.8 (Artificial Variables in Phase II) *If artificial variables form part of the basic sets of variables in the various iterations of Phase II, their values will never exceed zero.*

Proof. The proof follows from the discussion of the first way to maintain artificial variables at zero in Phase II. ∎

▷ **Exercise 3.16** Prove that the second way to maintain artificial variables at zero values is valid.

▷ **Exercise 3.17** Suppose Phase I terminates with a feasible solution with k artificials in the basis. Show that if we know in advance that there are no degenerate solutions after the removal of all redundant equations, then k is equal to the number of redundant equations.

▷ **Exercise 3.18** Suppose at the end of Phase I that one or more artificial variables remain in the basis at zero level. Assume that we can initiate Phase II by setting the cost coefficients c_j of all the artificials in the z equation at zero. Show that, as the iterations of the Simplex Algorithm are performed, we can ensure that the artificial variables remain at zero by the following procedure.

1. If for the incoming column s, we have one or more $\bar{a}_{is} \geq 0$ for i corresponding to an artificial variable, we perform the usual minimum ratio test.
2. On the other hand, if for all i corresponding to artificial variables we have $\bar{a}_{is} \leq 0$, then instead of performing the minimum ratio test, we pivot on any $r = i$ corresponding to an artificial variable.

3.4 BOUNDED VARIABLES

We now turn our attention to solving a linear program in *bounded variables*, that is,

$$\begin{array}{lll} \text{Minimize} & c^T x & \\ \text{subject to} & Ax = b, & A:\ m \times n, \\ & l \leq x \leq u. & \end{array} \tag{3.42}$$

As earlier, the independent variables will correspond to nonbasic variables, x_N, and the dependent ones to basics, x_B. Let the canonical form with respect to x_B be as before:

$$\begin{array}{rcl} (-z) + 0x_B + \bar{c}^T x_N & = & -\bar{z}_0 \\ Ix_B + \bar{A}x_N & = & \bar{b}, \end{array} \tag{3.43}$$

where subscript B (or $\mathcal{B}$) is the set of indices of basic variables, and subscript N (or $\mathcal{N}$) is the set of indices of the nonbasic variables.

Definition (Basic Solution): Assuming that at least one of the bounds for each x_j is finite, the special solution obtained by setting each nonbasic equal to either its lower bound or upper bound and then solving for the values of the basic variables will now be called a *basic solution.*

Definition (Degenerate Solution): A basic solution is *degenerate* if the values of one or more of the basic variables in the basic solution are at either their lower bound or upper bound.

▷ **Exercise 3.19** Assume that for some $j = j_0$, x_{j_0} in (3.42) has a lower bound of $-\infty$ and an upper bound of $+\infty$. Further assume that the lower bounds on all other variables are finite. Show how to convert such a problem into one where at least one of the bounds on each variable is finite by each of the following three ways:

1. By eliminating the variable x_{j_0}.
2. By replacing x_{j_0} by the difference of two nonnegative variables.
3. By making sure that the variable x_{j_0} is basic on the first iteration of Phase I and then never making it nonbasic thereafter.

Explain why method 3 is the same as method 1.

▷ **Exercise 3.20** Show that for problem (3.42) if each upper bound and lower bound is finite, then the linear program can never have an unbounded solution.

▷ **Exercise 3.21** Let $x^o = (x^o_B, x^o_N)$ be a basic solution as defined above. Assume that the lower and upper bounds for each x_j are finite. Make the following transformation:

1. If $x^o_j = u_j$, replace x_j by $u_j - y_j$.
2. If $x^o_j = l_j$, replace x_j by $l_j + y_j$.
3. If $l_j < x^o_j < u_j$, replace x_j by y_j.

Show that y_j is a basic solution as defined earlier on Page 66.

THEOREM 3.9 (Optimality Test for the Bounded Variable LP) *Let $x = (x^*_B, x^*_N)$ be a basic solution to (3.43) according to the above definition, i.e., $x^*_B = \bar{b} - \bar{A}x^*_N$ and each nonbasic component of x^*, for $j \in \mathcal{N}$, must satisfy $x^*_j = l_j$ or $x^*_j = u_j$. Then (x^*_B, x^*_N) is a minimal solution with total costs $z^* = \bar{z}_0 + \bar{c}^T x^*_N$ if*

$$\begin{aligned} & l_B \le x^*_B = \bar{b} - \bar{A}x^*_N \le u_B\,, && (a) \\ & \bar{c}_j \ge 0 \quad \textit{for } j \in \mathcal{L}, && (b) \\ & \bar{c}_j \le 0 \quad \textit{for } j \in \mathcal{U}, && (c) \end{aligned} \tag{3.44}$$

where

$$\begin{aligned} \mathcal{L} &= \{\, j \in \mathcal{N} \mid x^*_j = l_j \,\}, \\ \mathcal{U} &= \{\, j \in \mathcal{N} \mid x^*_j = u_j \,\}. \end{aligned} \tag{3.45}$$

Proof. Clearly, for the basic solution to be feasible (3.44a) must hold. The objective function value is given by

$$z^* = \bar{z}_0 + \bar{c}^T x^*_N = \bar{z}_0 + \sum_{j \in \mathcal{L}} \bar{c}_j x^*_j + \sum_{j \in \mathcal{U}} \bar{c}_j x^*_j. \tag{3.46}$$

For any feasible (x, z), we have

$$z = \bar{z}_0 + \sum_{j \in \mathcal{L}} \bar{c}_j x_j + \sum_{j \in \mathcal{U}} \bar{c}_j x_j. \tag{3.47}$$

Clearly, if $\bar{c}_j \geq 0$ for $j \in \mathcal{L}$, then the best we can do is to have the corresponding x_j be at their lower bounds, $x_j^* = l_j$. Similarly, if $\bar{c}_j \leq 0$ for $j \in \mathcal{U}$, then the best we can do is to have the corresponding x_j be at their upper bounds, $x_j^* = u_j$. Thus, the right-hand side of (3.46) is a lower bound for z. Since this lower bound is attained for the basic feasible solution (x_B^*, x_N^*), the minimum value of z in (3.47) is the minimum value of z attained for all feasible solutions, basic or otherwise. ∎

MODIFYING THE SIMPLEX ALGORITHM TO SOLVE THE BOUNDED VARIABLE PROBLEM

To see how to modify the Simplex Algorithm to solve the bounded variable LP, note the following:

1. The optimality conditions for a basic feasible solution $x = (x_B^o, x_N^o)$ are

$$\begin{aligned} \bar{c}_j \geq 0 &\quad \text{for } j \in \mathcal{L}, \\ \bar{c}_j \leq 0 &\quad \text{for } j \in \mathcal{U}. \end{aligned}$$

   (We are assuming, to simplify the discussion, that at least one bound on each x_j is finite.) If the optimality conditions are not satisfied, the incoming variable is selected by finding the index $j = s$ such that

$$s = \underset{\substack{\{j \in \mathcal{L}\} \\ \{k \in \mathcal{U}\}}}{\operatorname{argmin}} \{\bar{c}_j, -\bar{c}_k\}. \tag{3.48}$$

2. If not optimal, we try to bring the nonbasic variable with index $j = s$ into the basis. Clearly if $x_s^o = u_s$, we will try to decrease x_s, and if $x_s^o = l_s$, we will try to increase x_s in an attempt to bring it into the basis. Let δ be defined by

$$\delta = \begin{cases} 1 & \text{if } x_s^o = l_s, \\ -1 & \text{if } x_s^o = u_s. \end{cases} \tag{3.49}$$

   The nonbasic variable x_s will change as follows:

$$x_s = x_s^o + \delta\theta, \tag{3.50}$$

   where $\theta \geq 0$ will be determined as shown below so that x_s and the basic variables x_{j_i}, for $i = 1, \ldots, m$, all stay within their bounds.

   This implies that we must have $l_s \leq x_s^o + \delta\theta \leq u_s$ and $l_{j_i} \leq x_{j_i}^o - \delta\theta\bar{a}_{is} \leq u_{j_i}$ for $i = 1, \ldots, m$. The other nonbasic variables are fixed at $x_j = x_j^o$. From this it follows that θ is the smallest of three ratios

$$\theta_s = u_s - l_s, \tag{3.51}$$

$$\theta_l = \min_{\{i \mid \delta\bar{a}_{is} > 0\}} \left(\frac{x_{j_i}^o - l_{j_i}}{\delta\bar{a}_{is}}, \infty \right) \geq 0, \tag{3.52}$$

$$\theta_u = \min_{\{i \mid \delta\bar{a}_{is} < 0\}} \left(\frac{u_{j_i} - x_{j_i}^o}{-\delta\bar{a}_{is}}, \infty \right) \geq 0, \tag{3.53}$$

where δ is defined in equation (3.49). Note that θ_s is the maximum amount that x_s can change given its upper and lower bounds; θ_l reflects the maximum that x_s can change before one of the lower bounds on a basic variable is violated; and θ_u reflects the maximum that x_s can change before one of the upper bounds on a basic variable is violated. Let

$$r = \text{argmin}(\theta_s, \theta_l, \theta_u).$$

It may happen that $\theta = \theta_s$, in which case there is no change to the basis—the nonbasic variable x_s simply switches bounds and causes a feasible change in the values of the basic variables. The case $r = s$ occurs often in practice and requires less work because there is no change to the basis, and hence no pivoting is required.

3. The current value of x_B changes as follows:

$$x_B = x_B^o - \delta\theta \bar{A}_{\bullet s}, \tag{3.54}$$

where δ is given by (3.49).

▷ **Exercise 3.22** Compare the steps of the algorithm just described for the bounded variable problem, when the lower bounds l are equal to 0 and upper bounds u are equal to ∞, with the standard Simplex Algorithm 3.1 and show that the steps just discussed are identical.

▷ **Exercise 3.23** Write down the complete Simplex Algorithm for solving a linear program with bounded variables.

Example 3.2 (Solving a Linear Program with Bounded Variables) This example illustrates the use of the Simplex Algorithm as modified for a bounded variable linear program.

$$\begin{array}{lrcrcl} \text{Minimize} & -3x_1 & - & 2x_2 & = & z \\ \text{subject to} & 2x_1 & + & x_2 & \le & 10 \\ & 5x_1 & + & 3x_2 & \le & 27 \\ & \multicolumn{5}{c}{0 \le x_1 \le 4,\ 0 \le x_2 \le 5.} \end{array} \tag{3.55}$$

We add slacks x_3 and x_4 and set each of their lower bounds to 0 and upper bounds to $+\infty$. In this case we obtain an initial starting solution by setting x_1 and x_2 to their lower bounds of 0 and solving for the slack basic variables x_3 and x_4. In general, we may not be so lucky and may need to add artificial variables (and also possibly change signs of the equations).

At the start of iteration 0, the initial tableau is shown under iteration 0 in Table 3-3. Note that for nonbasic j the last row displays the bound status; if nonbasic j is at its lower bound the status is l, otherwise the status is u.

Notice that both x_1 and x_2 are at their lower bound with $\bar{c}_1 = -3$ and $\bar{c}_2 = -2$ (displayed in the z row). Therefore, by (3.48), $s = 1$. Thus, we should try to increase $x_s = x_1$ since it is at its lower bound. Applying (3.51)–(3.53), the three ratios are $\theta_s = 4$, $\theta_l = 5$, and $\theta_u = +\infty$, implying $r = s = 1$. Hence by the end of the initial iteration, x_1 has moved to its upper bound and remains nonbasic; since $\theta = \theta_s$, the basis is unchanged

Iteration 0 (Phase II)

			$-z$	x_1	x_2	x_3	x_4
Lower Bound			$-\infty$	0	0	0	0
Upper Bound			$+\infty$	4	5	$+\infty$	$+\infty$
Basic Variables			Variables				
			$-z$	x_1	x_2	x_3	x_4
$-z$	$=$	0	1	-3	-2	0	0
x_3	$=$	10		2	1	1	0
x_4	$=$	27		5	3	0	1
Status				l	l		

Iteration 1 (Phase II)

			$-z$	x_1	x_2	x_3	x_4
$-z$	$=$	12	1	-3	-2	0	0
x_3	$=$	2		2	**1**	1	0
x_4	$=$	7		5	3	0	1
Status				u	l		

Iteration 2 (Phase II)

			$-z$	x_1	x_2	x_3	x_4
$-z$	$=$	16	1	1	0	2	0
x_2	$=$	2		2	1	1	0
x_4	$=$	1		**-1**	0	-3	1
Status				u		l	

Iteration 3 (Phase II)

			$-z$	x_1	x_2	x_3	x_4
$-z$	$=$	17	1	0	0	-1	1
x_2	$=$	4		0	1	**-5**	2
x_1	$=$	3		1	0	3	-1
Status						l	l

Iteration 4 (Phase II)

			$-z$	x_1	x_2	x_3	x_4
$-z$	$=$	$17\frac{1}{5}$	1	0	$-\frac{1}{5}$	0	$\frac{3}{5}$
x_3	$=$	$\frac{1}{5}$		0	$-\frac{1}{5}$	1	$-\frac{2}{5}$
x_1	$=$	$\frac{12}{5}$		1	$-\frac{3}{5}$	0	$\frac{1}{5}$
Status					u		l

Table 3-3: Tableau Form Example of a Bounded Variable LP

and no pivoting is required. However, the new basic values for the basic variables need to be computed by (3.54):

$$\begin{pmatrix} -z \\ x_3 \\ x_4 \end{pmatrix} = \begin{pmatrix} 0 \\ 10 \\ 27 \end{pmatrix} - 4 \times \begin{pmatrix} -3 \\ 2 \\ 5 \end{pmatrix} = \begin{pmatrix} 12 \\ 2 \\ 7 \end{pmatrix}.$$

At the start of iteration 1, x_1 is at its upper bound and x_2 is at its lower bound, with $\bar{c}_1 = -3$ and $\bar{c}_2 = -2$, and therefore by (3.48), $s = 2$. Thus, we should try to increase $x_s = x_2$ since it is at its lower bound. Applying (3.51)–(3.53), the three ratios are $\theta_s = 5$, $\theta_l = 2$, and $\theta_u = +\infty$, implying $r = l = 1$, where $j_r = j_1 = 3$. Hence, $x_s = x_2$ enters the basis at value $x_2 = \theta = 2$, and x_3 leaves the basis to be at its lower bound. We compute the change of basic variables x_B and $x_s = x_2$ by (3.54) and (3.50) respectively:

$$\begin{pmatrix} -z \\ x_3 \\ x_4 \end{pmatrix} = \begin{pmatrix} 12 \\ 2 \\ 7 \end{pmatrix} - 2 \times \begin{pmatrix} -2 \\ 1 \\ 3 \end{pmatrix} = \begin{pmatrix} 16 \\ 0 \\ 1 \end{pmatrix},$$

$$x_2 = l_2 + \delta\theta = 0 + 2 = 2, \qquad \text{replacing } x_3 \text{ in the basis.}$$

The values of $(-z)$, x_2, and x_4 are posted to the left of the double line at the start of iteration 2. Next we pivot on the boldface term **1** and record the resulting values in the tableau for the start of iteration 2.

At the start of iteration 2, x_1 is at its upper bound and x_3 is at its lower bound, with $\bar{c}_1 = 1$ and $\bar{c}_3 = 2$. Therefore by (3.48), $s = 1$. Thus, we should try to decrease $x_s = x_1$ since it is at its upper bound; in this case $\delta = -1$. Applying (3.51)–(3.53), the three ratios are $\theta_s = 4$, $\theta_l = 1$, and $\theta_u = 3/2$, implying $r = l = 2$, where $j_r = j_2 = 4$. Hence, $x_s = x_1$ enters the basis at value $x_1 = u_1 - \theta = 3$, and x_4 leaves the basis at its lower bound. We compute the change of basic variables x_B and $x_s = x_1$ by (3.54) and (3.50) respectively:

$$\begin{pmatrix} -z \\ x_2 \\ x_4 \end{pmatrix} = \begin{pmatrix} 16 \\ 2 \\ 1 \end{pmatrix} - (-1) \times 1 \times \begin{pmatrix} 1 \\ 2 \\ -1 \end{pmatrix} = \begin{pmatrix} 17 \\ 4 \\ 0 \end{pmatrix},$$

$$x_1 = u_1 + \delta\theta = 4 + (-1) \times 1 = 3, \qquad \text{replacing } x_4 \text{ in the basis.}$$

The values of $(-z)$, x_2, and x_1 are posted to the left of the double line at the start of iteration 3. Next we pivot on the boldface term **−1** and record the resulting values in the tableau for the start of iteration 3.

At the start of iteration 3, x_3 is at its lower bound and x_4 is also at its lower bound, with $\bar{c}_3 = -1$ and $\bar{c}_4 = 1$. Therefore by (3.48), $s = 3$. Thus, we should try to increase $x_s = x_3$ since it is at its lower bound. Applying (3.51)–(3.53), the three ratios are $\theta_s = +\infty$, $\theta_l = 1$, and $\theta_u = 1/5$, implying $r = u = 1$, where $j_r = j_1 = 2$. Hence, $x_s = x_3$ enters the basis at value $x_3 = l_3 + \theta = 1/5$, and x_2 leaves the basis at its upper bound. We compute the change of basic variables x_B and $x_s = x_3$ by (3.54) and (3.50) respectively:

$$\begin{pmatrix} -z \\ x_2 \\ x_1 \end{pmatrix} = \begin{pmatrix} 17 \\ 4 \\ 3 \end{pmatrix} - \frac{1}{5} \times \begin{pmatrix} -1 \\ -5 \\ 3 \end{pmatrix} = \begin{pmatrix} 17\frac{1}{5} \\ 5 \\ \frac{12}{5} \end{pmatrix},$$

$$x_3 = l_3 + \delta\theta = 0 + 1/5 = 1/5, \qquad \text{replacing } x_2 \text{ in the basis.}$$

The values of $(-z)$, x_3, and x_1 are posted to the left of the double line at the start of iteration 4. Next we pivot on the boldface term $-\mathbf{5}$ and record the resulting values in the tableau for the start of iteration 4.

At the start of iteration 4, the tableau is optimal by Theorem 3.9, since x_2 is at its upper bound and $\bar{c}_2$ is negative; and, x_4 is at its lower bound and $\bar{c}_4$ is positive. The optimal solution is readily read off from the tableau: The basic values are $z = -17\frac{1}{5}$, $x_3 = 1/5$, and $x_1 = 12/5$; the nonbasic values are $x_2 = 5$ since x_2 has status u, and $x_4 = 0$ since x_4 has status l.

3.5 REVISED SIMPLEX METHOD

The Revised Simplex Method is not a different method but is a different way to carry out each computational step of the Simplex Method. The revised method offers several advantages over the Simplex Algorithm in tableau form.

1. Considerable savings in computations are possible if the fraction of nonzero coefficients is less than $1 - (2m/n)$. This is typically the case in practice.

2. Less data is recorded from one iteration to the next, which permits a larger problem to be solved when the memory capacity of an electronic computer is limited. Furthermore, considerable savings in computations are possible because not all the entries in the tableau need to be updated.

3. A weakness of the simplex algorithm in tableau form is that the inverse of the basis in explicit form is part of the full tableau. The explicit representation of the inverse can be numerically unstable. Since the inverse is only needed to solve certain systems of equations, the need to have an inverse can be bypassed in the Revised Simplex Algorithm by solving the system of equations by numerically stable methods.

3.5.1 MOTIVATION

While each iteration of the Simplex Method requires that a whole new tableau be computed and recorded, it may be observed that only the modified cost row and the column corresponding to the variable entering the basic set play any role in the decision process. That is, in order to carry out the steps of the Simplex Algorithm (see Algorithm 3.1),

1. we first look at the reduced costs $(\bar{c})$ to determine whether an improvement in the solution is possible, and if so, then we determine which column s to bring into the basis (see Steps 1–3 of the Simplex Algorithm);

2. next we look at the pivot column $\bar{A}_{\bullet s}$ and the right hand side $\bar{b}$ to determine the pivot row r and the variable that leaves the basis (see Steps 4–5 of the Simplex Algorithm).

Since $\bar{b}$ can be easily updated, at each iteration all we need are the reduced costs and the updated column corresponding to the incoming variable. It turns out that we can obtain this information directly if we have available a way to solve a system of equations whose coefficient matrix is the current basis matrix B or its transpose. Any efficient way to update the solution procedures for the next iteration will do. For the purpose of illustrating the Revised Simplex Algorithm on small examples, we will use an explicit representation of B^{-1} and a way to update it. In general, as we have just noted, computing and using an inverse can be a numerically unstable process and also, for large problems an explicit representation of B^{-1} may not be computationally efficient in terms of speed and storage.

If we have at hand an explicit representation of B^{-1}, it is easy to derive the required quantities $\bar{c}$ and $\bar{A}_{\bullet s}$ from the original data. Assume that an initial feasible solution is available for a given linear program in standard form (3.1). Suppose, for convenience of discussion, that the columns of the coefficient matrix A have been ordered so that the basis columns (represented by the matrix $B = B^t$ for iteration t) are the first m columns of A and the nonbasic columns (represented by the matrix $N = N^t$) are the last $(n - m)$ columns of A. That is,

$$\begin{aligned} (-z) + c_B^T x_B + c_N^T x_N &= 0 \quad (a) \\ B x_B + N x_N &= b \quad (b). \end{aligned} \tag{3.56}$$

Algebraically, the canonical form of (3.56) for iteration t is given by

$$\begin{aligned} (-z) \qquad + (c_N^T - c_B^T B^{-1} N) x_N &= -c_B^T B^{-1} b \quad (a) \\ I x_B + \qquad B^{-1} N x_N &= B^{-1} b. \quad (b) \end{aligned} \tag{3.57}$$

This can be seen by multiplying (3.56b) by B^{-1} on the left to obtain (3.57b). If now we multiply (3.57b) on the left by c_B^T and subtract from (3.56a), we obtain (3.57a). But this is not the best way to carry out the computations.

To compute $\bar{c}$ efficiently we first compute

$$\pi^T = c_B^T B^{-1}, \tag{3.58}$$

and then

$$\bar{c}^T = c_N^T - \pi^T N, \tag{3.59}$$

followed by $s = \text{argmin}_j\, \bar{c}_j$. Then if $\bar{c}_s < 0$, we compute

$$\bar{A}_{\bullet s} = B^{-1} A_{\bullet s}. \tag{3.60}$$

The value $\bar{b} = B^{-1} b$ is not explicitly computed; instead it is updated in the Revised Simplex Tableau when a pivot is performed.

Definition (Price): The vector π is called the *price* vector (or *simplex multipliers*). We will see in Section 7.1 why the economists refer to π as the price vector.

THEOREM 3.10 (Uniqueness of π) *At any iteration t, the simplex multipliers π are unique.*

▷ **Exercise 3.24** Prove Theorem 3.10.

Definition (Pricing Out): The operation of multiplying π times the nonbasic column $A_{\bullet j}$ for nonbasic j in the determination of $N^T\pi$ is called *pricing out* the columns j.

Definition (Representation in Terms of the Basis): The linear combination of the columns in the basis that yields the incoming column $A_{\bullet s}$ is called the *representation of the sth activity in terms of the basic set of activities.* It is easy to see that this representation is the updated column $A_{\bullet s}$, see (3.60), because in matrix notation (3.60) can be rewritten as:

$$B\bar{A}_{\bullet s} = A_{\bullet s}. \tag{3.61}$$

See Section A.6 in Appendix A where the notion of a basis in a vector space is discussed.

Both pricing out a column and representation of an activity in terms of a basis can be done very efficiently for matrices having many zero elements (so called *sparse matrices*).

Next we shall show how to easily find B^{-1} and π in the tableau form. For convenience of discussion we reorder the variables as before in (3.56) so that the first m columns of A are $B = B^t$. Assume also that now we add m artificial variables x_a to (3.56b) to obtain (3.62):

$$\begin{array}{rl} (-z) + c_B^T x_B + c_N^T x_N \qquad\qquad = 0 & (a) \\ Bx_B + Nx_N + Ix_a = b. & (b) \end{array} \tag{3.62}$$

Once again multiplying (3.62b) by B^{-1}, we get (3.63b), and subtracting the product of c_B^T times (3.63b) from (3.62a), we get (3.63a), which together give the canonical form for iteration t:

$$\begin{array}{rlll} (-z) & + (c_N^T - c_B^T B^{-1}N)x_N - c_B^T B^{-1}x_a = -c_B^T B^{-1}b & (a) \\ Ix_B + & B^{-1}Nx_N + B^{-1}x_a = B^{-1}b. & (b) \end{array} \tag{3.63}$$

Thus, $\pi^T = c_B^T B^{-1}$ and B^{-1} for any iteration t can be directly read off from the tableau by examining the columns corresponding to x_a.

Furthermore, if we were to change the basis by pivoting in a new column from $\bar{A} = B^{-1}N$, then the new basis inverse, say $\bar{B}^{-1}$, would still be available in the columns corresponding to x_a.

▷ **Exercise 3.25** Prove that no matter which column we bring into the basis, the new basis inverse will be available in the columns corresponding to x_a.

The above discussion implies that we can reduce the size of our tableau form by keeping only the $(-z)$ column, the columns corresponding to x_a, and the updated right-hand side. With these we can generate the reduced cost $\bar{c}$ and the incoming column $\bar{A}_{\bullet s}$. The reduced size tableau is shown below in Table 3-4. Thus, we never

$(-z)$	x_a	RHS	x_s	
1	$-\pi^T$	$-\pi^T b$	$\bar{c}_s$	
	B^{-1}	$\bar{b}$	$\bar{A}_{\bullet s}$	$\longleftarrow$ Pivot on $\bar{A}_{rs}$

Table 3-4: Revised Simplex Tableau Iteration t

explicitly compute $\bar{A}$. This is what gives considerable savings in storage space. To update the revised simplex tableau, Table 3-4 is pivoted on in the last column on element $\bar{a}_{rs}$. Thus the $\bar{b}$ for the next iteration is computed by the pivot step.

▷ **Exercise 3.26** In Phase I we replace the objective (3.62a) by the infeasibility form

$$(-w) + e^T x_a = 0.$$

Show that canonical form at the start of Phase I is

$$\begin{array}{rcrcrcl} (-w) & - & e^T A x & + & 0 x_a & = & 0 \\ & & A x & + & I x_a & = & b. \end{array}$$

Write down the steps of the Revised Simplex Algorithm for Phase I. Next write down the steps of the Revised Simplex Method.

3.5.2 REVISED SIMPLEX METHOD ILLUSTRATED

Example 3.3 (Revised Simplex Method) We shall illustrate the Revised Simplex Method on the same Example 3.1 used to illustrate the Simplex Method in tableau form. Namely, find min z, $x \geq 0$ such that

$$\begin{array}{rcrcrcrcrcl} 2x_1 & + & 1x_2 & + & 2x_3 & + & x_4 & + & 4x_5 & = & z \\ 4x_1 & + & 2x_2 & + & 13x_3 & + & 3x_4 & + & x_5 & = & 17 \\ x_1 & + & x_2 & + & 5x_3 & + & x_4 & + & x_5 & = & 7. \end{array} \tag{3.64}$$

For ease of exposition we will display the entire tableau form but only show the quantities that are determined explicitly or obtained by pivoting.

The detached coefficient form for (3.64) is shown in Table 3-5. The coefficients that appear in the top row are $d_j = -e^T A_{\bullet j}$, where $e = (1, 1, \ldots, 1)^T$. The Phase I reduced costs $\bar{d}_j$ that will be computed on iterations 0, 1, and 2 of Phase I of the Revised Simplex Method are stored in the $-w$ line of each iteration in Table 3-6, as will be the reduced costs $\bar{c}_j$ for iterations 2 and 3 of Phase II in Table 3-7. Actually, in practice, in the Revised Simplex Method, the $\bar{d}_j$ and $\bar{c}_j$ are computed but not stored; only the minimum value $\bar{d}_s$ or $\bar{c}_s$ is stored at the beginning of iteration t. Table 3-6 shows the recorded data at the beginning of iteration t of the Revised Simplex Method. The column labeled "Basic

Detached Coefficients of Original System

Basic Variables		Original Variables						Artificial Variables		Constants
	$-w$	$-z$	x_1	x_2	x_3	x_4	x_5	x_6	x_7	
$-w$	1		−5	−3	−18	−4	−2			−24
$-z$		1	2	1	2	1	4			0
x_6			4	2	13	3	1	1		17
x_7			1	1	5	1	1		1	7

Table 3-5: Detached Coefficients

Variables" shows the names of the variables in the order $(j_1, \ldots, j_m)$ in the basis. The *relevant* columns of the canonical form correspond to the artificial variables and contain the negative of the simplex multipliers and the basis inverse. The simplex multipliers π are the negative coefficients of the artificials in the $-w$ row in Phase I; in Phase II these are in the $-z$ row. The column after the last artificial variable is the modified right hand side. These are obtained by pivoting.

To compute $\bar{d}_j$ and $\bar{c}_j$ directly from the original system Table 3-5, we will need the simplex multipliers associated with iteration t. The numbers shown in italic (or boldface) are generated directly from the original system. Recall that $\bar{d}_j = d_j - \pi^T A_{\bullet j}$ and $\bar{c}_j = c_j - \pi^T A_{\bullet j}$ can be computed from the original data. If the solution is not optimal, we determine the incoming column index s. The inverse of the basis is used to compute $\bar{A}_{\bullet s}$ directly from the original data by the formula $\bar{A}_{\bullet s} = B^{-1} A_{\bullet s}$. The pivoting transforms column s to a unit column vector, which is not recorded.

On iteration 0, the Phase I basis inverse is the identity matrix (see Table 3-6), and hence the entries shown in the table are the same as the corresponding entries from the original data. Since the reduced cost $\bar{d}_3 = -\mathbf{18}$ is the smallest, we bring in the corresponding x_3 into the basis on the next iteration. We are now in a position to determine which variable leaves the basis; we compute $\bar{A}_{\bullet 3} = B^{-1} A_{\bullet 3}$. Since $B^{-1} = I$ for this iteration, $\bar{A}_{\bullet 3} = A_{\bullet 3}$ can be read off directly from the original data and entered in the corresponding column of Table 3-6 for iteration 0. The updated column $\bar{A}_{\bullet 3}$ and $\bar{b}$ allow us to locate the pivot position, which is $\bar{A}_{13}$. Pivoting on $\bar{A}_{13}$ drives x_6 out of the basis, replacing it by x_3, and generates the artificial columns and updated $\bar{b}$ of the next iteration, $t = 1$.

On iteration 1, the basis inverse is

$$B^{-1} = \begin{pmatrix} \frac{1}{13} & 0 \\ -\frac{5}{13} & 1 \end{pmatrix}$$

in the columns corresponding to x_6 and x_7 in Table 3-6. The simplex multipliers are $\pi_1 = -18/13$ and $\pi_2 = 0$ which are the negative of the entries in the first row in columns corresponding to x_6 and x_7. Thus the $\bar{d}_j$ can be computed by using (3.59) as $d_j - \pi^T A_{\bullet j}$ for iteration 1 and so forth.

3.5.3 REVISED SIMPLEX METHOD ALGORITHM

The algorithm is described using B^{-1}; in practice the inverse is not computed, but instead the system of equations involving B is solved directly by an LU factorization

Iteration 0 (Phase I)

Basic Variables	OBJ	Original Variables					Artificial Variables		RHS
	$-w$	x_1	x_2	x_3	x_4	x_5	x_6	x_7	
				$\bar{A}_{\bullet 3} = B^{-1}A_{\bullet 3}$			$-\pi_1$	$-\pi_2$	
$-w$	*1*	*−5*	*−3*	*−18*	*−4*	*−2*	0	0	−24
x_6				**13**			1		17
x_7				5				1	7
							B^{-1}		

Iteration 1 (Phase I) $\bar{d}_j = d_j - \pi^T A_{\bullet j}$

Basic Variables	$-w$	x_1	x_2	x_3	x_4	x_5	x_6	x_7	RHS
						$B^{-1}A_{\bullet 5} = \bar{A}_{\bullet 5}$	$-\pi_1$	$-\pi_2$	
$-w$	1	$\frac{7}{13}$	$-\frac{3}{13}$	*0*	$\frac{2}{13}$	$-\frac{8}{13}$	$\frac{18}{13}$		$-\frac{6}{13}$
x_3						$\frac{1}{13}$	$\frac{1}{13}$		$\frac{17}{13}$
x_7						$\mathbf{\frac{8}{13}}$	$-\frac{5}{13}$	1	$\frac{6}{13}$
							B^{-1}		

Iteration 2 (Phase I Optimal) $\bar{d}_j = d_j - \pi^T A_{\bullet j} \geq 0$

Basic Variables	$-w$	x_1	x_2	x_3	x_4	x_5	x_6	x_7	RHS
$-w$	1	*0*	*0*	*0*	*0*	*0*	1	1	0
x_3							$\frac{1}{8}$	$-\frac{1}{8}$	$\frac{5}{4}$
x_5							$-\frac{5}{8}$	$\frac{13}{8}$	$\frac{3}{4}$
							B^{-1}		

Table 3-6: Revised Simplex Method: Tableau Form Example

Iteration 2 (Phase II Start: Introduce z)

Basic Variables	OBJ	Original Variables					Artificial Variables		RHS
	$-z$	x_1	x_2	x_3	x_4	x_5	x_6	x_7	
$-z$	1	2	1	2	1	4	0	0	0
x_3							$\frac{1}{8}$	$-\frac{1}{8}$	$\frac{5}{4}$
x_5							$-\frac{5}{8}$	$\frac{13}{8}$	$\frac{3}{4}$

B^{-1} (columns x_6, x_7)

Iteration 2 (Phase II: Updated z) $\quad \bar{c}_j = c_j - \pi^T A_{\bullet j}$

$\bar{A}_{\bullet 2} = B^{-1} A_{\bullet 2}$ $\quad -\pi^T = -c_B^T B^{-1}$

Basic Variables	$-z$	x_1	x_2	x_3	x_4	x_5	x_6	x_7	RHS
$-z$	1	$\frac{19}{4}$	$-\frac{3}{4}$		$\frac{3}{2}$		$\frac{9}{4}$	$-\frac{25}{4}$	$-\frac{11}{2}$
x_3			$\frac{1}{8}$				$\frac{1}{8}$	$-\frac{1}{8}$	$\frac{5}{4}$
x_5			$\mathbf{\frac{3}{8}}$				$-\frac{5}{8}$	$\frac{13}{8}$	$\frac{3}{4}$

B^{-1} (columns x_6, x_7)

Iteration 3 (Phase II Optimal) $\quad \bar{c}_j = c_j - \pi^T A_{\bullet j} \geq 0$

Basic Variables	$-z$	x_1	x_2	x_3	x_4	x_5	x_6 ($-\pi_1$)	x_7 ($-\pi_2$)	RHS
$-z$	1	3	0	0	1	2	1	$\frac{19}{2}$	-4
x_3							$\frac{1}{3}$	$-\frac{2}{3}$	1
x_2							$-\frac{5}{3}$	$\frac{13}{3}$	2

B^{-1} (columns x_6, x_7)

Table 3-7: Revised Simplex Method: Tableau Form Example (Continued)

(based on Gaussian elimination).

Algorithm 3.3 (Revised Simplex Algorithm) Assume a linear program in standard form and that $x_B = \bar{b}$ is a basic feasible solution and that columns with indices $j_1, \ldots, j_m$ for a feasible basis B whose inverse B^{-1} is known.

1. *Simplex Multipliers.* Determine the simplex multipliers
$$\pi = (B^{-1})^T c_B.$$
2. *Reduced Costs.* Determine the reduced costs
$$\bar{c}_j = c_j - \pi^T A_{\bullet j} \qquad \text{for } j \text{ nonbasic.}$$
3. *Smallest Reduced Cost.* Same as Step 1 of the Simplex Algorithm 3.1.
4. *Test for Optimality.* Same as Step 2 of the Simplex Algorithm 3.1.
5. *Incoming Variable.* Same as Step 3 of the Simplex Algorithm 3.1.
6. *Determine* $\bar{A}_{\bullet s}$, the representation of $A_{\bullet s}$ in terms of the basis B.
7. *Test for unbounded* z. Same as Step 4 of the Simplex Algorithm 3.1.
8. *Outgoing Variable.* Same as Step 5 of the Simplex Algorithm 3.1.
9. Pivot on $\bar{a}_{rs}$ in the matrix $\big(\bar{b}, B^{-1}, \bar{A}_{\bullet s}\big)$ to obtain $\big(\text{updated } \bar{b}, \text{updated } B^{-1}, e_r\big)$, where e_r is the unit vector with 1 in row r and zeros elsewhere.
10. Set $j_r = e_r$ and return to Step 1 with updated $\bar{b}$ and B^{-1}.

▷ **Exercise 3.27** Show that pivoting on $\bar{a}_{rs}$ involves multiplying $\bar{b}$, B^{-1}, and $\bar{A}_{\bullet s}$ by the inverse of the elementary matrix
$$E_r = I + (\bar{A}_{\bullet s} - e_r)e_r^T,$$
which is
$$E_r^{-1} = I - \frac{(\bar{A}_{\bullet s} - e_r)e_r^T}{\bar{a}_{rs}}.$$

3.5.4 COMPUTATIONAL REMARKS

In the standard Simplex Method we modify on each iteration the entire tableau of $(m+1)(n+1)$ entries. Not counting the identity part of the standard simplex tableau, the number of operations (multiplication and addition pairs) on each iteration is
$$\big((n-m)+1\big)(m+1) = (n-2m)(m+1) + (m+1)^2. \tag{3.65}$$

In the Revised Simplex Method, we never explicitly compute all of $\bar{A}$. We only compute $\bar{A}_{\bullet s}$ and $\bar{c}$; and update $-\pi$, and $\bar{B}^{-1}$ by pivoting. Thus we modify only $(m+1)(m+1)$ entries of the tableau (including the computation of π). Thus, starting with an $m \times n$ system in a feasible canonical form, the total number of operations required per iteration is
$$f(n-m)(m+1) + fm(m+1) + (m+1)^2 = fn(m+1) + (m+1)^2, \tag{3.66}$$

where f is the fraction of nonzero coefficients in the original tableau, which we assume, on the average, is the same as that in the column entering the basis. The three terms on the left are the numbers of operations used (a) in "pricing out," (b) in representing the new column, and (c) in pivoting.

Comparing (3.66) and (3.65) it is easy to see that the Revised Simplex Method requires less work than the standard Simplex Method if f, the fraction of nonzeros, satisfies

$$f < 1 - 2m/n.$$

From the above it is clear that for the Revised Simplex Method to require less work than the standard Simplex Method we must have $n > 2m$.

3.6 NOTES & SELECTED BIBLIOGRAPHY

Since the invention of the Simplex Method by one of the authors (George Dantzig) in 1947, numerous papers have appeared; they are far too many to reference all the titles of the papers and presentations of linear programming.

On first glance it may appear that most problems will be nondegenerate. After all, what is the probability of four planes in three space meeting in a point (for example)! It turns out that although it appears that the probability of a linear program being degenerate is zero, almost every problem encountered in practice is highly degenerate. Degeneracy is the rule, not the exception! For this reason, the choice of the variable to leave the basis in the case of a tie has been and continues to be the subject of much investigation because of the theoretical possibility that a poor choice could lead to a repetition of the same basic solution after a number of iterations. In *Linear Programming 2*, several examples are given where using the standard rules results in a sequence of pivots that repeats, called *cycling in the Simplex Algorithm*. The choice of pivots under degeneracy and near degeneracy is examined in detail later, in *Linear Programming 2*. There proofs are provided for finite termination of the simplex algorithm under various degeneracy-resolving schemes such Dantzig's inductive method, Wolfe's rule, Bland's rule, Harris's procedure, and the Gill, Murray, Saunders, Wright anticycling procedure. A proof of convergence in a finite number of steps under the random choice rule can be found in Dantzig [1963]; see also *Linear Programming 2*.

Another question is concerned with the number of iterations required to solve a linear program using the Simplex Method. Examples have been contrived by Klee and Minty [1972] that require in the worst case $(2^m - 1)$ iterations. Consider the linear program

$$\begin{array}{lll} \text{Minimize} & \displaystyle\sum_{j=1}^{m} -10^{m-j} x_j & \\ \text{subject to} & \displaystyle\left(2\sum_{j=1}^{i-1} 10^{i-j} x_j\right) + x_i + z_i = 100^{i-1}, & \text{for } i = 1, \dots, m, \\ & x_j \geq 0, \ z_j \geq 0, \quad \text{for } j = 1, \dots, m. & \end{array} \tag{3.67}$$

If we apply the Simplex Method (using the largest coefficient rule) to solve the problem (3.67), then it can be shown that the Simplex Method performs $(2^m - 1)$ iterations before finding the optimal solution. If examples such as these are representative of the "real-world," the number of iterations would be too high for the Simplex Method to be a practical algorithm. Experience solving thousands and thousands of practical problems shows that the method solves them in surprisingly few iterations. To explain why this is so requires some way to characterize the special properties of problems encountered in practice. So far nobody has been able to do so, and the best that one has been able to do along these lines is to use some hypothesized probabilistic model for generating the class of linear programs to be solved and proving theorems about the *expected number* of iterations. See, for example, Borgwardt [1982a, 1982b, 1987a, 1987b] and Smale [1982]).

In *Linear Programming 2* we describe other, more complex, rules for selecting an incoming column, including those that are scale invariant, that are more computationally efficient for large problems.

In this chapter, we described a Phase I procedure that uses a full set of artificial variables to obtain a starting basic feasible solution; it was first proposed by Dantzig [1951a]. Another technique, labeled the Big-M method, was first suggested by Dantzig and subsequently by others. (See Problems 3.28, 3.29, 3.30, 3.31, and 3.32). The technique was developed in the context of the standard form, but is also directly applicable to a linear program with bounded variables. After adding a full set of artificial variables, the objective function is modified by adding to it the sum of the artificial variables each multiplied by a large cost M. Then the problem is solved with the new objective function. Provided the cost M is chosen large enough, it is clear that the artificial variables will be driven to zero when minimizing. The problems with this method are (1) M has to be chosen to be large enough, and (2) a large value of M can cause numerical problems when computing the multipliers and reduced costs. In *Linear Programming 2* we discuss different methods for finding an initial feasible solution.

The Revised Simplex Method was first proposed by Dantzig & Orchard-Hays [1953].

Many excellent books, too numerous to mention, are available on linear programming. Among them are Bazarra, Jarvis, & Sherali [1990], Bradley, Hax, & Magnanti [1977], Brickman [1989], Chvátal V. [1983], Dantzig [1963], Gass [1985], Hillier & Lieberman [1995], Murty [1983], and Nering & Tucker [1993].

3.7 PROBLEMS

3.1 First solve the following using the Simplex Method. Next solve it using the Revised Simplex Method. Finally, solve it using the `DTZG Simplex Primal` software option.

$$\begin{array}{lrcrcrcrcl}
\text{Maximize} & 3x_1 & + & x_2 & + & 5x_3 & + & 4x_4 & = & z \\
\text{subject to} & 3x_1 & - & 3x_2 & + & 2x_3 & + & 8x_4 & \leq & 50 \\
 & 4x_1 & + & 6x_2 & - & 4x_3 & - & 4x_4 & \leq & 40 \\
 & 4x_1 & - & 2x_2 & + & x_3 & + & 3x_4 & \leq & 20
\end{array}$$

and $x_1 \geq 0,\ x_2 \geq 0,\ x_3 \geq 0,\ x_4 \geq 0.$

3.2 Consider:

$$\begin{array}{lrcrcrcrcl} \text{Minimize} & 3x_1 & & & - & x_3 & & & = & z \\ \text{subject to} & x_1 & + & x_2 & + & x_3 & + & x_4 & = & 4 \\ & -2x_1 & + & x_2 & - & x_3 & & & = & 1 \\ & & & 3x_2 & + & x_3 & + & x_4 & = & 9 \end{array}$$

and $x_1 \geq 0,\ x_2 \geq 0,\ x_3 \geq 0,\ x_4 \geq 0$.

(a) Solve the above linear program using the Simplex Method. Be sure to carry the part of the tableau corresponding to the artificial variables all the way to the final tableau.

(b) What is the index set (call it $\mathcal{B}$) of the optimal basis? What is $A_{\bullet\mathcal{B}}$? Where is $[A_{\bullet\mathcal{B}}]^{-1}$ in the final tableau?

(c) Compute $\pi = ([A_{\bullet\mathcal{B}}]^{-1})^T c_{\mathcal{B}}$. Where is π in the final tableau? Verify through direct computation that $\bar{c} = c - A^T\pi$.

3.3 Solve the following bounded variable linear program by hand.

$$\begin{array}{lrcrcl} \text{Minimize} & -2x_1 & - & x_2 & = & z \\ \text{subject to} & 3x_1 & + & x_2 & \leq & 9 \\ & 2x_1 & - & 2x_2 & \leq & 3 \end{array}$$

and $0 \leq x_1 \leq 1,\ 0 \leq x_2 \leq 8$.

3.4 Read each of the following statements carefully and decide whether it is true or false. Briefly justify your answer.

(a) If a tie occurs in the pivot row choice during a pivot step while solving an LP by the Simplex Method, the basic feasible solution obtained after this pivot step is degenerate.

(b) In solving an LP by the Simplex Method, a different basic feasible solution is generated after every pivot step.

(c) The total number of optimal solutions of an LP is always finite since the total number of different bases is finite.

3.5 A farm is comprised of 240 acres of cropland. The acreage to be devoted to corn production and the acreage to be used for oats production are the decision variables. Profit per acre of corn production is \$40 and the profit per acre of oats production is \$30. An additional resource restriction is that the total labor hours available during the production period is 320. Each acre of land in corn production uses 2 hours of labor during the production period, whereas production of oats requires only 1 hour. Formulate an LP to maximize the farm's profit. Solve it.

3.6 Suppose we are solving the problem

$$\begin{array}{ll} \text{Minimize} & c^Tx \\ \text{subject to} & Ax = b \\ & x \geq 0. \end{array}$$

and we arrive at the following Phase II tableau:

$(-z)$	0	1	0	c_1	14
0	1	1	0	a_1	b_1
0	0	-2	1	a_2	b_2

(a) Identity the current basic solution and give conditions that assure it is a basic feasible solution.

(b) Give conditions that assure that the basic solution is an optimal basic feasible solution.

(c) Give conditions that assure that the basic solution is the *unique* optimal basic feasible solution.

(d) Give conditions that guarantee that the objective value is unbounded below.

(e) Assuming the conditions in (d) hold, exhibit a feasible ray on which the objective value goes to $-\infty$ and exhibit a set of infeasibility multipliers for the dual problem. Note that the *ray* generated by $q \in \Re^n$ is the set of points $\{ x \mid x = \theta q \}$ as the scalar parameter θ varies from 0 to $+\infty$.

(f) Assuming the conditions of (a) hold, give all conditions under which you would perform a pivot on the element a_1.

3.7 Late one night, while trying to make up this problem set, your trusty course assistant decided to give you the following linear program to solve:

x_1	x_2	x_3	x_4	x_5	x_6	x_7	x_8	x_9	x_{10}	x_{11}	RHS
1	0	0	0	0	0	1	1	1	1	1	4
0	1	0	0	0	0	2	2	−1	−3	5	6
0	0	1	0	0	0	2	2	3	0	0	4
0	0	0	1	0	0	−3	0	4	5	6	6
0	0	0	0	1	0	−9	3	−3	0	−1	9
0	0	0	0	0	1	−4	0	−2	−1	5	4
0	0	0	0	0	0	−5	−8	−5	−6	−7	0

After pivoting with the Simplex Algorithm until he got an optimal solution, he then—klutz that he is—spilled his can of Coke over the final tableau (not a suggested practice). Unfortunately, by Murphy's Law of Selective Gravitation (i.e., objects fall where they do the most damage) the spilled Coke dissolved the right-hand side (RHS) of the final tableau, leaving only:

x_1	x_2	x_3	x_4	x_5	x_6	x_7	x_8	x_9	x_{10}	x_{11}	RHS
43/24	1/24	−15/16	−1/3	0	0	1	0	−115/48	0	1	–
3/8	1/8	−5/16	0	0	0	0	0	−11/16	0	1	–
−43/24	1/8	23/16	1/3	0	0	0	1	187/48	0	0	–
5/8	−1/8	−3/16	0	0	0	0	0	3/16	1	0	–
175/8	5/8	−209/16	−4	1	0	0	0	−591/16	0	0	–
71/12	−7/12	−19/8	−4/3	0	1	0	0	−191/24	0	0	–
1	0	7/2	1	0	0	0	0	21/2	0	0	–

Luckily the course assistant was able to fill in the missing right-hand side (optimal basic feasible solution and final objective value) without doing any further pivoting. How did he do it? What is the missing right hand side?

3.8 Spaceman Spiff hurtles through space toward planet Bog. Cautiously, he steps out of his spacecraft and explores the dunes. Suddenly, the Great Giztnad pounces! Spaceman Spiff turns his hydroponic laser against it, but it has no effect! The Great Giztnad rears its gruesome head and speaks: "Solve this problem using the Simplex Method before I return, or I shall boil you in cosmic oil! Hee, hee, hee!"

x_1	x_2	x_3	x_4	x_5	x_6	x_7	x_8	x_9	x_{10}	RHS
1	0	0	0	0	1	1	2	1	3	3
0	1	0	0	0	−1	1	2	3	4	2
0	0	1	0	0	2	5	−4	0	2	4
0	0	0	1	0	1	3	1	1	2	1
0	0	0	0	1	5	4	3	−2	1	3
0	0	0	0	0	−3	−2	2	−1	−2	

Using what he learned in OR340, Spiff quickly applied the Simplex Method and arrived at the final tableau

x_1	x_2	x_3	x_4	x_5	x_6	x_7	x_8	x_9	x_{10}	RHS
1	0	0	−1	0	0	−2	1	0	1	2
0	1	0	−13/7	4/7	0	−16/7	13/7	0	6/7	13/7
0	0	1	−4/7	−2/7	0	15/7	−38/7	0	4/7	18/7
0	0	0	5/7	−1/7	0	11/7	2/7	1	9/7	2/7
0	0	0	2/7	1/7	1	10/7	5/7	0	5/7	5/7
0	0	0	11/7	2/7	0	27/7	31/7	0	10/7	17/7

"I have solved the problem, your Vileness," Spaceman Spiff announces. "Goodbye, oh hideous master of–"

"Not so fast!" booms the Great Giztnad. "You have missed a column. Those numbers farther off to the right are the RHS. Also, you copied some of the numbers incorrectly. I'll give you ten more minutes. Solve this quickly, before I lose my patience and eat you raw!"

x_1	x_2	x_3	x_4	x_5	x_6	x_7	x_8	x_9	x_{10}	x_{11}	RHS
1	0	0	0	0	1	1	2	1	3	3*	90*
0	1	0	0	0	−1	1	2	3	4	2*	155*
0	0	1	0	0	2	5	−4	0	2	4*	62*
0	0	0	1	0	1	3	1	1	7*	1*	70*
0	0	0	0	1	5	4	3	−2	7*	3*	70*
0	0	0	0	0	−3	−2	2	−1	−2	−2*	

(* represents changes from the original tableau).

Show what the resulting final tableau and optimal solution will be, and *explain* why it is optimal. Hint: You do not need to redo the Simplex Method.

3.9 Consider the following linear program:

$$\begin{array}{lrcrcrcrcl} \text{Minimize} & 2x_1 & - & x_2 & + & x_3 & + & 5x_4 & = & z \\ \text{subject to} & x_1 & + & x_2 & + & x_3 & + & x_4 & = & 4 \\ & 2x_1 & + & 3x_2 & - & 4x_3 & + & 2x_4 & \le & 5 \\ & x_1 & + & 2x_2 & - & 5x_3 & + & x_4 & \ge & 2 \\ & \multicolumn{9}{l}{x_j \ge 0,\ j = 1, \ldots, 4.} \end{array}$$

(a) Use the Simplex Method (with no more than 2 artificial variables) to show that it is infeasible.

(b) Use the **DTZG Simplex Primal** software option on the problem.

3.10 Use the **DTZG Simplex Primal** software option to demonstrate that the following system of inequalities in nonnegative variables is infeasible.

$$\begin{aligned} 4x_1 \qquad\quad + x_3 - x_4 &\geq 4 \\ 4x_1 + 4x_2 + 4x_3 + 5x_4 &\leq 4 \\ 5x_1 + 3x_2 + x_3 + 2x_4 &\leq 9 \\ x_j \geq 0,\ j = 1, \ldots, 4. \end{aligned}$$

3.11 As we shall see later (see Chapter 5)

$$\begin{array}{lrcl} \text{Maximize} & 4y_1 - 4y_2 - 5y_3 & = & v \\ \text{subject to} & 4y_1 - 4y_2 - 5y_3 & \leq & 0 \\ & 4y_2 - 3y_3 & \leq & 0 \\ & y_1 - 4y_2 - y_3 & \leq & 0 \\ & y_1 - 5y_2 - 2y_3 & \leq & 0 \\ & y_j \geq 0,\ j = 1, \ldots, 3 & & \end{array}$$

is the dual of the system in Problem 3.10, assuming it has a zero coefficient objective. Use the **DTZG Simplex Primal** software option to demonstrate that it is unbounded. How could you have deduced that it is unbounded simply by looking at it.

3.12 Solve the following linear program by hand and also by the **DTZG Simplex Primal** software option to demonstrate that it is unbounded.

$$\begin{array}{lrcl} \text{Minimize} & 1x_1 - 2x_2 + x_3 + 3x_4 & = & z \\ \text{subject to} & 2x_1 - x_2 + x_3 - x_4 & \leq & 10 \\ & -5x_1 + 2x_2 - 2x_3 + x_4 & \leq & 20 \\ & 3x_1 - 4x_2 + 4x_3 - 2x_4 & \leq & 30 \\ & x_j \geq 0,\ j = 1, \ldots, 4. & & \end{array}$$

Write down the class of feasible solutions that cause $z \to -\infty$.

3.13 Consider the following linear program.

$$\begin{array}{lrcl} \text{Minimize} & -x_1 - 2x_2 + x_3 + 2x_4 + 3x_5 & = & z \\ \text{subject to} & 2x_1 - x_2 - x_3 - x_4 + 2x_5 & = & 0 \\ & 2x_1 - x_2 + 2x_3 - x_4 + x_5 & = & 0 \\ & x_1 + x_2 + x_3 + x_4 + x_5 & = & 0 \\ & x_j \geq 0,\ j = 1, \ldots, 5. & & \end{array}$$

(a) Apply Phase I of the Simplex Method to obtain a feasible solution to the linear program.

(b) At the end of Phase I, use each of the three methods of handling artificials in the basis and proceed to Phase II.

3.14 *Kuhn's Example of Cycling.*

$$\begin{array}{lrcl} \text{Minimize} & -2x_1 - 3x_2 + x_3 + 12x_4 & = & z \\ \text{subject to} & -2x_1 - 9x_2 + x_3 + 9x_4 & \leq & 0 \\ & \frac{1}{3}x_1 + x_2 - \frac{1}{3}x_3 - 2x_4 & \leq & 0 \end{array}$$

and $x_1 \geq 0$, $x_2 \geq 0$, $x_3 \geq 0$, $x_4 \geq 0$.

Add slack variables x_5 and x_6 and let the starting feasible basis contain x_5 and x_6. Assume that

- the index s of the incoming variable is chosen as the nonbasic variable with the most negative reduced cost;
- the index j_r of the outgoing variable is determined by looking at all $\bar{a}_{is} > 0$ as specified in the Simplex Algorithm and choosing $r = i$ as the smallest index such that $\bar{a}_{is} > 0$.

(a) Solve the problem by hand and show that it cycles after iteration 6.

(b) Re-solve the problem applying Bland's rule to show that the Simplex Algorithm converges to an optimal solution.

(c) Re-solve the problem using the Random Choice Rule to show that the Simplex Algorithm converges to an optimal solution. (For example use dice to make your selections.)

3.15 *Beale's Example of Cycling.*

$$\begin{array}{lrcrcrcrcl} \text{Minimize} & -3x_1/4 & + & 150x_2 & - & x_3/50 & + & 6x_4 & = & z \\ \text{subject to} & x_1/4 & - & 60x_2 & - & x_3/25 & + & 9x_4 & \leq & 0 \\ & x_1/2 & - & 90x_2 & - & x_3/50 & + & 3x_4 & \leq & 0 \\ & & & & & x_3 & & & \leq & 1 \end{array}$$

and $x_1 \geq 0,\ x_2 \geq 0,\ x_3 \geq 0,\ x_4 \geq 0$.

Add slack variables x_5, x_6, and x_7 and let the starting feasible basis contain x_5, x_6, and x_7. Assume that

- the index s of the incoming variable is chosen as the nonbasic variable with the most negative reduced cost;
- the index j_r of the outgoing variable is determined by looking at all $\bar{a}_{is} > 0$ as specified in the Simplex Algorithm and choosing $r = i$ as the smallest index such that $\bar{a}_{is} > 0$.

(a) Solve the problem by hand and show that it cycles.

(b) Re-solve the problem applying Bland's rule to show that the Simplex Algorithm converges to an optimal solution.

(c) Re-solve the problem using the Random Choice Rule to show that the Simplex Algorithm converges to an optimal solution. (For example use dice to make your selections.)

3.16 Consider the following 2-equation, 5-variable linear program.

$$\begin{array}{lrcrcrcrcrcl} \text{Minimize} & x_1 & + & 7x_2 & + & 5x_3 & + & x_4 & + & 6x_5 & = & z \\ \text{subject to} & 1x_1 & + & 2x_2 & + & 2x_3 & + & 5x_4 & + & 4x_5 & = & 10 \\ & 3x_1 & + & 6x_2 & - & 2x_3 & + & x_4 & - & x_5 & = & 30 \\ & \multicolumn{11}{l}{x_j \geq 0,\ j = 1, \ldots, 5.} \end{array}$$

(a) Apply Phase I of the Simplex Method to obtain a feasible solution to the linear program.

(b) If any artificials remain in the basis at the end of Phase I, use each of the three methods of handling artificials in the basis.

(c) Apply Phase II to obtain an optimal solution.

3.17 Consider the following 3-equation, 5-variable linear program.

$$\begin{array}{lrcl} \text{Minimize} & 2x_1 + 3x_2 + 4x_3 + x_4 + x_5 & = & z \\ \text{subject to} & 2x_1 + x_2 + x_3 + x_4 + x_5 & = & 10 \\ & 2x_1 \qquad + 3x_3 + 2x_4 \qquad & = & 6 \\ & x_2 - 2x_3 - x_4 + x_5 & = & 4 \\ & x_j \geq 0,\ j = 1, \ldots, 5. & & \end{array}$$

(a) Apply Phase I of the Simplex Method to obtain a feasible solution to the linear program.

(b) Show that an artificial remains in the basis at the end of Phase I because the constraints are redundant.

(c) Apply Phase II to obtain an optimal solution.

3.18 Consider the following 3-equation, 4-variable linear program.

$$\begin{array}{lrcl} \text{Maximize} & x_1 & = & z \\ \text{subject to} & x_1 + 3x_2 + 4x_3 + x_4 & = & 20 \\ & 2x_1 \qquad + x_3 \qquad & = & 5 \\ & -7x_1 + 3x_2 \qquad + x_4 & = & 0 \\ & x_j \geq 0,\ j = 1, \ldots, 4. & & \end{array}$$

(a) Apply Phase I of the Simplex Method to obtain a feasible solution to the linear program.

(b) Show that an artificial remains in the basis at the end of Phase I because the constraints are redundant.

(c) Apply Phase II to solve the problem.

3.19 Consider the following two variable linear program.

$$\begin{array}{lrcl} \text{Maximize} & x_1 + x_2 & = & z \\ \text{subject to} & x_1 - x_2 & \geq & 1 \\ & x_1 + x_2 & \leq & 3 \\ & 2x_1 - x_2 & \leq & 3 \\ & x_1 \geq 0,\ x_2 \geq 0. & & \end{array}$$

(a) Plot the feasible region.

(b) Solve the problem graphically.

(c) Show graphically that the optimal solution is degenerate.

(d) On the figure, indicate which constraint can be dropped to obtain a non-degenerate optimal solution.

3.20 The purpose of this exercise is to examine the effect of changes to an optimal solution as a result of changes to data.

$$\begin{array}{ll} \text{Minimize} & 4x_1 + 3x_2 + 2x_3 + 1x_4 = z \\ \text{subject to} & 2x_1 - 3x_2 + x_3 + 2x_4 = 10 \\ & 1x_1 + 4x_2 - 2x_3 + 3x_4 \geq 16 \\ & x_j \geq 0, \ j = 1, \ldots, 4. \end{array}$$

(a) Solve the linear program using the `DTZG Simplex Primal` software option.
(b) Change the cost on x_4 from 1 to 4 and re-solve the problem. Change it to 8 and re-solve the problem. How does the solution change in each case?
(c) Change the coefficient of x_2 in the second equation from $a_{22} = 4$ to $a_{22} = 5$ and re-solve the problem. How does the solution change?
(d) Decrease the right hand side $b_1 = 10$ on the first equation to 8 and re-solve the problem. Next increase it to $b_1 = 12$ and re-solve. Finally, increase it to $b_1 = 20$ and re-solve. Examine how the solution changes in each case.
(e) Add a new activity 5 with level $x_5 \geq 0$ to the problem with $c_5 = -1$, $a_{15} = -2$, $a_{25} = -3$. Re-solve the problem and indicate how the solution changes. How could you have predicted this based on your original run?
(f) Add a new activity 6 with level $x_6 \geq 0$ to the problem with $c_6 = -2$, $a_{16} = 2$, $a_{16} = 3$. Re-solve the problem and indicate how the solution changes. How could you have predicted this based on your original run?
(g) Add in turn the following rows to the original problem and examine how the solution changes:
- $x_1 + x_2 + x_3 + x_4 \geq 4$.
- $2x_1 + 2x_2 + 4x_3 + x_4 \leq 8$.
- $x_1 + x_2 + x_3 + x_4 = 6$.

3.21 Consider the linear program

$$\begin{array}{ll} \text{Minimize} & 2x_1 + x_2 + 3x_3 = z \\ \text{subject to} & x_1 + 2x_2 + x_3 \leq 6 \\ & 2x_1 + x_3 \leq 4 \\ & x_j \geq 0, \ j = 1, \ldots, 3. \end{array}$$

(a) Add slack variables x_4 and x_5 to the inequality constraints. Starting with x_4 and x_5 as basic variables, solve the linear program by hand using the Revised Simplex Method.
(b) Multiply the second constraint by 2 and re-solve the problem. What is the effect on the multipliers π_1 and π_2? What is the effect on the optimal solution?

3.22 Solve the following linear program by the Revised Simplex Method by hand, starting with a full artificial basis.

$$\begin{array}{ll} \text{Minimize} & x_1 - 3x_2 + x_3 = z \\ \text{subject to} & 2x_1 + x_2 + x_3 = 6 \\ & x_1 + x_2 - x_3 = 2 \\ & x_j \geq 0, \ j = 1, \ldots, 3. \end{array}$$

Change the first inequality to be a '$\leq$' inequality. Re-solve the problem; this time add a slack variable x_4 to the first constraint and an artificial variable x_5 to the second constraint and minimize $w = x_6$.

3.23 Consider the following bounded variable linear program:

$$\begin{array}{ll} \text{Minimize} & 4x_1 + 2x_2 + 3x_3 = z \\ \text{subject to} & x_1 + 3x_2 + x_3 = 5 \\ & 2x_1 \qquad\quad - x_3 = 1 \\ & 0 \leq x_1 \leq 1,\ 0 \leq x_2 \leq \infty,\ 1 \leq x_3 \leq \infty. \end{array}$$

(a) Solve the problem by hand.

(b) Solve it using the **DTZG Simplex Primal** software option to verify your solution.

3.24 Consider the following bounded 2-equation, 4-variable linear program:

$$\begin{array}{ll} \text{Minimize} & 2x_1 + x_2 + 3x_3 + 2x_4 = z \\ \text{subject to} & x_1 + x_2 + 3x_3 - x_4 = 7 \\ & x_1 + 2x_2 + 2x_3 + x_4 = 5 \\ & x_1 \qquad + x_3 \qquad = 3 \\ & 0 \leq x_1 \leq 2,\ 0 \leq x_2 \leq \infty,\ 1 \leq x_3 \leq 4,\ -4 \leq x_4 \leq -1. \end{array}$$

(a) Solve the problem by hand.

(b) Solve it using the **DTZG Simplex Primal** software option to verify your solution.

3.25 Apply Phase I of the Simplex Method by hand to show that the following linear program is infeasible:

$$\begin{array}{ll} \text{Maximize} & x_1 - 3x_2 + 2x_3 = z \\ \text{subject to} & x_1 + 2x_2 + 3x_3 \leq 5 \\ & 2x_1 + 3x_2 + 2x_3 \leq 4 \\ & 2 \leq x_1 \leq 4,\ -\infty \leq x_2 \leq -1,\ 3 \leq x_3 \leq 8. \end{array}$$

3.26 Consider the following bounded variable linear program:

$$\begin{array}{ll} \text{Minimize} & x_1 - 3x_2 + 2x_3 = z \\ \text{subject to} & x_1 + 2x_2 + x_3 \leq 2 \\ & 2x_1 + x_2 + 4x_3 \leq 4 \\ & 0 \leq x_1 \leq 2,\ -\infty \leq x_2 \leq 0,\ -2 \leq x_3 \leq 2. \end{array}$$

(a) Solve the problem by hand and show that it has a class of solutions that cause the objective function to go to $-\infty$.

(b) Solve it using the **DTZG Simplex Primal** software option to verify your solution.

3.27 Consider the following 2-equation, 4-variable linear program.

$$\begin{array}{ll} \text{Minimize} & 9x_1 + 2x_2 + 4x_3 + 8x_4 = z \\ \text{subject to} & 2x_1 + x_2 + x_3 + 3x_4 \geq 20 \\ & 3x_1 - x_2 + x_3 - x_4 \geq 10 \\ & x_j \geq 0,\ j = 1, \ldots, 4. \end{array}$$

(a) Solve the problem by hand or by the **DTZG Simplex Primal** software option.

(b) Show that the linear program has multiple optimal solutions.

(c) Find every basic feasible optimal solution.

3.28 *Big-M Method.* Suppose that we are trying to find the solution to the following linear program:

$$\begin{array}{lrcrcl} \text{Minimize} & -2x_1 & - & 3x_2 & = & z \\ \text{subject to} & x_1 & + & 2x_2 & \leq & 4 \\ & x_1 & + & x_2 & = & 3 \end{array}$$
$$x_1 \geq 0,\ x_2 \geq 0.$$

To initiate the Simplex Method we add a slack $x_3 \geq 0$ to the first constraint and an artificial variable $x_4 \geq 0$ to the second constraint and then construct a Phase I objective $w = x_4$ and minimize w. With the Big-M method we solve the linear program in one pass by solving the problem:

$$\begin{array}{lrcrcrcrcl} \text{Minimize} & -2x_1 & - & 3x_2 & + & 0x_3 & + & Mx_4 & = & z \\ \text{subject to} & x_1 & + & 2x_2 & + & x_3 & & & = & 4 \\ & x_1 & + & x_2 & & & + & x_4 & = & 3 \end{array}$$
$$x_1 \geq 0,\ x_2 \geq 0,\ x_3 \geq 0,\ x_4 \geq 0,$$

where M is a large, prespecified positive number, and (x_3, x_4) are chosen to form the starting basic feasible solution.

(a) Solve the above linear program with $M = 100$.

(b) Plot the steps of the algorithm in (x_1, x_2) space.

(c) Re-solve the original problem using the Phase I/Phase II approach.

(d) Plot the steps of the Phase I/Phase II algorithm in (x_1, x_2) space and compare with the Big-M method.

3.29 Consider the following linear program:

$$\begin{array}{lrcrcl} \text{Minimize} & x_1 & + & x_2 & = & z \\ \text{subject to} & 2x_1 & + & x_2 & = & 4 \\ & 3x_1 & + & 2x_2 & \geq & 5 \end{array}$$
$$x_1 \geq 0,\ x_2 \geq 0.$$

(a) Solve the above linear program by the Big-M method of Problem 3.28 with $M = 100$.

(b) Plot the steps of the algorithm in (x_1, x_2) space.

3.30 Solve the following linear program by the Big-M method described in Problem 3.28

$$\begin{array}{lrcrcrcl} \text{Minimize} & -x_1 & - & 2x_2 & + & 3x_3 & = & z \\ \text{subject to} & 2x_1 & - & x_2 & + & 4x_3 & = & 5 \\ & 3x_1 & + & 2x_2 & - & x_3 & \leq & 4 \end{array}$$
$$x_1 \geq 0,\ x_2 \geq 0,\ x_3 \geq 0.$$

3.31 Using the Big-M method described in Problem 3.28 show that the following linear program is infeasible.

$$\begin{array}{lrcl} \text{Minimize} & -2x_1 + 3x_2 - 3x_3 & = & z \\ \text{subject to} & 4x_1 + x_2 + x_3 & = & 5 \\ & x_1 + x_2 + x_3 & \geq & 6 \\ & x_1 & \geq & 1 \end{array}$$

$$x_1 \geq 0,\ x_2 \geq 0,\ x_3 \geq 0.$$

3.32 Suppose that you apply the Big-M method to solve a linear program and obtain an unbounded solution. Does this imply that the original problem has an unbounded solution?

3.33 Suppose that the following LP problem is feasible:

$$\begin{array}{lll} \text{Minimize} & c^T x & \\ \text{subject to} & Ax \geq b, & A: m \times n, \\ & x \geq 0. & \end{array}$$

Show that $z^* \to -\infty$ if and only if there exists an $\bar{x} \neq 0$ such that $\bar{x} \geq 0$, $A\bar{x} \geq 0$, $c^T\bar{x} < 0$.

3.34 Show that the system of equations

$$\begin{array}{rcl} x_1 + x_2 - x_3 + x_4 & = & 1 \\ -4x_2 + x_3 + x_5 & = & 2 \\ -3x_2 + x_3 + x_6 & = & 3 \end{array}$$

has an unbounded solution.

3.35 Given a linear program $Ax = b$, $x \geq 0$, and $c^T x = \min$. Prove that you can

(a) add a set of the columns to produce new columns with new nonnegative variables, and
(b) add or subtract equations

without affecting the optimal solution.

3.36 Show that a one-equation linear program can be solved in at most one iteration. Specifically, show how to choose the entering variable. What are your conclusions on different cases that might arise. You can assume that the problem is given in canonical form with respect to the variable x_1 as follows:

$$\begin{array}{lrcl} \text{Minimize} & c_2x_2 + c_3x_3 + \cdots + c_nx_n & = & z \\ \text{subject to} & x_1 + a_2x_2 + a_3x_3 + \cdots + a_nx_n & = & b, \end{array}$$

where $b > 0$.

3.37 *Dantzig [1963]*. Show, by changing units of any activity k whose $\bar{c}_k < 0$, that x_k can be chosen by the rule of $\bar{c}_s = \min \bar{c}_j$ to be the candidate to enter the next basic set. Can you suggest another selection rule that might be better; does it involve more work?

3.38 Consider an LP in canonical form (3.11).

(a) The standard Simplex Method determines the incoming column by selecting the column that maximizes the decrease in the objective function per unit increase of the incoming variable. In terms of the given canonical form, indicate how the standard Simplex Method determines the pivot row and pivot column.

(b) A pivoting method is scale-invariant if the selection of pivot row and column remains unchanged under any positive scaling of some (or all) of the variables. A positive scaling of a variable x_j is of the form $x'_j = d_j x_j$ where $d_j > 0$. Is the standard Simplex Method scale-invariant? Justify your answer.

(c) An alternative pivoting method is to choose the pivot row and column so as to maximize the decrease in the objective function at each pivot step while maintaining feasibility. In terms of the canonical form, describe how this method determines the pivot row and pivot column.

(d) Is the method in (c) scale-invariant? Justify your answer.

3.39 Suppose that Phase I of the Simplex Method is applied to the system $Ax = b$, $x \geq 0$, and the procedure results in infeasibility. Show that the method has generated infeasibility multipliers. How can we derive the infeasibility multipliers from the final tableau?

3.40 Suppose that for an LP in standard form, the system $Ax = b$ has rank $r < m$, i.e., it has $m - r$ redundant equations.

(a) Show that there will be at least $k = m - r$ artificial variables left in the tableau at the end of Phase I.

(b) Show that if these k artificial variables are dropped from the tableau, then the subsystem of equations associated with these artificial variables also has $m-r$ redundant equations and that if $k = m-r$, then the associated $m-r$ equations are vacuous.

3.41 *T. Robacker in Dantzig [1963].* In some applications it often happens that many variables initially in the basic set for some starting canonical form remain until the final canonical form, so that their corresponding rows in the successive tableaux of the Simplex Method, though continuously modified, have never been used for pivoting. Devise a technique for generating rows only as needed for pivoting and thereby avoiding needless work.

3.42 *Bazarra, Jarvis, & Sherali [1990].* Consider the following two problems where $Ax = b$, $x \geq 0$ form a bounded region.

$$\begin{array}{lrcll} \text{Minimize} & x_n & = & z & \\ \text{subject to} & Ax & = & b, & \qquad A: m \times n, \\ & x & \geq & 0, & \end{array}$$

and

$$\begin{array}{lrcll} \text{Maximize} & x_n & = & z & \\ \text{subject to} & Ax & = & b, & \qquad A: m \times n, \\ & x & \geq & 0. & \end{array}$$

Let the optimal objective values of the two problems be x'_n and x''_n respectively. If x_n is a number in the interval $[x'_n, x''_n]$, show that there exists a feasible point whose nth component is equal to x_n.

3.43 Suppose that one equation of a linear program in standard form has one positive coefficient, say that of x_k, while all remaining coefficients of the equation are nonpositive and the constant b is positive. This implies that $x_k > 0$ for any solution, whatever the values of the remaining $x_j \geq 0$. Pivot on any nonzero term in x_k to eliminate x_k from the remaining equations and set aside the one equation involving x_k. Prove that the resulting linear program in one fewer equation and one fewer variable can be solved to find the optimal solution of the original problem.

3.44 If it is known in advance that a solution cannot be optimal unless it involves a variable x_k at positive value, show that this variable can be eliminated and the reduced system with one fewer equation and variable solved in its place.

3.45 Suppose that we assume that x_1, which has a nonzero coefficient in row 1, is in the optimal solution of a linear program and eliminate it from rows $2, \ldots, m$ and from the objective function. Next we solve the linear program without row 1 to obtain an optimal solution. Then we substitute the values of $x_2, x_3, \ldots, x_m$ into the original row 1 to obtain the value of x_1. Show that

(a) If $x_1 \geq 0$ then the solution $x_1, x_2, \ldots, x_m$ is optimal.
(b) If $x_1 < 0$ then its value must be zero in an optimal solution.

3.46 Show that if a linear program is feasible and the set of feasible solutions is bounded, the linear program cannot have a homogeneous solution.

3.47 Suppose that the linear program

$$\begin{array}{lll} \text{Minimize} & c^T x & \\ \text{subject to} & Ax = b, & A: \, m \times n, \\ & x \geq 0 & \end{array}$$

has a finite optimal solution. Show that if the right-hand side is changed from b to some b', the resulting linear program is either infeasible or has a finite optimal feasible solution. Thus, by changing the right-hand side, we cannot make a linear program, with a finite optimal solution, into a linear program that is unbounded below.

3.48 *Ph.D. Comprehensive Exam, June 13, 1968, at Stanford.* In a Leontief substitution model, there are m items that can be produced within the system. Let the constant b_i, $i = 1, \ldots, m$, denote the annual amount to be delivered by the system to satisfy external demands.

Each process produces one and only one item j for $j = 1, \ldots, m$. Let the index j_k identify the kth alternative process for producing item j, and let the unknown x_{j_k} denote the number of units of item j produced annually by process k, where $k = 1, \ldots, n$. In order to produce one unit of item j by process k, it requires a_{ij_k} units of input of item i, and a cost of c_{j_k} is incurred. The coefficients $a_{ij_k} \geq 0$, and for each j_k pair

$$\sum_{i=1}^{m} a_{ij_k} < 1.$$

(a) Formulate the linear programming model as one of meeting the annual delivery requirements at minimum cost.
(b) Suppose that $b_i > 0$, for $i = 1, \ldots, m$. Prove that in *any* basic feasible solution to the linear programming problem formulated in part (a), exactly one of the k alternative processes for item j will be operated at positive intensity.
(c) Suppose that $b_i \geq 0$, for $i = 1, \ldots, m$. Prove that there exists a basic feasible linear programming solution.

(d) Prove that if we have an optimal solution with shadow prices (multipliers) π_i^o for a particular set of delivery requirements b_i^o, the same shadow prices will also be optimal for any set of delivery requirements such that $b_i \geq 0$.

3.49 *Ph.D. Comprehensive Exam, November 1995, at Stanford.* Consider the linear program:

$$\begin{array}{lrcll} \text{Minimize} & \displaystyle\sum_{j=1}^{n} c_j x_j & = & z & \\ \text{subject to} & \displaystyle\sum_{j=1}^{n} P_j x_j & = & b, & P_j \in \Re^m \\ & x_j & \geq & 0, & \text{for } j = 1, \ldots, n. \end{array}$$

Suppose that $z = z^*$, $x_j = x_j^*$ for $j = 1, \ldots, n$ is a *nondegenerate* optimal basic feasible solution. A new activity vector is being considered for augmenting the problem to

$$\begin{array}{lrcll} \text{Minimize} & \displaystyle\sum_{j=1}^{n} c_j x_j + c_{n+1} x_{n+1} & = & z & \\ \text{subject to} & \displaystyle\sum_{j=1}^{n} P_j x_j + P_{n+1} x_{n+1} & = & b, & P_j \in \Re^m \\ & x_j \geq 0, \quad \text{for } j = 1, \ldots, n. & & & \end{array}$$

(a) How would you test whether the column $n+1$ is a candidate for improving the solution?

(b) Assume that column $n+1$ passes the test and x_{n+1} enters the basic set, displacing x_r. Let $z = \hat{z}$, $x_j = \hat{x}_j$ for $j = 1, \ldots, n+1$ be the updated basic feasible solution. Prove that it is a strict improvement, i.e., $\hat{z} < z^*$.

(c) Suppose the augmented problem is now iterated to a new optimal solution. Prove that x_{n+1}, once in, never leaves the basic set.

CHAPTER 4

INTERIOR-POINT METHODS

An interior-point algorithm is one that improves a feasible interior solution point of the linear program by steps through the interior, rather than one that improves by steps around the boundary of the feasible region, as the classical Simplex Algorithm does.

Just as there are many variants of the Simplex Method (which we refer to as pivot-step algorithms), so there are many variants of interior methods such as projective, affine, and path-following. Some interior-point methods inscribe an ellipsoidal ball in the feasible region with its center at the current iterate, or first transform the feasible space and then inscribe a hypersphere with the current iterate at the center. Next an improving direction is found by joining the current iterate to the point on the boundary of the ellipsoid or sphere that maximizes (or minimizes) the linear objective function (obtained by solving a linear system of equations). A point is then selected on the improving direction line as the next current iterate. Sometimes this iterate is found along a line that is a linear combination of the improving direction and some other direction.

During the period 1979–1996, there has been intensive interest in the development of interior-point methods. These methods are related to classical least-square methods used in numerical analysis for making fits to data or fitting simpler functional forms to more complicated ones. Therefore interior research can tap a vast literature of approximation theory. A theoretical breakthrough came in 1979; the Russian mathematician L.G. Khachian discovered an ellipsoid algorithm whose running time in its worst case was significantly lower than that of the Simplex Algorithm in its worst case—its iterates are not always required to be feasible. Other theoretical results quickly followed, notably that of N. Karmarkar who discovered an interior-point algorithm whose running time performance in its worst case was significantly lower than that of Kachiyan's. This in turn was followed by more theoretical results by others improving on the worst-case performance.

It soon became apparent that these worst-case running times were many, many times greater than the actual running times of the Simplex Method on practical problems and totally misleading as to what algorithm to choose for solving practical problems. It appears that the problems encountered in practical applications belong to a very specialized class (that is yet to be characterized in a simple way). Or to put it in another way, the class of problems simply described by $Ax = b$, $x \geq 0$, $c^T x = \min$ is much too broad, and the worst-case examples drawn from this very broad class are far different from anything ever encountered in practical applications or likely ever to be encountered in the future.

Attempts to characterize in a simple way the class (or classes) of practical problems from which one could derive a theoretical explanation of the excellent performance times of various algorithms in practice have, in general, failed. In special cases, such as the shortest path problem, the worst-case performance and performance on actual problems are comparable, and the algorithms are very efficient. There has been progress proving that average performance on classes of *randomly generated* problems using a parametric variant of the Simplex Method resembles that obtained on practical problems, but no one claims that these randomly generated problems belong to the class of practical problems.

Because the theoretical results are totally misleading as to what algorithm to choose to solve a practical problem, a different approach is used. The linear programming profession has accumulated a large collection of test problems drawn from practical sources. These are used to compare running times of various proposed algorithms. The general goal of these efforts is to develop algorithms that surpass the performance of the Simplex Method on this collection of problems. For example, Karmarkar claimed that on very large problems his technique is significantly faster. As of this writing, as far as the authors of this book can ascertain, there appears to be no algorithm that is a clear winner, i.e., solves all (or almost all of) the test problems faster than all the other proposed methods. On problems with many bounding hyperplanes in the neighborhood of the optimum point, an interior-method will probably do better. On problems with relatively few boundary planes (which is often the case in practice) an exterior method will be hard to beat. For this reason, it is likely that the commercial software of the future will be some sort of a hybrid, because one does not know which kind of problem is being solved or because one wishes to obtain an extreme-point solution. Many specialized, efficient codes have been proposed for solving structured linear programs such as network problems, staircase problems, and block-angular systems.

Here we illustrate the primal affine method which is the same as Dikin's method, and which has the distinction of having been rediscovered by many.

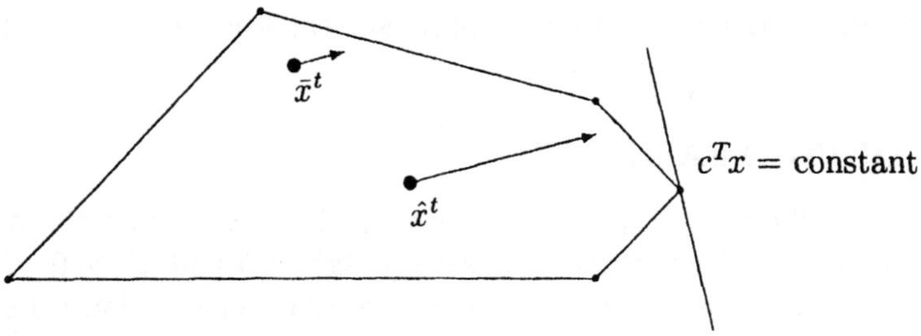

Figure 4-1: Comparison of a Move from a Point $\hat{x}^t$ Near the Center Versus a Point $\bar{x}^t$ Near the Boundary.

4.1 BASIC CONCEPTS

In interior-point methods for solving linear programs, we are given, or must be able to find, an initial starting interior feasible point from which we proceed to an improved interior feasible point, and so on, eventually stopping by some criterion.

Our development here of an interior-point method will be for solving a linear program in standard form:

$$\begin{array}{lrcl} \text{Minimize} & c^Tx & = & z \\ \text{subject to} & Ax & = & b, \qquad A: m \times n, \\ & x & \geq & 0. \end{array} \tag{4.1}$$

The rationale for the approach is based on the following observations. When minimizing, one is first tempted to move from the current solution x^t in the direction of ***steepest descent*** of the objective function (i.e., in the negative gradient of the objective function, which is the same as moving orthogonal to the objective hyperplane c^Tx = constant). If the current iterate x^t is an interior point, so that $x^t > 0$, such a move will in general violate the constraints $Ax = b$. To adjust for this, one typically moves instead in the direction given by the projection of the negative gradient of the objective onto the hyperplanes $Ax = b$. However, if x^t is close to the boundary hyperplanes, as $x^t = \bar{x}^t$ is in Figure 4-1, very little improvement will occur. On the other hand, if the current iterate happens to be near the "center," such as $x^t = \hat{x}^t$ in Figure 4-1, there could be a big improvement.

The key ideas behind interior-point methods are as follows, assuming an initial feasible interior point is available and that all moves satisfy $Ax = b$:

1. Try to move through the interior in directions that show promise of moving quickly to the optimal solution.

2. Recognize that if we move in a direction that sets the new point too "close" to the boundary, this will be an obstacle that will impede our moving quickly to an optimal solution. One way around this is to transform the feasible region so that the current feasible interior point is at the center of the transformed

feasible region. Once a movement has been made, the new interior point is transformed back to the original space, and the whole process is repeated with the new point as the center.

MAINTAINING FEASIBILITY

It is clear that the feasible point must satisfy $Ax = b$. Hence the only feasible interior points $x = x^t$ are those that satisfy $Ax^t = b$ and $x^t > 0$. As a result, whenever we move from a point (in the original space or transformed space) x^t to $x^{t+1} = x^t + \alpha p^t$, with $\alpha > 0$, we need to satisfy $Ax^{t+1} = b$ and $x^{t+1} > 0$. This implies that

$$Ax^{t+1} = A(x^t + \alpha p^t) = b + \alpha A p^t. \tag{4.2}$$

Since $\alpha \neq 0$, the above equation implies that in order to stay feasible, p^t must satisfy

$$Ap^t = 0. \tag{4.3}$$

In other words, $p^t \neq 0$ must be orthogonal to the rows of A. A vector direction p^t that satisfies $Ap^t = 0$ is said to belong to the *null space* of A. It is easy to see that the matrix

$$P = I - A^T(AA^T)^{-1}A, \tag{4.4}$$

called the *projection matrix*, will transform any vector v into $Pv = p$, and p will be in the null space of A because $AP = 0$ is the zero matrix. Hence, in order to maintain feasibility, every move from x^t to x^{t+1} must be in the null space of A, i.e., $p^t = Pv$ for some vector v.

STEEPEST DESCENT DIRECTION

The next thing to note is that the direction of maximum decrease or steepest descent is the direction that is the negative of the gradient of the objective function z. Hence, the direction of steepest descent is $-c$. However, the condition that we must maintain feasibility implies that we must move in the direction of steepest descent projected into the null space of A. That is,

$$p^t = -Pc. \tag{4.5}$$

Such a direction is called a *projected gradient direction.* To see that such a direction is a direction of decrease, let $x^{t+1} = x^t + \alpha p^t$, with $\alpha > 0$ and $p^t = -Pc$. Then

$$\begin{aligned} c^T x^{t+1} &= c^T(x^t + \alpha p^t) \\ &= c^T x^t - \alpha c^T P c \\ &\leq c^T x^t, \end{aligned} \tag{4.6}$$

where the last inequality follows because P has the property that $P = P^T P$, and therefore

$$c^T P c = c^T P^T P c = (Pc)^T (Pc) \geq 0.$$

▷ **Exercise 4.1** Prove $P^T P = P$, where P is defined by (4.4).

STAYING IN THE INTERIOR

We plan to stay in the interior of $x \geq 0$ by limiting the amount of movement $\alpha > 0$ in the direction p^t. The quantity α, called the ***steplength***, is chosen to ensure feasibility of the nonnegativity constraints. However, we do not want to move all the way to the boundary because we want to avoid the need to move to the boundary on subsequent iterations. Hence we move some portion of the way towards the boundary. Based on the experience of others on practical problems, moving 50% to 90% of the way is a simple good rule for the primal affine/Dikin's method that we are about to describe.

CENTERING

What is left is the procedure for transforming the space so that an interior point x^t is the "center" point in the transformed space. The basic idea is very simple—we change the scaling of the solution so that each component is equidistant from the constraint boundaries in the transformed feasible region; for example, after rescaling, $x^t = e$, where $e = (1, 1, \ldots, 1)^T$. Let $D = \mathrm{Diag}\,(x^t)$ be the diagonal matrix having the components of the current iterate x^t as the diagonal elements. The simplest such scheme is to rescale x by setting $x = D\hat{x}$, so that $\hat{x}^t = e$. Substituting $x = D\hat{x}$ in (4.1) we obtain

$$\begin{array}{lrll} \text{Minimize} & \hat{c}^T\hat{x} & = z & \\ \text{subject to} & \widehat{A}\hat{x} & = b, & \widehat{A}: m \times n, \\ & \hat{x} & \geq 0, & \end{array} \tag{4.7}$$

where $\hat{c} = Dc$ and $\widehat{A} = AD$. As a result of this transformation, clearly the projection matrix for the system (4.7) is now defined by

$$\widehat{P} = I - \widehat{A}^T(\widehat{A}\widehat{A}^T)^{-1}\widehat{A}. \tag{4.8}$$

Hence at each iteration t, we rescale x^t to $\hat{x}^t = e$ and move to $\hat{x}^{t+1}$ in the negative projected gradient direction $-\widehat{P}\hat{c}$; thus

$$\hat{x}^{t+1} = e - \alpha\widehat{P}\hat{c}, \tag{4.9}$$

where $e = (1, 1, \ldots, 1)^T$ and $\widehat{P}$ and $\hat{c}$ are defined as above. Finally, we rescale back to the original space by setting $x^{t+1} = D\hat{x}^{t+1}$. Next we repeat the iterative process replacing D by $D = \mathrm{Diag}\,(x^{t+1})$.

STOPPING RULE

The only item left out in our discussion is the stopping rule. The simple rule typically followed is to *stop* with an approximate optimal solution when the difference between iterates x^{t+1} and x^t is "deemed" sufficiently small in the original space.

4.2 PRIMAL AFFINE / DIKIN'S METHOD

Now we are prepared to write down the specific steps of the primal affine algorithm and illustrate it with an example. The name *affine method* results from the following definition:

> *Definition (Affine Transformation)*: Let F be an $n \times n$ nonsingular matrix and let $d \in \Re^n$. The transformation $y = Fx + d$ is called an *affine transformation* of x into y.

In the literature, an interior-point method that uses an affine transformation to obtain the search direction is called an *affine* method.

The algorithm is described with $\widehat{P}$ computed explicitly. In practice $\widehat{P}$ is never computed; instead, $p = -\widehat{P}\hat{c}$ is computed directly using a QR factorization or Cholesky factorization.

Algorithm 4.1 (Primal Affine/Dikin's Method) Assume that an initial feasible interior point solution $x = x^o > 0$ satisfying $Ax^o = b$ is known. The steps of the algorithm are as follows:

1. *Initialize Counter:* Set $t \leftarrow 0$.
2. *Create D:* Set $D = \mathrm{Diag}(x^t)$.
3. *Compute Centering Transformation:* Compute $\widehat{A} = AD$, $\hat{c} = Dc$.
4. *Determine the Projection Matrix:* Compute $\widehat{P} = I - \widehat{A}^T(\widehat{A}\widehat{A}^T)^{-1}\widehat{A}$.
5. *Compute Steepest Descent Direction:* Set $p^t = -\widehat{P}\hat{c}$.
6. Set $\theta = -\min_j p_j^t$.
7. *Test for Unbounded Objective.* If $\theta \leq 0.0$ report the objective as being unbounded and stop.
8. *Obtain $\hat{x}^{t+1}$:* Compute
$$\hat{x}^{t+1} = e + \left(\frac{\alpha}{\theta}\right) p^t,$$
where $e = (1, 1, \ldots, 1)^T$ and α is set to a number strictly between 0 and 1. Typically, α is set to be between 0.5 and 0.95 for primal affine methods (in most implementations it is set between 0.9 and 0.95).
9. *Transform Back to Original Space:* Compute $x^{t+1} = D\hat{x}^{t+1}$.
10. *Termination Check:* If x^{t+1} is "close" to x^t, report x^{t+1} as "close" to optimal and stop.
11. Set $t \leftarrow t + 1$ and go to Step 2.

Example 4.1 (Illustration of the Primal Affine Method) Consider the two-variable example

$$\begin{array}{lrcrcl} \text{Minimize} & -2x_1 & - & x_2 & = & z \\ \text{subject to} & x_1 & + & x_2 & \leq & 5 \\ & 2x_1 & + & 3x_2 & \leq & 12 \end{array}$$

and $x_1 \geq 0,\ x_2 \geq 0$.

An initial starting feasible interior solution $x_1^o = 1$, $x_2^o = 2$ is given.

We start by transforming this into the standard form by adding slack variables x_3 and x_4 to obtain

$$\begin{array}{lrcl} \text{Minimize} & -2x_1 - x_2 & = & z \\ \text{subject to} & x_1 + x_2 + x_3 & = & 5 \\ & 2x_1 + 3x_2 \qquad + x_4 & = & 12 \end{array}$$

and $x_j \geq 0$ for $j = 1, \ldots, 4$.

The corresponding feasible interior-point starting solution is

$$x^o = (1 \quad 2 \quad 2 \quad 4)^T. \tag{4.10}$$

The objective value at x^o is $z_o = -4$.

We start the first iteration by setting the scaling matrix D to be

$$D = \begin{pmatrix} 1 & & & \\ & 2 & & \\ & & 2 & \\ & & & 4 \end{pmatrix}. \tag{4.11}$$

The rescaled $\widehat{A}$ matrix and objective function coefficient $\hat{c}$ are computed as

$$\widehat{A} = AD = \begin{pmatrix} 1 & 1 & 1 & 0 \\ 2 & 3 & 0 & 1 \end{pmatrix} \begin{pmatrix} 1 & & & \\ & 2 & & \\ & & 2 & \\ & & & 4 \end{pmatrix} = \begin{pmatrix} 1 & 2 & 2 & 0 \\ 2 & 6 & 0 & 4 \end{pmatrix}, \tag{4.12}$$

$$\hat{c} = Dc = \begin{pmatrix} 1 & & & \\ & 2 & & \\ & & 2 & \\ & & & 4 \end{pmatrix} \begin{pmatrix} -2 \\ -1 \\ 0 \\ 0 \end{pmatrix} = \begin{pmatrix} -2 \\ -2 \\ 0 \\ 0 \end{pmatrix}. \tag{4.13}$$

Next we compute the projection matrix $\widehat{P}$:

$$\begin{aligned} \widehat{P} &= I - \widehat{A}^T(\widehat{A}\widehat{A}^T)^{-1}\widehat{A} \\ &= \begin{pmatrix} 1 & & & \\ & 1 & & \\ & & 1 & \\ & & & 1 \end{pmatrix} - \begin{pmatrix} 1 & 2 \\ 2 & 6 \\ 2 & 0 \\ 0 & 4 \end{pmatrix} \left[\begin{pmatrix} 1 & 2 & 2 & 0 \\ 2 & 6 & 0 & 4 \end{pmatrix} \begin{pmatrix} 1 & 2 \\ 2 & 6 \\ 2 & 0 \\ 0 & 4 \end{pmatrix} \right]^{-1} \begin{pmatrix} 1 & 2 & 2 & 0 \\ 2 & 6 & 0 & 4 \end{pmatrix} \\ &= \begin{pmatrix} .8831 & -.2597 & -.1818 & -.0519 \\ -.2597 & .3117 & -.1818 & -.3377 \\ -.1818 & -.1818 & .2727 & .3636 \\ -.0519 & -.3777 & .3636 & .5325 \end{pmatrix}. \end{aligned} \tag{4.14}$$

Hence the projected gradient is

$$p^o = -\widehat{P}\hat{c} = -\begin{pmatrix} .8831 & -.2597 & -.1818 & -.0519 \\ -.2597 & .3117 & -.1818 & -.3377 \\ -.1818 & -.1818 & .2727 & .3636 \\ -.0519 & -.3777 & .3636 & .5325 \end{pmatrix} \begin{pmatrix} -2 \\ -2 \\ 0 \\ 0 \end{pmatrix} = \begin{pmatrix} 1.2468 \\ 0.1039 \\ -0.7273 \\ -0.7792 \end{pmatrix}. \tag{4.15}$$

Next we pick θ by

$$\theta = -\min_j p_j^o = .7792. \tag{4.16}$$

Next we rescale the current iterate to $\hat{x}^o = D^{-1}x^o = e$ and move to $\hat{x}^1$ in the transformed space:

$$\hat{x}^1 = \begin{pmatrix} 1 \\ 1 \\ 1 \\ 1 \end{pmatrix} + \frac{\alpha}{\theta} p^o = \begin{pmatrix} 1 \\ 1 \\ 1 \\ 1 \end{pmatrix} + \frac{0.5}{.7792} \begin{pmatrix} 1.2468 \\ 0.1039 \\ -0.7273 \\ -0.7792 \end{pmatrix} = \begin{pmatrix} 1.8000 \\ 1.0667 \\ 0.5333 \\ 0.5000 \end{pmatrix}. \tag{4.17}$$

Transforming this point to the original space we obtain

$$x^1 = D\hat{x}^1 = \begin{pmatrix} 1 & & & \\ & 2 & & \\ & & 2 & \\ & & & 4 \end{pmatrix} \begin{pmatrix} 1.8000 \\ 1.0667 \\ 0.5333 \\ 0.5000 \end{pmatrix} = \begin{pmatrix} 1.8000 \\ 2.1333 \\ 1.0667 \\ 2.0000 \end{pmatrix}. \tag{4.18}$$

The objective value at x^1 is $z_1 = -5.7333$. This completes the first iteration of the method. Because x^1 is quite different from x^o we proceed to the next iteration.

We update the scaling matrix D to $\mathrm{Diag}\,(x^{t+1})$:

$$D = \begin{pmatrix} 1.8000 & & & \\ & 2.1333 & & \\ & & 1.0667 & \\ & & & 2.0000 \end{pmatrix}. \tag{4.19}$$

The rescaled $\widehat{A}$ matrix and objective function coefficient $\hat{c}$ are computed as

$$\begin{aligned} \widehat{A} = AD &= \begin{pmatrix} 1 & 1 & 1 & 0 \\ 2 & 3 & 0 & 1 \end{pmatrix} \begin{pmatrix} 1.8000 & & & \\ & 2.1333 & & \\ & & 1.0667 & \\ & & & 2.0000 \end{pmatrix} \\ &= \begin{pmatrix} 1.8000 & 2.1333 & 1.0667 & 0.0000 \\ 3.6000 & 6.4000 & 0.0000 & 2.0000 \end{pmatrix}, \end{aligned} \tag{4.20}$$

$$\hat{c} = Dc = \begin{pmatrix} 1.8000 & & & \\ & 2.1333 & & \\ & & 1.0667 & \\ & & & 2.0000 \end{pmatrix} \begin{pmatrix} -2 \\ -1 \\ 0 \\ 0 \end{pmatrix} = \begin{pmatrix} -3.6000 \\ -2.1333 \\ 0.0000 \\ 0.0000 \end{pmatrix}. \tag{4.21}$$

Next we need to compute the projection matrix $\widehat{P}$. This is

$$\begin{aligned} \widehat{P} &= I - \widehat{A}^T\left(\widehat{A}\widehat{A}^T\right)^{-1}\widehat{A} \\ &= \begin{pmatrix} .6203 & -.3718 & -.3031 & .0733 \\ -.3718 & .2885 & .0505 & -.2539 \\ -.3031 & .0505 & .4106 & .3841 \\ .0733 & -.2539 & .3841 & .6806 \end{pmatrix}. \end{aligned} \tag{4.22}$$

Hence the projected gradient is

$$p^1 = \widehat{P}\hat{c} = \begin{pmatrix} .6203 & -.3718 & -.3031 & .0733 \\ -.3718 & .2885 & .0505 & -.2539 \\ -.3031 & .0505 & .4106 & .3841 \\ .0733 & -.2539 & .3841 & .6806 \end{pmatrix} \begin{pmatrix} -3.6000 \\ -2.1333 \\ 0.0000 \\ 0.0000 \end{pmatrix} = \begin{pmatrix} 1.4399 \\ -0.7231 \\ -0.9836 \\ -0.2779 \end{pmatrix}. \tag{4.23}$$

Next we pick θ by

$$\theta = -\min_j p_j^1 = 0.9836 \tag{4.24}$$

Next we rescale the current iterate to $\hat{x}^1 = D^{-1}x^1 = e$ and move to $\hat{x}^2$ in the transformed space:

$$\hat{x}^2 = \begin{pmatrix} 1 \\ 1 \\ 1 \\ 1 \end{pmatrix} + \frac{\alpha}{\theta} p = \begin{pmatrix} 1 \\ 1 \\ 1 \\ 1 \end{pmatrix} + \frac{0.5}{0.9836} \begin{pmatrix} 1.4399 \\ -0.7231 \\ -0.9836 \\ -0.2779 \end{pmatrix} = \begin{pmatrix} 1.7320 \\ 0.6324 \\ 0.5000 \\ 1.8588 \end{pmatrix}. \tag{4.25}$$

Transforming this point to the original space, we obtain

$$x^2 = D\hat{x}^2 = \begin{pmatrix} 1.8000 & & & \\ & 2.1333 & & \\ & & 1.0667 & \\ & & & 2.0000 \end{pmatrix} \begin{pmatrix} 1.7320 \\ 0.6324 \\ 0.5000 \\ 1.8588 \end{pmatrix} = \begin{pmatrix} 3.1175 \\ 1.3492 \\ 0.5333 \\ 1.7175 \end{pmatrix}. \tag{4.26}$$

The objective value is $z_2 = -7.5842$. This completes the second iteration of the method. Because x^2 is quite different from x^1 we proceed to the next iteration. We leave the rest of the iterations as an exercise for the reader.

It is clear that the solution of the above problem using the primal affine method will take many more iterations than the Simplex Method. However, the experience with very large practical problems indicates that interior-point methods tend to solve faster than the Simplex Method.

▷ **Exercise 4.2** Complete the steps of the algorithm on Example 4.1 using the `Primal Affine / Dikin` software option.

4.3 INITIAL SOLUTION

So far we have assumed that an initial feasible interior-point solution is available. If an initial feasible interior solution is not available, we can easily generate one by picking an arbitrary $x^o > 0$, say $x^o = e$ where $e = (1, 1, \dots, 1)^T$, and setting up the following linear program with one artificial variable $x_a \geq 0$ and M (called "Big-M" in the literature) a large scalar constant:

$$\begin{array}{lrcl} \text{Minimize} & c^Tx + \quad Mx_a & = & z \\ \text{subject to} & Ax + (b - Ax^0)x_a & = & b \qquad x \geq 0,\ x_a \geq 0. \end{array} \tag{4.27}$$

Then $x = x^0 > 0$ and $x_a = 1$ is clearly a feasible solution to (4.27).

Example 4.2 (Initial Feasible Point) Consider the two-variable example described in Example 4.2 with the slacks put in:

$$\begin{array}{lrcrcrcrcl} \text{Minimize} & -2x_1 & - & x_2 & & & & & = & z \\ \text{subject to} & x_1 & + & x_2 & + & x_3 & & & = & 5 \\ & 2x_1 & + & 3x_2 & & & + & x_4 & = & 12 \end{array}$$

and $x_j \geq 0$ for $j = 1, \dots, 4$.

We pick an arbitrary

$$x^o = (1 \quad 1 \quad 1 \quad 1)^T.$$

Then we compute the quantity $b - Ax^o$:

$$b - Ax^o = \begin{pmatrix} 5 \\ 12 \end{pmatrix} - \begin{pmatrix} 1 & 1 & 1 & 0 \\ 2 & 3 & 0 & 1 \end{pmatrix} \begin{pmatrix} 1 \\ 1 \\ 1 \\ 1 \end{pmatrix} = \begin{pmatrix} 2 \\ 6 \end{pmatrix}.$$

Hence we create the following linear program with $M = 1000$:

$$\begin{array}{lrcrcrcrcrcl} \text{Minimize} & -2x_1 & - & x_2 & & & & & + & 1000x_a & = & z \\ \text{subject to} & x_1 & + & x_2 & + & x_3 & & & + & 2x_a & = & 5 \\ & 2x_1 & + & 3x_2 & & & + & x_4 & + & 6x_a & = & 12 \end{array}$$

and $x_j \geq 0$ for $j = 1, \ldots, 4$ and $x_a \geq 0$.

Then a starting feasible solution is $x^o = e$, where $e = (1, 1, \ldots, 1)^T$, and $x_a = 1$.

▷ **Exercise 4.3** Solve the problem in Example 4.2 using the **Primal Affine / Dikin** software option.

See Problems 4.4 and 4.5 and Section 4.4 for a further discussion of the Big-M Method.

4.4 NOTES & SELECTED BIBLIOGRAPHY

Interior-point methods are not recent, they have been around for a very long time. For example, in chronological order, von Neumann [1947], Dantzig [1995] (based on an idea of von Neumann, 1948), Hoffman, Mannos, Sokolowsky, & Wiegmann [1953], Tompkins [1955, 1957], Frisch [1957], Dikin [1967]. (Fiacco & McMormick [1968] further developed Frisch's Barrier Method approach to nonlinear programming.) None of these earlier methods, including Khachian's [1979] ellipsoidal polynomial-time method, turned out to be competitive in speed to the Simplex Method on practical problems.

Interior-point methods became quite a popular way to solve linear programs in the late 1970s and early 1980s. However interior-point (and polynomial-time) methods have been around for a very long time. For example, in chronological order, von Neumann [1947], Hoffman, Mannos, Sokolowsky, & Wiegmann [1953], Tompkins [1955, 1957], Frisch [1957], Dikin [1967], and Khachian [1979]. (Fiacco & McMormick [1968] further developed Frisch's Barrier Method approach to nonlinear programming.) None of these earlier methods turned out to be competitive in speed on practical problems to the Simplex Method.

In 1979, Khachian presented an algorithm, based on a nonlinear geometry of shrinking ellipsoids, with a worst-case polynomial-time bound of $O(n^6 L_B^2)$ (where L_B is the number of bits required to represent the input data on a computer). Given an open set of inequalities of the form $Ax < b$, where A is $m \times n$ with $m \geq 2$, $n \geq 2$, Khachian's algorithm either finds a feasible point if the system is nonempty or demonstrates that no feasible point exists. Assuming that the inequalities have a feasible solution, the method starts by drawing a ball that is large enough to contain a sufficiently large volume of the feasible space defined by the inequalities $Ax < b$. If the center of the ball is within the open set of inequalities, a feasible solution has been found and the algorithm terminates. If a feasible solution is not obtained, the method proceeds to the next iteration by constructing an ellipsoid of smaller volume that contains the feasible space of the inequalities contained in the previously drawn ball. If the center of the ellipsoid is in the feasible space of $Ax < b$, we have found a feasible solution; otherwise the method proceeds to the next iteration by constructing another ellipsoid of smaller volume, and so on. The theory developed by Khachian states that if a feasible solution exists, then the center of some ellipsoid will lie in the feasible space within a number of iterations bounded by some polynomial expression in the data. Although Khachian's ellipsoid method has nice theoretical properties, unfortunately it performs poorly in practice. First, the number of iterations tend to be very large; and second, the amount of computation associated with each iteration is much more than that with the Simplex Method. Khachian's work specialized to linear programming is based on earlier work done by Shor [1971a, 1971b, 1972a, 1972b, 1975, 1977a, 1977b] for the more general case of convex programming. Other work that was influenced by Shor and preceded Khachian was due to Judin & Nemirovskii [1976a,b,c]. Predating all this was an article by Levin [1965] for convex programming.

In 1984, Karmarkar presented his interior point ellipsoid method with a worst-case polynomial-time bound of $O(n^{3.5}L_B^2)$, where L_B, as defined above, is the number of bits required to represent the input data on a computer. Claims by Karmarkar that his method is much faster (in some cases 50 times faster) than the Simplex Method stimulated improvements in the simplex-based algorithms and the development of alternative interior-point methods. Up to 1996, no method has been devised to our knowledge that is superior for all problems encountered in practice. Since the publication of Karmarkar's [1984] paper there have as of 1996 been well over a thousand papers published on interior-point methods. See Kranich [1991] for a bibliography, M. Wright [1992] for a review of interior-point methods. Also see Lustig, Marsten, & Shanno [1994] for a review of the computational aspects of interior-point methods.

The primal affine method presented in this chapter (which is the same as Dikin's method) was rediscovered by Barnes [1986] and Vanderbei, Meketon, & Freedman [1986]. Dikin [1974] proved convergence of his method under primal nondegeneracy. Proofs of convergence of Dikin's iterates can also be found in Adler, Resende, Veiga, & Karmakar [1989], Barnes [1986], Dantzig [1988a], Dantzig & Ye [1990], Monma & Morton [1987], Vanderbei, Meketon, & Freedman [1986], and, under somewhat weaker assumptions, in Vanderbei & Lagarias [1988].

A projected Newton Barrier method is described and related to Karmarkar's method in Gill, Murray, Saunders, Tomlin, & Wright [1986]. Dual approaches of both these methods have also been developed, see Adler, Resende, Veiga, & Karmarkar [1989] and Renegar [1988]. Megiddo [1988, 1986] first devised the theory for primal-dual interior-methods (see *Linear Programming 2* for details of the method) which has performed very well in practice. Lustig, Marsten, & Shanno [1990, 1991a, 1991b, 1992a, 1992b, 1992c, 1994] implemented and reported promising results for various versions of a primal-dual algorithm. For a review of methods with a focus on computational results, see, for example, Gill, Murray, & Saunders [1988] and Lustig, Marsten, & Shanno [1994].

For a description of various interior-point methods see *Linear Programming 2*. Also, see *Linear Programming 2* for a discussion on what to do in the event that M is not sufficiently large that the vector x_a goes to zero. Computational techniques (QR factorization, Cholesky factorization) for solving linear progams by interior-point methods are dicsussed in *Linear Programming 2*.

4.5 PROBLEMS

4.1 Consider the linear program

$$\begin{array}{lrcl} \text{Minimize} & 2x_1 + x_2 & = & z \\ \text{subject to} & x_1 + x_2 & = & 2 \\ & x_1 \geq 0, & x_2 \geq 0. & \end{array}$$

(a) Plot the feasible region.

(b) Starting with an initial interior feasible solution $x_1 = 1$, $x_2 = 1$, solve the problem by hand by the primal affine method described in this chapter. Plot each iteration and clearly show the direction of steepest descent that was used.

(c) Solve the problem by the `Primal Affine / Dikin` software option.

(d) Solve it by the `DTZG Simplex Primal` software option and compare the solution obtained to that obtained by the primal affine method.

4.2 Consider the linear program

$$\begin{array}{lrcl} \text{Maximize} & x_1 + x_2 & = & z \\ \text{subject to} & 2x_1 + x_2 & \leq & 3 \\ & x_1 & \leq & 1 \\ & x_1 \geq 0, & x_2 \geq 0. & \end{array}$$

(a) Plot the feasible region.

(b) Starting with an initial interior feasible solution $x_1 = 0.5$, $x_2 = 0.5$, solve the problem by hand by the primal affine method described in this chapter. Plot each iteration.

(c) Solve the problem by the `Primal Affine / Dikin` software option.

(d) Solve it by the `DTZG Simplex Primal` software option and compare the solution obtained to that obtained by the primal affine method.

4.3 Consider the linear program

$$\begin{array}{lrcl} \text{Minimize} & x_1 - 2x_2 & = & z \\ \text{subject to} & x_1 + 2x_2 & = & 5 \\ & x_1 & \geq & 1 \\ & x_1 \geq 0, & x_2 \geq 0. & \end{array}$$

(a) Plot the feasible region.

(b) Starting with an initial interior feasible solution $x_1 = 2$, $x_2 = 1.5$, solve the problem by hand by the primal affine method described in this chapter. Plot each iteration.

(c) Solve the problem by the `Primal Affine / Dikin` software option.

(d) Solve it by the `DTZG Simplex Primal` software option and compare the solution obtained to that obtained by the primal affine method.

4.4 Use the primal affine method to solve

$$\begin{array}{ll} \text{Minimize} & 2x_1 + 3x_2 = z \\ \text{subject to} & 2x_1 + x_2 \le 6 \\ & x_1 \ge 1 \\ & x_1 \ge 0,\ x_2 \ge 0 \end{array}$$

as follows:

(a) Plot the feasible region of the problem.
(b) Use the Big-M method of Section 4.3 to generate an initial feasible interior-point solution.
(c) Plot the path taken by the iterates of the primal-affine method.

4.5 Apply the Big-M method of Section 4.3 to generate an initial feasible solution to the linear program

$$\begin{array}{ll} \text{Minimize} & x_1 + x_2 = z \\ \text{subject to} & x_1 + 2x_2 \ge 2 \\ & 3x_1 + 2x_2 \le 1 \\ & x_1 \ge 0,\ x_2 \ge 0. \end{array}$$

(a) Apply the primal affine method of this chapter and demonstrate that the problem is infeasible.
(b) Solve the problem using the **Primal Affine / Dikin** software option.
(c) When in practice, the Big-M method is applied to large practical problems, how do you think the choice of M makes a difference in determining whether the problem is infeasible.

4.6 Consider the following linear program:

$$\begin{array}{ll} \text{Minimize} & -x_1 = z \\ \text{subject to} & x_1 + x_2 \ge 4 \\ & x_1 \ge 0,\ x_2 \ge 0. \end{array}$$

(a) Use the Big-M method of Section 4.3 to generate an initial feasible interior-point solution.
(b) Apply the primal affine method of this chapter and demonstrate that the problem is unbounded.

4.7 Solve the following problem by the primal affine method (software option **Primal Affine / Dikin**) described in this chapter:

$$\begin{array}{ll} \text{Minimize} & 3x_1 + 2x_2 + 4x_3 = z \\ \text{subject to} & x_1 + x_2 + x_3 = 1 \\ & x_1 + 3x_2 - x_3 = 0 \\ \multicolumn{2}{l}{\text{and } x_1 \ge 0, x_2 \ge 0, x_3 \ge 0.} \end{array} \tag{4.28}$$

(a) Use the Big-M method of Section 4.3 to generate an initial feasible interior-point solution.

(b) Solve the problem by a two-phase procedure (similar to the Simplex Method) as follows. Start by introducing an artificial variable as for the Big-M method.

- Use a Phase I objective of the form min x_a to solve the problem.
- Do Phase II. That is, drop the artificial variable and restart the method with the feasible solution obtained at the end of Phase I and the real objective of the problem.

(c) Compare the Big-M method with the two-phase method.

4.8 Solve the following linear program with a redundant constraint by hand with the primal-affine method.

$$\begin{array}{lrcrcl} \text{Minimize} & 3x_1 & - & 2x_2 & = & z \\ \text{subject to} & x_1 & + & x_2 & = & 1 \\ & 2x_1 & + & 2x_2 & = & 2 \end{array} \tag{4.29}$$

and $x_1 \geq 0, x_2 \geq 0$.

4.9 Consider the following linear program with a redundant constraint:

$$\begin{array}{lrcrcrcrcrcl} \text{Minimize} & x_1 & + & x_2 & + & x_3 & + & x_4 & + & x_5 & = & z \\ \text{subject to} & x_1 & + & 2x_2 & + & x_3 & + & x_4 & + & 2x_5 & = & 5 \\ & x_1 & - & x_2 & & & & & & & = & 1 \\ & 3x_1 & & & + & x_3 & + & x_4 & + & 2x_5 & = & 7 \\ & \multicolumn{11}{l}{x_j \geq 0, \ j = 1, \ldots, 5.} \end{array}$$

(a) Solve the linear program by hand using primal-affine method. (You may want to use mathematical software or spreadsheets to help with matrix multiplications and solving of systems of equations.)

(b) Apply the **Primal Affine / Dikin** software option to solve the problem. Explore the solution of the problem using different values of α (consider at least 5 different values between 0.5 and 0.99, with 0.95 being one of the values). Comment on your results.

4.10 Solve the following linear program:

$$\begin{array}{lrcrcl} \text{Minimize} & -x_1 & & & = & z \\ \text{subject to} & x_1 & + & x_2 & \geq & 4 \\ & \multicolumn{5}{l}{x_1 \geq 0, \ x_2 \geq 0.} \end{array}$$

(a) Plot the feasible region.

(b) Apply the primal-affine method and display the iterates on your plot.

4.11 Consider the following linear program due to Klee & Minty [1972]:

$$\begin{array}{lll} \text{Minimize} & \displaystyle\sum_{j=1}^{m} -10^{m-j} x_j & \\ \text{subject to} & \displaystyle\left(2\sum_{j=1}^{i-1} 10^{i-j} x_j\right) + x_i + z_i = 100^{i-1}, & \text{for } i = 1, \ldots, m \\ & x_j \geq 0, \ z_j \geq 0, \quad \text{for } j = 1, \ldots, m. & \end{array} \tag{4.30}$$

If we apply the Simplex Method (using the largest coefficient rule) to solve the problem (4.30), then it can be shown that the Simplex Method performs $(2^m - 1)$ iterations before finding the optimal solution. Apply the **`Primal Affine / Dikin`** software option for $m \geq 8$ to solve the problem. Explore the solution of the problem using different values of α (consider at least 5 different values between 0.5 and 0.99). Comment on your results.

4.12 Solve the following problem using the **`Primal Affine / Dikin`** software option:

$$\begin{array}{ll} \text{Minimize} & c^T x \\ \text{subject to} & Ax = b \\ & x \geq 0, \end{array}$$

where A, b, and c are defined as follows:

$$\begin{aligned} A_{ij} &= \frac{1}{i+j-1}, \text{ for } i, j = 1, \ldots, 10, \\ b_i &= \sum_{j=1}^{10} A_{ij}, \text{ for } i = 1, \ldots, 10, \\ c_j &= 1, \text{ for } j = 1, \ldots, 10. \end{aligned}$$

Comment on your solution.

4.13 Use the **`Primal Affine / Dikin`** software option to solve the transportation problem (1.4) on page 5.

4.14 How would you modify the primal affine method to handle upper and lower bounds on the variables, assuming that on each variable at least one bound is finite. Discuss how you would handle the case when a variable is unrestricted in value. Hint: Replace each unrestricted variable by the difference of two nonnegative variables (see Section 6.2).

CHAPTER 5

DUALITY

Earlier, see (1.6) and (1.8), we defined the dual of a linear program in standard form, min $z = c^T x$, $Ax = b$, $x \geq 0$, as max $v = b^T \pi$, $A^T \pi \leq c$. The original problem in relation to its dual is called the *primal*. It is easy to prove, by rewriting the dual in standard form, that the dual of the dual is the primal; see Exercise 5.2. Feasible solutions to the primal and to the dual may appear to have little relation to one another. Actually, there is a strong relationship between their objective values; moreover, their optimal basic feasible solutions are such that it is possible to use one to obtain the other readily. Commercial software is written using the primal Simplex Method; as a result, it may sometimes be more convenient to use the dual to obtain the solution to a linear programming problem than the primal.

5.1 DUAL AND PRIMAL PROBLEMS

5.1.1 VON NEUMANN SYMMETRIC FORM

Given the linear program in the form

$$\text{PRIMAL: Find } \min z \text{ such that } Ax \geq b,\ x \geq 0,\ c^T x = z, \tag{5.1}$$

von Neumann defined its dual as

$$\text{DUAL: Find } \max v \text{ such that } A^T y \leq c,\ y \geq 0,\ b^T y = v. \tag{5.2}$$

This definition is equivalent to the one given above when the primal problem is stated in standard form; see also Section 5.1.4.

▷ **Exercise 5.1** Based on the above definition of a dual, prove that the dual of a dual is a primal problem.

	Primal						
	Variables	$x_1 \geq 0$	$x_2 \geq 0$	$\cdots$	$x_n \geq 0$	Relation	Constants
	$y_1 \geq 0$	a_{11}	a_{12}	$\cdots$	a_{1n}	$\geq$	b_1
	$y_2 \geq 0$	a_{21}	a_{22}	$\cdots$	a_{2n}	$\geq$	b_2
Dual	$\vdots$	$\vdots$	$\vdots$	$\vdots$	$\vdots$	$\vdots$	$\vdots$
	$y_m \geq 0$	a_{m1}	a_{m2}	$\cdots$	a_{mn}	$\geq$	b_m
	Relation	$\leq$	$\leq$	$\cdots$	$\leq$		$\leq$ Max v
	Constants	c_1	c_2	$\cdots$	c_n	$\geq$ Min z	

Table 5-1: Tucker Diagram

5.1.2 TUCKER DIAGRAM

To see more clearly the connection between the primal and dual problems, we shall use Tucker's detached coefficient array, Table 5-1. The primal problem reads across, the dual problem down. A simple way to remember the direction of inequality is to write the primal inequalities $\geq$ to correspond to the z-form being always $\geq \min z$, and to write the dual inequalities $\leq$ to correspond to the v-form being always $\leq \max v$.

5.1.3 DUALS OF MIXED SYSTEMS

It is always possible to obtain the dual of a system consisting of a mixture of equations, inequalities (in either direction), nonnegative variables, or variables unrestricted in sign by changing the system to an equivalent inequality system in the form (5.1). However, an easier way is to apply certain rules, displayed in Table 5-2, which we will now discuss. Both the primal and dual systems can be viewed as consisting of a set of variables with their sign restrictions and a set of linear equations and inequalities, such that the variables of the primal are in one-to-one correspondence with the equations and inequalities of the dual, and such that the equations and inequalities of the primal are in one-to-one correspondence with the variables of the dual. When the primal relation is a linear inequality ($\geq$), the corresponding dual variable of the dual is nonnegative; note in Table 5-2 that if the primal relation is an equation, the corresponding dual variable will be unrestricted in sign.

As an illustration of these rules, consider the following example:

Example 5.1 (Dual of a Mixed Primal System) Suppose we have the following mixed primal system find min z, $x_1 \geq 0$, $x_2 \leq 0$, x_3 unrestricted, subject to

$$\begin{array}{rcrcrcl} x_1 & + & x_2 & + & x_3 & = & z \text{ (min)} \\ x_1 & - & 3x_2 & + & 4x_3 & = & 5 \\ x_1 & - & 2x_2 & & & \geq & 3 \\ & & 2x_2 & - & x_3 & \leq & 4 \end{array} \tag{5.3}$$

Primal	*Dual*
Minimize Primal Objective	Maximize Dual Objective
Objective Coefficients	RHS of Dual
RHS of Primal	Objective Coefficients
Coefficient Matrix	Transposed Coefficient Matrix
Primal Relation:	Dual Variable:
(ith) Inequality: $\geq$	$y_i \geq 0$
(ith) Inequality: $\leq$	$y_i \leq 0$
(ith) Equation: $=$	y_i unrestricted in sign
Primal Variable:	Dual Relation:
$x_j \geq 0$	(jth) Inequality: $\leq$
$x_j \leq 0$	(jth) Inequality: $\geq$
x_j unrestricted in sign	(jth) Equation: $=$

Table 5-2: Correspondence Rules Between Primal and Dual LPs

The above primal system in detached coefficient form is displayed by reading across Table 5-3. Its dual, obtained by applying the rules in Table 5-2, is displayed by reading down in Table 5-3.

To see why this is the case, suppose we rewrite system (5.3) in its equivalent von Neumann inequality form:

$$\begin{array}{rrrrl} x_1 + & (-x_2') + & (x_3' - x_3'') & \geq & \min z \\ x_1 - & 3(-x_2') + & 4(x_3' - x_3'') & \geq & 5 \qquad (x_1 \geq 0, x_2' \geq 0, x_3' \geq 0, x_3'' \geq 0) \\ -\big[x_1 - & 3(-x_2') + & 4(x_3' - x_3'')\big] & \geq & -5 \\ x_1 - & 2(-x_2') & & \geq & 3 \\ & -\big[2(-x_2') - & (x_3' - x_3'')\big] & \geq & -4. \end{array} \tag{5.4}$$

We write $x_2' = -x_2 \geq 0$; this changes the sign of x_2. Also, we have written $x_3 = x_3' - x_3''$ as *the difference of two nonnegative variables* (see Section 6.2), and we have written the first equation of (5.3) as *equivalent to two inequalities*, $x_1 - 3x_2 + 4x_3 \geq 5$ and $x_1 - 3x_2 + 4x_3 \leq 5$. The relationship between the primal of (5.4) and its dual is displayed in Table 5-4.

Here it is convenient to let $y_1' \geq 0$ and $y_1'' \geq 0$ be the dual variables corresponding to the first two inequalities. Since coefficients of y_1' and y_1'' differ only in sign in every inequality, we may set $y_1' - y_1'' = y_1$, where y_1 can have either sign. Note next that the coefficients in the inequalities of the dual corresponding to x_3' and x_3'' differ only in sign, which implies the equation

$$4(y_1' - y_1'') - y_3 = 1, \quad \text{or} \quad 4y_1 - y_3 = 1.$$

From these observations it is clear that Table 5-3 is stating more concisely the relations of Table 5-4.

	Primal					
	Variables	$x_1 \geq 0$	$x_2 \leq 0$	x_3	Relation	Constants
	y_1	1	−3	4	=	5
	$y_2 \geq 0$	1	−2		$\geq$	3
Dual	$y_3 \leq 0$		2	−1	$\leq$	4
	Relation	$\leq$	$\geq$	=		$\leq$ Max v
	Constants	1	1	1	$\geq$ Min z	

Table 5-3: Example of a Dual of a Mixed System

	Primal						
	Variables	$x_1 \geq 0$	$x_2' \geq 0$	$x_3' \geq 0$	$x_3'' \geq 0$	Relation	Constants
	$y_1' \geq 0$	1	3	4	−4	$\geq$	5
	$y_1'' \geq 0$	−1	−3	−4	4	$\geq$	−5
Dual	$y_2 \geq 0$	1	2			$\geq$	3
	$-y_3 \geq 0$		2	1	−1	$\geq$	−4
	Relation	$\leq$	$\leq$	$\leq$	$\leq$		$\leq$ Max v
	Constants	1	−1	1	−1	$\geq$ Min z	

Table 5-4: Example of a Dual of a Mixed System (Continued)

5.1.4 THE DUAL OF THE STANDARD FORM

Applying the correspondence rules of Table 5-2, the dual of the standard form is easily obtained; see Table 5-5. The dual variables, which in this case are all unrestricted in sign, are denoted by π_i (instead of y_i) to conform with the notation used in earlier chapters (See Equation 1.8).

Thus, the primal problem for the standard linear program given in Table 5-5 is to

$$\begin{array}{lll} \text{Minimize} & c^Tx = z & \\ \text{subject to} & Ax = b, & A:\ m \times n, \\ & x \geq 0, & \end{array} \tag{5.5}$$

and the dual problem for the standard linear program is to

$$\begin{array}{lll} \text{Maximize} & b^T\pi = v & \\ \text{subject to} & A^T\pi \leq c, & A:\ m \times n, \end{array} \tag{5.6}$$

where π_i is unrestricted in sign.

	Primal						
Dual	Variables	$x_1 \geq 0$	$x_2 \geq 0$	$\cdots$	$x_n \geq 0$	Relation	Constants
	π_1	a_{11}	a_{12}	$\cdots$	a_{1n}	$=$	b_1
	π_2	a_{21}	a_{22}	$\cdots$	a_{2n}	$=$	b_2
	$\vdots$	$\vdots$	$\vdots$	$\vdots$	$\vdots$	$\vdots$	$\vdots$ (Dual Obj)
	π_m	a_{m1}	a_{m2}	$\cdots$	a_{mn}	$=$	b_m
	Relation	$\leq$	$\leq$	$\cdots$	$\leq$		$\leq$ Max v
	Constants	c_1	c_2 (Primal Objective)	$\cdots$	c_n	$\geq$ Min z	

Table 5-5: Primal/Dual System for Standard Form

▷ **Exercise 5.2** Show that the dual of a dual of a primal linear program in standard form is itself the primal linear program in standard form.

5.1.5 PRIMAL-DUAL FEASIBLE-INFEASIBLE CASES

All four combinations of feasibility and infeasibility of the primal and dual systems are possible as shown in the examples below:

Example 5.2 (Primal Feasible, Dual Feasible, and Min $z =$ Max v)

$$\text{Primal:}\quad \begin{aligned} x_1 &= z \text{ (Min)} \\ x_1 &= 5 \\ x_1 &\geq 0 \end{aligned} \qquad\qquad \text{Dual:}\quad \begin{aligned} 5\pi_1 &= v \text{ (Max)} \\ \pi_1 &\leq 1 \end{aligned}$$

Example 5.3 (Primal Feasible, Dual Infeasible, and Min $z \to -\infty$)

$$\text{Primal:}\quad \begin{aligned} -x_1 - x_2 &= z \text{ (Min)} \\ x_1 - x_2 &= 5 \\ x_1 \geq 0,\ x_2 &\geq 0 \end{aligned} \qquad\qquad \text{Dual:}\quad \begin{aligned} 5\pi_1 &= v \text{ (Max)} \\ \pi_1 &\leq -1 \\ -\pi_1 &\leq -1 \end{aligned}$$

Example 5.4 (Primal Infeasible, Dual Feasible, and Max $v \to \infty$)

$$\text{Primal:}\quad \begin{aligned} x_1 &= z \text{ (Min)} \\ x_1 &= -5 \\ x_1 &\geq 0 \end{aligned} \qquad\qquad \text{Dual:}\quad \begin{aligned} -5\pi_1 &= v \text{ (Max)} \\ \pi_1 &\leq 1 \end{aligned}$$

Example 5.5 (Primal Infeasible, Dual Infeasible)

$$\text{Primal:}\quad \begin{aligned} -x_1 - x_2 &= z \text{ (Min)} \\ x_1 - x_2 &= +5 \\ x_1 - x_2 &= -5 \\ x_1 \geq 0,\ x_2 &\geq 0 \end{aligned} \qquad\qquad \text{Dual:}\quad \begin{aligned} 5\pi_1 - 5\pi_2 &= v \text{ (Max)} \\ \pi_1 + \pi_2 &\leq -1 \\ -\pi_1 - \pi_2 &\leq -1 \end{aligned}$$

▷ **Exercise 5.3** Verify the above four examples.

5.2 DUALITY THEOREMS

The *Duality Theorem* is a statement about the range of possible z values for the primal versus the range of possible v values for the dual. This is depicted graphically in Figure 5-1 for the case where the primal and dual are both feasible.

Von Neumann stated but did not prove the Duality Theorem: *If the primal* (5.1) *and dual* (5.2) *as stated above have feasible solutions, then there exist optimal feasible solutions to both the primal and the dual that are equal.* Here we simply state and illustrate the Strong Duality Theorem and the Weak Duality Theorem.

Assuming that primal and dual solutions exist, the weaker form of the Duality Theorem is

THEOREM 5.1 (Weak Duality Theorem) *If x^o is any feasible solution to the primal* (5.1) *and y^o is any feasible solution to the dual* (5.2), *then*

$$y^{oT}b = v^o \le z^o = c^T x^o. \tag{5.7}$$

Example 5.6 (Illustration of Weak Duality) Consider the primal problem

$$\begin{array}{rcrcl} 5x_1 & + & 3x_2 & = & z \text{ (min)} \\ x_1 & + & 3x_2 & \ge & -2 \\ 2x_1 & + & 4x_2 & \ge & -2 \\ \multicolumn{5}{l}{x_1 \ge 0,\ x_2 \ge 0} \end{array} \tag{5.8}$$

and its dual

$$\begin{array}{rcrcl} -2y_1 & - & 2y_2 & = & v \text{ (max)} \\ y_1 & + & 2y_2 & \le & 5 \\ 3y_1 & + & 4y_2 & \le & 3 \\ \multicolumn{5}{l}{y_1 \ge 0,\ y_2 \ge 0.} \end{array} \tag{5.9}$$

A feasible solution to the primal problem is $x_1 = 1$, $x_2 = 1$, with $z = 8$. A feasible solution to the dual is $y_1 = 0.1$, $y_2 = 0.1$, with $v = -0.4$. As expected, according to the Weak Duality Theorem, we have $-0.4 = v \le z = 8$.

Note: The bounds on the objective values are for any pair of feasible solutions. When the Simplex Method is used to solve a linear program, each iteration in Phase II ensures primal feasibility but not dual feasibility. In fact when dual feasibility is attained, an optimal solution is obtained. In the intermediate steps of Phase II, however, the primal and dual objectives are equal to each other, because $v = \pi^T b = \pi^T B x = c_B^T x_B = z$, where π is dual infeasible

COROLLARY 5.2 (Bounds on the Objectives) *Every feasible solution y^o to the dual yields a lower bound $y^{oT}b$ to values of z^o for feasible solutions x^o to the primal. Conversely, every feasible solution x^o to the primal yields an upper bound $c^T x^o$ to values of v^o for feasible solutions y^o to the dual.*

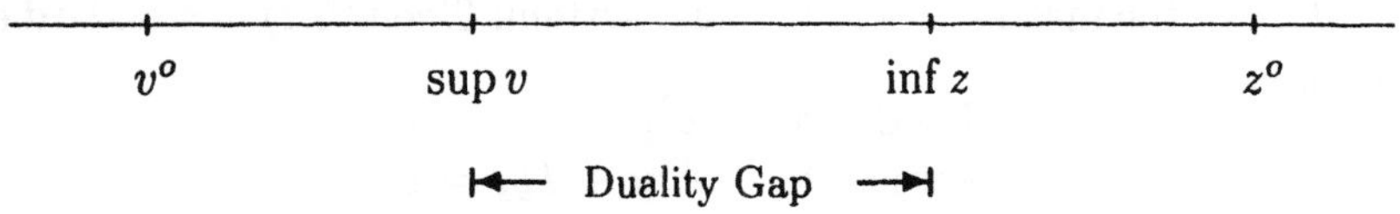

Figure 5-1: Illustration of the Duality Gap

We can depict the relationship by plotting the points v^o and z^o on a line, as shown in Figure 5-1.

We are now ready to formally state Von Neumann's Duality Theorem which in words states that if feasible solutions to the primal and dual exist then the *duality gap* (depicted in Figure 5-1) *is zero and that* $\sup v$ *is actually attained for some choice of* y *and* $\inf z$ *is attained for some choice of* x.

THEOREM 5.3 (Strong Duality Theorem) *If the primal system* min $z = c^T x$, $Ax \geq b$, $x \geq 0$ *has a feasible solution and the dual system* max $v = b^T y$, $A^T y \leq c$, $y \geq 0$ *has a feasible solution, then there exist optimal feasible solutions* $x = x^*$ *and* $y = y^*$ *to the primal and dual systems such that*

$$b^T y^* = \max v = \min z = c^T x^*. \tag{5.10}$$

Example 5.7 (Illustration of the Duality Theorem) An optimal solution to the primal problem (5.8) is $x_1 = 0$, $x_2 = 0$, with $z = 0$. A optimal solution to the dual (5.9) is $y_1 = 0.0$, $y_2 = 0.0$, with $v = 0.0$. As expected, according to the Duality Theorem, we have $0.0 = v = z = 0.0$.

5.3 COMPLEMENTARY SLACKNESS

An important property of primal/dual systems is complementary slackness. Let $x_j \geq 0$ be any feasible solution satisfying (5.1) and $y_i \geq 0$ be any feasible solution satisfying (5.2); we assume here that there exist feasible solutions. We rewrite the former in standard equality form:

$$\begin{array}{lllll} \text{Minimize} & c^T x & & = z & \\ \text{subject to} & Ax & - \; I x_s & = b & \\ & x & & \geq 0, & \end{array} \tag{5.11}$$

where $x_s = (x_{n+1}, x_{n+2}, \ldots, x_{n+m})^T \geq 0$ are variables that measure the extent of inequality, or *negative slack*, between the left- and right-hand sides of the inequalities.

It will be convenient also to let $y_s = (y_{m+1}, y_{m+2}, \dots, y_{m+n}) \geq 0$ measure the *positive slack* in the inequalities of the dual system. Then (5.2) in standard equality form becomes

$$\begin{array}{lll} \text{Maximize} & b^T y & = v \\ \text{subject to} & A^T y + I y_s & = c \\ & y & \geq 0, \end{array} \tag{5.12}$$

where $y_s = (y_{m+1}, y_{m+2}, \dots, y_{m+n})^T \geq 0$.

THEOREM 5.4 (Complementary Slackness) *For optimal feasible solutions of the primal* (5.1) *and dual* (5.2) *systems, whenever the kth relation of either system is slack, the kth variable of its dual is zero; if the kth variable is positive in either system, the kth relation of its dual is tight, i.e.,*

$$x_k y_{m+k} = 0, \quad k = 1, \dots, n, \qquad \textit{and} \qquad y_k x_{m+k} = 0, \quad k = 1, \dots, m. \tag{5.13}$$

Example 5.8 (Illustration of Complementary Slackness) At the optimal solution of (5.8), see Example 5.7, we have $n = 2$, $m = 2$, and

$$x = (x_1, x_2) = (0, 0)^T, \quad x_s = (x_3, x_4)^T = (2, 2)^T,$$

and

$$y = (y_1, y_2) = (0, 0)^T, \quad y_s = (y_3, y_4)^T = (5, 3)^T.$$

Clearly

$$x_1 y_3 = 0, \ x_2 y_4 = 0, \qquad y_1 x_3 = 0, \ y_2 x_4 = 0.$$

▷ **Exercise 5.4** Create an example with at least 3 variables that has an obvious feasible solution to the primal and an obvious lower bound for z. State the dual. Use the `DTZG Simplex Primal` software option to verify the Duality Theorem by solving both systems.

5.4 OBTAINING A DUAL SOLUTION

It turns out that when a primal linear program is solved, the dual solution can also be obtained very easily. We shall illustrate this through an example of a linear program in standard form to which a full set of artificial variables has been added.

Example 5.9 (Dual Solution from an Optimal Tableau) Consider the problem of Example 3.1: Find min z, $x \geq 0$ such that

$$\begin{array}{rcrcrcrcrcl} 2x_1 & + & 1x_2 & + & 2x_3 & + & x_4 & + & 4x_5 & = & z \\ 4x_1 & + & 2x_2 & + & 13x_3 & + & 3x_4 & + & x_5 & = & 17 \\ x_1 & + & x_2 & + & 5x_3 & + & x_4 & + & x_5 & = & 7. \end{array} \tag{5.14}$$

The optimal tableau for this problem, found in 3 iterations, is displayed in Table 5-6. Recall from (3.63) in the discussion of the Revised Simplex Method that the reduced costs corresponding to the artificial variables are $-c_B^T B^{-1}$, which are the same as the negative of the simplex multipliers, or dual variables, π. Hence the dual variables π can be read off directly from the optimal simplex tableau, which contains the artificial variables. In this case they are $\pi_1 = -1$, $\pi_2 = -19/2$. The optimal dual objective value $v = 4$ is the same as the optimal primal objective value $z = 4$ by the Duality Theorem.

Optimal Tableau										
Basic Variables	OBJ	Original Variables					Artificial Variables		RHS
	$-z$	x_1	x_2	x_3	x_4	x_5	x_6	x_7	
$-z$	1	3			1	2	1	$\frac{19}{2}$	-4
x_3		$\frac{2}{3}$		1	$\frac{1}{3}$	$-\frac{1}{3}$	$\frac{1}{3}$	$-\frac{2}{3}$	1
x_2		$-\frac{7}{3}$	1		$-\frac{2}{3}$	$\frac{8}{3}$	$-\frac{5}{3}$	$\frac{13}{3}$	2

Table 5-6: Simplex Method: Optimal Tableau

▷ **Exercise 5.5** Explain why $\pi_1 = -1$, $\pi_2 = -19/2$ satisfies $A^T\pi$ and hence is a feasible dual solution. Explain why $x_2 = 2$, $x_3 = 1$, $x_1 = x_4 = x_5 = x_6 = x_7 = 0$ is a feasible primal solution and why we have $c^Tx = b^T\pi$.

Another important property for primal/dual systems is

THEOREM 5.5 (Primal/Dual Optimality Criteria) *Let $(x_1^*, \ldots, x_n^*, z^*)$ be a feasible solution to a primal linear program in standard form and $(\pi_1^*, \ldots, \pi_m^*, v^*)$ be a feasible solution to its dual, satisfying*

$$\bar{c}^* = c - A^T\pi^* \geq 0, \qquad b^T\pi^* = v^*. \tag{5.15}$$

Then a necessary and sufficient condition for optimality of both solutions is

$$\bar{c}_j^* = 0, \quad \textit{for} \quad x_j^* > 0. \tag{5.16}$$

Example 5.10 (Illustration of Primal/Dual Optimality) In the optimal tableau displayed in Table 5-6, we have

$$\bar{c}^* = \begin{pmatrix} 3 \\ 0 \\ 0 \\ 1 \\ 2 \\ 1 \\ 19/2 \end{pmatrix}, \qquad x^* = \begin{pmatrix} 0 \\ 2 \\ 1 \\ 0 \\ 0 \\ 0 \\ 0 \end{pmatrix}.$$

It is clear that $\bar{c}_j^* = 0$ for $x_j^* > 0$.

5.5 NOTES & SELECTED BIBLIOGRAPHY

As noted in this chapter, associated with every linear programming problem is another linear programming problem called the *dual.* The fundamental notion of duality and the term were introduced by John von Neumann (in conversations with George Dantzig in October 1947) and appear implicitly in a working paper he wrote a few weeks later (von Neumann [1947]). George Dantzig's report, *A Theorem on Linear Inequalities*, dated 5 January 1948, contains the first known rigorous (unpublished) proof of the Duality Theorem. Subsequently Gale, Kuhn, & Tucker [1951] independently formulated the Duality Theorem, which they proved by means of a classical lemma due to Farkas [1902]. This theorem, known as *Farkas's Lemma* (see *Linear Programming 2*), first appeared as a lemma in Farkas's 1902 paper. A constructive proof of the Duality Theorem using the Simplex Method can be found in Dantzig [1963]. J. Abadie in verbal communications [1965] with one of the authors showed how to use the Infeasibility Theorem to prove von Neumann's Strong Duality Theorem. We shall formally prove the Duality Theorem in *Linear Programming 2*, using the Infeasibility Theorem 2.3.

It was Tobias Dantzig, mathematician and author, well known for his books popularizing the history of mathematics, who suggested around 1955 to his son George the term *primal* as the natural antonym to dual since both primal and dual derive from the Latin.

A systematic presentation of theoretical properties of dual linear programs can be found in Gale [1956] and Goldman & Tucker [1956a,b]. A review of von Neumann's contributions can be found in Kuhn & Tucker [1958]. Today everyone cites von Neumann as the originator of the Duality Theorem and credits Gale, Kuhn, & Tucker as the publishers of the first rigorous proof.

As already noted, there are several important duality-type results, known as "Either Or" theorems of the alternatives, that predate the linear programming era: Farkas [1902], Gordan [1873], Motzkin [1936], Stiemke [1915], and Ville [1938]. The earliest known result on feasibility is one concerning *homogeneous systems*, Gordan [1873]: "Either a linear homogeneous system of equations $Ax = 0$ possesses a nontrivial solution in nonnegative variables or there exists an equation, formed by taking some linear combination of the equations, that has all positive coefficients."

A natural question to ask is why not use the classical method of Lagrange multipliers to solve the linear programming problem. If we were to do so we would be required to find optimal multipliers (or prices π), which, if they exist, must satisfy a "dual" system with the property that the $\bar{c}_j$ (or relative cost factors) and optimal x_j satisfy $\bar{c}_j x_j = 0$ for $j = 1, \ldots, n$. The latter leads to 2^n possible cases of either $\bar{c}_j = 0$ or $x_j = 0$. It is here that this classical approach breaks down, for it is not practical to consider all 2^n possible cases for large n. In a certain sense, however, the Simplex Method can be viewed as a systematic way to eliminate most of these cases and to consider only a few. Indeed, it immediately restricts the number of cases by considering only those with $n - m$ of the $x_j = 0$ at one time and such that the coefficient matrix of the remaining m variables is nonsingular; moreover, the unique value of these variables is positive (under nondegeneracy). The conditions $\bar{c}_j x_j = 0$ tell us that $\bar{c}_j = 0$ for $x_j > 0$ and this determines uniquely π_i and the remaining $\bar{c}_j$. If it turns out that not all $\bar{c}_j \geq 0$, the case is dropped and a special new one is examined on the next iteration, and so on.

5.6 PROBLEMS

5.1 Set up the dual of the problem of finding x_j:

$$\begin{array}{lrcl} \text{Minimize} & x_1 + x_2 & = & z \\ \text{subject to} & x_1 + 2x_2 & \geq & 3 \\ & x_1 - 2x_2 & \geq & -4 \\ & x_1 + 7x_2 & \leq & 6, \end{array}$$

where $x_1 \geq 0$ and x_2 is unrestricted in sign. Determine the simplex multipliers of the optimal solution of the primal and verify that it satisfies the dual and gives the same value for the objective form.

5.2 Find the dual of

$$\begin{array}{ll} \text{Minimize} & x_1 + 2x_2 + x_3 = z \\ \text{subject to} & x_1 + x_2 + x_3 = 1 \\ & |x_1| \leq 4,\ x_2 \geq 0,\ x_3 \geq 0. \end{array}$$

5.3 Use the **DTZG Simplex Primal** software option to show that the following linear program is infeasible.

$$\begin{array}{ll} \text{Minimize} & -x_1 - 3x_2 + 5x_3 + 2x_4 = z \\ \text{subject to} & x_1 + 2x_2 - x_3 + x_4 \leq 4 \\ & 2x_2 - x_3 \geq 8 \\ & x_j \geq 0,\ j = 1, \ldots, 4. \end{array}$$

Write down its dual and use the **DTZG Simplex Primal** software option to show that the dual is unbounded.

5.4 Consider the linear program:

$$\begin{array}{ll} \text{Minimize} & -2x_1 - x_2 = z \\ \text{subject to} & 2x_1 - x_2 \geq 4 \\ & x_1 \leq 1 \\ & x_1 \geq 0,\ x_2 \geq 0. \end{array}$$

(a) Plot the feasible region of the primal and show that it is empty.

(b) Plot the feasible region of its dual and show that a class of solutions exists for which the dual objective $\rightarrow \infty$.

5.5 Show that the linear program

$$\begin{array}{ll} \text{Minimize} & x_1 + x_2 = z \\ \text{subject to} & 2x_1 + 3x_2 \geq 3 \\ & 3x_1 - 2x_2 \geq 4 \\ & x_1 \leq 1 \\ & x_1 \geq 0,\ x_2 \geq 0 \end{array}$$

is infeasible. Write down its dual and show that a class of solutions exists for which the dual objective $\rightarrow \infty$.

5.6 Consider the linear program:

$$\begin{array}{lrcrcl} \text{Minimize} & -x_1 & + & x_2 & = & z \\ \text{subject to} & 2x_1 & + & 3x_2 & \geq & 4 \\ & x_1 & - & x_2 & \geq & 1 \\ & \multicolumn{5}{l}{x_1 \geq 0,\ x_2 \geq 0.} \end{array}$$

(a) Plot the feasible region of the primal and show that it is possible to construct a class of solutions for which the primal objective $z \to -\infty$.

(b) Plot the feasible region of its dual and show that it is empty.

5.7 Use the **`DTZG Simplex Primal`** software option to show that the linear program

$$\begin{array}{lrcrcrcl} \text{Minimize} & x_1 & - & 2x_2 & + & x_3 & = & z \\ \text{subject to} & & & x_2 & - & x_3 & \geq & 4 \\ & x_1 & + & x_2 & + & x_3 & \geq & 1 \\ & \multicolumn{7}{l}{x_1 \geq 0,\ x_2 \geq 0,\ x_3 \geq 0} \end{array}$$

has a class of solutions for which the objective $z \to -\infty$. Write down the dual and show that it is infeasible.

5.8 Consider the linear program

$$\begin{array}{lrcrcrcl} \text{Minimize} & x_1 & + & 2x_2 & - & 3x_3 & = & z \\ \text{subject to} & 2x_1 & + & 2x_2 & + & x_3 & \geq & 10 \\ & 5x_1 & - & x_2 & - & x_3 & \leq & 5 \\ & \multicolumn{7}{l}{x_1 \geq 0,\ x_2 \geq 0,\ x_3 \geq 0.} \end{array}$$

(a) Use the **`DTZG Simplex Primal`** software option to solve the problem.

(b) Find its dual and use the **`DTZG Simplex Primal`** software option to solve it.

(c) Verify that the solution at each stage of Phase II satisfies the Weak Duality Theorem.

(d) Verify that complementary slackness holds at the primal and dual optima.

5.9 Consider the linear program

$$\begin{array}{lrcrcl} \text{Minimize} & 2x_1 & + & x_2 & = & z \\ \text{subject to} & 2x_1 & + & x_2 & \leq & 6 \\ & x_1 & + & 3x_2 & \geq & 3 \\ & \multicolumn{5}{l}{x_1 \geq 0,\ x_2 \geq 0,\ x_3 \geq 0.} \end{array}$$

(a) Plot the feasible region and solve the problem graphically.

(b) Find its dual and solve it graphically.

(c) Verify that each and every basic feasible solution (corner point) satisfies the Weak Duality Theorem.

(d) Verify that complementary slackness holds at the primal and dual optima.

5.10 *Dantzig [1963]*. Show that neither the primal nor the dual of the system

$$\begin{array}{ll} \text{Minimize} & x_1 - 2x_2 = z \\ \text{subject to} & x_1 - x_2 \geq 2 \\ & -x_1 + x_2 \geq -1 \\ & x_1 \geq 0,\ x_2 \geq 0 \end{array}$$

has a feasible solution.

5.11 Write down the dual of the linear program

$$\begin{array}{ll} \text{Minimize} & x_1 - 3x_2 + x_3 = z \\ \text{subject to} & 2x_1 - 3x_2 + x_3 \geq 4 \\ & x_1 \qquad\quad - x_3 \leq 2 \\ & \qquad - x_2 + x_3 \leq 1 \\ & x_1 \geq 0,\ x_2 \geq 0,\ x_3 \geq 0. \end{array}$$

Use the **DTZG Simplex Primal** software option to demonstrate that neither the primal nor its dual has a feasible solution.

5.12 Prove that the optimal dual solution is never unique if the optimal primal basic solution is degenerate and the optimal dual solution is not.

5.13 Prove in general that an equation in the primal corresponds to an unrestricted variable in the dual and a variable unrestricted in sign corresponds to an equation.

5.14 Show that the dual of the linear program

$$\begin{array}{ll} \text{Minimize} & b^T y = z \\ \text{subject to} & A^T y = 0, \\ & e^T y = 1, \\ & y \geq 0, \end{array}$$

where $e = (1, 1, \ldots, 1)^T$, can be used to find a feasible solution to $Ax \leq b$. Note that the x variables are not restricted in sign.

5.15 Consider an LP in standard form (3.1). Let $\mathcal{B}$ be the index set corresponding to the basic variables (i.e., $A_{\bullet\mathcal{B}}$ is the submatrix consisting of the basic columns). In terms of the original data, c, A, b, and $\mathcal{B}$,

(a) What is π (the simplex multipliers)? Describe the pivot selection rules.

(b) Show that if x is optimal then π is dual feasible. (By x optimal, we mean that x prices out optimally in our tableau.)

(c) Show that if x is an optimal feasible solution to the LP, then π is optimal for the dual.

(d) Assume that our current tableau indicates that the problem is unbounded. Show that there exists some vector $0 \neq \bar{x} \geq 0$ such that $A\bar{x} = 0$ and $c^T\bar{x} < 0$.

5.16 Let A be a skew symmetric matrix (that is, $A^T = -A$). Show that the following linear program is its own dual.

$$\begin{array}{ll} \text{Minimize} & c^T x \\ \text{subject to} & Ax \geq -c \\ & x \geq 0. \end{array}$$

If x^* is an optimal solution to the above linear program, what is the value of $c^T x^*$? Why?

5.17 Consider the linear program

$$\begin{array}{ll} \text{Minimize} & c^T x = z \\ \text{subject to} & Ax \le b \\ & x \ge 0, \end{array}$$

where

$$A = \begin{pmatrix} 2 & 1 & -4 \\ -1 & -1 & 1 \\ 1 & 0 & 0 \end{pmatrix}, \quad b = \begin{pmatrix} 2 \\ -1 \\ 2 \end{pmatrix}, \quad c = \begin{pmatrix} -2 & 1 & 0 \end{pmatrix}.$$

The final tableau, where s_1, s_2, and s_3 are slacks, is

$(-z)$	x_1	x_2	x_3	s_1	s_2	s_3	
1	0	1	0	0	0	2	4
0	0	−3	0	1	4	2	2
0	1	0	0	0	0	1	2
0	0	−1	1	0	1	1	1

(a) What is an optimal solution x^*? What is an optimal solution π^* for the dual? What is the optimal objective value z^*? What is the optimal basis for the primal and what is its inverse?

(b) Suppose the right-hand side is changed to $\hat{b} = (1, -1, 2)^T$. Does the current basis remain feasible?

(c) Suppose the objective coefficients are changed to $\hat{c} = (-2, 1, 2)^T$. Does the current solution remain optimal?

5.18 Consider the linear program

$$\begin{array}{ll} \text{Minimize} & a^T x - b^T y = z \\ \text{subject to} & x - y = c \\ & x \le 0 \\ & y \le 0, \end{array}$$

where a, b, c are n-vectors.

(a) Exhibit a feasible solution.

(b) Since the problem is feasible, either there exists an optimal solution or there exists a class of solutions with unbounded objective value. Find the dual and using it, derive necessary and sufficient conditions on a and b for when the primal has an optimal solution.

(c) Find the (obvious) optimal solution to the dual when it exists and use it and complementary slackness to find a basic optimal solution to the primal.

(d) Eliminate y from the LP (by substituting $y = x - c$) and re-derive your conditions for a bounded optimum for the primal without referring to the dual problem. (Hint: This should agree with your answer in (b).)

5.19 What is the dual linear program for

$$\begin{array}{ll} \text{Minimize} & c^T x \\ \text{subject to} & Ay = x \\ & y \le b. \end{array}$$

5.20 What is the dual linear program for

$$\begin{array}{lrcl} \text{Minimize} & & c^T x & \\ \text{subject to} & b_l \leq & Ax & \leq b_u \\ & l \leq & x & \leq u, \end{array}$$

where A is an $m \times n$ matrix. What would be the dual if we were maximizing instead of minimizing?

5.21 Suppose that a linear program in standard form (5.5) has a finite optimal value. Show that if the right-hand side is changed from b to some b', the resulting linear program is either infeasible or has a finite optimal value. Thus, by changing the right-hand side b we cannot obtain an LP that is unbounded below.

5.22 In practical applications of linear programming, the scaling of coefficients is important for numerical reasons and for knowing how to interpret model results. Consider a typical linear program and its associated dual:

$$\begin{array}{lrcl} \text{Minimize} & c^T x & = & z \\ \text{subject to} & Ax & \geq & b, \\ & x & \geq & 0. \end{array} \qquad \begin{array}{lrcl} \text{Maximize} & b^T y & = & v \\ \text{subject to} & A^T y & \leq & c, \\ & y & \geq & 0. \end{array}$$

What is the effect of each of the actions (a), (b), (c), (d) below

(a) Multiplying a row of $[A\ b]$ by a factor γ;

(b) Multiplying the right-hand side b by a factor $\gamma > 0$;

(c) Multiplying a column of $\begin{bmatrix} c \\ A \end{bmatrix}$ by a factor $\gamma > 0$;

(d) Multiplying the objective row c by a factor γ (consider both positive and negative γ)

upon

- the optimal objective value;
- the optimal choice of decision variables and their optimal activity levels;
- the optimal values of the dual variables.

CHAPTER 6

EQUIVALENT FORMULATIONS

Real world problems tend to be "dirty," that is, there is a certain softness—a lack of precision—in their definition that often permits them to have two or more ways of being expressed mathematically. These different expressions may not be equivalent in the mathematical sense that there is an exact one-to-one correspondence between their solutions. However, from the point of view of a person looking for an answer, either is an equally satisfactory statement of the practical problem to be solved. One of these could turn out to be completely amenable to mathematical analysis and solution while another could be computationally hopeless. Only through detailed knowledge of the application and knowledge of what is intractable can one decide which one of these two acceptable formulations is best to use. Linear programming models have been successful because many large problems can be on the one hand satisfactorily formulated by planners as linear programs and can be on the other hand solved using computers.

6.1 RESTRICTED VARIABLES

Typically, the variables x_j have a lower bound of 0. In practice, however, the variables x_j may have specified upper or lower bounds, or both. For example, $x_j \geq -10$. In this case the user could define a new variable $x'_j = x_j + 10$ and substitute $x'_j - 10$ for x_j into the equations and solve the resulting system as a "standard" LP. If x_j has both a finite upper and lower bound this simple substitution trick is not sufficient to reduce the system to an equivalent standard form; an additional slack variable and equation are needed.

It would appear that having specified lower bounds increases the work required by the user to set up the problem as a standard linear program. Fortunately, most linear programming software makes it easy for the user to state the bounds. In Chapter 3 we discussed special techniques for efficiently solving linear programs whose variables have both upper and lower bounds.

6.2 UNRESTRICTED (FREE) VARIABLES

An interesting special case occurs when x_j has neither a specified finite upper or lower bound. This is called the *unrestricted* case. Except for possible numerical difficulties, we could eliminate any unrestricted variable by using one of the equations that it is in with a nonzero coefficient, to solve for it and substitute its expression in terms of the other variables into the other equations. Another approach is to replace each such variable x_j by the difference of two nonnegative variables,

$$x_j = x_j^+ - x_j^-, \text{ where } x_j^+ \geq 0, \ \ x_j^- \geq 0.$$

The variable x_j^+ is called the *positive part* of x_j and the variable x_j^- is called the *negative part* of x_j. The variables x_j^+ and x_j^- are defined as follows:

$$x_j^+ = \begin{cases} +x_j & \text{if } x_j \geq 0 \\ 0 & \text{if } x_j < 0 \end{cases}, \qquad x_j^- = \begin{cases} 0 & \text{if } x_j \geq 0 \\ -x_j & \text{if } x_j < 0 \end{cases}.$$

We shall show later that there exists an optimal solution for which either $x_j^+ = 0$ or $x_j^- = 0$ or both equal zero.

Having no specified bounds appears to increase the work required by the user to set up the problem as well as to increase the number of variables. Fortunately, special techniques, such as those discussed in Chapter 3, not only allow the user to specify the problem in a "user-friendly" way, but also in a way that does not increase the size of the problem or the computational effort required to specify and solve such problems.

▷ **Exercise 6.1** Suppose that you have an n-dimensional linear program with $1 \leq k \leq n$ unrestricted variables. That is,

$$\begin{array}{lrl} \text{Minimize} & \sum_{j=1}^{n} c_j x_j & = z \\ \text{subject to} & \sum_{j=1}^{n} a_{ij} x_j & \geq b_i \quad \text{for } i = 1, \ldots, m. \end{array}$$

Show that it is possible to modify the linear program to one that has all nonnegative variables by replacing the unrestricted variables x_j by restricted variables $\bar{x}_j \geq 0$ and one additional unrestricted variable x_{n+1}.

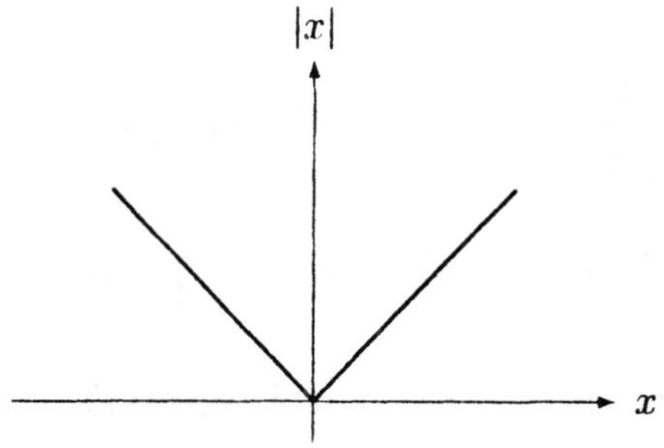

Figure 6-1: Absolute Value Function

▷ **Exercise 6.2** Consider the problem Min x subject to $x = 3$. Since x is an unrestricted variable, we replace it by $x = u - v$, where $u \geq 0$, $v \geq 0$ and we now have

$$\begin{array}{lrcrcl} \text{Minimize} & -u & + & v & = & z \\ \text{subject to} & u & - & v & = & 3. \end{array}$$

Assume that a basic feasible solution u and simplex multiplier $\pi = -1$ have been determined. Checking for optimality, the v column prices out to zero. Hence the basic feasible solution is optimal. However, due to slight round-off error, it apparently prices out -0.0000000000001 and the software therefore chooses to introduce v as the incoming variable. Discuss what happens next and what safeguards must be put into the simplex algorithm software to prevent this from happening.

6.3 ABSOLUTE VALUES

An absolute value variable $|x|$ is defined as a function of x as follows:

$$|x| = \begin{cases} x \text{ if } x \geq 0, \\ -x \text{ if } x < 0. \end{cases}$$

The plot of the function is illustrated in Figure 6-1. When the absolute value variables appear in a general system of equations, it is not possible to reformulate the problem as a linear program. However, when such functions appear only as terms in the objective function, it is sometimes possible to reformulate the model as a linear program.

Consider the following problem, where the objective is to minimize the weighted sum of absolute values subject to linear constraints, where the weights c_j are nonnegative.

$$\begin{array}{llll} \text{Minimize} & \displaystyle\sum_{j=1}^{n} c_j|x_j| = z, & \text{where } c_j \geq 0 \text{ for } j = 1,\dots,n, & \\ \text{subject to} & \displaystyle\sum_{j=1}^{n} a_{ij}x_j = b_i, & \text{for } i = 1,\dots,m. & \end{array} \tag{6.1}$$

We can replace in the equations each variable x_j by the difference of its positive and negative parts, that is,

$$x_j = x_j^+ - x_j^-, \qquad \text{where } x_j^+ \geq 0, \;\; x_j^- \geq 0.$$

In the objective function, where the functions $|x_j|$ appear, $|x_j|$ can be replaced by $x_j^+ + x_j^-$, the sum of the positive and negative parts, provided that $x_j^+ x_j^- = 0$. This results in the following modified problem:

$$\begin{array}{lll} \text{Minimize} & \displaystyle\sum_{j=1}^{n} c_j(x_j^+ + x_j^-) = z & \\ \text{subject to} & \displaystyle\sum_{j=1}^{n} a_{ij}(x_j^+ - x_j^-) = b_i & \text{for } i = 1,\dots,m, \\ & x_j^+ \geq 0, \; x_j^- \geq 0 & \text{for } j = 1,\dots,n, \\ & x_j^+ x_j^- = 0 & \text{for } j = 1,\dots,n. \end{array} \tag{6.2}$$

The above problem is not a linear program because of relations $x_j^+ x_j^- = 0$ in (6.2). However, if $c_j \geq 0$ for all j, it can be shown (see Lemma 6.1) that even if conditions $x_j^+ x_j^- = 0$ in (6.2) are dropped, either x_j^+ or x_j^- will be zero at an optimum solution. Therefore, in the case $c_j \geq 0$ for all j, the above problem reduces to a linear program.

LEMMA 6.1 (Optimality Property) *In the problem defined by* (6.2), *if each* $c_j \geq 0$, *then there exists a finite optimum solution to*

$$\begin{array}{lll} \text{Minimize} & \displaystyle\sum_{j=1}^{n} c_j(x_j^+ + x_j^-) = z & \\ \text{subject to} & \displaystyle\sum_{j=1}^{n} a_{ij}(x_j^+ - x_j^-) = b_i & \text{for } i = 1,\dots,m, \\ & x_j^+ \geq 0, \; x_j^- \geq 0 & \text{for } j = 1,\dots,n, \end{array} \tag{6.3}$$

such that either x_j^+ *or* x_j^- *is zero provided there is a feasible solution.*

Proof. Let $\hat{x}^+ = (\hat{x}_1^+, \hat{x}_2^+, \dots, \hat{x}_n^+)^T$ and $\hat{x}^- = (\hat{x}_1^-, \hat{x}_2^-, \dots, \hat{x}_n^-)^T$ be a feasible solution to problem (6.3). For $j = 1,\dots,n$ define $v_j = \min(\hat{x}_j^+, \hat{x}_j^-)$ and use these v_j's to define

$$\begin{aligned} \bar{x}_j^+ &= \hat{x}_j^+ - v_j, \\ \bar{x}_j^- &= \hat{x}_j^- - v_j. \end{aligned} \tag{6.4}$$

The $\bar{x}_j^+$, $\bar{x}_j^-$ are clearly also a feasible solution to problem (6.3) with the additional property that either $\bar{x}_j^+ = 0$ or $\bar{x}_j^- = 0$. Therefore, they are also a feasible solution to the problem defined by (6.2). Furthermore, if each $c_j \geq 0$, then

$$\begin{aligned} c^T(\bar{x}^+ + \bar{x}^-) &= c^T(\hat{x}^+ + \hat{x}^-) - 2\sum_{j=1}^{n} c_j v_j \\ &\leq c^T(\hat{x}^+ + \hat{x}^-). \end{aligned}$$

Thus, no matter what the solution, we can always find an equal or better solution with either the positive part or negative part of each x_j equal to zero. This completes the proof. ∎

The lemma shows that if problem (6.1) has an optimum solution, then it can be found by solving the linear program (6.3) and setting $x_j = x_j^+ - x_j^-$.

▷ **Exercise 6.3** Assume in problem (6.1) that some $c_j < 0$ and that feasible solutions exist. Construct an example where the minimum z is finite. Show for such an example that if we try to solve by means of (6.3) we will obtain a class of feasible solutions to (6.3) where the minimum z of (6.3) tends to $-\infty$.

▷ **Exercise 6.4** Generalize Lemma 6.1 where (6.1) is replaced by

$$\begin{array}{lll} \text{Minimize} & \displaystyle\sum_{j\in S} c_j|x_j| + \sum_{j\notin S} c_j x_j & = z \\ \text{subject to} & \displaystyle\sum_{j=1}^{n} a_{ij}x_j & = b_i \text{ for } i = 1,\ldots,m, \\ & x_j \geq 0 & \text{for } j \notin S, \\ & c_j \geq 0 & \text{for } j \in S. \end{array} \tag{6.5}$$

▷ **Exercise 6.5** Extend the discussion in this section to list the conditions under which the following problem can be formulated as a linear program.

$$\begin{array}{lll} \text{Minimize} & \displaystyle\sum_{i=1}^{p} f_i \left| \sum_{j=1}^{n} c_j^i x_j \right| & = z \\ \text{subject to} & \displaystyle\sum_{j=1}^{n} a_{ij}x_j & = b_i \text{ for } i = 1,\ldots,m. \end{array}$$

6.4 GOAL PROGRAMMING

So far, all the linear programming examples have had a simple objective that could be incorporated into a single objective function. In some practical situations there may be no clear single objective to be minimized but rather several objectives that we would like to minimize at the same time, which, in general, is nonsense because the objectives are incompatible. Nevertheless, humans to do this all the time; and there are a few situations where the objectives are not incompatible and it may make sense to do so, as we shall see.

The technique, referred to as *goal programming*, endeavors to satisfy multiple objectives simultaneously. As noted, this may not always be possible because the various objectives may conflict with each other. One possibility is to give all the objectives roughly equal priorities but to give different weights measuring their relative importance to one another and to add the different objectives together.

▷ **Exercise 6.6** Consider the product mix problem of Section 1.4.1. Assuming the same production constraints, modify the objective function so that profit from each of the desks is 10, 20, 18, and 40 respectively. As before, the objective is to maximize profit. However, it may also be important to try to maximize the production of Desk 1, which sells at a lower price than the other desks, in order to satisfy a community requirement of always having some of the cheaper desks available. State the two objective functions that you would want to consider in this case. Assign a relative weighting to these two objectives in order to formulate a goal program and optimize the linear program associated with this goal program.

Other possibilities are when weights are assigned to deviations from goals or when the objectives have priorities associated with them. These are described in the next two subsections.

WEIGHTED GOAL PROGRAMMING

In general, suppose that there are K objectives, $z_1, z_2, \ldots, z_K$, with respective goal values $g_1, g_2, \ldots, g_K$. In *weighted goal programming*, a composite objective function is created of the form

$$z = \sum_{k=1}^{K} \phi_k(g_k - z_k),$$

where the function ϕ_k is defined by

$$\phi_k(y) = \begin{cases} \alpha_k y & \text{if } y \geq 0, \\ \beta_k y & \text{if } y < 0, \end{cases}$$

where $\alpha_k \geq 0$ is the penalty (or weight) of overshooting the goal, and $\beta_k \geq 0$ is the penalty (or weight) of undershooting the goal. For example the objectives for a manufacturing corporation may to be to maximize profit while attempting to keep the capital expenditures within a specified budget. Note that the budget restriction is not a rigid constraint since it need not be satisfied. In this case we call the right hand side of the budget restriction a goal. Depending on how important it is to stay within budget, we can attach a penalty cost to the violation of the budget goal.

The goal programming problem as defined above falls under a class of problems called *separable nonlinear programming*. When the constraints are linear, we can easily convert the problem into a linear programming problem as long as the weights α_k and β_k are nonnegative for all $k = 1, \ldots, K$. Define

$$g_k - z_k = u_k^+ - u_k^-, \quad u_k^+ \geq 0, \quad u_k^- \geq 0, \quad k = 1, \ldots, K.$$

Each ϕ_k can then be replaced by the linear function

$$\phi_k(g_k - z_k) = \alpha_k u_k^+ + \beta_k u_k^-,$$

together with the additional constraints $u_k^+ \geq 0$ and $u_k^- \geq 0$. The resulting problem is a linear program provided α_k and β_k are nonnegative for all $k = 1, \ldots, K$.

PRIORITY GOAL PROGRAMMING

In *Priority Goal Programming* the objectives fall into different priority classes. We assume that no two goals have equal priority. In this case we first satisfy the highest priority objectives and then if different solutions tie for the optimum, the tie is resolved by the next set of priorities, and so on.

We will describe two methods for solving such problems. The first solves a number of linear programs *sequentially*. First, order the objectives so that z_1 has the highest priority, z_2 has the next highest priority, and so on. The objective of the first linear program is the linear form z_1 with highest priority. The resulting linear program is solved. Nonbasic variables with reduced costs different from zero are dropped (this technique for identifying alternative optimal solutions was discussed in Section 3.2.4) and the next priority objective form z_2 is made into the new objective function. (Alternatively, we could optimize using z_2 and setting the upper and lower bounds on the nonbasic variables with positive reduced costs to be 0.) After the new linear program is solved, the process is continued until all the priorities have been converted into objective forms. The process also terminates if a unique solution is found. There is a technique that was discussed in Section 3.2.4 for finding out whether or not an optimal solution is unique.

▷ **Exercise 6.7** Another technique would be to make the first objective z_1 a constraint by setting the value of z_1 equal to its optimal value and then using the next priority objective form z_2 as the new objective function. Discuss what numerical difficulties may be encountered. Suggest a method that may possibly help in getting around such difficulties.

▷ **Exercise 6.8** Solve, using the software, Exercise 6.6 by the method of Exercise 6.7. Show that if the optimal value of z_1 is not correctly inputted into the second linear program it can cause numerical difficulties if set too small. Graph the dual problem and identify the corner solutions that are tied for minimum costs. Which corner solution was selected and why?

The second method is to form a single composite objective function by assigning weights to the linear objective forms $z_1, z_2, \ldots, z_K$ and summing. For example, the weights on the two objectives in Exercise 6.6 might be $w_1 = 10^3$ for profits (if it has the highest priority) and $w_2 = 1$ for the production of desk 1. It can be shown that theoretically if a sufficiently high weight w_1 is given to one profit goal relative to the other goal, it will give the same solution as the first method. How to assign such sufficiently high weights is an interesting problem in itself, that in general is as difficult to solve as the original problem.

▷ **Exercise 6.9** Solve, using the software, Exercise 6.6 by the second method with various relative weights and find approximately how high w_1 must be relative to w_2 to give the same answer as the first method.

6.5 MINIMIZING THE MAXIMUM OF LINEAR FUNCTIONS

In goal programming we tried to satisfy multiple objectives simultaneously according to some preassigned goals. In other situations there may be several different disasters that could happen, and we may wish to minimize the cost of the worst disaster. Let there be K different least cost functions z_k that measure the cost of disaster k:

$$z_k = \sum_{j=1}^{n} c_j^k x_j, \quad k = 1, \ldots, K. \tag{6.6}$$

In this case we want to

$$\begin{array}{lrcl} \text{Minimize} & \multicolumn{3}{l}{\text{Maximum}\{z_1, z_2, \ldots, z_K\}} \\ \text{subject to} & z_k & = & \sum_{j=1}^{n} c_j^k x_j, \quad k = 1, \ldots, K, \\ & Ax & = & b, \\ & x & \geq & 0, \end{array} \tag{6.7}$$

where $A \in \Re^{m \times n}$, $b \in \Re^m$.

The problem as defined above does not appear to be a linear program but can be made into one by defining a new variable

$$z \geq z_k, \quad \text{for } k = 1, \ldots, K, \tag{6.8}$$

and minimizing z. We may rewrite (6.8) as

$$z - \sum_{j=1}^{n} c_j^k x_j \geq 0, \quad \text{for } k = 1, \ldots, K.$$

Thus we can reformulate problem (6.7) as the linear program

$$\begin{array}{lrl} \text{Minimize} & z & \\ \text{subject to} & z - \sum_{j=1}^{n} c_j^k x_j \geq 0 & \text{for } k = 1, \ldots, K \\ & Ax = b & \\ & x \geq 0. & \end{array} \tag{6.9}$$

▷ **Exercise 6.10** A department store chain may have stores in several cities and wishes to avoid closing the least profitable one by maximizing its net revenues. Find the linear programming problem that maximizes the minimum of the different revenue functions $z_k = \sum_{j=1}^{n} c_j^k x_j$, $k = 1, \ldots, K$, subject to $Ax = b$, $x \geq 0$, where $A \in \Re^{m \times n}$, $b \in \Re^m$.

Example 6.1 (Finding a Center) Some of the recent developments in solving linear programs are based on interior-point methods. Some of these techniques require an initial solution point either at the "*center*" or an initial solution point close to the center. Various definitions of a center are used. For our purposes we define it as the center of the largest sphere that can be inscribed inside the constraints.

The following approach can be used to generate this center. Let $Ax \leq b$, $x \geq 0$, define m constraints in $\Re^n$ and let $A_{i\bullet}x = b_i$ define the boundary of the ith constraint. The boundary of a constraint is called a *hyperplane*. The problem of finding a center with respect to these hyperplanes can be stated as the minimum of the maximum of m linear functions. In analytic geometry we learned that the distance from any point $\hat{x}$ to a plane $A_{i\bullet}x = b_i$ is given by

$$\theta_i = \frac{b_i - A_{i\bullet}\hat{x}}{||A_{i\bullet}||},$$

where $||A_{i\bullet}|| = \left(\sum_{j=1}^{n} A_{ij}^2\right)^{1/2}$. It is convenient to divide each inequality $A_{i\bullet}x \leq b_i$ by $||A_{i\bullet}||$. This results in a new set of inequalities $\bar{A}x \leq \bar{b}$, where $\bar{A}_{i\bullet} = A_{i\bullet}/||A_{i\bullet}||$ and $\bar{b}_i = b_i/||A_{i\bullet}||$. The distance from any point $\hat{x}$ to a hyperplane $A_{i\bullet}x = b_i$ is then given by

$$\theta_i = \bar{b}_i - \bar{A}_{i\bullet}\hat{x}.$$

The problem of finding the center point can then be stated as

$$\begin{array}{lrll} \text{Minimize} & \text{Maximum}\{\theta_1, \theta_2, \ldots, \theta_m\} & & \\ \text{subject to} & \bar{A}_{i\bullet}x + \theta_i = & b_i & \text{for } i = 1, \ldots, m \\ & x_j \quad\quad & \geq 0 & \text{for } j = 1, \ldots, n \\ & \theta & \geq 0 & \text{for } i = 1, \ldots, m. \end{array} \tag{6.10}$$

▷ **Exercise 6.11** Set up (6.10) as a linear program. Discuss why it was formulated with $\theta_i \geq 0$ for $i = 1, \ldots, m$. Discuss what happens if the system $Ax \leq b$, $x \geq 0$, is infeasible. Construct a two-variable problem with five hyperplanes (actually lines in two dimensional space). Solve your problem using the software. Draw a picture and verify that you have obtained a correct solution.

No.	t	y
(1)	50	10
(2)	100	55
(3)	150	118
(4)	200	207
(5)	250	308

Table 6-1: Experimental Points

6.6 CURVE FITTING

Suppose physical theory says that the two variable quantities t (for time) and y (for distance) are functionally related by

$$y = f(t),$$

where $f(t) = x_0 + x_1 t + x_2 t^2$, for example. The values of the coefficients x_0, x_1, x_2 are not known. A number of experiments are run with different values of $t = t_i$, and the values of $y = y_i$ observed. This results in m data points

$$(t_i, y_i), \quad i = 1, \dots, m.$$

For example, the experiment may result in the data points shown in Table 6-1. Since the observed y_i are subject to error, the best that we can hope for is to find values for x_0, x_1, x_2 such that the resulting curve $f(t)$ comes closest in some sense to fitting the observed data. More generally, in many applications, the functional form to be approximated is a polynomial in the independent variable t:

$$f(t) \approx x_0 + x_1 t + x_2 t^2 + \cdots + x_n t^n.$$

Typically, because the observed data are subject to error, the experiment is repeated many times, resulting in m data points (t_i, y_i), where m is much greater than $n+1$; we shall assume this for the rest of this section. If there were no experimental errors and if the underlying functional form $y = f(t)$ is correct, then the problem reduces to using the m data points to generate m equations and solving the overdetermined but consistent system for the unknowns $x_0, x_1, \dots, x_n$. Typically there is error, and since the number of data points m is greater than the number of unknowns $n+1$, the problem will turn out to be overdetermined and not consistent, in which case it is not possible to solve for the x_j's exactly. Thus, we try to find x_j's such that the error in the fit of the curve to the real-world data is minimized in some sense. There are three commonly used "best-fit" models for choosing $x_0, x_1, x_2, \dots, x_n$:

1. *The* 1-*norm.* Minimize the sum of the absolute errors of the fit:

$$\text{Minimize} \quad z_1 = \sum_{i=1}^{m} \left| y_i - \sum_{j=0}^{n} x_j t_i^j \right|.$$

2. *The* ∞*-norm.* Minimize the maximum absolute value of the errors of the fit:

$$\text{Minimize} \quad z_\infty = \max_{1\le i\le m} \left| y_i - \sum_{j=0}^{n} x_j t_i^j \right|.$$

3. *The* 2*-norm.* Minimize the sum of the squares of the errors of the fit:

$$\text{Minimize} \quad z_2 = \sum_{i=1}^{m} \Big(y_i - \sum_{j=0}^{n} x_j t_i^j \Big)^2.$$

In general, each of the above models will generate a different "best fit" solution. Often in practice the curve found is almost the same whatever the norm chosen, so that the choice of what is the appropriate norm to use is usually based on the *ease* with which one can solve the resulting model.

MINIMIZING THE 1-NORM

In this case, the problem is

$$\text{Minimize} \quad z_1 = \sum_{i=1}^{m} \left| y_i - \sum_{j=0}^{n} x_j t_i^j \right|. \tag{6.11}$$

Using the techniques outlined in Section 6.3, we write

$$v_i^+ - v_i^- = y_i - \sum_{j=0}^{n} x_j t_i^j, \quad \text{for } i = 1, \dots, m,$$

where $v_i^+ \ge 0$ and $v_i^- \ge 0$ for $i = 1, \dots, m$. The problem (6.11) can be reformulated as the following equivalent linear program:

$$\begin{array}{lllll} \text{Minimize} & \displaystyle\sum_{i=1}^{m} (v_i^+ + v_i^-) & & & = z \\ \text{subject to} & v_i^+ & - \; v_i^- & + \displaystyle\sum_{j=0}^{n} x_j t_i^j & = y_i \\ & v_i^+ & & & \ge 0 \\ & & v_i^- & & \ge 0 \\ & \text{for } i = 1, \dots, m. & & & \end{array} \tag{6.12}$$

▷ **Exercise 6.12** Why is not necessary to specify $v_i^+ v_i^- = 0$ in problem (6.12)?

One of the advantages of using linear programming to solve the 1-norm model is that additional constraints on the parameters $x_0, x_1, x_2, \dots, x_n$ can easily be incorporated into the model.

▷ **Exercise 6.13** Suppose that we want to minimize the 1-norm for the example displayed in Table 6-1, assuming a quadratic functional form, i.e., $y = x_0 + x_1 t + x_2 t^2$. Use the `DTZG Simplex Primal` software option to solve the problem separately under the following three assumptions:

1. x_0, x_1, x_2 are unrestricted in sign.
2. Assume that the theory says that $x_0 \geq 0, x_1 \geq 0, x_2 \geq 0$.
3. Assume that $x_2 \geq 2x_1$ but otherwise the x_i are unrestricted in sign.

Compare the three solutions obtained.

▷ **Exercise 6.14** Suppose that it is known that the errors in the experiment are always greater to some known value α. In this case we look for a *one-sided fit*, i.e.,

$$\begin{array}{ll} \text{Minimize} & z_1 = \sum_{i=1}^{m} e_i, \\ \text{subject to} & y_i - f(t_i) = e_i, \quad e_i \geq \alpha, \quad i = 1, \ldots, m. \end{array}$$

Redo Exercise 6.13 assuming a one-sided fit.

MINIMIZING THE ∞-NORM

In this case, the problem is:

$$\text{Minimize} \quad z_\infty = \max_{1 \leq i \leq m} \left| y_i - \sum_{j=0}^{n} x_j t_i^j \right|. \tag{6.13}$$

Using the techniques outlined in Section 6.3, we write

$$v_i^+ - v_i^- = y_i - \sum_{j=0}^{n} x_j t_i^j \quad \text{for } i = 1, \ldots, m,$$

where $v_i^+ \geq 0$ and $v_i^- \geq 0$ for $i = 1, \ldots, m$. Combining this with the techniques from Section 6.5, we define a variable z such that

$$z \geq v_i^+ + v_i^- \quad \text{for } i = 1, \ldots, m.$$

The problem (6.13) can then be reformulated as the following equivalent linear program:

$$\begin{array}{lrcl} \text{Minimize} & z & & \\ & z - v_i^+ - v_i^- & \geq & 0 \\ & v_i^+ - v_i^- + \sum_{j=0}^{n} x_j t_i^j & = & y_i \\ & v_i^+ & \geq & 0 \\ & v_i^- & \geq & 0 \\ \text{for } i = 1, \ldots, m & & & \end{array} \tag{6.14}$$

▷ **Exercise 6.15** Solve Exercise 6.13 assuming the ∞-norm in place of the 1-norm. Compare your solution with that found by minimizing the 1-norm.

MINIMIZING THE 2-NORM

In the third case, the problem is

$$\text{Minimize} \quad z_2 = \sum_{i=1}^{m} \left(y_i - \sum_{j=0}^{n} x_j t_i^j \right)^2. \tag{6.15}$$

This problem is known as the *least squares problem.* It is not equivalent to a linear program. The techniques that are used to solve this problem reduce to solving $n+1$ equations in $n+1$ unknowns. Statistical software packages for fitting curves to data typically use the 2-norm. Techniques for solving such least-squares problems usually are based on the QR factorization; see Section 6.8 for references.

▷ **Exercise 6.16** Solve Part (1) of Exercise 6.13 assuming the 2-norm in place of the 1-norm. Plot and compare your solution with those found by minimizing the 1-norm and the ∞-norm.

6.7 PIECEWISE LINEAR APPROXIMATIONS

In this section we consider linear programs having piecewise linear functions as objectives. If the function being minimized is convex (or is concave, if being maximized), convergence is guaranteed to a global minimum.

6.7.1 CONVEX/CONCAVE FUNCTIONS

Convergence proofs for many algorithms often assume convexity of the objective function and constraints.

> *Definition (Convex/Concave Functions)*: Let $f(x)$ be a real-valued function defined over the points $x = (x_1, x_2, \ldots, x_n)^T \in \Re^n$. Then $f(x)$ is said to be a *convex function* if and only if for any two points $x = (x_1, x_2, \ldots, x_n)^T$ and $y = (y_1, y_2, \ldots, y_n)^T$ we have
>
> $$f\big(\lambda x + (1-\lambda)y\big) \le \lambda f(x) + (1-\lambda) f(y) \tag{6.16}$$
>
> for all values of λ satisfying $0 \le \lambda \le 1$. It is said to be a *strictly convex function* if the inequality (6.16) is a strict inequality, i.e., $\le$ is replaced by $<$, for all $0 < \lambda < 1$. The function $f(x)$ is said to be *concave* (or *strictly concave*) if the above inequality (6.16) holds with $\ge$ instead of $\le$ (or $>$ instead of $<$).

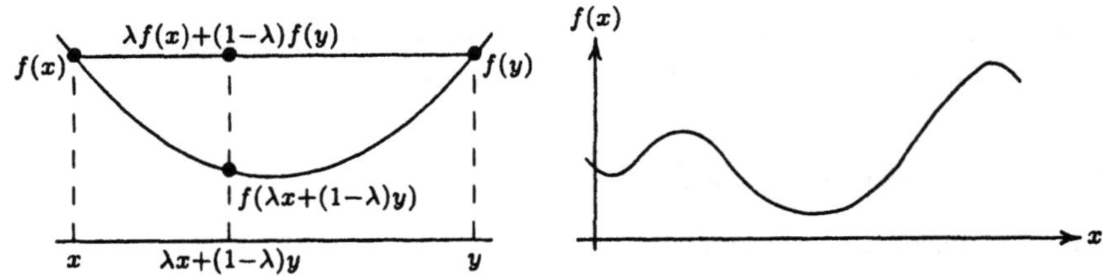

Figure 6-2: Examples of Convex and General Functions

A convex function is illustrated on the left in Figure 6-2. From the figure it is easy to see a geometric interpretation of a convex function. That is, if you take any two points on the graph of a convex function and join them with a chord, then the chord will lie on or above the graph of the function between the points. Note that this is not true for every two points for the function on the right in Figure 6-2.

Convex functions have the following properties.

1. The negative of a convex function is a concave function.

2. A nonnegative linear combination of r convex functions is another convex function. A similar result holds for concave functions.

3. A linear function (sometimes referred to as an *affine function*) is both convex and concave.

4. In calculus, a function in one variable that is twice continuously differentiable is defined to be convex if and only if its second derivative is greater than or equal to zero everywhere. It is defined to be strictly convex if and only if its second derivative is strictly greater than zero everywhere. It is defined to be concave (or strictly concave) if and only if its second derivative is less than or equal to zero (or strictly less than zero) everywhere. Our definitions can be shown to be equivalent in the case that our functions are twice continuously differentiable.

5. In calculus, a function in n variables that is twice continuously differentiable is defined to be convex if and only if its matrix of partial second derivatives, called the Hessian matrix, is positive semidefinite everywhere. It is strictly convex if and only if its Hessian matrix is positive definite everywhere. (A matrix M is positive definite if $x^TMx > 0$ for all $x \neq 0$, see Section A.13.) It is concave (or strictly concave) if its Hessian matrix is negative semidefinite (or negative definite) everywhere. Again our definitions can be shown to be equivalent in the case that our functions are twice continuously differentiable.

▷ **Exercise 6.17** Prove properties 1, 2, and 3.

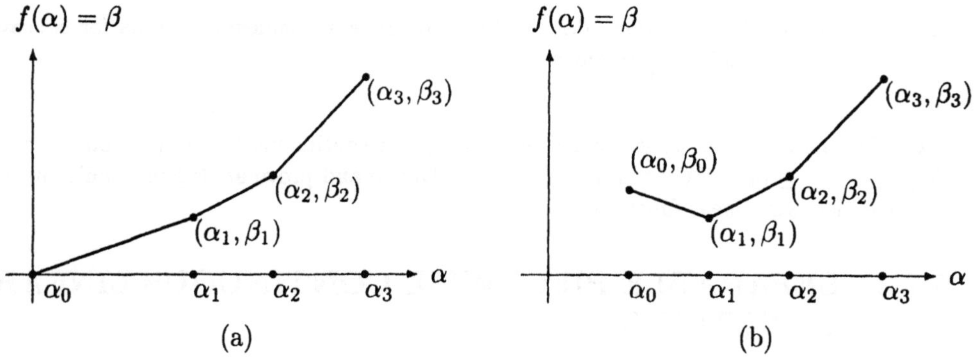

Figure 6-3: Piecewise Continuous Linear Functions

▷ **Exercise 6.18** Prove properties 4 and 5.

6.7.2 PIECEWISE CONTINUOUS LINEAR FUNCTIONS

Definition (Piecewise Continuous Linear): A function $\beta = f(\alpha)$ is said to be ***piecewise continuous linear*** on the interval $\alpha_0 \le \alpha \le \alpha_t$ if and only if the range over which it is defined can be partitioned into successive closed intervals:

$$f(\alpha) = d_i + c_i\alpha, \qquad \text{for } \alpha_i \le \alpha \le \alpha_{i+1},\ i = 0, 1, \ldots, t-1, \tag{6.17}$$

where d_i and c_i are known constants and where we require for continuity,

$$d_i + c_i\alpha_{i+1} = d_{i+1} + c_{i+1}\alpha_{i+1} \text{ for } i = 0, 1, \ldots, t-1. \tag{6.18}$$

Definition (Breakpoints): The points $\alpha_1, \alpha_2, \ldots, \alpha_{t-1}$ are called ***breakpoints***. It is not required that the endpoints α_0 or α_t be finite; thus $\alpha_0 = -\infty$ or $\alpha_t = \infty$ or both are allowed.

An example of a piecewise continuous linear function is the absolute value function illustrated in Figure 6-1, where $\alpha_0 = -\infty$, $\alpha_1 = 0$, $\alpha_2 = \infty$. Other examples of piecewise linear functions are shown in Figure 6-3.

▷ **Exercise 6.19** Draw a function that is piecewise linear but not continuous.

LEMMA 6.2 (Convexity of a Piecewise Continuous Linear Function) *A continuous piecewise linear function is convex if and only if its slope is nondecreasing with respect to α, that is, $c_0 \le c_1 \le \cdots \le c_{t-1}$.*

▷ **Exercise 6.20** Prove Lemma 6.2. Define and prove an analogous lemma for concave piecewise continuous linear functions.

▷ **Exercise 6.21** Show that for a convex piecewise continuous linear function we have $d_0 \geq d_1 \geq \cdots \geq d_{t-1}$. What is true if $\alpha_0 < 0$? Define and prove analogous conditions for concave piecewise linear functions.

6.7.3 SEPARABLE PIECEWISE CONTINUOUS LINEAR FUNCTIONS

Definition (separable piecewise continuous linear function): A real-valued function $f(x)$ defined over $x = (x_1, x_2, \ldots, x_n)^T \in \Re^n$ is said to be a ***separable piecewise continuous linear function*** if and only if it can be written as

$$f(x) = f_1(x_1) + f_2(x_2) + \cdots + f_n(x_n),$$

where each component function $f_j(x_j)$ depends only on one variable, x_j, and is piecewise continuous linear with respect to x_j as described above.

For linear programs in standard form we assume that $f(x) = c^T x$, that is, $f(x)$, the objective function, is separable in the variables x_j and linear with respect to each variable x_j. In this section we generalize the discussion to include functions where the contribution of variable x_j to the cost function $f(x)$ varies in a linear or piecewise continuous linear manner. If all the $f_j(x_j)$ are piecewise continuous linear convex functions in the case we are minimizing, or all piecewise continuous linear concave functions in the case we are maximizing, the problem can be modeled as a linear program. Since a convex function of one variable can be approximated by a piecewise continuous linear convex function, we can use the results developed here to approximate the minimization of a general separable convex function by a linear program.

Suppose that we wish to solve the following problem:

$$\begin{array}{llll} \text{Minimize} & \displaystyle\sum_{j=1}^{n} f_j(x_j) = f(x) & & \\ \text{subject to} & \qquad Ax = b, & A:\ m \times n, & \qquad (6.19) \\ & \qquad\; x \geq 0, & & \end{array}$$

where each $f_j(x_j)$ is either a linear or piecewise continuous linear convex function. Let $f_k(x_k)$ be a piecewise continuous linear convex function of the form

$$f_k(x_k) = d_k^r + c_k^r x_k, \qquad \text{for } \alpha_k^r \leq x_k \leq \alpha_k^{r+1}, \quad r = 0, 1, \ldots, t_k - 1,$$

where, assuming $x_k \geq 0$, we have $\alpha_k^o = 0$, $\alpha_k^{t_k} = \infty$, and $d_k^o = 0$. Note that the continuity assumption requires that $d_k^r \geq d_k^{r+1}$ for $r = 0, 1, \ldots, t_k - 1$, and the convexity assumption requires that $c_k^r \leq c_k^{r+1}$ $r = 0, 1, \ldots, t_k - 1$.

To formulate this problem as a linear program in standard form with upper and lower bounds on the variables, define new variables y_k^r for $r = 1, \ldots, t_k$ such that

$$\begin{aligned} 0 \leq\ & y_k^1 \leq \alpha_k^1, \\ 0 \leq\ & y_k^2 \leq \alpha_k^2 - \alpha_k^1, \\ & \cdots \\ 0 \leq\ & y_k^{t_k-1} \leq \alpha_k^{t_k-1} - \alpha_k^{t_k-2}, \\ 0 \leq\ & y_k^{t_k}. \end{aligned} \tag{6.20}$$

Next, in $Ax = b$, replace each variable x_k by

$$x_k = \sum_{r=1}^{t_k} y_k^r. \tag{6.21}$$

Then $f_k(x_k)$ can be replaced by (6.22) in the objective function in (6.19):

$$f_k(x_k) = \sum_{r=0}^{t_k-1} c_k^r y_k^{r+1} + \gamma_k, \tag{6.22}$$

where

$$\gamma_k = \sum_{r=0}^{t_k-1} d_k^r$$

is a constant. Denoting the optimal values of y_k^r by $\hat{y}_k^r$, we *claim* that substituting the values $\hat{y}_k^r$ into (6.21) and (6.22) will yield the optimal values for x_k and $f_k(x_k)$, because the functions $f_k(x_k)$ are all convex.

▷ **Exercise 6.22** Prove the above claim for the simple case of

$$f(x) = f_1(x_1) + c_2x_2 + c_3x_3 + \cdots + c_nx_n,$$

where $f_1(x_1)$ is continuous and piecewise linear. Extend your arguments to the more general case of $f(x) = \sum_{j=1}^{n} f_j(x_j)$, where each component $f_j(x_j)$ is continuous and piecewise linear. Hint: In order for our claim to hold we need y_k^r to satisfy an additional property at a minimum, namely,

$$y_k^r > 0 \implies y_k^i = \alpha_k^i - \alpha_k^{i-1} \quad \text{for} \quad i < r.$$

Show that this holds when we are minimizing a piecewise continuous linear convex function.

A similar technique can be used when maximizing a separable concave piecewise continuous linear function.

6.8 NOTES & SELECTED BIBLIOGRAPHY

Goal Programming is a special case of solving multiple objectives in general. Additional information on goal programming can be found in Davis & McKeown [1981], Hillier & Lieberman [1995], Ignizio [1976], Lee [1972], Murty [1983], and Steuer [1985]. For applications of goal programming to financial management see, for example, Sponk [1981]. Techniques for formulating linear programs in practice can be found in Bradley, Hax, & Magnanti [1977], Shapiro [1984], and in Tilanus, DeGans, & Lenstra (eds.) [1986].

Curve fitting is used extensively, especially in statistical applications. In statistics, the method described in Section 6.6 is referred to as least-squares or multiple linear regression. Many excellent references are available, for example, Chambers [1977], Feller [1957, 1969], Kennedy & Gentle [1980], and Mood, Graybill, & Boes [1974]. See *Linear Programming 2* for a discussion on the use of the QR factorization to solve the linear least-squares problem.

The proof of Exercise 6.1 is originally due to Tucker (see, Dantzig [1963]). Problem 6.10 on page 165 introduces the concept of a fixed charge problem, see Dantzig & Hirsch [1954]. Problem 6.12 on page 166 briefly introduces the concept of game theory, the mathematical formulation of which can be found in Borel [1921, 1924, 1927], von Neumann [1928, 1937, 1948a, 1953], and in the now famous book *Theory of Games and Economic Behavior* by von Neumann & Morgenstern [1944].

6.9 PROBLEMS

6.1 Consider the following linear program:

$$\begin{array}{lrcrcrcl} \text{Minimize} & -2x_1 & + & x_2 & + & x_3 & = & z \\ \text{subject to} & x_1 & + & x_2 & + & x_3 & = & 4 \\ & 2x_1 & & & - & x_3 & \geq & 3 \\ & \multicolumn{7}{l}{-\infty \leq x_1 \leq \infty,\ x_2 \geq 0, x_3 \geq 0.} \end{array}$$

(a) Replace the variable x_1 by the difference of two nonnegative variables and solve the resulting linear program by hand.

(b) Solve the linear program using the `DTZG Simplex Primal` software option.

6.2 Formulate the following problem as a linear program in standard form:

$$\begin{array}{lrcrcrcrcl} \text{Minimize} & x_1 & - & 3x_2 & + & x_3 & - & x_4 & = & z \\ \text{subject to} & x_1 & - & 2x_2 & - & 3x_3 & + & x_4 & \leq & 10 \\ & 2x_1 & + & x_2 & + & x_3 & - & 2x_4 & \leq & 6 \\ & \multicolumn{9}{l}{x_1,\ x_2 \text{ unrestricted}, x_3 \geq 0,\ x_4 \geq 0.} \end{array}$$

(a) Replace each of the unrestricted variables x_1 and x_2 by the difference of two nonnegative variables and solve the resulting linear program by hand.

(b) Solve the linear program using the `DTZG Simplex Primal` software option.

6.3 Formulate the following problem as a linear program and solve it using the **DTZG Simplex Primal** software option.

$$\begin{array}{lrcl} \text{Minimize} & 2|x_1| + 3x_2 + 4x_3 & = & z \\ \text{subject to} & x_1 + x_2 + 2x_3 & \leq & 8 \\ & 2x_1 - x_2 + x_3 & \geq & 4 \\ & \multicolumn{3}{l}{x_1 \text{ unrestricted}, \ x_2 \geq 0, \ x_3 \geq 0.} \end{array}$$

6.4 Formulate the following problem as a linear program and solve it using the **DTZG Simplex Primal** software option.

$$\begin{array}{lrcl} \text{Minimize} & -6x_1 - 8x_2 - 5x_3 - 4x_4 & = & z \\ \text{subject to} & 3x_1 + 3x_2 + 8x_3 + 2x_4 & = & 50 \\ & |2x_1 + 3x_2 + 2x_3 + 2x_4| & \leq & 20 \\ & \multicolumn{3}{l}{x_j \leq 40, \ \text{for } j = 1, \ldots, 4.} \end{array}$$

What happens if the bounds $x_j \leq 40$ are dropped.

6.5 Where possible, reformulate the following problems as linear programs in standard form, $\min c^T x$, $Ax = b$, $x \geq 0$.

(a)

$$\begin{array}{lrcl} \text{Minimize} & c^T x & = & z, \quad \text{with } c > 0 \\ \text{subject to} & x_j & = & \max(u_j, v_j), \quad j = 1, \ldots, n. \end{array}$$

(b)

$$\begin{array}{lrcl} \text{Minimize} & c^T x & = & z \\ \text{subject to} & A_1 x & \geq & 0 \\ & A_2 x & \leq & 0. \end{array}$$

(c)

$$\begin{array}{lrcl} \text{Minimize} & c^T x & = & z \\ \text{subject to} & Ax & = & b \\ & x & \geq & 0 \\ & \multicolumn{3}{l}{x_j \text{ integer for } j = 1, \ldots, n.} \end{array}$$

(d)

$$\begin{array}{lrcl} \text{Minimize} & c^T x & = & z \\ \text{subject to} & Ax & = & b \\ & x & \geq & 0 \\ & x & \neq & 0. \end{array}$$

6.6 *Excesses and Shortages.* Problems that occur in practice often have costs associated with coming below or exceeding certain requirements. For example, if production of an item is below the contracted amount, a *shortage* cost is incurred either because of the contract stipulation or because the demand is met by having to buy the item from an expensive outside source. On the other hand, if production exceeds the demand, then a holding cost is incurred for having to store the *excess* production.

Suppose that your company manufactures n different items x_j, each of which can be sold at a unit profit of p_j. Assume that the production constraints are $Ax = b$. Except for item 1, assume that all other items, $x_2, x_3, \dots, x_n$, can be sold easily on the open market. Suppose that your contractual agreements are such that you must deliver d_1 units of item 1. If you are unable to deliver exactly d_1 units of item 1, a penalty cost, or shortage cost, of s is incurred. On the other hand, in the hopes of selling item 1 in the next time period, if you produce more than the demand, you incur a holding cost, or excess cost, of e. Develop a linear programming model to maximize profit. Clearly state the conditions under which the linear programming model succeeds in solving this problem.

6.7 Use the **DTZG Simplex Primal** software to find the maximum of the minimum value of x_j for $j = 1, \dots, 3$ that satisfy the system of inequalities

$$\begin{aligned} 2x_1 + x_2 + 2x_3 &\leq 15 \\ x_1 + 3x_2 + 4x_3 &\leq 25 \\ 3x_1 + x_2 + 5x_3 &\leq 30 \\ x_j \geq 0,\ j &= 1, 2, 3. \end{aligned}$$

6.8 Use the **DTZG Simplex Primal** software option to find the maximum of the minimum value of x_j for $j = 1, \dots, 3$ that satisfy the system of inequalities

$$\begin{aligned} 2y_1 + y_2 + 3y_3 &\geq 5 \\ y_1 + 3y_2 + 1y_3 &\geq 10 \\ 2y_1 + 4y_2 + 5y_3 &\geq 20 \\ y_j \geq 0,\ j &= 1, 2, 3. \end{aligned}$$

6.9 *Optical Scanning Problem (Avi-Itzhak, H., Van Mieghen, J., & Rub, L. [1993]).* An optical scanner compares the preprogrammed pattern of each letter, described by a matrix of pixels $50 \times 50 = 2500$ recorded as a 2500 dimensional vector normalized to be of unit length, with that of an unknown letter observed by scanning a matrix of pixels $50 \times 50 = 2500$. The unknown letter described by the vector x will be said to be the same as the preprogrammed letter described by the vector a if the correlation

$$\frac{a^T x}{||x||} \geq k,$$

where k is a constant typically chosen to be greater than say 0.95.

(a) Suppose that you are designing software for a scanner to be used to differentiate between the several kinds of letters of an alphabet and the typefaces of each letter. Then each unknown would need to be compared with several preprogrammed versions of each letter, which unfortunately turns out to take too long to do and requires too much storage. Instead, we would like to use some sort of average representation of each letter in order to reduce the processing time and storage. If we let $a^i = A_{i\bullet}$, $i = 1, \dots, m$, represent m different representations of a given letter, then we might be interested in choosing as the "average" representation of typefaces of a letter as $\beta = \beta^*$, of unit length, that maximizes the minimum correlation:

$$\rho = \max_{\beta} \left\{ \min_{i=1,\dots,m} \frac{A_{i\bullet}\beta}{||\beta||} \right\}, \qquad ||\beta|| = 1. \tag{6.23}$$

Show that this is equivalent to solving the following quadratic program (i.e., an optimization problem where the objective function is a quadratic and the constraints are linear):

$$\begin{array}{lll} \text{Minimize} & y^T y = z & \\ \text{subject to} & A_{i\bullet} y \geq 1, & i = 1, \ldots, m, \end{array} \tag{6.24}$$

and setting $\beta = y/||y||$. Hint: Show that (6.24) is equivalent to finding β that solves

$$\begin{array}{lll} \text{Maximize} & \rho = u & \\ \text{subject to} & \rho \leq \dfrac{A_{i\bullet}\beta}{||\beta||}, & i = 1, \ldots, m. \end{array}$$

(b) If software for solving a quadratic program is unavailable, show how you can apply piecewise linear approximations to solve the quadratic program (6.24) approximately as a bounded variable program. Hint: Show that $y^T y$ is a separable quadratic form.

6.10 *Fixed Charge Problem.* In many practical applications there is the underlying notion of a fixed charge. For example, a refinery may want to know whether building an additional refinery would result in better serving the current set of customers. In such situations the cost is characterized by

$$c = \begin{cases} \alpha x + \beta & \text{if } x > 0, \\ 0 & \text{if } x = 0, \end{cases}$$

where β is the fixed charge.

(a) Show that we may model this by writing the cost form as

$$c = \alpha x + \beta y$$

and including the constraints

$$y = \{0, 1\}$$

and

$$x - yU \leq 0,$$

where U is an upper bound on x.

(b) The inclusion of integer variables in the formulation requires a special-purpose solver. However, if only one or two variables have an associated fixed charge, it is possible to use the `DTZG Simplex Primal` option to solve the problem. Discuss how you would do this.

(c) Consider the product mix problem of Section 1.4.1. Suppose that management has the possibility of a special order on a new desk 5 that requires 12 carpentry hours, 50 finishing hours, and generates a profit of \$60. Unfortunately, it requires some additional equipment for finishing purposes and the cost of this equipment is \$500. Should desk 5 be manufactured. What if the cost of the new equipment was \$1,000.

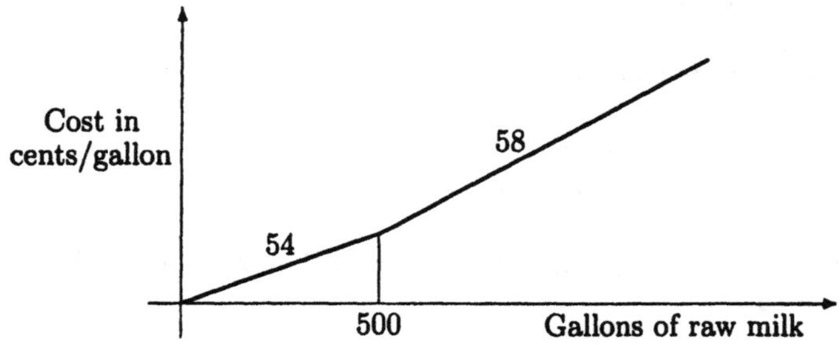

Figure 6-4: Purchase Price of Raw Milk in Region A

6.11 (a) Consider the linear program

$$\begin{array}{lrcl} \text{Minimize} & 1x_1 + 2x_2 + 3x_3 & = & z \\ \text{subject to} & 2x_1 - x_2 + 3x_3 & = & 3 \\ & 0 \leq x_j \leq 2, \text{ for } j = 1,2,3, & & \end{array}$$

where all variables must take integer values. Show how to convert this to a binary integer program where all the variables are either 0 or 1 and the coefficients and right-hand sides are -1, 0, or $+1$.

(b) Generalize to a linear program where all the variables must take integral values and all the coefficients and right-hand side are integers.

6.12 *Game Theory.* In a zero-sum matrix game, the row player, to find his optimal mixed strategy, must solve the linear program

$$\begin{array}{lrcl} \text{Maximize} & & L & \\ \text{subject to} & A^T x & \geq & Le \\ & e^T x & = & 1, \end{array}$$

and the column player, to find her optimal strategy, must solve the linear program

$$\begin{array}{lrcl} \text{Minimize} & & M & \\ \text{subject to} & Ay & \leq & M\hat{e} \\ & \hat{e}^T x & = & 1, \end{array}$$

where A is $m \times n$, $e = (1, 1, \ldots, 1)^T$ is of dimension m, and $\hat{e} = (1, \ldots, 1)^T$ is of dimension n. Prove that these two programs are duals of each other and Max $L =$ Min $M = v$. Note: v is called the value of the game.

6.13 *Ph.D. Comprehensive Exam, September 23, 1972, at Stanford.* Happy Milk Distributors (HMD) purchases raw milk from farmers in two regions: A and B. Prices, butterfat content, and separation properties of the raw milk differ between the two regions. HMD processes the raw milk to produce cream and milk to desired specifications for distribution to the consumers.

Region A Raw Milk. The purchase price in 1972 dollars of raw milk in Region A is indicated in Figure 6-4. For example, to purchase 700 gallons would cost $54(500) + 58(200)$ cents. There is no upper bound on the amount that can be purchased. Raw milk from Region A has 25% butterfat and when separated (at

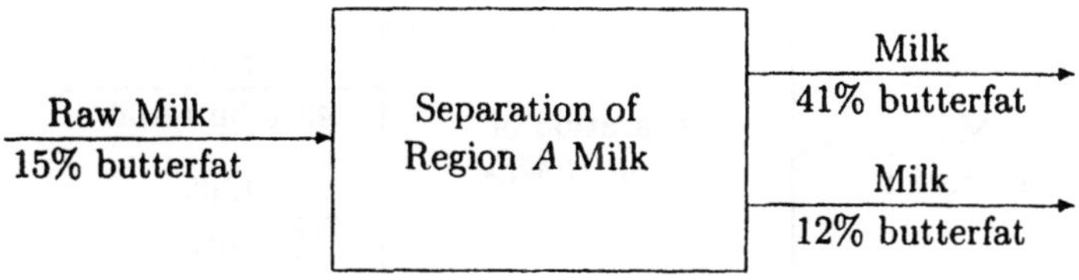

Figure 6-5: Separation of Region A Milk

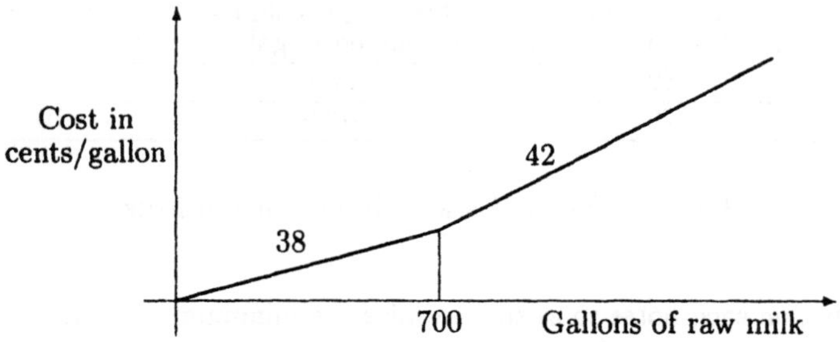

Figure 6-6: Purchase Price of Raw Milk in Region B

5 cents per gallon) yields two "milks," one with 41% butterfat and another with 12% butterfat; this is shown in Figure 6-5. The volume of milk is conserved in *all* separation processing.

Region B Raw Milk. The purchase price in 1972 dollars (as for Region A raw milk) is illustrated in Figure 6-6. Raw milk from Region B has 15% butterfat and when separated (at 7 cents per gallon) yields two "milks," one with 43% butterfat and another with 5% butterfat; this is shown in Figure 6-7. The volume of milk is conserved in *all* separation processing.

Production Process. After the milk is purchased and collected at the plant, it is either mixed directly or separated and then mixed. Mixing, to produce cream and milk to specifications, is done at no cost. For example, some of the raw milk from Region A may be separated and then mixed, and some of it may be mixed directly (i.e., without having been first separated).

Demand and Selling Price. The demand and selling price are described in Table 6-2. For example, all the cream produced must have at least 40% butterfat; it sells at \$1.50 per gallon; no more than 250 gallons of the cream produced can be sold.

The Problem. Assuming disposal is free, formulate a linear program that when solved on a computer would enable HMD to maximize its profits.

6.14 *Vajda [1961].* Suppose that you visit the racecourse with B dollars available for betting. Assume further that there is only one race on this day and that N horses are competing. The betting works as follows: If you bet one dollar on horse k, then you get $\alpha_k > 0$ dollars if it comes in first and 0 dollars otherwise. Formulate a linear program to determine how much of a total of B dollars should

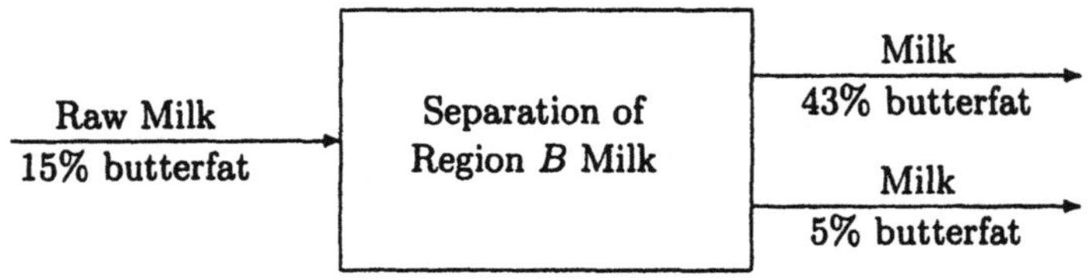

Figure 6-7: Separation of Region B Milk

	Minimum required percentage of butterfat	Maximum volume demanded in gallons	Selling price in cents per gallon
Cream	40	250	150
Milk	20	2000	70

Table 6-2: Demand and Selling Price of Milk

you bet on each horse so as to maximize the minimum net gain, irrespective of which horse comes in first.

6.15 (a) Solve Problem 6.14 for $N = 2$, $\alpha_1 = 1$, $\alpha_2 = 2$, and $B = 1$.

(b) Observe that the bets are inversely proportional to α_1 and α_2. Prove that, in general, the optimal solution is to bet inversely proportional to the return on a bet α_k. Hint: Show that the bets x_j, for $j = 1, \ldots, N$, must be basic variables in an optimal solution and that they price out as optimal.

6.16 Suppose an experiment results in the data points shown in Table 6-3. It is hypothesized that the relation is a cubic of the form

$$y = x_0 + x_1 t + x_2 t^2 + x_3 t^3.$$

Find the parameters x_0, x_1, x_2, x_3 that give the best fit to the data. Solve the problem using each of the three models described in this chapter. (Note: For the least squares problem, simply set up and solve the normal equations $A^T Ax = A^T b$.)

6.17 *Davis & McKeown [1981].* Sigma Paper Company, Inc., is about to build a new plant. The labor requirements to build the plant are 2,000 nonprofessionals and

t	y
1	6
2	18
3	50
4	101
5	177
6	296
7	447
8	642

Table 6-3: Experimental Points for a Cubic

850 professionals. Because of labor market shortages in both categories, costs for recruiting women and minorities are greater than for others. Specifically, the cost for recruiting minorities (men or women) is \$740 for nonprofessionals and \$1,560 for professionals. For recruiting nonprofessional women the cost is \$850 and for recruiting professional women the cost is \$1,450. Otherwise the recruiting costs for men average \$570 for each nonprofessional position and \$1,290 for each professional position. The company has budgeted \$2.4 million for recruiting purposes, and the management has established the following goals in order of priority.

Goal 1: At least 45% of the new labor force should be women.

Goal 2: Minorities should constitute at least 40% of the labor force. Furthermore, a minority woman is counted both as a minority and as a woman employee.

Goal 3: The cost of recruiting should be minimized.

Goal 4: The recruiting cost should be limited to \$300,000.

(a) Use the goal programming technique to formulate a linear programming model for this problem.

(b) Solve the problem using the software provided with the book.

6.18 *Adapted from Hillier & Lieberman [1995].* Consider a preemptive goal programming problem with three priority levels, one goal for each priority level, and just two activities to contribute towards these goals as shown in the table below:

Unit Contribution			
	Activity		
Priority Level	a	b	Goal
One	1	2	≤ 20
Two	1	1	$= 15$
Three	2	1	≥ 40

(a) Use the goal programming technique to formulate a linear programming model for this problem.

(b) Solve the problem using the software provided with the book.

(c) Use the logic of preemptive goal programming to solve the problem graphically by focusing on just the two decision variables. Clearly explain the logic used.

(d) Use the sequential programming technique to solve this problem with the software provided with the book. After using the goal programming technique to formulate the linear programming model at each stage, solve the model graphically by focusing on just the two decision variables. Identify all optimal solutions obtained for each stage.

6.19 Use the weighted goal programming method to solve Problem 6.18.

6.20 Estimate the failure rate x_i of an item as a function of its age i. An experiment is run with K new items put in use at time 0. Time is divided into unit intervals and the number still working at the beginning of each interval is observed. If we let f_i be the number of items observed to be working at the start of time interval i, then $f_i - f_{i+1}$ are the number of items that failed during time interval i

and $a_i = (f_i - f_{i+1})/f_i$ is the observed failure rate at the age of i time units. Suppose that from engineering considerations we know that the failure rate in a large population increases with age; this is the Increasing Failure Rate (IFR) property. The actual failure rates in the experiment with only K items may not be increasing due to random fluctuations in the observations. The solution x to the problem is required to satisfy the IFR property and at the same time be one that is closest to the observed failure rates a in the sense of minimizing the maximum absolute values of the difference. Formulate this problem.

6.21 *Ubhaya [1974a, 1974b] in Murty [1983].* Let $w = (w_1, w_2, \ldots, w_n)^T > 0$, a vector of positive weights, be given. For each vector $y \in \Re^n$, define the function $L(y)$ by

$$L(y) = \max\{\, w_i|y_i| \,:\, i = 1, \ldots, n \,\}.$$

Given a vector $a \in \Re^n$, formulate a linear program to find a vector $x \in \Re^n$ to

$$\begin{array}{ll} \text{Minimize} & L(a - x) \\ \text{subject to} & x_i - x_{i+1} \le 0, \quad i = 1, \ldots, n-1. \end{array}$$

(a) Show that this is a generalization of Problem 6.20, where we have attached weights w_i to the deviations.
(b) How does the formulation change if, in addition, each x_i is required to be nonnegative, and also less than or equal to one?
(c) Solve the problem for $n = 3$, $w = (0.3, 0.6, 0.1)^T$, and $a = (0.20, 0.45, 0.35)^T$.

6.22 Solve the calculus problem

$$\begin{array}{ll} \text{Minimize} & x_1^2 + x_2^2 + x_3^2 + 2x_1 + 4x_2 + 8x_3 = z \\ \text{subject to} & x_1 + x_2 + x_3 = 6 \end{array}$$

by using linear programming software and piecewise continuous linear approximations to the quadratic functions (with upper and lower bounds on the variables introduced to make the approximations). What properties of the approximates ensures the convexity of the approximations.

6.23 Consider the following problem, which has a convex objective function and linear inequality constraints

$$\begin{array}{lrcrcl} \text{Minimize} & x_1^2 & + & 2x_2^2 & = & z \\ \text{subject to} & 4x_1 & + & 5x_2 & \le & 20 \\ & x_1 & + & x_2 & \ge & 2 \end{array}$$

Solve the problem using piecewise continuous linear approximations to the convex function. Use 4, 8, and 12 pieces. Compare the results under the different approximations.

CHAPTER 7

PRICE MECHANISM AND SENSITIVITY ANALYSIS

The term *sensitivity analysis*, sometimes called *post-optimality analysis*, refers to an analysis of the effect on the optimal solution of changes in the input-output coefficients, cost coefficients, and constant terms. Such analysis can often be more important in practice than finding the optimal solution. It is a very important part of solving linear programs in practice. Most practical problems have input data whose values are not known with certainty. For example, the costs of raw materials may change after the model is solved, or the costs used in the model may only be a guess as to what they will be in the future; the right-hand-side constraints may have to be changed because more or less of the resources are available in the market than previously estimated or were earlier authorized by management to buy; or the coefficients may have to be changed because product specifications have been changed.

In addition to the optimal tableau giving us the point of most profitable operation, it is possible to obtain from it a wealth of information concerning a wide range of operations in the neighborhood of this optimum by performing a sensitivity analysis. As noted, in many applications, the information obtained is often more valuable than the specification of the optimum solution itself.

Sensitivity analysis is important for several reasons:

1. Stability (robustness) of the optimum solution under changes of parameters may be highly desirable. For example, using the old optimum solution point; a slight variation of a parameter in one direction may result in a large unfavorable difference in the objective function relative to the new minimum, while a large variation in the parameter in another direction may result in only a small difference. In an industrial situation where there are certain inherent

variabilities in processes and materials not taken account of in the model, it may be desirable to move away from the optimum solution in order to achieve a solution less sensitive to such changes in the input parameters.

2. Values of the input-output coefficients, objective function coefficients, and/or constraint constants may be to some extent *controllable* at some cost; in this case we want to know the effects that would result from changing these values and what the cost would be to make these changes.

3. Even if the input-output and objective function coefficients and constraint constants are *not* controllable, their values may be only approximate; thus it is still important to know for what ranges of their values the solution is still optimal. If it turns out that the optimal solution is extremely sensitive to their values, it may become necessary to seek better estimates.

The problem of finding optimal solutions to linear programs whose coefficients and right-hand sides are uncertain is called *stochastic programming* or *planning under uncertainty.* The idea is to find solutions that hedge against various combinations of contingencies that might arise in the future and at the same time are solutions that are minimal in some expected-value sense.

7.1 THE PRICE MECHANISM OF THE SIMPLEX METHOD

Recall that in previous chapters we often referred to the simplex multipliers as *prices.* In this section we shall show how this viewpoint of multipliers arises naturally and we shall provide an economic interpretation of the Simplex Method. In fact, it will turn out that an important part of sensitivity analysis is being able to interpret the price mechanism of the Simplex Method.

For the discussion we shall once again consider the primal problem in standard form:

$$\begin{array}{lrl} \text{Minimize} & c^T x = z & \\ \text{subject to} & Ax = b, & A:\ m \times n, \\ & x \geq 0; & \end{array} \tag{7.1}$$

and its dual

$$\begin{array}{lrl} \text{Maximize} & b^T \pi = v & \\ \text{subject to} & A^T \pi \leq c, & A:\ m \times n. \end{array} \tag{7.2}$$

For the purpose of our discussion, suppose that an optimal solution is available to the primal-dual system shown above. Let $x^* = (x_B^*, x_N^*)$ be the optimal primal solution, and let π^* be the corresponding dual solution (or multipliers). Let the optimal primal objective value be

$$z^* = c^T x^* = c_B^T x_B^*, \tag{7.3}$$

and the optimal dual objective value be

$$v^* = b^T \pi^*, \tag{7.4}$$

where $v^* = z^*$ and the optimal basic indices are denoted by B. Let $\bar{c} = c - A^T\pi^* \geq 0$ be the optimal reduced costs.

7.1.1 MARGINAL VALUES OR SHADOW PRICES

The *price* of an item is its exchange ratio relative to some standard item. If the item measured by the objective function is taken as a standard, *then the price π_i^* of item i is the change it induces in the optimal value of the objective z per unit change of b_i, for infinitesimally small changes of b_i.*

> *Definition (Price, Marginal Value, Shadow Price)*: The *price*, or *marginal value*, or *shadow price* of a constraint i is defined to be the rate of the change in the objective function as a result of a change in the value of b_i, the right-hand side of constraint i.

At an optimal solution, the primal objective and dual objective are equal. Thus,

$$z^* = v^* = b^T \pi^*$$

is a function of the right-hand sides b, and hence this relation can be used to show that the price associated with b_i at the optimum is π_i^* if the basic solution is *nondegenerate.* If so, then for small changes in any b_i, the basis, and hence the simplex multipliers, will remain constant. Under nondegeneracy the change in value of z per change of b_i for small changes in b_i is obtained by partially differentiating z with respect to the right-hand side b_i, i.e.,

$$\frac{\partial z}{\partial b_i} = \pi_i^*, \tag{7.5}$$

and thus, by the above definition, π_i^* can be interpreted as the price that, when associated with the right-hand side, is referred to as the *shadow price*, a term attributed to Paul Samuelson.

Example 7.1 (Shadow Prices Under Nondegeneracy) Consider the linear program to find min z, $x \geq 0$ such that

$$\begin{array}{rcrcrcrcl}
-x_1 & - & 2x_2 & - & 3x_3 & + & x_4 & = & z \\
x_1 & & & & & + & 2x_4 & = & 6 \\
& & x_2 & & & + & 3x_4 & = & 2 \\
& & & & x_3 & - & x_4 & = & 1.
\end{array}$$

It is easy to verify that the optimal basic solution is

$$x_1 = 6,\ x_2 = 2,\ x_3 = 1,\ x_4 = 0,\ z = -13.$$

The shadow prices associated with the optimal basis (columns 1, 1, 3) are

$$\pi_1^o = -1,\ \pi_2^o = -2,\ \pi_3^o = -3,$$

because

$$z = b_1\pi_1 + b_2\pi_2 + b_3\pi_3, \qquad \text{where } b = (6, 2, 1),$$

implies

$$\frac{\partial z}{\partial b_1} = \pi_1 = -1, \quad \frac{\partial z}{\partial b_2} = \pi_2 = -2, \quad \frac{\partial z}{\partial b_3} = \pi_3 = -3.$$

Note that these do not change for small values of b_i.

If the basic solution is *degenerate*, then the situation is not so straightforward. However, it is easy to see from the above discussion that the objective function is piecewise linear in the parameter b_i. A price interpretation can still be obtained by examining positive and negative changes to b_i. In the case of degeneracy of a single basic variable, a change in a particular b_i will *either* result in no change in the value of the degenerate basic variable, in which case there is no change in the shadow prices, *or* it will result in no change in prices if b_i is changed in one direction, but will result in a complete set of new prices if b_i is changed in the other direction.

In general the two sets of prices can be obtained by applying parametric techniques; see Section 7.8.

Example 7.2 (Shadow Prices Under Degeneracy) Suppose that in Example 7.1, we replace $b_3 = 1$ by $b_3 = 0$; then $\partial z/\partial b_3^+ = -3$ if the change in b_3 is positive. But if the change in b_3 is negative, then x_3 drops out of the basis and x_4 becomes basic and the shadow prices are $\pi_1^o = -1$, $\pi_2^o = -2$, $\pi_3^o = \partial z/\partial b_3^- = -9$.

7.1.2 ECONOMIC INTERPRETATION OF THE SIMPLEX METHOD

When applied to the primal problem (7.1), the goal of the Simplex Method is to determine a basic feasible solution that uses the available resources in the most cost-effective manner. The objective function of the dual (7.2) at iteration t is the total cost:

$$v = \pi^T b = \sum_{i=1}^{m} \pi_i b_i, \tag{7.6}$$

where π_i are the simplex multipliers associated with the basis B. As we will see, π_i (or π_i^* at the optimal solution) can be interpreted as implicit revenues to offset direct costs (i.e., the shadow price) per unit of resource i given the primal basic variables at iteration t. Thus $\pi_i b_i$ can be interpreted as the offsets to the direct costs $\sum_{j=1}^n c_j x_j$ of having b_i units of resource i available in the primal problem.

With this interpretation of the dual variables π and dual objective function v we can examine each row j of the dual problem corresponding to column j of the primal problem. Each unit of activity j in the primal problem consumes a_{ij} units of resource i. Using shadow prices, the left-hand side, $\sum_{i=1}^m a_{ij}\pi_i$, of the jth constraint of the dual problem is the *implicit indirect* cost of the mix of resources that would be consumed and/or produced by one unit of activity j. On the other hand, the right hand side, c_j, of the jth constraint of the dual problem is the *direct* cost of one unit of activity j. That is, the jth dual constraint, $\sum_{i=1}^{m} a_{ij}\pi_i \leq c_j$, can be interpreted as saying that the implicit indirect costs of the resources consumed by activity j less the implicit indirect costs of the resources produced by activity j must not exceed the direct costs c_j; and if these costs are strictly less than c_j, it does not pay to engage in activity j. On the other hand, if $\sum_{i=1}^{m} a_{ij}\pi_i > c_j$, it means that it does pay to engage in activity j.

In other words, the dual constraints associated with the nonbasic variables x_N may be satisfied, feasible, or infeasible. That is, letting

$$\bar{c} = c_N - N^T\pi, \tag{7.7}$$

where π solves $B^T\pi = c_B$, at any iteration t we could have $\bar{c}_j \geq 0$ (the dual constraint j is feasible) or $\bar{c}_j < 0$ (the dual constraint j is infeasible). From an economic standpoint, $\bar{c}_j < 0$ implies that activity j can use its resources more economically than any combination of activities in the basis, whose net input-output vector is the same as activity j, implying an improved solution is possible. If, on the other hand, $\bar{c}_j \geq 0$, then the resources used by activity j are already being used by the activities in the basis in a more cost-effective way elsewhere. *The prices of the dual problem are selected so as to maximize the implicit indirect costs of the resources consumed by all the activities.*

The complementary slackness conditions of dual optimality can be given the following economic interpretation: Whenever an activity j is "operated" at a strictly positive level, i.e., x_j basic and $x_j > 0$, the marginal value of the resources it consumes ($\sum_{i=1}^{m} a_{ij}\pi_i$) per unit level of this activity must exactly equal the cost c_j and all nonbasic activities must "operate" at a zero level.

When the primal-dual system is expressed in the von Neumann symmetric form (see Section 5.1), if a slack variable is strictly positive in an optimal solution, this implies that the corresponding dual variable, π_i, is equal to 0. That is, the resource i for which the primal slack variable is positive is a free resource, i.e., the marginal value of obtaining the resource is zero. If the slack variable corresponding to the difference between the direct and indirect costs of activity j is positive, this implies that the corresponding primal activity level is zero.

7.1.3 THE MANAGER OF A MACHINE TOOL PLANT

The example of this section is based on material supplied by Clopper Almon Jr. to one of the authors. Consider the dilemma of a manager of a machine tool plant, say in an economy that has just been socialized by a revolution (for example, Russia just after the 1917 revolution). The central planners have allocated to this manager *input* quantities $+b_1, \ldots, +b_k$ of materials (which we designate by $1, \ldots, k$) and have instructed this manager to produce *output* quantities $-b_{k+1}, \ldots, -b_m$ of the machine tools (which we designate by $k+1, \ldots, m$). The $b_1, \ldots, b_k$, being inputs, are nonnegative, and $b_{k+1}, \ldots, b_m$, being outputs, are nonpositive by our conventions.

The planners further direct the manager that he shall use as little labor as possible to meet his required production goals and that he must pay the workers with labor certificates as wages, one certificate for each hour of labor. The central planners have declared that the prices of items of the old free market economy are no longer to be used but have not provided any new prices to guide the manager.

The manager has at his disposal many production activities, say n of them, each of which he can describe by a column vector, $A_{\bullet j} = (a_{1j}, \dots, a_{mj})^T$. If the jth process inputs a_{ij} units of the ith item per unit level of operation, a_{ij} is positive. If, on the other hand, the jth process outputs a_{ij} units of item i per unit level of operation, a_{ij} is negative. The jth process also requires c_j units of labor per unit level of operation. The manager's problem then is to find levels of operation for all the processes, $x_1, x_2, \dots, x_n$, that satisfy

$$\begin{array}{ccccccccc} a_{11}x_1 & + & a_{12}x_2 & + & \cdots & + & a_{1n}x_n & = & b_1 \\ a_{21}x_1 & + & a_{22}x_2 & + & \cdots & + & a_{2n}x_n & = & b_2 \\ \vdots & & \vdots & & \vdots & & \vdots & & \vdots \\ a_{m1}x_1 & + & a_{m2}x_2 & + & \cdots & + & a_{mn}x_n & = & b_m \end{array} \tag{7.8}$$

and minimize the total amount of labor used,

$$c_1x_1 + c_2x_2 + \cdots + c_nx_n = z \text{ (min)}.$$

The components of x must, of course, be nonnegative. In matrix notation,

$$Ax = b, \;\; x \geq 0, \;\; c^Tx = z.$$

The manager knows of m old reliable processes (activities), namely $1, \dots, m$, which he is sure he can use in some combination to produce the required outputs from the given inputs. However, the labor requirements may be excessive. Thus, he knows he can find nonnegative $x_1, x_2, \dots, x_m$ such that

$$\begin{array}{ccccccccc} a_{11}x_1 & + & a_{12}x_2 & + & \cdots & + & a_{1m}x_m & = & b_1 \\ a_{21}x_1 & + & a_{22}x_2 & + & \cdots & + & a_{2m}x_m & = & b_2 \\ \vdots & & \vdots & & \vdots & & \vdots & & \vdots \\ a_{m1}x_1 & + & a_{m2}x_2 & + & \cdots & + & a_{mm}x_m & = & b_m \end{array} \tag{7.9}$$

or in matrix notation,

$$Bx_B = b.$$

We shall assume that B is a feasible basis for (7.8).

This manager knows, alas, that his workers are prone to be extravagant with materials, using far more inputs or producing far fewer outputs for each activity j than that specified in (7.8). Unless he can keep this tendency in check, he knows he will fail to meet his quotas, and he knows that failure to meet quotas is dangerous to his health. Before the revolution, he deducted the cost of the excess use of materials from the worker's wages; but now that all prices have been swept away, he lacks a monetary measure for computing the cost of materials or the value of items produced. Suddenly, in a stroke of genius, it occurs to him that he can make up his own prices in terms of labor certificates, charge the operators of each process for the materials they use, credit them for their products, and give them the difference as their pay. (This sounds like a pretty good idea, it might even work in our free market economy.)

Being a fair man, he wants to set prices such that the efficient workers can take home a certificate for each hour worked. That is, he wants to set the prices, $\pi_1, \pi_2, \ldots, \pi_m$, for raw material and products produced such that the net yield on a unit level of each basic activity j is equal to the amount of labor c_j that it requires:

$$\begin{array}{ccccccccc} \pi_1 a_{11} & + & \pi_2 a_{12} & + & \cdots & + & \pi_m a_{1m} & = & c_1 \\ \pi_1 a_{21} & + & \pi_2 a_{22} & + & \cdots & + & \pi_m a_{2m} & = & c_2 \\ \vdots & & \vdots & & \vdots & & \vdots & & \vdots \\ \pi_1 a_{m1} & + & \pi_2 a_{m2} & + & \cdots & + & \pi_m a_{mm} & = & c_m \end{array} \tag{7.10}$$

or in matrix notation,

$$B^T \pi = c_B.$$

The manager now uses his PC to solve (7.9) for x_B and verify that $x_B \geq 0$, and to solve (7.10) for π. Common sense tells the manager that by using his pricing device he would have to pay out exactly as many labor certificates as he would if he paid the labor by the hour and all labor worked efficiently. Indeed, this is easily proved since the total cost, using his calculated prices for basic activities, is, noting $x_N = 0$,

$$\pi^T b = \pi^T (B x_B) = (\pi^T B) x_B = c_B^T x_B = c^T x,$$

where $c^T x$ is the cost of paying wages directly.

The manager harbors one qualm about his pricing device, however. He remembers that there are other processes besides the m he is planning to use and suspects that his lazy, wily workers will try to substitute one or more of these in place of some combination of the basic ones that he is planning to use, and he could end up issuing more labor certificate hours than actual hours worked. In order to discover whether such activities exist he "prices out" each activity j, that is,

$$c'_j = \pi_1 a_{1j} + \pi_2 a_{2j} + \cdots + \pi_m a_{mj}, \tag{7.11}$$

and sees whether any of them is greater than the direct wages c_j. On looking over the list of processes in (7.8), the manager finds several for which the inequality $c'_j > c_j$ holds. Denoting the excess wages of the jth process by $\bar{c}_j$, the manager determines

$$\bar{c}_j = c_j - (\pi_1 a_{1j} + \pi_2 a_{2j} + \cdots + \pi_m a_{mj})$$

and singles out process s, the one offering the most excess wages:

$$\bar{c}_s = \min_j \bar{c}_j < 0. \tag{7.12}$$

Before devising repressive measures to keep the workers from using processes that yield excess wages, the manager, a meditative sort of fellow, who worries about his health, pauses to reflect on the meaning of these excess wages. Having always had a bent for mathematics, he soon discovers a relation that mathematically we express by saying that the vector of coefficients $\bar{a}_{ij}$, for any activity j in the canonical form, can be interpreted as weights to form a linear combination of the *original* vectors of the basic activities that has the *same* net input and output coefficients for all items as those of activity j, except possibly the cost coefficient c_j. In particular, he finds that he can represent activity s, the one yielding the most excess wages, as a linear combination of his "old reliable" activities as follows:

$$\begin{pmatrix} a_{11} \\ a_{21} \\ \vdots \\ a_{m1} \end{pmatrix} \bar{a}_{1s} + \begin{pmatrix} a_{12} \\ a_{22} \\ \vdots \\ a_{m2} \end{pmatrix} \bar{a}_{2s} + \cdots + \begin{pmatrix} a_{1m} \\ a_{2m} \\ \vdots \\ a_{mm} \end{pmatrix} \bar{a}_{ms} = \begin{pmatrix} a_{1s} \\ a_{2s} \\ \vdots \\ a_{ms} \end{pmatrix}, \tag{7.13}$$

where $\bar{a}_{is}$ are the coefficients of x_s in the canonical form. In words, (7.13) tells him that x_s units of activity s can be *simulated* by a combination of the basic set of activities $(1, \ldots, m)$; i.e., by $\bar{a}_{1s}x_s, \bar{a}_{2s}x_s, \ldots, \bar{a}_{ms}x_s$ units. Thus, if the workers introduce x_s units of activity s, the levels of the basic activities must be *adjusted* by these amounts (up or down, depending on sign) if the material constraints and output quotas are to remain satisfied. Now, the labor cost of simulating one unit of activity s by the m old reliables is

$$c_1\bar{a}_{1s} + c_2\bar{a}_{2s} + \cdots + c_m\bar{a}_{ms}.$$

This amount is precisely what the manager would pay for the various inputs and outputs of one unit of the real activity s if he were to use the prices π_i. For considering the vector equation (7.13) as m equations and multiplying the first equation through by π_1, the second by π_2, etc. and summing, one obtains immediately from (7.10)

$$c_1\bar{a}_{1s} + c_2\bar{a}_{2s} + \cdots + c_m\bar{a}_{ms} = \pi_1 a_{1s} + \pi_2 a_{2s} + \cdots + \pi_m a_{ms}. \tag{7.14}$$

It is now readily shown that the fact that the process s yields excess wages means to the manager that it takes less labor to operate s directly than to simulate it with some combination of the m old activities. This is clear from (7.11), (7.12), and (7.14), which yield

$$c_1\bar{a}_{1s} + \cdots + c_r\bar{a}_{rs} + \cdots + c_m\bar{a}_{ms} > c_s. \tag{7.15}$$

Hence, he reasons, s must be in a sense more efficient than at least one of these old processes. Recalling that the central planners instructed him to use as little labor as possible, the manager decides to use activity s in place of one of the original m. He soon discovers that if he wishes to avoid the nonsensical situation of planning to use some activity at a negative level, the process r to be replaced by process s must be chosen so that

$$\frac{\bar{b}_r}{\bar{a}_{rs}} = \min_{\{i|\bar{a}_{is}>0\}} \frac{\bar{b}_i}{\bar{a}_{is}}, \qquad (\bar{a}_{rs} > 0).$$

Because $\bar{a}_{rs} > 0$, it follows from (7.15) that

$$\frac{c_1\bar{a}_{1s} + \cdots + c_{r-1}\bar{a}_{r-1,s} + c_{r+1}\bar{a}_{r+1,s} \cdots + c_m\bar{a}_{ms} - c_s}{-\bar{a}_{rs}} < c_r. \tag{7.16}$$

The coefficients $\bar{a}_{js}/(-\bar{a}_{rs})$ of c_j for $j = 1, \ldots, r-1, r+1, \ldots, m$ and the coefficient $-1/(-\bar{a}_{rs})$ of c_s in (7.16) are precisely the weights required to simulate activity r out of the activities in the new basis, as can be seen by re-solving (7.13) for column r in terms of the others. But, for example, $c_1\bar{a}_{1s}/(-\bar{a}_{rs})$ is the labor cost of the first activity in the simulation of activity r, so that the left-hand side of (7.16) represents the total labor cost of simulating a unit level of activity r by the activities in the new basis, while c_r is the labor cost of one unit of the real activity r. Hence (7.16) shows that activity r is indeed less efficient in its use of labor than those in the new basis.

In summary, the manager now knows that if there exist processes for which his pricing device yields more labor certificates than are actually required, then he had better substitute one of these more efficient processes for one in the original set and thereby bring about a more efficient use of labor. Since the central planners instructed him to use as little labor as possible, it is clearly wise for him to plan production using activity s instead of one of the m he had originally intended to use, to readjust the levels of use of the remaining ones, and to change the prices.

Having learned this lesson, the manager proceeds again to look for more efficient processes that provide his wily workers with excess wages by being more efficient than one in his new basis. If there are any, he makes the substitution, readjusts prices, and again looks for more efficient processes, and so on until he finds a set of prices π^* under which no process prices out as more efficient. Fortunately for him, it turns out (as we know) that in a finite number of steps he will find such a set of prices.

Let us pause for a moment to ponder the meaning of one of these prices, say π_i. Suppose we introduce into the manager's A matrix in equation (7.8) a fictitious activity that consists simply in increasing his allotment of item i if $b_i > 0$ or in decreasing his production quota on i if $b_i < 0$. Such an activity will be represented by a column that has all zeros except for a one in the ith row. Thus the labor cost of simulating this activity by a combination of those of the final basis is, by (7.11), precisely π_i. Thus π_i is the *labor value*, the labor that can be displaced by one additional unit of item i.

The manager has now achieved his objective of finding a set of prices to charge for raw materials and to pay for finished goods that will keep his workers from wasting inputs and yet offer no possibilities of his paying out more labor certificates than actual hours worked. But he now begins to wonder whether he is truly safe from the central planners' criticism for the amount of labor he uses. He begins by specifying explicitly what he intends to do. His operating plan consists of a set of activity levels $x^* = (x_1^*, x_2^*, \dots, x_n^*)^T$ satisfying

$$\begin{aligned} c^T x^* &= z^*, \\ Ax^* &= b, \\ x^* &\geq 0 \end{aligned} \tag{7.17}$$

and a set of prices $\pi^* = (\pi_1^*, \pi_2^*, \dots, \pi_m^*)^T$ satisfying $B^T\pi^* = c_B$ with the property that

$$\bar{c}_j = c_j - \sum_{i=1}^{m} \pi_i^* a_{ij} > 0 \quad \Longrightarrow \quad x_j^* = 0. \tag{7.18}$$

We shall now prove that the manager's operating plan has minimized his labor costs. Writing $\bar{c} = c - A^T\pi^*$, we have from (7.17) that

$$\bar{c}^T x^* = (c - A^T\pi^*)^T x^* = z^* - b^T\pi^*, \tag{7.19}$$

where, by (7.17), z^* is the total labor requirement of the manager's operating plan. But because of (7.18), $\bar{c}^T x^* = 0$, and therefore

$$z^* = b^T\pi^*. \tag{7.20}$$

Now let $x = (x_1, x_2, \dots, x_n)^T$ be *any other feasible operating plan*, and let z be *its labor requirements*; then

$$\begin{aligned} c^T x &= z, \\ Ax &= b, \\ x &\geq 0. \end{aligned} \tag{7.21}$$

It follows by multiplying $Ax = b$ by π^*, subtracting from $c^T x = z$, and noting (7.20), that

$$(c - A^T\pi^*)^T x = z - b^T\pi^* = z - z^*, \qquad \text{or} \qquad \sum_{j=1}^{n} \bar{c}_j x_j = z - z^*. \tag{7.22}$$

But the left member is the sum of nonnegative terms and therefore $z \geq z^*$. Hence, no other feasible operating plan exists whose labor requirement is less than the one found by the manager.

At this point, we can imagine the manager's delight at his genius, for *as a by-product of his search for prices that will cause his workers to work efficiently, he has also discovered those processes that minimize his labor requirements.* Without explicitly trying, he has solved his assigned task of keeping his use of labor to a *minimum*!

Let us review the requirements satisfied by the prices found by the manager. *First*, there will be no excess wages in any activity; that is,

$$A^T\pi \leq c. \tag{7.23}$$

Second, the total amount of wages to be paid for all activities should be the same whether they are paid directly or by use of the pricing device; that is

$$z^* = c^T x^* = b^T \pi, \tag{7.24}$$

where x^* is an optimal solution to (7.8).

Let us now show that these prices π themselves represent the optimal solution to another linear programming problem—specifically, to the dual problem of our manager's original production problem. By multiplying the jth equation of (7.23) by x_j^* and summing, we find that

$$\pi_1 \sum_{j=1}^{n} a_{1j} x_j^* + \pi_2 \sum_{j=1}^{n} a_{2j} x_j^* + \cdots + \pi_m \sum_{j=1}^{n} a_{mj} x_j^* \leq \sum_{j=1}^{n} c_j x_j^*. \tag{7.25}$$

Substituting from (7.8)

$$b_i = \sum_{j=1}^{n} a_{1j} x_j^*, \qquad i = 1, \ldots, m, \tag{7.26}$$

gives

$$\pi_1 b_1 + \pi_2 b_2 + \cdots + \pi_m b_m \leq c_1 x_1^o + c_2 x_2^o + \cdots + c_n x_n^o.$$

Thus, $\pi^T b \leq c^T x^*$ for any π that satisfies (7.23). The prices π^* found by the manager give $b^T \pi^* = c^T x^*$, and thus π^* maximizes $b^T \pi$, subject to the constraints (7.23). Hence, $\pi = \pi^*$ may be viewed as an optimal solution to the *dual* linear programming problem, namely,

$$\begin{aligned} b^T \pi &= v \text{ (max)}, \\ A^T \pi &\leq c. \end{aligned} \tag{7.27}$$

The relation $b^T \pi^* = c^T x^*$, where π^* is an optimal solution to (7.27) and x^* an optimal solution to (7.8), agrees with the *Strong Duality Theorem* (see Theorem 5.3).

▷ **Exercise 7.1** Interpret the economic meaning of maximizing $b^T \pi$ in the case of the tool plant manager.

7.1.4 THE AMBITIOUS INDUSTRIALIST

In this section we shall present a situation where the primal formulation and the dual formulation both are problems that a planner would like to have solved. Consider a defense plant that has just been built for the government. The plant has been designed to produce certain definite amounts $-b_i$, $i = k+1, \ldots, m$, of certain defense items and to use only certain definite amounts, $+b_i$, $i = 1, \ldots, k$, of certain scarce materials that will be *provided without cost* by other government plants. The consulting engineers who designed the plant provided the government with a list of the various processes available in the plant and their input and output coefficients. Somewhat confused by this mass of data, the civil servants who were supposed to operate the plant decide to call in a private industrialist to consult on how they should plan their production. The industrialist realizes that it would be good training for his men and a feather in his cap if he could contract to actually operate the plant. Accordingly, once he gets the information and studies the data, he proposes a flat fee for which he will manage the plant, turn over to the government the required amounts of output, and use no more than the allotted quantities of the scarce materials. The civil service men declare that all other things being equal, they think it would be best for the government to operate the plant, but if he can convince them that his proposal is a good one (meaning that if the government operates the plant, it is unlikely it could do so less expensively), they will accept his offer.

The industrialist takes the data back to his office, gets out his linear programming book titled *Linear Programming 1: Introduction*, and uses input-output coefficient data to form a matrix A and a labor cost vector c.

To determine the minimum fee for which he can afford to operate the defense plant, the industrialist has only to solve the following linear program:

$$\begin{aligned} c^T x &= z \text{ (min)}, \\ Ax &= b, \\ x &\geq 0. \end{aligned} \tag{7.28}$$

Using the software provided, he quickly solves the problem on his PC and prints out the results using his printer. The results are that z^* is the minimum cost and x^* is the vector of optimal process utilization levels. His first thought is to explain the linear programming technique to his civil service friends, show them the final tableau, and thereby convince them that they can do no better than to accept his offer and pay him z^*. But then he realizes that this plan will give away his secret; the civil servants will have no further need for him. They will take his vector of operating levels x^* to optimally operate the plant themselves. To prevent this from happening, he must find a way to convince the government that z^* is minimal without giving away his plan x^*.

To this end, he decides to invent a system of prices that he will offer to pay for the materials, provided he is paid certain prices for the outputs. He wants these prices to be such that there are no profits on any individual activity, for if there were profits, the government would spot them and know that they could find a way to run the plant with lower labor cost. On the other hand, given these restraints, he wants to make as much money as possible. That is, he wants his price vector π to satisfy

$$\begin{aligned} \pi^T b &= v \text{ (max)}, \\ A^T \pi &\leq c. \end{aligned} \tag{7.29}$$

He recognizes this problem as the dual of the one he just solved and immediately produces the dual solution: optimal $\pi = \pi^*$, the simplex multipliers from the last iteration of the Simplex Method used to solve the primal problem, and maximal $v = v^*$. Fortunately, he notes with relief, $v^* = z^*$.

With these results under his arm, the industrialist goes back to see the civil servants and presents his offer in price terms. The bureaucrats check to be sure that every one of the inequalities (7.29) is satisfied, and, of course, calculate the total cost using these prices: $b^T\pi^* = v^*$. The industrialist then invites them to consider any program x satisfying (7.28). Its cost to them, if they operate the plant themselves, is c^Tx. But replacing π by π^* in (7.29) and multiplying both sides by any *feasible* x yields

$$(\pi^*)^T Ax \le c^T x \tag{7.30}$$

or, by (7.28),

$$(\pi^*)^T b \le c^T x. \tag{7.31}$$

Hence,

$$v^* \le c^T x, \tag{7.32}$$

so that the cost of any feasible program that the bureaucrats could devise will be at least v^*. This argument convinces the civil servants that they can do no better than to accept the industrialist's flat fee offer of v^*. With one last hope of operating the plant themselves, they try to pry out of him just how much of each process he intends to operate; but he feigns ignorance of such details and is soon happily on his way with a signed contract in his pocket.

7.1.5 SIGN CONVENTION ON PRICES

Economists use the sign convention that the flow of items produced by an activity are positive and the flow of items consumed by an activity are negative. They also assume the value of the cost item has a price of unity and that costs (money paid out) are being minimized. With this in mind, let us introduce into the linear program a fictitious "procurement" activity $j = n + i$ that increases the allotment of item i; its coefficients are zero except for unity in row i and c_{n+i} in the cost row.

How low must the cost c_{n+i} be before it pays to increase the allotment of item i? Pricing out this activity, we see that it pays if

$$c_{n+i} < \pi_i.$$

Hence, π_i is the break-even cost of the item i procurement activity.

If an item is produced by an activity $a_{ij} > 0$ and if the item has value, then *the flow of money* $\pi_i a_{ij} > 0$ *is toward the activity.* Similarly, for each unit of activity j, the input $a_{ij} < 0$ would induce a payment of $\pi_i a_{ij} < 0$. In other words, *the flow of money is out.*

The total flow of money into the activity by the price device is given by pricing it out, that is,

$$\sum_{i=1}^{m} \pi_i a_{ij}.$$

If this value exceeds c_j, the amount that would be received by the alternative of direct payment, then this activity (or some other with the same property) will be used in lieu of a basic activity now in use. This in turn will generate a new set of prices, etc.

Basic Variables	Admissible Variables (Including Slacks)							Constants
	$-z$	x_1	x_2	x_3	x_4	x_5	x_6	
$-z$	1	−12	−20	−18	−40	0	0	0
x_5	0	4	9	7	10	1	0	6
x_6	0	1	1	3	40	0	1	4

Table 7-1: Initial Tableau for Sensitivity Analysis

7.2 INTRODUCING A NEW VARIABLE

After an optimal solution $x = x^*$ has been determined, suppose that we want to examine the effect on the optimal solution of introducing a new variable x_{n+1} with cost coefficient c_{n+1} and input-output coefficients $A_{\bullet n+1}$.

LP IN STANDARD FORM

The augmented problem is then

$$\begin{array}{lrcl} \text{Minimize} & c^T x + c_{n+1}x_{n+1} & = & z \\ \text{subject to} & Ax + A_{\bullet n+1}x_{n+1} & = & b \\ & x & \geq & 0 \\ & x_{n+1} & \geq & 0. \end{array} \tag{7.33}$$

Introducing a new variable into (7.1) as shown in (7.33) does not alter feasibility since we can make it nonbasic and set its value to be at its lower bound of 0. Thus, the current solution is still feasible. However, the solution need not be optimal since the reduced cost $\bar{c}_{n+1}$ corresponding to the new variable x_{n+1} may be negative. In order to check for optimality we compute

$$\bar{c}_{n+1} = c_{n+1} - A_{\bullet n+1}^T \pi^*. \tag{7.34}$$

If $\bar{c}_{n+1} \geq 0$, the old solution with $x_{n+1} = 0$ is optimal. If, on the other hand, $\bar{c}_{n+1} < 0$, then we know that we can improve on the solution by bringing the variable x_{n+1} into the basis. It may be necessary to perform a series of pivot steps before optimality is regained.

Example 7.3 (A New Column Introduction) Consider the product mix problem as stated in Section 1.4.1:

$$\begin{array}{rcrcrcrcrcrcll} -12x_1 & - & 20x_2 & - & 18x_3 & - & 40x_4 & & & & & = z\ (\text{min}) & \\ 4x_1 & + & 9x_2 & + & 7x_3 & + & 10x_4 & + & x_5 & & & = 6 & (\text{carpentry}) \\ x_1 & + & x_2 & + & 3x_3 & + & 40x_4 & & & + & x_6 & = 4 & (\text{finishing}) \end{array} \tag{7.35}$$

This is shown in simplex tableau form in Table 7-1. Since this is already in canonical form, addition of artificial variables is unnecessary, and we can proceed directly to Phase II of

Basic Variables	Admissible Variables (Including Slacks)							Constants
	$-z$	x_1	x_2	x_3	x_4	x_5	x_6	
$-z$	1	0	20/3	10/3	0	44/15	4/15	56/3
x_1	0	1	7/3	5/3	0	4/15	−1/15	4/3
x_4	0	0	−1/30	1/30	1	−1/150	2/75	1/15

Table 7-2: Optimal Tableau for Sensitivity Analysis

the Simplex Method. After several iterations we arrive at the optimum solution as shown in the final tableau in Table 7-2.

From the information contained in this tableau (Table 7-2) we see that the optimum product mix for the problem as stated in thousands of units is at the rate of 4/3 thousand desks of type 1 and 1/15 thousand desks of type 4 per time period for a total profit rate of $z = \$56/3$ thousand per period.

Suppose a new desk called Type 7 has been designed that will require 6 man-hours of carpentry shop time and 2 man hours of finishing department labor per desk. Based on an estimated profit of $18 per desk, we would like to know whether it would pay to produce this desk.

Note that the negatives of the values of the simplex multipliers, 1, 44/15, 4/15, for the last iteration can be obtained from the top row vector of the inverse of the final basis (columns corresponding to $(-z)$, x_5, and x_6). This yields, after pricing out,

$$\bar{c}_7 = -18 + \frac{44}{15}(6) + \frac{4}{15}(2) = \frac{2}{15}.$$

Since $\bar{c}_7 > 0$, it does *not* pay to produce this desk.

▷ **Exercise 7.2** If the economic sign convention of Section 7.1.5 is followed, show that the signs of the bottom two equations of (7.35) are reversed and the optimal prices are $\pi_1 = 44/15$ and $\pi_2 = 4/15$, corresponding to the top row of the inverse in Table 7-1.

BOUNDED VARIABLE LP

Next, let us consider the more general case when x_{n+1} has upper and lower bounds specified, i.e., $l_{n+1} \leq x_{n+1} \leq u_{n+1}$, where the lower bound is not necessarily 0 and the upper bound is not necessarily ∞. In this case, besides regaining optimality, we also have to be concerned with feasibility. There are three cases to consider:

1. The lower bound $l_{n+1} = -\infty$ and the upper bound $u_{n+1} = \infty$. In this case $x_{n+1} = 0$ provides a feasible solution. However, the solution (if not degenerate) is not optimal if $\bar{c}_{n+1} \neq 0$ because it is profitable to increase x_{n+1} if $\bar{c}_{n+1} < 0$ and to decrease x_{n+1} if $\bar{c}_{n+1} > 0$.

2. The lower and upper bounds are both finite. We first look at easy cases such as setting set x_{n+1} at its lower bound and checking to see whether the solution found by adjusting the basic variables is feasible. If not, we try again setting

x_{n+1} at its upper bound. If this also does not work, we check to see whether 0 is bracketed by the bounds, in which case we set $x_{n+1} = 0$ to obtain a feasible solution. See Exercise 7.3.

If a feasible solution is not obtained at either bound of x_{n+1} or at $x_{n+1} = 0$, we set $x_{n+1} = 0$ and perform Phase I of the Simplex Method by creating a Phase I objective and a modified problem for which this solution is feasible. If $u_{n+1} < 0$, we set the Phase I objective function to minimize $w = x_{n+1}$ and temporarily set $u_{n+1} = \infty$. If $0 < l_{n+1}$, we set the Phase I objective function to be $w = -x_{n+1}$ and temporarily set $l_{n+1} = -\infty$. If by checking at each iteration of the Simplex Algorithm to see whether x_{n+1} is feasible with respect to the original bounds a feasible solution is found, we terminate Phase I, reset the original bounds, and continue with Phase II. If Phase I terminates optimal with $w > 0$, we report the original problem as infeasible. This method is called *minimizing the sum of infeasibilities.*

If a feasible solution is obtained, we compute the value of $\bar{c}_{n+1}$ using equation (7.34). If $x_{n+1} = l_{n+1}$ and $\bar{c}_{n+1} \geq 0$ the solution is optimal, else if $\bar{c}_{n+1} < 0$, we perform one or more iterations of the Simplex Method. If $x_{n+1} = u_{n+1}$ and $\bar{c}_{n+1} \leq 0$ the solution is optimal, else if $\bar{c}_{n+1} > 0$, we perform one or more iterations of the Simplex Method. If $x_{n+1} = 0$ is feasible and $\bar{c}_{n+1} = 0$ the solution is optimal, else if $\bar{c}_{n+1} \neq 0$, we perform one or more iterations of the Simplex Method.

3. Either the lower bound or the upper bound is infinite but not both; see Exercise 7.4.

▷ **Exercise 7.3** Show how we can take advantage of the sign of $\bar{c}_{n+1}$ to possibly reduce the number of computations in Case 2 above.

▷ **Exercise 7.4** For Case 3 above, complete the detailed steps for the case where either the lower bound or the upper bound is infinite but not both.

7.3 INTRODUCING A NEW CONSTRAINT

We assume that the constraint being introduced is of one of the following forms:

$$\begin{aligned} A_{m+1\bullet}x &= b_{m+1}, \\ A_{m+1\bullet}x &\leq b_{m+1}, \\ A_{m+1\bullet}x &\geq b_{m+1}, \end{aligned} \tag{7.36}$$

where $b_{m+1} \geq 0$. By setting different bounds on the slack variable x_{n+1} we can write any of the constraints (7.36) in the form

$$A_{m+1\bullet}x + x_{n+1} = b_{m+1}. \tag{7.37}$$

Thus if the constraint reads "$= b_{m+1}$," then $0 \le x_{n+1} \le 0$, if it reads "$\le b_{m+1}$," then $0 \le x_{n+1} \le \infty$, and if it reads "$\ge b_{m+1}$," then $-\infty \le x_{n+1} \le 0$. The cost c_{n+1} associated with x_{n+1} is zero.

The current solution $x = x^*$ is feasible, optimal, and satisfies $Ax^* = b$ for the original problem. If it also satisfies the added constraint (7.37) with $x_{n+1} = x^*_{n+1}$, the solution (x^*, x^*_{n+1}) is also optimal for the augmented problem, as can be easily seen by keeping the old optimal prices and setting the price on the extra constraint to be equal to zero.

But what if the solution $x = x^*$ is infeasible after the addition of the constraint? That is, for the equality constraint, $x_{n+1} \ne 0$; for the $\le$ constraint, $x_{n+1} < 0$; or for the $\ge$ constraint, $x_{n+1} > 0$. Then we can solve the infeasibility problem by minimizing a Phase I objective, and a modified problem is constructed as follows.

1. For the "$= b_{m+1}$" case, if $x^*_{n+1} > 0$, then set $w = x_{n+1}$ and set $u_{n+1} = \infty$. On the other hand, if $x^*_{n+1} < 0$, then set $w = -x_{n+1}$ and set $l_{n+1} = -\infty$.

2. For the "$\le b_{m+1}$" case, if $x^*_{n+1} < 0$, then set $w = -x_{n+1}$ and set $l_{n+1} = -\infty$.

3. For the "$\ge b_{m+1}$" case, if $x^*_{n+1} > 0$, then set $w = x_{n+1}$ and set $u_{n+1} - \infty$.

If by checking at each iteration of the Simplex Algorithm to see whether x_{n+1} is feasible with respect to the original bounds a feasible solution is found, we terminate Phase I, reset the original bounds, and continue with Phase II. If Phase I terminates optimal with $w > 0$, we report the original problem as infeasible. This method is called *minimizing the sum of infeasibilities*.

The new multipliers (π, π_{m+1}) can be easily computed from

$$\begin{pmatrix} B & 0 \\ v_B^T & 1 \end{pmatrix}^T \begin{pmatrix} \pi \\ \pi_{m+1} \end{pmatrix} = \begin{pmatrix} 0 \\ \theta \end{pmatrix} \quad \text{or} \quad \begin{pmatrix} B^T & v_B \\ 0 & 1 \end{pmatrix} \begin{pmatrix} \pi \\ \pi_{m+1} \end{pmatrix} = \begin{pmatrix} 0 \\ \theta \end{pmatrix} \tag{7.38}$$

where $\theta = 1$ if $w = x_{n+1}$ and $\theta = -1$ if $w = -x_{n+1}$, and, letting $j_1, j_2, \ldots, j_m$ be indices of the variables in the optimal basis B of the original problem, $v_B = (a_{m+1,j_1}, a_{m+1,j_2}, \ldots, a_{m+1,j_m})^T$. It is easy to see that the solution is given by

$$\pi_{m+1} = \theta \quad \text{and} \quad \pi = \theta (B^{-1})^T v_B. \tag{7.39}$$

The reduced costs can be easily obtained from

$$\begin{aligned} \bar{d} &= d_N - \begin{pmatrix} N^T & v_N \end{pmatrix} \begin{pmatrix} \pi \\ \pi_{m+1} \end{pmatrix} \\ &= -\begin{pmatrix} N^T & v_N \end{pmatrix} \begin{pmatrix} \pi \\ \pi_{m+1} \end{pmatrix}, \end{aligned} \tag{7.40}$$

where $v_N = (a_{m+1,j_{m+1}}, \; a_{m+1,j_{m+2}}, \; \ldots, \; a_{m+1,j_n})^T$ and $j_{m+1}, \ldots, j_n$ are the indices of the nonbasic variables.

▷ **Exercise 7.5** Show that

$$\begin{pmatrix} B & 0 \\ v^T & 1 \end{pmatrix}^{-1} = \begin{pmatrix} B^{-1} & 0 \\ -v^T B^{-1} & 1 \end{pmatrix}.$$

Derive the inverse form by using Equation (A.21) on Page 330.

7.4 COST RANGING

Often costs are uncertain in industrial applications. For example, in the oil industry, the spot prices for purchasing crude oils for making gasoline may have changed after the linear program has been run, and we are concerned with whether the changes in price are sufficient to require us to rerun the program. It is clear that changing the costs does not affect feasibility, and thus we only need to be concerned with optimality. Cost ranging refers to the determination of the range of costs for a variable x_j such that the solution stays optimal. We are going to consider only two cases: the effect of changing the cost of one nonbasic variable and the effect of changing the cost for one of the basic variables.

THE EFFECT OF NONBASIC-VARIABLE COST CHANGE

Consider the nonbasic variable x_s with cost c_s. For optimality to hold we need the reduced costs to be nonnegative. That is, we need

$$c_s \geq A_{\bullet s}^T \pi^*. \tag{7.41}$$

Thus, the range of costs c_s for the nonbasic variable x_s over which optimality of the basis B is preserved is $c_s \geq A_{\bullet j}^T \pi^*$.

Example 7.4 (Change in Nonbasic Cost) In Example 7.3 on page 184, how much would the profit for desk Type 7 have to change before it becomes worthwhile to produce it? It follows from (7.41) that if the cost coefficient c_j for any nonbasic activity is decreased by the value of its relative cost factor $\bar{c}_j$ in the optimal solution, it becomes a candidate to enter the basis. In this case, desk Type 7 becomes a candidate for production if its *profit* per unit can be increased by $\bar{c}_7 = 2/15$.

▷ **Exercise 7.6** How much must the profit on desk Type 2 be increased to bring it into an optimum solution? How much would you have to raise the selling price on desk Type 3 in order to make its production optimal? How would you modify the model if as the result of a market survey you have determined that the amount that can be sold is a function of selling price and that there is an upper bound on the amount that can be sold?

THE EFFECT OF BASIC-VARIABLE COST CHANGE

The determination of the range of costs over which optimality of the basis is preserved is slightly more complicated in this case. For convenience, let us suppose that the cost of optimal basic activity j_r is changed from c_{j_r} to γ and we are interested in how much γ must increase or decrease before the basis B is no longer optimal. The basic cost vector c_B changes to

$$c_B(\gamma) = c_B + (\gamma - c_{j_r})e_r, \tag{7.42}$$

where e_r is the rth column of the identity matrix. Then the new multipliers π are now a function of γ, which we write as $\pi(\gamma)$. By definition, $B^T\pi(\gamma) = c_B(\gamma)$, that is,

$$B^T\pi(\gamma) = c_B + (\gamma - c_{j_r})e_r. \tag{7.43}$$

Then for the current solution to remain optimal, we need

$$\tilde{c}_j(\gamma) = c_j - \pi(\gamma)^T A_{\bullet j} \geq 0 \qquad \text{for all nonbasic } j. \tag{7.44}$$

The inequalities (7.44) are all linear in γ, because multiplying (7.43) on the left by $(B^T)^{-1}$, and noting that $(B^T)^{-1}) = (B^{-1})^T$, we have

$$\pi(\gamma) = \pi^* + (\gamma - c_{j_r})(B^{-1})^T e_r. \tag{7.45}$$

From these we can compute a range of values on γ that maintain optimality of the solution.

Therefore, the new reduced costs are given by

$$\begin{aligned}
\tilde{c}_j(\gamma) &= c_j - A_{\bullet j}^T\pi(\gamma) = c_j - A_{\bullet j}^T\pi^* - (\gamma - c_{j_r})A_{\bullet j}^T(B^{-1})^T e_r \\
&= c_j - A_{\bullet j}^T\pi^* - (\gamma - c_{j_r})(B^{-1}A_{\bullet j})^T e_r \\
&= \bar{c}_j - (\gamma - c_{j_r})\bar{A}_{\bullet j}^T e_r \\
&= \bar{c}_j - (\gamma - c_{j_r})\bar{a}_{rj}.
\end{aligned} \tag{7.46}$$

As long as $\tilde{c}_j(\gamma) \geq 0$ for all nonbasic j, the solution will remain optimal. The range of γ for which the solution is optimal can be computed by setting $\tilde{c}_j(\gamma) \geq 0$ for each nonbasic j and solving for γ. Thus, the range on γ is given by

$$c_{j_r} + \max_{\bar{a}_{rj}<0} \frac{\bar{c}_j}{\bar{a}_{rj}} \leq \gamma \leq c_{j_r} + \min_{\bar{a}_{rj}>0} \frac{\bar{c}_j}{\bar{a}_{rj}}. \tag{7.47}$$

As long as γ lies in the range defined by (7.47), the current basis B remains optimal; however, the optimal value of the objective function changes according to

$$z(\gamma) = c_B^T(\gamma)x_B^* = c_B^T x_B^* + (\gamma - c_{j_r})e_r^T x_B^* = z^* + (\gamma - c_{j_r})x_{j_r}, \tag{7.48}$$

where x_B^* is the current optimal solution.

Example 7.5 (Change in Basic Cost) For what range of cost of a particular basic activity $j_r = 1$ in the desk production Example 7.3 does the present optimal solution still remain optimal? The present solution will remain optimal as long as cost coefficient $\gamma = c_1$ satisfies Equation (7.47). Hence, based on (7.47) we use Table 7-2 to first compute

$$\min_{\bar{a}_{rj}>0} \frac{\bar{c}_j}{\bar{a}_{rj}} = \min\left\{\frac{\bar{c}_2}{\bar{a}_{12}}, \frac{\bar{c}_3}{\bar{a}_{13}}, \frac{\bar{c}_5}{\bar{a}_{15}}\right\} = \left\{\frac{20/3}{7/3}, \frac{10/3}{5/3}, \frac{44/15}{4/15}\right\} = 2,$$
$$\max_{\bar{a}_{rj}<0} \frac{\bar{c}_j}{\bar{a}_{rj}} = \max\left\{\frac{\bar{c}_6}{\bar{a}_{16}}\right\} = \left\{\frac{4/15}{-1/15}\right\} = -4.$$

Thus from (7.47) we see that the present solution remains optimal for c_1 in the range $-12 - 4 \leq c_1 \leq -12 + 2$, where -12 was the original value of c_1.

▷ **Exercise 7.7** For what range of profit for desk Type 4 is the present solution (see Table 7-2) still optimal? Determine what activity enters the solution if c_1 is decreased to -20, increased to $-19/2$. What activity leaves the solution in each case?

▷ **Exercise 7.8** Construct an example by changing b_1 in the original problem in Table 7-1 to show that if the profit for desk Type 1 is decreased to $19/2$, i.e., c_1 is increased to $-19/2$, then desk Type 1 is not the one that leaves the solution.

▷ **Exercise 7.9** Given an initial optimal basic feasible solution, prove the theorem that reducing the cost of a basic activity will not necessarily cause it to be dropped from the optimal solution. Verify this by showing, in the example of Exercise 7.7, that no matter how much c_1 is decreased, activity $j = 1$ will still remain in the basis.

▷ **Exercise 7.10** Modify the cost ranging analysis for the case when x has both upper and lower bounds associated with it.

7.5 CHANGES IN THE RIGHT-HAND SIDE

It is important to be able to examine the effect of changes to the right-hand side of the constraints, especially those that specify what resources are available. For example, the availability of a resource may be constrained as a result of a company's policy decision to place a quota on imports of an item from a foreign company.

Assume that we have an optimal solution to a linear program in standard form. Let the right-hand side of the rth constraint be given by a parameter θ with a particular value of θ being the current b_r. Let

$$b(\theta) = b + (\theta - b_r)e_r, \tag{7.49}$$

where e_r is the rth column of the identity matrix. If the value of θ does not make the current basis infeasible, then the revised basic solution is still optimal. However, if it is infeasible, this will result in a change of basis requiring that the new problem be solved by a Phase I / Phase II procedure or by the Dual-Simplex Method. (The Dual-Simplex Method is a pivot procedure that uses different rules as to which column leaves the basis and which enters the basis. It turns out in reality to be a clever way to solve the dual of the original problem by the Simplex Method without having to transpose the matrix.) The range of values θ for the right hand side of the rth constraint, for which the current basis remains feasible and thus optimal, is obtained by ensuring that the solution $x_B(\theta)$ to

$$Bx_B(\theta) = b(\theta) \tag{7.50}$$

is such that $x_B(\theta) \geq 0$. From (7.49) and (7.50) we get

$$x_B(\theta) = B^{-1}b + (\theta - b_r)B^{-1}e_r = x_B^* + (\theta - b_r)B^{-1}e_r, \tag{7.51}$$

where x_B^* is the optimal solution with $\theta = b_r$. For feasibility we need $x_B(\theta) \geq 0$, thus the range of θ is

$$b_r + \max_{\beta_{ir}>0} \frac{-(x_B^*)_i}{\beta_{ir}} \leq \theta \leq b_r + \min_{\beta_{ir}<0} \frac{-(x_B^*)_i}{\beta_{ir}}, \tag{7.52}$$

where β_{ir} is element (i,r) of B^{-1}.

As long as θ lies in the range defined by (7.52), the current basis B remains optimal; however, the optimal values of the optimal basic solution (7.51) change, and the optimal value of the objective function changes according to

$$z(\theta) = c_B^T x_B(\theta) = c_B^T x_B^* + (\theta - b_r)c_B^T Be_r = z^* + (\theta - b_r)\pi_r^*, \tag{7.53}$$

where z^* is the current optimal objective value and π^* are the simplex multipliers of the current optimal solution.

▷ **Exercise 7.11** Derive the relation (7.53) alternatively by noting that $z(\theta) = b(\theta)^T\pi^*$.

Example 7.6 (Change in Capacity) What is the effect of increasing finishing department capacity in Example 7.3? The present solution will remain optimal as long as the right-hand side $\theta = b_2$ satisfies (7.52). Hence, based on (7.52) we use Table 7-2 to first compute

$$\max_{\beta_{ir}>0} \frac{-(x_B^*)_i}{\beta_{ir}} = \max\left\{\frac{-x_4}{\beta_{22}}\right\} = \left\{\frac{-1/15}{2/75}\right\} = -2.5,$$

$$\min_{\beta_{ir}<0} \frac{-(x_B^*)_i}{\beta_{ir}} = \min\left\{\frac{-x_1}{\beta_{12}}\right\} = \left\{\frac{-4/3}{-1/15}\right\} = 20.$$

Thus from (7.52), we see that the present solution remains feasible and optimal for b_2 in the range $4 - 2.5 \leq b_2 \leq 4 + 20$, or $1.5 \leq b_2 \leq 24$, where 20 was the original value of b_2. Thus, the answer to our question is that we can increase finishing department capacity up to 20,000 hours. The net profit per hour of increase is $-\pi_2^* = \$4/15$.

▷ **Exercise 7.12** If finishing department capacity in Example 7.3 has been increased by 20,000+ϵ hours, where ϵ is infinitesimally small, what is the resulting product mix? Which basic activity has dropped out of the solution?

▷ **Exercise 7.13** In Example 7.3, equipment needed to increase the carpentry capacity by 10% can be rented for $5,000. Also, overtime hours up to 20% of the rated capacity of either carpentry or finishing can be obtained at a premium of $1.50 per hour. Above this figure, the premium is estimated to be about $3.00 per hour because of loss of productive efficiency. Is it better to rent or to use overtime with increased capacity.

▷ **Exercise 7.14** Show that if a slack variable corresponding to b_i is basic in the optimal solution with value $\bar{x}_{n+i}$, then the corresponding constraint constant $b_i = b_i^o$ in the initial tableau can take on any value $b_i \geq b_i^o - \bar{x}_{j_i}$, with no change in the values of the objective function or the other basic variables in the optimal solution. In this range, is b_i actually constraining the solution?

▷ **Exercise 7.15** Modify the analysis of this section for changing the right hand side for the case when x has both upper and lower bounds associated with it.

7.6 CHANGES IN THE COEFFICIENT MATRIX

When blending scrap metal, it may turn out that the actual percentage of, say, chromium in the scrap is not according to specifications in the jth column in the coefficient matrix, and it is important to determine the extent to which the variation in chromium would affect the optimal blend. In general, when the coefficient matrix is large, it is not practical to examine the effects of changes in the coefficients by solving the problem again from scratch. We are going to assume that we have an optimal solution to a linear program in standard form, and we will consider only two cases: the effect of changes in a coefficient in a nonbasic column and the effect of changes in a coefficient in a basic column.

THE EFFECT OF NONBASIC-COLUMN COEFFICIENT CHANGES

Changing the coefficients of nonbasic columns does not affect the feasibility, only optimality because the nonbasic variables are all set at their lower bound of 0. Let j be a nonbasic column in the current optimal solution and let a_{kj} be replaced by a parameter α_{kj} subject to change. This α_{kj} has current value a_{kj}. The solution stays optimal provided the reduced cost for column j is nonnegative. Thus, we wish to determine the range of values for α_{kj} for which

$$c_j - \big(A_{\bullet j} + (\alpha_{kj} - a_{kj})e_k\big)^T \pi^* \geq 0, \tag{7.54}$$

where π^* are the simplex multipliers of the current optimal solution. Hence

$$\pi_k^* \alpha_{kj} \leq \bar{c}_j + \pi_k^* a_{kj}. \tag{7.55}$$

Therefore, the current optimal solution remains optimal as long as (7.55) holds. If $\pi_k^* = 0$, the value of α_{kj} does not affect optimality. Indeed, if $\pi_k^* = 0$, the optimal solution is not affected by the change of the coefficient in row k of any nonbasic column j.

▷ **Exercise 7.16** Modify the above analysis for the case when a_{ij} is changed for a nonbasic column j whose corresponding variable x_j has both upper and lower bounds associated with it.

Example 7.7 (Change in Nonbasic Coefficient) Recall in Example 7.3 on page 184, a new desk Type 7 with coefficients $c_7 = -18$, $a_{17} = 6$, and $a_{27} = 2$ was introduced and found to be unprofitable. How much would the carpentry shop labor requirement of $a_{17} = 6$ for desk Type 7 have to change for it to become more profitable to produce? Replacing the original value of $a_{17} = 6$ by the parameter α_{17} in the $\bar{c}_j$ calculation, we have

$$\bar{c}_7 = c_7 - \pi_1^* \alpha_{17} - \pi_2^* a_{27} = -18 + \frac{44}{15}\alpha_{17} + \frac{4}{15}\alpha_{27} = \frac{44}{15}\alpha_{17} - \frac{262}{15}, \tag{7.56}$$

where $\pi_1^* = -44/15$, $\pi_2^* = -4/15$ are the current optimal prices from Table 7-2. In order for activity $j = 7$ to become a candidate to enter the optimal basis, we require $\bar{c}_7 \leq 0$ or $\alpha_{17} \leq 131/22 = 5\frac{21}{22}$. Therefore, the carpentry shop labor requirement for desk Type 7 must drop by at least $6 - 5\frac{21}{22} = \frac{1}{22}$.

▷ **Exercise 7.17** In Example 7.3, to what value would the carpentry shop hours for desk Type 2 have to be reduced to make it competitive?

Example 7.8 (Basis Change When Nonbasic Coefficients Change) Suppose that we are not really sure of either the labor requirements *or* profit for desk Type 2. We would like to determine a formula for these parameters that may be used to determine whether it pays to change the production mix and produce some desks of Type 2. For activity $j = 2$ to become a candidate to enter the solution, its coefficients c_2, a_{12}, a_{22} must satisfy

$$\begin{aligned} \bar{c}_2 &= c_2 - \pi_1^* a_{12} - \pi_{22}^* a_{22} \\ &= c_2 + \frac{44}{15}a_{12} + \frac{4}{15}a_{22} \leq 0, \end{aligned} \tag{7.57}$$

where $\pi_1^* = -44/15$, $\pi_2^* = -4/15$ are the current optimal prices from Table 7-2. If, for example, it turns out that $a_{12} = 8$, $a_{22} = 2$, $c_2 = -25$, then substitution of these values into (7.57) gives $\bar{c}_2 = -1$, which implies that it pays to produce some desks of Type 2. In the general case, the formula is given by Equation (7.57).

EFFECT OF BASIC-COLUMN COEFFICIENT CHANGES

Changing the coefficients of basic columns can affect both the feasibility and the optimality. For convenience we relabel the basic columns to be $i = 1, \ldots, m$. Let i be a basic column in the current optimal solution x^*, let a_{ki} be replaced by a parameter α_{ki} subject to change, and let the modified optimal basis B be denoted by

$$\bar{B} = B(\alpha_{ki}).$$

In particular, $B(a_{ki}) = B$. The modified optimal basic feasible solution will be denoted by $x_B(\alpha_{ki})$ (or equivalently, by $x_{\bar{B}}$) and the modified multipliers will be denoted by $\pi(\alpha_{ki})$. Our goal is to determine the range of α_{ki} over which the basis $\bar{B}$ remains optimal, i.e., $x_{\bar{B}} \geq 0$, $x_N = 0$, and $\bar{c} \geq 0$. We will first determine this range for the following example.

Example 7.9 (Change in a Basic Coefficient) What happens if the carpentry shop requirement for desk Type 1 in Example 7.3 changes? To evaluate the effect of varying a_{11} we replace a_{11} by α_{11}. First we note that the current optimal basis and its inverse from Tables 7-1 and 7-2 are

$$B = \begin{pmatrix} 4 & 10 \\ 1 & 40 \end{pmatrix} \quad \text{and} \quad B^{-1} = \frac{1}{150} \begin{pmatrix} 40 & -10 \\ -1 & 4 \end{pmatrix}. \tag{7.58}$$

Replacing B_{11} by α_{11} note that

$$\bar{B} = \begin{pmatrix} \alpha_{11} & 10 \\ 1 & 40 \end{pmatrix} \quad \text{and} \quad \bar{B}^{-1} = \frac{1}{40\alpha_{11} - 10} \begin{pmatrix} 40 & -10 \\ -1 & \alpha_{11} \end{pmatrix}. \tag{7.59}$$

The formula for $\bar{B}^{-1}$ can be obtained by using Exercise A.13 on page 330.

▷ **Exercise 7.18** Verify that $\bar{B}^{-1}\bar{B} = I$.

Note that in order for $\bar{B}$ to serve as a basis, $\bar{B}^{-1}$ must exist. In this case this means that $1/(40\alpha_{11} - 10) \neq 0$, or $\alpha_{11} \neq 1/4$. We will first determine the range of α_{11} over which the adjusted basic solution is primal feasible. Thus, α_{11} must be such that

$$x_{\bar{B}} = \bar{B}^{-1}b = \bar{B}^{-1} \begin{pmatrix} 6 \\ 4 \end{pmatrix} = \frac{1}{40\alpha_{11} - 10} \begin{pmatrix} 200 \\ -6 + 4\alpha_{11} \end{pmatrix} \geq 0. \tag{7.60}$$

This implies that $40\alpha_{11} - 10 \geq 0$, or $\alpha_{11} \geq 1/4$ and $-6 + 4\alpha_{11} \geq 0$, or $\alpha \geq 6/4$. Together these imply that for primal feasibility we require

$$\alpha_{11} \geq 6/4. \tag{7.61}$$

Next we determine the range of α_{11} over which the simplex multipliers π are dual feasible. This requires ensuring that the reduced costs $\bar{c}_j$, as a function of α_{11}, remain nonnegative for all the nonbasic variables j. We first compute the new multipliers as a function of α_{11}, i.e., $\pi(\alpha_{11})$ as

$$\begin{aligned} \pi(\alpha_{11}) &= (\bar{B}^T)^{-1}c_B = \frac{1}{40\alpha_{11} - 10} \begin{pmatrix} 40 & -1 \\ -10 & \alpha_{11} \end{pmatrix} \begin{pmatrix} -12 \\ -40 \end{pmatrix} \\ &= \frac{4}{4\alpha_{11} - 1} \begin{pmatrix} -11 \\ 3 - \alpha_{11} \end{pmatrix}. \end{aligned} \tag{7.62}$$

Using these multipliers the reduced costs can be computed as

$$\bar{c}_N = \begin{pmatrix} c_2 \\ c_3 \\ c_5 \\ c_6 \end{pmatrix} - N^T\pi(\alpha_{11}) = \begin{pmatrix} -20 \\ -18 \\ 0 \\ 0 \end{pmatrix} + \frac{4}{4\alpha_{11}-1}\begin{pmatrix} 96+\alpha_{11} \\ 68+3\alpha_{11} \\ 11 \\ -3+\alpha_{11} \end{pmatrix} \geq 0. \tag{7.63}$$

Recalling that $40a_{11} - 10 > 0$ by (7.60), we have by multiplying by $4a_{11} - 1$ that the range of α_{11} over which the simplex multipliers stay dual feasible is given by

$$3 \leq \alpha_{11} \leq \frac{29}{6}. \tag{7.64}$$

Since this range happens also to preserve primal feasibility, see (7.61), we conclude that as long as $3 \leq \alpha_{11} \leq 29/6$, the adjusted primal basic solution will remain optimal.

▷ **Exercise 7.19** Suppose in Example 7.3, that the carpentry shop hours needed to manufacture desk Type 1 should have been $4\frac{1}{2}$ hours instead of 4 hours per desk. How would the solution change? What would be the effect if it turned out to be 5 hours instead of 4 hours per desk?

▷ **Exercise 7.20** What is the effect on the optimal solution and the value of the objective of simultaneous changes to carpentry shop requirements, finishing department requirements, and profit for desk Type 1 in Example 7.3?

Next we derive formulae for ranges on α_{ki} that preserve both primal and dual feasibility. In order to compute this range, we note that $a_{ki} = B_{ki}$ and

$$\bar{B} = B + (\alpha_{ki} - B_{ki})e_k e_i^T = B\big(I + (\alpha_{ki} - B_{ki})B^{-1}e_k e_i^T\big), \tag{7.65}$$

where e_r is the rth column of the identity matrix for $r = i, k$. The inverse of the new basis $\bar{B}$ is given by (see Equation (A.21) on Page 330)

$$\bar{B}^{-1} = \left(I - \frac{(\alpha_{ki} - B_{ki})}{1 + (\alpha_{ki} - B_{ki})\beta_{ik}} B^{-1}e_k e_i^T\right)B^{-1}, \tag{7.66}$$

providing $1 + (\alpha_{ki} - B_{ki})\beta_{ik} \neq 0$, where β_{ik} is element (i, k) of B^{-1}.

▷ **Exercise 7.21** In (7.65) and (7.66) multiply $\bar{B}^{-1}$ by $\bar{B}$ and verify that $\bar{B}^{-1}\bar{B} = I$.

Letting

$$\phi = \frac{1}{B_{ik}^{-1} + 1/(\alpha_{ki} - B_{ki})}, \tag{7.67}$$

we note that $\bar{B}^{-1}$ can be written as

$$\bar{B}^{-1} = \big(I - \phi B^{-1}e_k e_i^T\big)B^{-1}. \tag{7.68}$$

1. *Primal Feasibility.* For primal feasibility we need $x_B(\alpha_{ki}) \geq 0$ where we have $\bar{B}x_B(\alpha_{ki}) = b$ and $x_B(a_{ki}) = x_B^*$. Then for feasibility we need to restrict α_{ki} so that
$$x_B(\alpha_{ki}) = \bar{B}^{-1}b \geq 0. \tag{7.69}$$
Substituting (7.68) into (7.69), we obtain
$$\begin{aligned} x_B(\alpha_{ki}) &= \bar{B}^{-1}b = B^{-1}b - \phi B^{-1}e_k e_i^T B^{-1}b \\ &= x_B^* - \phi B^{-1}e_k e_i^T x_B^* \\ &= x_B^* - \phi [x_B^*]_i B_{\bullet k}^{-1} \\ &\geq 0. \end{aligned} \tag{7.70}$$
This then provides the following range on ϕ:
$$\max_{\{q \in \mathcal{B} \mid [x_B^*]_i \beta_{qk} < 0\}} \frac{x_q^*}{[x_B^*]_i \beta_{qk}} \leq \phi \leq \min_{\{q \in \mathcal{B} \mid [x_B^*]_i \beta_{qk} > 0\}} \frac{x_q^*}{[x_B^*]_i \beta_{qk}}, \tag{7.71}$$
where β_{qk} is element (q, k) of B^{-1}.

2. *Dual Feasibility.* In addition to primal feasibility we also need dual feasibility. Hence we restrict α_{ki} so that for all nonbasic j
$$\bar{c}_j(\alpha_{ki}) = c_j - A_{\bullet j}^T \pi(\alpha_{ki}) \geq 0, \tag{7.72}$$
where $\pi(\alpha_{ki})$, a function of α_{ki}, is computed from $\bar{B}^T\pi(\alpha_{ki}) = c_B$ and $\bar{B}$ is a function of α_{ki}.

   Next we determine the range on ϕ (and thus α_{ki}) for the current solution to remain dual feasible. The new multipliers $\pi(\alpha_{ki})$ are determined by:
$$\begin{aligned} \pi(\alpha_{ki}) &= (\bar{B}^{-1})^T c_B = (B^{-1})^T c_B - \phi (B^{-1})^T e_i e_k^T (B^{-1})^T c_B \\ &= \pi^* - \phi (B^{-1})^T e_i e_k^T \pi^* \\ &= \pi^* - \phi \pi_k^* (B^{-1})^T e_i. \end{aligned} \tag{7.73}$$
Using the above, the new reduced costs $\bar{c}_N(\alpha_{ki})$, as a function of α_{ki}, are
$$\begin{aligned} \bar{c}_N(\alpha_{ki}) &= c_N - N^T\pi(\alpha_{ki}) = c_N - N^T\pi^* + \phi\pi_i^*(B^{-1}N)^T e_k \\ &= \bar{c}_N - \phi\pi_i^*(\bar{A}_{k\bullet})^T. \end{aligned} \tag{7.74}$$
For dual feasibility we need $\bar{c}_N(\alpha_{ki}) \geq 0$, which provides the following range on ϕ:
$$\max_{\{j \in \mathcal{N} \mid \pi_i^* \bar{A}_{kj} < 0\}} \frac{\bar{c}_j}{\pi_i^* \bar{A}_{kj}} \leq \phi \leq \min_{\{j \in \mathcal{N} \mid \pi_i^* \bar{A}_{kj} > 0\}} \frac{\bar{c}_j}{\pi_i^* \bar{A}_{kj}}. \tag{7.75}$$

Relations (7.71) and (7.75) together determine the range of values of α_{ki} that maintain feasibility and optimality of the current solution.

▷ **Exercise 7.22** First, from (7.71) derive the range on α_{ki} for primal feasibility. Next, from (7.75) derive the range on α_{ki} for dual feasibility. Use these two ranges to derive a formula for the range on α_{ki} that preserves optimality of the current basis.

If the value of α_{ki} changes so that the basis is no longer primal and dual feasible for some α_{ki}, there are three cases to consider:

1. If for a particular value of α_{ki}, the primal solution is not feasible but the dual solution is feasible, we can apply the Dual-Simplex Method.
2. If for a particular value of α_{ki}, the primal solution is feasible but not optimal, we can apply the primal Simplex Method.
3. If for a particular value of α_{ki}, both the primal solution and dual solution are feasible, we revert back to Phase I of the primal Simplex Algorithm.

▷ **Exercise 7.23** Modify the above analysis, for the case when a_{ki} is changed for a basic column i when the variables x have both upper and lower bounds associated with it.

▷ **Exercise 7.24** Prove that although x_B and $\bar{c}_N$ are nonlinear functions of $\alpha_{kj} - a_{kj}$, the ranges of α_{jk} are determined by simple linear inequalities in ϕ, which in turn implies that the range of values of ϕ are determined by simple linear inequalities in α_{kj}.

▷ **Exercise 7.25** Under what conditions can the value of an input-output coefficient for a basic activity be changed without any effect on the optimal value of the objective?

In the course of the discussion above, we have proved the following:

THEOREM 7.1 (Change in Basic Coefficient) *Let basis element $a_{ki} = B_{ki}$ be changed to α_{ki}, and let the (i,k) element of B^{-1} be denoted by β_{ik}. For small changes $\alpha_{ki} - a_{ki}$, the changes to x_B, π, and $\bar{c}$ are nonlinear in $\alpha_{ki} - a_{ki}$ but are linear in*

$$\phi = \frac{\alpha_{ki} - a_{ki}}{1 + (\alpha_{ki} - a_{ki})\beta_{ik}}. \tag{7.76}$$

▷ **Exercise 7.26** Show that ϕ is a monotonically increasing function of $\alpha_{ki} - a_{ki}$:

$$\phi = \frac{1}{\beta_{ik} + 1/(\alpha_{ki} - a_{ki})}. \tag{7.77}$$

▷ **Exercise 7.27** In Theorem 7.1, under what conditions would the changes to x_B, π, and $\bar{c}$ be linear in α_{ki}.

THEOREM 7.2 (Convexity of Variation of $B_{\bullet j}$) *Given a general linear program, the domain of all possible variations of a column $B_{\bullet j}$ for which the given basis remains optimal is convex in the space of the components of $B_{\bullet j}$.*

▷ **Exercise 7.28** Prove Theorem 7.2.

7.7 THE SUBSTITUTION EFFECT OF NONBASIC ACTIVITIES ON BASIC ACTIVITIES

The Simplex Method tells us how many units of the entering activity $j = s$ can be brought into the solution and what will be its effect upon the quantities of the other basic activities, namely,

$$\max x_s = \min_{\{\,i \mid \bar{a}_{is}>0\,\}} \left(\frac{\bar{b}_i}{\bar{a}_{is}} \right). \tag{7.78}$$

The rows of an optimal tableau (see, for example, Table 7-2) express the values of the basic variables in terms of the nonbasic variables, while a column in the optimal tableau expresses a nonbasic activity in terms of the set of basic activities. Thus, in the latter case, the coefficients of the jth column vector, i.e., $\bar{a}_{ij}$, can be interpreted as the amounts of basic columns, j_i, to be used as "substitution" factors; see Equation (7.13) in the discussion of the price mechanism of the Simplex Method.

Observe that the relative cost factor $\bar{c}_j$ for each variable can be calculated from the "substitution" factors $\bar{a}_{ij}$ and the original cost coefficients c_j, namely,

$$\bar{c}_j = c_j - \sum_{i=1}^{m} \bar{a}_{ij} c_{j_i} \tag{7.79}$$

where j_i is the index of the ith basic activity. This is an alternative way to do the "pricing out" calculations $\bar{c}_j = c_j - \sum_{i=1}^{m} \pi_i a_{ij_i}$.

▷ **Exercise 7.29** Prove that $\sum_{i=1}^{m} \bar{a}_{ij} c_{j_i} = \sum_{i=1}^{m} \pi_i a_{ij_i}$.

Example 7.10 (Substitution Effect) From (7.78), when $\max x_2$ units of activity $j = 2$ are brought into the solution, then the corresponding ith basic activity drops out. However, it does not pay to increase the level of activity $j = 2$ and adjust the values of the basic activities because for each unit of activity $j = 2$ (see Table 7-2) that we bring into the solution, we would have to remove 7/3 units of basic activity $j = 1$ and add 1/30 units of basic activity $j = 4$ for a resulting net increase of 20/3 units in the objective function.

An alternative way to do the calculations is to use Equation (7.79). For each unit of $j = 2$ added, the following changes in the basic variables result:

Variable	Quantity Change	Cost per Unit	Cost Change
x_1	$-7/3$	-12	$+28$
x_4	$1/30$	-40	$-4/3$
x_2	$+1$	-20	-20
Relative cost factor: $\bar{c}_2$			$+20/3$

▷ **Exercise 7.30** In Example 7.3, if the profit on desk Type 2 is increased by exactly $\bar{c}_2 = \frac{20}{3}$ per desk, show that up to $571\frac{3}{7}$ desks of Type 2 can be produced per period without reducing total profit. What is the resulting product mix?

7.8 NOTES AND SELECTED BIBLIOGRAPHY

Most of the examples in this chapter are adapted from Dantzig [1963]. In particular, the material in Section 7.1.3 was contributed by Clopper Almon, Jr. and Example 7.9 is based on a private communication of William Blattner.

Sensitivity analysis has many uses in industry. This is the reason why many books on linear programming extensively cover sensitivity analysis; see, for example, Bradley, Hax, & Magnanti [1977], Chvátal [1983], Dantzig [1963], Gass [1985], Hadley [1972], Murty [1983], and Shapiro [1979].

Many of the variants discussed in *Linear Programming 2* can be used for sensitivity analysis; for example, the Dual-Simplex Method and parametric programming. Obtaining prices when the basic solution is degenerate can be obtained through the use of the parametric techniques described in *Linear Programming 2*. For further details on the method of minimizing the sum of infeasibilities, see *Linear Programming 2*.

Planning under uncertainty, which is the problem of finding optimal solutions to linear programs whose coefficients and right-hand sides are uncertain, is discussed in *Linear Programming 3*.

7.9 PROBLEMS

7.1 Given a linear program in standard form with cost form $z = c^T x$, suppose that x_1 is basic in an optimal solution. Prove that x_1 remains basic if its cost coefficient c_1 is reduced.

7.2 Let $\pi = (\pi_1, \pi_2, \dots, \pi_m)^T$ be optimal simplex multipliers for a linear program in standard form. Suppose that the basic feasible solution is nondegenerate. Prove:

(a) $\partial z / \partial b_i = \pi_i$, i.e., a small change δ in b_i will change z by $\pi_i \delta$. Thus show that the price π_i represents the change in the total costs z per infinitesimal change in the availability of item i.

(b) $\partial z / \partial x_j = \bar{c}_j$ for j nonbasic, i.e., an increase by a small change δ of x_j by adjusting the values of the basic variables will change z by $\bar{c}_j \delta$.

(c) What is the breakeven cost k of a procurement activity that produces δ units of item i at a cost of $k\delta$? Note: The procurement activity has no other inputs or outputs.

7.3 *Dantzig [1963]*. Show that:

Variety	*Man-days per acre*	*Cost \$/acre*	*Profit \$/acre*
Cabernet Sauvignon	20	115	70
Sauvignon Blanc	10	90	50
Chardonnay	15	200	120

Table 7-3: Data for the Grape Grower's Dilemma

(a) If in an optimal solution there are surpluses of certain items, their prices are zero.

(b) The price of an item is zero if there is no cost associated with the activity of storing it and there is a surplus of the item.

(c) The above case might lead to excessive use of the raw material inputs, unless the central planners place some cost, in terms of labor, of using surplus excess raw material.

(d) It would be better, in general, not to use slacks but to introduce activities for procurement of additional inputs and to place a labor cost on these as well.

7.4 *Dantzig [1963]*. Which of the various properties associated with the duality theorems of Chapter 5 explains why the manager of the tool plant discovered the process that minimizes his labor requirements in the course of developing a pricing system?

7.5 A grape grower has just purchased 1,000 acres of vineyards. Due to the quality of the soil and the excellent climate in the region, he can sell all that he can grow of cabernet sauvignon, chardonnay, and sauvignon blanc grapes. He would like to determine how much of each variety to grow on the 1,000 acres, given various costs, profits, and manpower limitations as shown in Table 7-3. Suppose that he has a budget of \$100,000 and has staff available to provide 8,000 man-days.

(a) Formulate the problem as a linear program.

(b) Solve it using the `DTZG Simplex Primal` software option.

(c) Being curious about other possibilities in the future, the grape grower would like to know whether he should grow another variety merlot. This would require 12 man-days/acre, cost \$80 per acre, and produce a profit of \$55/acre. Without rerunning the problem, determine whether it makes sense to try to grow merlot. If it does make sense, re-solve the linear program with the new grape variety included in the formulation.

(d) In order to obtain some initial cash, the grape grower decides to sell futures (at a lower profit) of the yield from 25 acres of sauvignon blanc and 150 acres of cabernet sauvignon grapes. How does this change the optimal solution?

7.6 Suppose that the solution of a linear program of the form

$$\min z = c^T x \quad \text{such that} \quad Ax \leq b, \quad x \geq 0,$$

results in the following terminal simplex tableau:

				slacks			
$(-z)$	x_1	x_2	x_3	s_1	s_2	s_3	RHS
1	10	0	0	3	2	0	2
0	−2	1	0	−2	1	0	1
0	−4	0	1	1	1	0	1
0	7	0	0	4	2	1	3

(a) Consider a modification (ranging) of coefficient c_3 of the form $c_3 + \delta$, where δ is a scalar. For what range of δ (positive and negative) does the current basis remain optimal?

(b) Similarly, change c_1 to $c_1 + \delta$. What happens to the solution as $\delta \to +\infty$ and as $\delta \to -\infty$?

(c) Modify the right-hand side to be $b + \Delta$, where $\Delta = (\delta_1, \delta_2, \delta_3)^T$. Give a system of inequalities that defines the joint range of variation in $\delta_1, \delta_2, \delta_3$ over which the current basis remains feasible.

7.7 Consider the linear program

$$\begin{array}{ll} \text{Minimize} & 2x_1 + x_2 - x_3 = z \\ \text{subject to} & x_1 + 2x_2 + 3x_3 = 12 \\ & -x_1 + 4x_2 - x_3 \leq 20 \\ & x_1 \geq 0,\ x_2 \geq 0,\ x_3 \geq 0. \end{array}$$

(a) Solve the linear program by the **DTZG Simplex Primal** software option to obtain the optimal solution x^*.

(b) By hand, compute ranges on each right-hand side for which the solution x^* stays optimal. How does the objective value change over the two ranges? Clearly show your calculations.

(c) By hand compute ranges on the cost coefficients for which the solution x^* stays optimal. How does the objective value change over these ranges? Clearly show your calculations.

(d) By hand compute ranges on each of the matrix coefficients for which the solution x^* stays optimal. How does the objective value change over these ranges? Clearly show your calculations.

(e) Rerun the **DTZG Simplex Primal** software option together with the **Sensitivity Analysis** software option and verify your hand calculations.

(f) Add a new column $\begin{pmatrix} 1 \\ -3 \\ 2 \end{pmatrix}$ corresponding to a new variable x_4. Without re-solving the problem, determine whether it enters the basis. If it does, re-solve the linear program with this new column.

(g) Add the new row $x_1 + x_2 + x_3 \geq 4$. Without re-solving the problem, determine whether it causes the optimal solution to change. If it does, re-solve the linear program with this row included.

7.8 Consider the linear program

$$\begin{array}{ll} \text{Minimize} & -x_1 + 3x_2 + x_3 - 2x_4 = z \\ \text{subject to} & x_1 + 2x_2 + 3x_3 - 4x_4 \leq 40 \\ & 2x_1 - 3x_2 + x_3 + x_4 \leq 50 \\ & x_1 \geq 0,\ x_2 \geq 0,\ x_3 \geq 0,\ x_4 \geq 0. \end{array}$$

(a) Solve the linear program with **DTZG Simplex Primal** software option to obtain the optimal solution x^*.

(b) By hand, compute ranges on each right hand side for which the solution x^* stays optimal. How does the objective value change over the two ranges? Clearly show your calculations.

(c) By hand compute ranges on the cost coefficients for which the solution x^* stays optimal. How does the objective value change over these ranges? Clearly show your calculations.

(d) By hand compute ranges on each of the matrix coefficients for which the solution x^* stays optimal. How does the objective value change over these ranges? Clearly show your calculations.

(e) Re-run the **DTZG Simplex Primal** software option together with the **Sensitivity Analysis** software option and verify your hand calculations.

(f) Add a new column $\begin{pmatrix} -6 \\ 2 \\ -3 \end{pmatrix}$ corresponding to a new variable x_5. Without re-solving the problem, determine whether it enters the basis. If it does, re-solve the linear program with this new column.

(g) Add the new row $x_1 - x_2 + x_3 = 10$. Without re-solving the problem, determine whether it causes the optimal solution to change. If it does, re-solve the linear program with this row included.

7.9 *Ph.D. Comprehensive Exam, June 13, 1968, at Stanford.* Consider the linear program, find x and min z such that

$$Ax = b, \quad x \geq 0, \quad c^T x = z \text{ (Min)}.$$

Assume that $x^o = (x_1^o, x_2^o, \ldots, x_n^o)^T$ is an optimal solution to the problem. Suppose further that a constraint $a^T x \leq \theta$ has been omitted from the original problem and that a feasible program results from adjoining this constraint.

(a) Prove that if $a^T x^o \leq \theta$, then x^o is still optimal for the augmented problem.

(b) Prove that if $a^T x^o > \theta$, then either there exists no feasible solution $a^T x^o \leq b$ or there exists an optimal solution x^* such that $a^T x^* = \theta$.

7.10 *Dantzig [1963].* Given an optimal basic feasible solution and the corresponding system in canonical form, show that $\bar{c}_j$ represents the change necessary in the unit cost of the jth nonbasic activity before it would be a candidate to enter the basis. If the other coefficients as well as cost coefficients can vary, show that

$$\bar{c}_j = c_j - \sum_{i=1}^{m} \pi_i a_{ij}$$

is the amount of change where π_i are the prices associated with the basic set of variables.

7.11 Let B be an optimal basis for a linear program in standard form. Prove or disprove (by means of a counter example) the claim that if some of the basic costs c_B are decreased, B will remain optimal.

7.12 *Dantzig [1963].* Develop a formula for the change in cost c_j of a basic activity before it is a candidate for being dropped from the basis. Which activity would enter in its place?

7.13 Suppose we have an optimal solution for a linear program and the cost coefficient c_s of nonbasic variable x_s is reduced sufficiently so that x_s in positive amount enters the basic set. Prove that in any optimal solution to the linear program, x_s is a basic variable.

7.14 Modify the analysis in Section 7.6 for the case when *all* the coefficients in a column $A_{\bullet j}$ change. Consider the cases where j is a basic column or a nonbasic column when the linear program is in standard form with upper and lower bounds on the variables.

7.15 *Ph.D. Comprehensive Exam, September 26, 1992, at Stanford.* Consider a linear program in standard form:

$$\begin{array}{lrcll} \text{Minimize} & c^T x & & & \\ \text{subject to} & Ax & = & b, & \quad A: m \times n, \\ & x & \geq & 0. & \end{array}$$

Assume that every basic solution is nondegenerate. Let

$$\begin{aligned} \mathcal{K} &= \{\, x \mid Ax = b, x \geq 0 \,\}, \\ \mathcal{K}' &= \{\, x \mid Ax = b, x \geq 0, x_n = 0 \,\}, \end{aligned}$$

where x_n is a scalar variable. Assume that the sets $\mathcal{K}$ and $\mathcal{K}'$ are nonempty and bounded.

Suppose that we allow only the cost coefficient c_n to vary. Show that there exists a constant γ such that x_n is a nonbasic for $c_n > \gamma$, and x_n is basic for $c_n < \gamma$, in all optimal bases. Show how to compute the scalar γ. What happens if $c_n = \gamma$?

CHAPTER 8

TRANSPORTATION AND ASSIGNMENT PROBLEM

The general case of the transportation and assignment problem is the minimum-cost capacitated network-flow problem

$$
\begin{array}{lll}
\text{Minimize} & c^T x & \\
\text{subject to} & Ax = b, & A:\ m \times n, \\
 & l \le x \le u, &
\end{array}
\tag{8.1}
$$

where each column $A_{\bullet j}$ has at most one positive coefficient $+1$ and at most one negative coefficient -1. This matrix structure implies that every basis is *triangular* and that all basic solutions have integer values if the right-hand side and upper and lower bounds have integer values.

The *classical* transportation problem, see (8.3), was studied as early as 1941 by Hitchcock, who proposed a solution technique that has points in common with the Simplex Method. This case and the general case (which will be discussed in Chapter 9) are reducible to one another.

8.1 THE CLASSICAL TRANSPORTATION PROBLEM

The *classical transportation problem*, see (8.3), is to determine an optimal schedule of shipments that:

1. originate at sources where known quantities of a commodity are available;

2. are allocated and sent directly to their final destinations, where the total amount received must equal the known quantities required;

3. exhaust the supply and fulfill the demand; hence, total given demands must equal total given supplies;

4. and finally, the cost of each shipment from a source to a destination is proportional to the amount shipped, and the total cost is the sum of the individual costs; i.e., the costs satisfy a linear objective function.

8.1.1 MATHEMATICAL STATEMENT

Suppose, for example, that m warehouses (origins) contain various amounts of a commodity that must be allocated to n cities (destinations). Specifically, the ith warehouse must dispose of exactly the quantity $a_i \geq 0$, while the jth city must receive exactly the quantity $b_j \geq 0$. For this problem, it is assumed that the *total demand equals the total supply*, that is,

$$\sum_{i=1}^{m} a_i = \sum_{j=1}^{n} b_j = T, \qquad a_i \geq 0, \; b_j \geq 0. \tag{8.2}$$

The cost c_{ij} of shipping a unit quantity from i to j is given; typically $c_{ij} \geq 0$ but $c_{ij} < 0$ is also possible. The problem then is to determine the number of units to be shipped from i to j at an overall minimum cost.

The problem can be stated mathematically as

$$\begin{array}{lrll}
\text{Minimize} & \displaystyle\sum_{i=1}^{m}\sum_{j=1}^{n} c_{ij}x_{ij} & = z & \\
\text{subject to} & \displaystyle\sum_{j=1}^{n} x_{ij} & = a_i, & i = 1,\ldots,m \\
& \displaystyle\sum_{i=1}^{m} x_{ij} & = b_j, & j = 1,\ldots,n \\
& x_{ij} & \geq 0, & i = 1,\ldots,m, \; j = 1,\ldots,n.
\end{array} \tag{8.3}$$

Note that in this chapter, the symbols m and n denote the number of sources and demand centers respectively, and are not the symbols used to denote the number of constraints and variables for a general linear program. In this case *the number of equations is* $m + n$ *and the number of variables is* mn.

8.1.2 PROPERTIES OF THE SYSTEM

This problem has a very nice structure that can be exploited to compute an initial basic feasible solution very easily and then to compute improving solutions until an

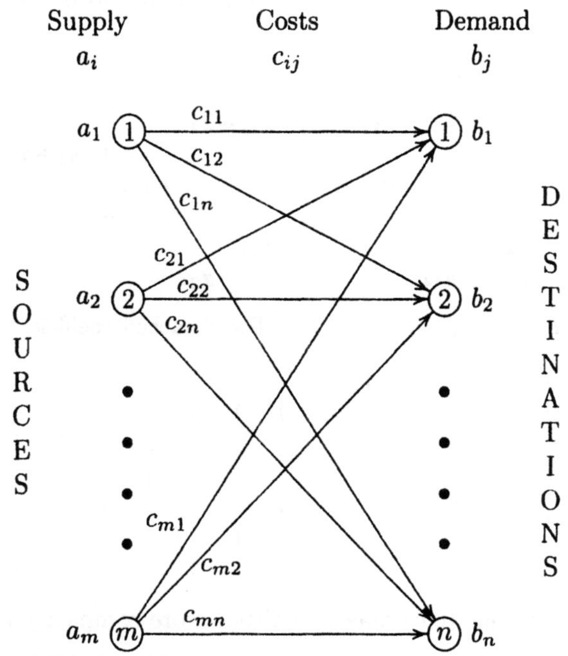

Figure 8-1: Network Representation of the Transportation Problem

optimal solution is determined. The nonzero pattern of coefficients (other than the objective function) is evident when the equations are displayed as shown below.

$$
\begin{array}{llllllll}
x_{11}+x_{12}+\cdots+x_{1n} & & & & & & & = a_1 \\
& & x_{21}+x_{22}+\cdots+x_{2n} & & & & & = a_2 \\
& & & & \ddots & & & \vdots \\
& & & & & x_{m1}+x_{m2}+\cdots+x_{mn} & & = a_m \\
x_{11} & & +x_{21} & & & x_{m1} & & = b_1 \\
& x_{12} & & +x_{22} & & & +x_{m2} & = b_2 \\
& \ddots & & \ddots & & & \ddots & \vdots \\
& & x_{1n} & & +x_{2n} & & +x_{mn} & = b_n
\end{array}
\qquad (8.4)
$$

The transportation problem is a special case of the more general network problem. A network representation of the classical transportation problem is shown in Figure 8-1.

Example 8.1 (Prototype Example) The example presented here is one that we shall use to illustrate several aspects of the transportation problem. Consider the following $m = 3$ (sources), $n = 4$ (destinations): Find

$$\min\ z = 7x_{11} + 2x_{12} + 5x_{13} + 4x_{14} + 3x_{21} + 5x_{22}$$

$$+ 4x_{23} + 1x_{24} + 2x_{31} + 1x_{32} + 3x_{33} + 4x_{34} \tag{8.5}$$

subject to

$$\begin{array}{lllllllllllll}
x_{11} + x_{12} + x_{13} + x_{14} & & & & & & & & & & & & = 3 \\
 & & & & x_{21} + x_{22} + x_{23} + x_{24} & & & & & & & & = 4 \\
 & & & & & & & & x_{31} + x_{32} + x_{33} + x_{34} & & & & = 5 \\
x_{11} & & & & +x_{21} & & & & +x_{31} & & & & = 3 \\
 & x_{12} & & & & +x_{22} & & & & +x_{32} & & & = 4 \\
 & & x_{13} & & & & +x_{23} & & & & +x_{33} & & = 2 \\
 & & & x_{14} & & & & +x_{24} & & & & +x_{34} & = 3
\end{array} \tag{8.6}$$

and $x_{ij} \geq 0$, for $i = 1, \ldots, 3$, $j = 1, \ldots, 4$. The detached coefficients have the following structure:

$$\begin{bmatrix}
1 & 1 & 1 & 1 & & & & & & & & \\
 & & & & 1 & 1 & 1 & 1 & & & & \\
 & & & & & & & & 1 & 1 & 1 & 1 \\
1 & & & & 1 & & & & 1 & & & \\
 & 1 & & & & 1 & & & & 1 & & \\
 & & 1 & & & & 1 & & & & 1 & \\
 & & & 1 & & & & 1 & & & & 1
\end{bmatrix}. \tag{8.7}$$

The system of equations (8.6) may be written more compactly as follows:

x_{11}	x_{12}	x_{13}	x_{14}	$= 3$
x_{21}	x_{22}	x_{23}	x_{24}	$= 4$
x_{31}	x_{32}	x_{33}	x_{34}	$= 5$
$\shortparallel$ 3	$\shortparallel$ 4	$\shortparallel$ 2	$\shortparallel$ 3	

▷ **Exercise 8.1** Draw the network representation for the prototype Example 8.1 (see Figure 8-1).

FEASIBILITY OF THE SYSTEM

It turns out that the assumption (8.2) that total supply equals total demand is *necessary* for feasibility of (8.3), because adding the first m equations in (8.3) results in

$$\sum_{i=1}^{m} \sum_{j=1}^{n} x_{ij} = \sum_{i=1}^{m} a_i$$

and adding the last n equations in (8.3) results in

$$\sum_{i=1}^{m} \sum_{j=1}^{n} x_{ij} = \sum_{j=1}^{m} b_j.$$

Therefore, $\sum_{i=1}^{m} a_i = \sum_{j=1}^{m} b_j$. In addition, the assumption (8.2) is *sufficient* for feasibility because

$$x_{ij} = \frac{a_i b_j}{T} \geq 0$$

is a feasible solution for (8.3).

▷ **Exercise 8.2** Verify that $x_{ij} = a_i b_j / T$ is a feasible solution for (8.3).

Example 8.2 (Illustration of Feasibility) In the system (8.6) of Example 8.1 suppose that we change the right-hand side of the first equation to 4. Now, subtracting the sum of the first three equations from the sum of the last four equations results in $0 = 1$, showing that the system is infeasible. This demonstrates that $\sum_{i=1}^{m} a_i = \sum_{j=1}^{n} b_j$ is a necessary condition for feasibility.

▷ **Exercise 8.3** Suppose that the total supply $\sum_{i=1}^{m} a_i$ available at the sources exceeds the sum of the demands $\sum_{i=1}^{n} b_j$ at the destinations. Show that we can make the system feasible by introducing a dummy destination to absorb the excess supply at no cost. If, on the other hand, $\sum_{i=1}^{m} a_i < \sum_{j=1}^{n} b_j$ show that we can introduce a dummy source to supply the excess requirement at, for example, the market price.

▷ **Exercise 8.4** In the case of excess supply, elaborate the model of Exercise 8.3 to include the possibility of storage at various external places at a transportation and holding cost, and also the possibility of storing at the supply site at a holding cost. Similarly, in the case of excess demand, elaborate the model of Exercise 8.3 to include the possibility of buying from various external suppliers.

RANK OF THE SYSTEM

The necessary and sufficient condition (8.2) for feasibility renders the system (8.4) dependent since it implies that the sum of the first m equations is the same as the sum of the last n. Therefore the rank of the system is at most $m + n - 1$.

LEMMA 8.1 (Rank of the Transportation Problem) *The rank of the system (8.3) is exactly $m + n - 1$. Furthermore, each equation is a linear combination of the other $m + n - 1$ equations, so that any one equation may be called redundant and may be discarded if it is convenient to do so.*

COROLLARY 8.2 (Number of Basic Variables) *There are exactly $m+n-1$ basic variables x_{ij}.*

Example 8.3 (Illustration of Rank) In the system (8.6) of Example 8.1 it is easy to verify that if we subtract the sum of the first three equations from the sum of the last four equations we obtain the zero equation $0 = 0$, implying that the system contains at least one redundant equation.

Dropping the first (or, for that matter, any other) equation will result in a full rank system, as we will now illustrate on Example 8.1. Clearly the last four equations are of full rank because they each contain exactly one of the variables x_{11}, x_{12}, x_{13}, x_{14}, implying that none of these equations can be generated by a linear combination of the other equations. The second and third equations contain no variables in common and cannot be generated from each other. Furthermore, the second equation cannot be generated from the third and the last four equations. Similarly, the third equation also cannot be generated from the second and the last four equations. This implies that the remaining system is of full rank.

▷ **Exercise 8.5** Prove Lemma 8.1 and Corollary 8.2.

BASIS TRIANGULARITY

A fundamental property of a transportation (or network flow) problem is that every basis is triangular.

Definition (Triangular Matrix): We give the following two equivalent definitions of a *triangular* matrix.

1. A square matrix is said to be *triangular* if it satisfies the following properties.
   (a) The matrix contains at least one row having exactly one nonzero element.
   (b) If the row with a single nonzero element and its column are deleted, the resulting matrix will once again have this same property.
2. Equivalently, we can define a square matrix to be *triangular* if its rows and columns can be permuted to be either an upper triangular or lower triangular matrix. See Section A.8 in the linear algebra Appendix A for a definition of upper and lower triangular matrices.

Example 8.4 (Basis Triangularity) A basis can be formed by using the columns corresponding to the variables x_{11}, x_{13}, x_{21}, x_{24}, x_{31}, and x_{32} (as we shall see later) for the transportation problem of Example 8.1. Below we display the system of equations after deleting the nonbasic variables and deleting the last equation as redundant.

$$\begin{array}{llllllll} x_{11} & + x_{13} & & & & & = 3 \\ & & x_{21} & + x_{24} & & & = 4 \\ & & & & x_{31} & + x_{32} & = 5 \\ x_{11} & & + x_{21} & & + x_{31} & & = 3 \\ & & & & & + x_{32} & = 4 \\ & x_{13} & & & & & = 2. \end{array} \tag{8.8}$$

It is easy to verify directly that this system is triangular by applying the first definition. It can also be verified by the second definition by rearranging the rows and columns to

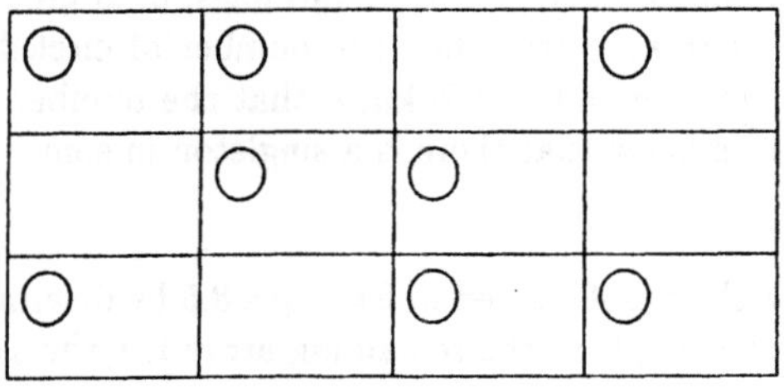

Figure 8-2: Illustration of the Basis Triangularity Theorem

obtain the lower triangular form

$$
\begin{array}{lllllllll}
x_{13} & & & & & & & = & 2 \\
& x_{32} & & & & & & = & 4 \\
x_{13} & & + \; x_{11} & & & & & = & 3 \\
& x_{32} & & + \; x_{31} & & & & = & 5 \\
& & x_{11} & + \; x_{31} & + \; x_{21} & & & = & 3 \\
& & & & x_{21} & + \; x_{24} & & = & 2.
\end{array}
\tag{8.9}
$$

▷ **Exercise 8.6** Find another basis for Example 8.1 and by reordering rows and columns, demonstrate its triangularity.

▷ **Exercise 8.7** Prove that the above two definitions of triangularity of a square matrix are equivalent.

▷ **Exercise 8.8** Show that an upper triangular matrix can be permuted to become a lower triangular matrix.

If a system has the property that every basis is triangular, then given a basic set of variables, it is easy to compute the associated basic solution. We can start by setting the values of all nonbasic variables to zero. The resulting system of equations will then involve only the basic variables. Since the system is triangular the basic variables can be easily evaluated by applying the first definition of triangularity defined above. The following is known as the *Fundamental Theorem for the Transportation Problem.*

THEOREM 8.3 (Basis Triangularity) *Every basis of the transportation problem* (8.3) *is triangular.*

Example 8.5 (Illustration of the Basis Triangularity Theorem) Suppose on the contrary that the set of basic variables are those circled in Figure 8-2 and they do not have the property that there is an equation (i.e., a row or column) with exactly one circled (basic) variable. Then every row and every column has two or more circled variables.

Because $n = 4 \geq m = 3$, the number of circled variables, 8, is greater than or equal to $2n \geq m + n = 7$. (If $m \geq n$ were true then the number of circled variables would be greater than or equal to $2m \geq m + n$.) We know that the number of basic variables is $m + n - 1 = 6$. Therefore we know that there is a singleton in some row or column.

▷ **Exercise 8.9** Complete the proof started in Example 8.5 by deleting the row or column having the singleton and showing that the remaining array has the same property.

INTEGRAL PROPERTY

The integer property of the solution is very important in practice. In the cannery Example 1.5 on page 4, x_{ij} represents the integer number of cases shipped from cannery i to warehouse j. An optimal solution with x_{ij} having a fractional value would be unacceptable. Fortunately, Theorem 8.4 tells us that this will not happen.

THEOREM 8.4 (Integral Property) *All the basic variables have integer values if the row and column totals a_i and b_j are integers.*

Example 8.6 (Integral Property) It is easy to solve the triangular system (8.9) to obtain the integer solution $x_{13} = 2$, $x_{32} = 4$, $x_{11} = 1$, $x_{31} = 1$, $x_{21} = 1$, and $x_{24} = 3$. The solution is integral as Theorem 8.4 stated. The example also illustrates that it is not possible to obtain fractional values when the right-hand sides of the equations have integer values, because the nonzero coefficients of $+1$ imply that all the variables are either set equal to the right-hand side or evaluated by simple subtractions.

▷ **Exercise 8.10** Apply the basis triangularity property to prove Theorem 8.4.

8.2 STANDARD TRANSPORTATION ARRAY

As noted earlier, the special structure of the transportation problem allows us to compactly represent the variables x_{ij} in an $m \times n$ array such that the sums across the rows correspond to the demand constraints and the sums across the columns correspond to the supply constraints. The ijth cell of this rectangular array corresponds to variable x_{ij}. We shall see later that this compact representation can be used very efficiently to solve the transportation problem by hand. A rectangular array suitable for solving such a transportation problem is shown in Figure 8-3 for a 3×5 case.

In Figure 8-3 the column of cells to the right of the double vertical lines is called the *marginal column*, and the row of cells below the double horizontal lines is called the *marginal row*. The rest of the cells are referred to as the rectangular array.

Typically, in the rectangular array, in hand calculations, the following is done:

1. The cost coefficient, c_{ij}, of the objective function is stored in the lower right corner of the ijth cell.

x_{11} c_{11}	x_{12} c_{12}	x_{13} c_{13}	x_{14} c_{14}	x_{15} c_{15}	a_1 u_1
x_{21} c_{21}	x_{22} c_{22}	x_{23} c_{23}	x_{24} c_{24}	x_{25} c_{25}	a_2 u_2
x_{31} c_{31}	x_{32} c_{32}	x_{33} c_{33}	x_{34} c_{34}	x_{35} c_{35}	a_3 u_3
b_1 v_1	b_2 v_2	b_3 v_3	b_4 v_4	b_5 v_5	

Figure 8-3: Example of a Standard Transportation Array

2. Each of the values of $m+n-1$ basic variables x_{ij} is stored in hand calculations in the upper left corner of the ijth cell for all (i, j). The zero nonbasic variables are left blank. *A zero (nonblank) entry in the ijth cell indicates that the value of the corresponding basic variable x_{ij} is equal to zero, implying a degenerate basic solution.*

3. For each of the first $i = 1, \ldots, m$ equations, the right-hand side availability a_i is stored in the upper left corner of the marginal column cell i, and its corresponding simplex multiplier u_i (to be discussed later) is stored in the cell's lower right corner.

4. For each of the last $j = 1, \ldots, n$ equations, the right-hand side demand b_j is stored in the upper left corner of the marginal row cell j and its corresponding simplex multiplier v_j (to be discussed later) is stored in the cell's lower right corner.

5. The reduced costs $\bar{c}_{ij}$ (to be discussed later) for the nonbasic variables are stored in the lower left corner of the cells corresponding to the nonbasic variables instead of the zero values of the basic variables.

As we have noted, each row and column of the array represents an equation. Thus, for feasibility, during the course of the algorithm, the sum of the x_{ij} entries in each row and column must equal the appropriate row or column total, a_i or b_j, that appears in the margins.

Example 8.7 (Transportation Array Illustrated) The transportation problem of Example 8.1 is shown in Figure 8-4 with the basic variables circled. During hand calculations the values of the basic variables will be stored rather than the symbols.

(x_{11}) 7	$\bar{c}_{12}$ 2	(x_{13}) 5	$\bar{c}_{14}$ 4	3 u_1
(x_{21}) 3	$\bar{c}_{22}$ 5	$\bar{c}_{23}$ 4	(x_{24}) 1	4 u_2
(x_{31}) 2	(x_{32}) 1	$\bar{c}_{33}$ 3	$\bar{c}_{34}$ 4	5 u_3
3 v_1	4 v_2	2 v_3	3 v_3	

Figure 8-4: Transportation Array Example

8.3 FINDING AN INITIAL SOLUTION

The fact that every basis in the classical transportation problem is triangular makes it easy to generate a starting basic feasible solution. We shall discuss five ways to get started, illustrating each on prototype Example 8.1.

Example 8.8 (Order from the Cheapest Source) This is not one of the five ways because this rule may lead to an infeasible solution, if it does lead to a feasible solution, it will be optimal. The rule is that a buyer at each destination j orders the total supply required b_j from the cheapest source $i_j = \text{argmin}_i\, c_{ij}$. If we apply this rule, then it will typically result in an infeasible starting solution. Applying the rule to the prototype transportation problem of Example 8.1, the amount that each buyer $j = 1, 2, 3, 4$ orders from a supplier is illustrated by the circled amounts in Figure 8-5.

In the figure, buyer $j = 1$ orders 3 units from the cheapest source $i = 3$; buyer $j = 2$ orders 4 units from the cheapest source $i = 3$; buyer $j = 3$ orders 2 units from the cheapest source $i = 3$; buyer $j = 4$ orders 3 units from the cheapest source $i = 2$. This solution is clearly infeasible because source $i = 3$ can only supply a total of 5 units, whereas 9 units have been ordered.

▷ **Exercise 8.11** Prove that if by good luck the orders from the cheapest source turns out to be a feasible solution, it is optimal.

8.3.1 TRIANGULARITY RULE

The simplest way to generate a starting basic feasible solution is by the following triangularity rule (algorithm).

Triangularity Rule: Choose arbitrarily any variable x_{pq} as the candidate for the first feasible basic variable. Make x_{pq} as large as possible without violating the row and column totals, i.e., set

$$x_{pq} = \min \{ a_p, b_q \}. \tag{8.10}$$

7	2	5	4	3
3	5	4	(3) 1	4
(3) 2	(4) 1	(2) 3	4	5
3	4	2	3	

Figure 8-5: Buy from the Cheapest Source

The next variable to be made basic is determined by this same procedure after *reducing the rectangular array* depending on which of the following three cases arises.

1. If $a_p < b_q$, then all the other variables in the pth row are given the value zero and designated as nonbasic. Next the pth row is deleted, and the value of b_q in column q is reduced to $(b_q - a_p)$.

2. If $a_p > b_q$, then all the other variables in the qth column are given the value zero and designated as nonbasic. Next the qth column is deleted and the value of a_p in row p is reduced to $(a_p - b_q)$.

3. If $a_p = b_q$, then randomly choose either the pth row or the qth column to be deleted, but *not both.* However, if several columns, but only one row, remain in the reduced array, then drop the qth column, and conversely, if several rows and one column remain in the reduced array, drop the pth row. If the pth row is deleted, the value of b_q in column q is reduced to 0. If the qth column is deleted, the value of a_p in row p is reduced to 0.

If after deletion of a row or column, there remains only one row or one column, then all remaining cells are basic and are evaluated in turn as equal to the residual amount in the row or column. On the last step exactly one row and one column remain, and both must be dropped after the last variable is evaluated. Thus, this Triangularity Rule will select as many variables for the basic set as there are rows plus columns, less one, i.e., $m + n - 1$.

▷ **Exercise 8.12** Show that every reduced array retains the property that the sum of the adjusted right-hand side entries in the marginal column is equal to the sum of the adjusted right-hand side entries in the marginal row. This implies that the last remaining variable acquires a value of the total for the single row that is equal to the total for the remaining single column in the final reduced array.

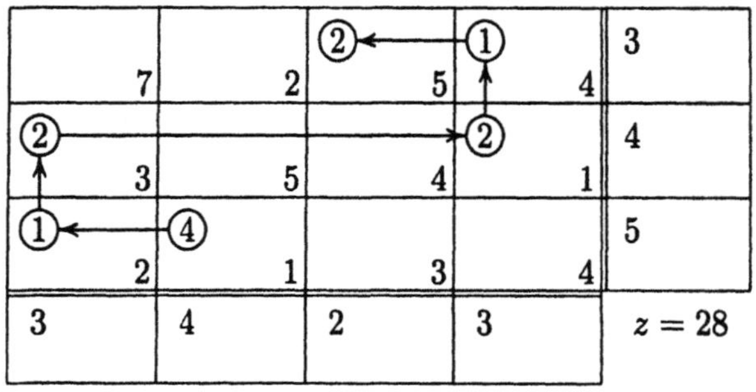

Figure 8-6: Illustration of the Triangularity Rule

Example 8.9 (Illustration of the Triangularity Rule) The sequence of steps to find a feasible solution for the prototype example 8.1 by the triangularity rule is illustrated in Figure 8-6.

We start by picking any variable (for example, in this case, the lowest cost variable) and trying to assign as much as possible to it so as not to violate the row or column total. We pick the variable in row 3, column 2 and assign it the value 4; this satisfies the column total and we ignore column 2 from now on. Next we could pick any x_{ij} from the remaining cells to increase. For example, look in the current row 3 for the lowest cost. The variable in row 3, column 1 has the lowest cost of 2, and we assign it the value 1 because with this assignment the row total is satisfied; and so forth. The objective value obtained for the starting feasible basic solution is $z = 28$.

Special Case: Northwest Corner Rule. The Northwest Corner Rule is a particular case of the Triangularity Rule described here. It starts off by selecting x_{11} (the Northwest corner variable) as the first variable to enter the basis. The variable x_{11} is assigned a value as large as possible without violating the row or column totals. Then the iterative procedure proceeds as follows. If x_{ij} was the last variable to be selected, the next variable to be selected is the one to the right of it, i.e., $x_{i,j+1}$ if the partial ith row total is still nonzero; otherwise, the next variable to be selected is the one below it, i.e., $x_{i+1,j}$. The same three cases can arise as discussed for the Triangularity Rule on Page 214 and are treated in the same way as discussed there.

Example 8.10 (Illustration of the Northwest Corner Rule) The sequence of steps to find a feasible solution for the prototype Example 8.1 by the Northwest Corner Rule is illustrated in Figure 8-7. The objective value obtained for the starting feasible basic solution is: $z = 59$.

The remaining rules described in this section are similar to the Triangularity Rule for choosing an initial feasible basic set and dropping other variables from further consideration for entering the basis. The main difference is in the sequence and reasons for choosing the basic variables.

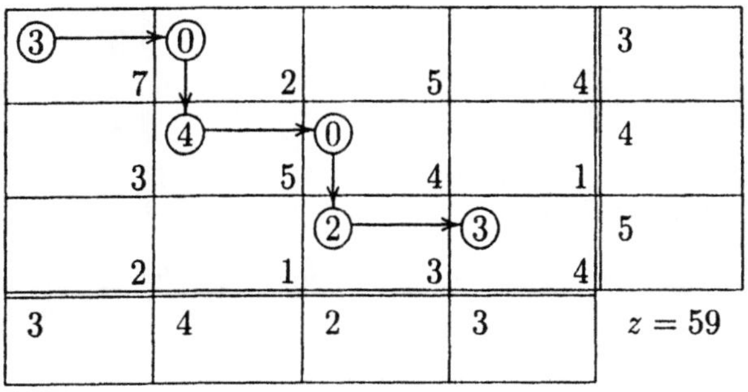

Figure 8-7: Illustration of the Northwest Corner Rule

The Northwest Corner Rule for finding an initial solution is very simple but often does not generate a good start because the rule does not take into account the values of the coefficients in the objective function. It is a special case of the Triangularity Rule which also may not generate a good start unless a low-cost cell is selected from among those remaining.

8.3.2 THE LEAST REMAINING COST RULE

Scan the costs c_{ij} for the *smallest* c_{ij} and choose the first basic variable x_{pq} such that

$$c_{pq} = \min_{(i,j)} c_{ij}. \tag{8.11}$$

Set the value of x_{pq} to be the minimum of its row or column total, and make the remaining variables in that row or column ineligible for further increases in the values of its variables (see the discussion for the Triangularity Rule on Page 214). For subsequent entries, find the smallest cost factor c_{ij} among the remaining cells and continue.

Example 8.11 (Illustration of the Least Remaining Cost Rule) The sequence of steps to find a feasible solution for the prototype Example 8.1 by the Least Remaining Cost Rule is illustrated in Figure 8-8. The objective value obtained for the starting feasible basic solution is $z = 29$.

8.3.3 VOGEL'S APPROXIMATION METHOD

Vogel's method is similar to the method described above; the main difference is in the way the basic variable is chosen from the eligible basic variables. The method has been popular because it turns out that in practical applications it often finds a solution that is close to the optimal. For each row i with eligible variables we compute the *cost difference* between the smallest cost and the next to smallest cost of eligible variables in the row; in a similar manner, we compute the *cost difference*

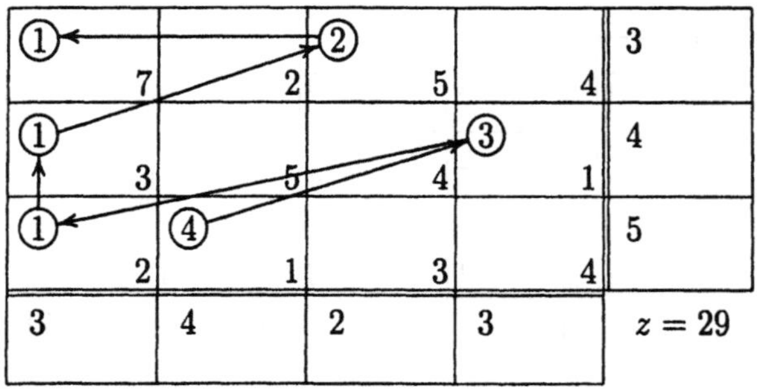

Figure 8-8: Illustration of the Least Remaining Cost Rule

between the smallest cost and the next to smallest cost of eligible variables in each column. The row or column with the largest cost difference is identified and the variable with the lowest cost is then picked from this row or column as the next basic variable. Ties for the largest cost difference can be broken in favor of the row or column having the smallest cost. Ties between those having the smallest cost can be broken at random.

The idea behind this approach is as follows. Suppose row k has the largest cost difference among all the rows and columns. By selecting the variable with the smallest cost from this row we avoid a future selection whereby the smallest cost variable becomes ineligible and the next to smallest cost variable comes into the basis.

This method clearly requires more computations than the other methods discussed so far. However, it turns out that in practice it is well worth this extra effort since often a very good starting basis is obtained.

Example 8.12 (Illustration of Vogel's Approximation Method) The sequence of steps to find a feasible solution for the prototype example 8.1 by Vogel's Approximation Method is illustrated in Figure 8-9. The columns Δc_i^t are the differences in the two smallest costs for row i on iteration t. The rows Δc_j^t are the differences in the two smallest costs for column j on iteration t. After the first four iterations, the differences are not relevant.

The objective value obtained for the starting feasible basic solution is $z = 23$.

8.3.4 RUSSEL'S APPROXIMATION METHOD

Like Vogel's method, Russel's method attempts to find a good or near-optimal starting basis for the transportation problem. This method requires more computations than Vogel's method and is claimed to work better on the average. However, there is no clear cut winner.

For each row and column containing variables eligible for entering the basis, we

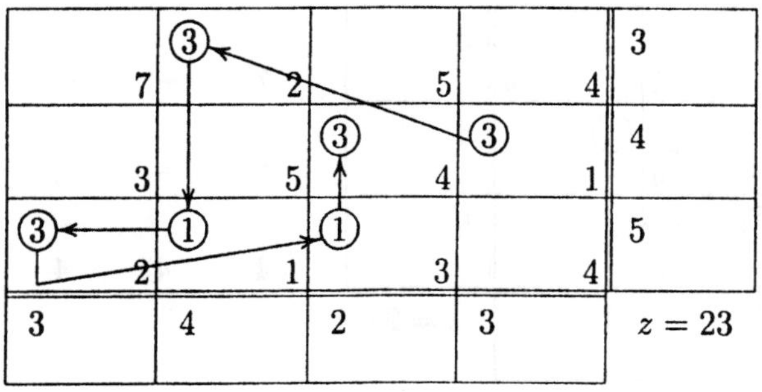

Δc_i^1	Δc_i^2	Δc_i^3	Δc_i^4
2	3*	–	–
2	1	1	1
1	1	1	1

Δc_j^1	1	1	1	3*
Δc_j^2	1	1	1	–
Δc_j^3	1	4*	1	–
Δc_j^4	1	–	1*	–

Figure 8-9: Illustration of Vogel's Approximation Method

first compute the maximum costs, i.e.,

$$\bar{u}_i = \max_j c_{ij} \qquad \text{and} \qquad \bar{v}_j = \max_i c_{ij}.$$

Then for all the variables eligible to enter the basis, we compute the quantities

$$\gamma_{ij} = c_{ij} - \bar{u}_i - \bar{v}_j.$$

Then the variable with the most negative γ_{ij} is selected as a candidate to enter the basis.

Example 8.13 (Illustration of Russel's Approximation Method) The sequence of steps to find a feasible solution for the prototype Example 8.1 by Russel's Approximation Method is illustrated in Figure 8-10. The columns $\bar{u}_i^t$ are the largest costs in row i on iteration t. The rows $\bar{v}_j^t$ are the largest costs in column j on iteration t. After the first three iterations, these values are not relevant.

The objective value obtained for the starting feasible basic solution is $z = 25$.

8.3.5 COST PREPROCESSING

The general idea is to replace the original rectangular array of c_{ij} by another array γ_{ij} with the property that the new problem has the same optimal basic feasible solution but is more convenient for finding a good initial basic solution. After the preprocessing any of the above rules described earlier may be applied. The preprocessing steps are:

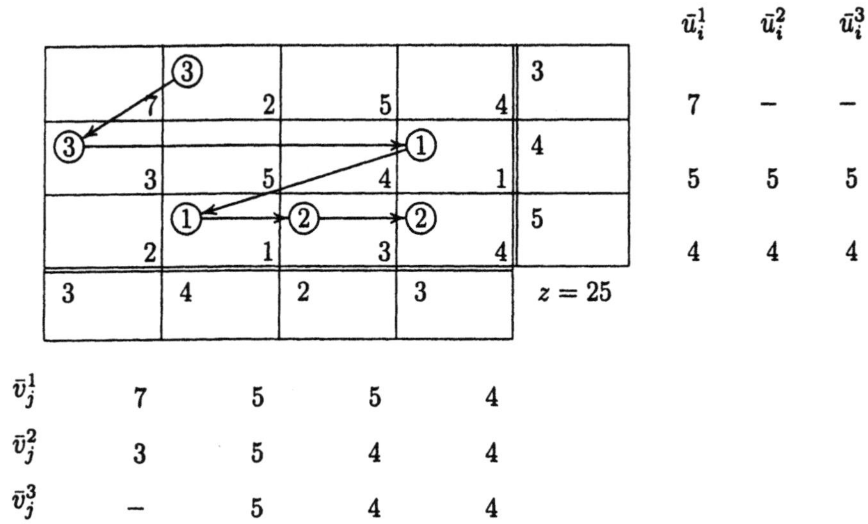

Figure 8-10: Illustration of Russel's Approximation Method

1. For $i = 1, \ldots, m$ replace c_{ij} by $\delta_{ij} = c_{ij} - \min_{l=1,\ldots,n} c_{il}$.

2. For $j = 1, \ldots, n$ replace δ_{ij} by $\gamma_{ij} = c_{ij} - \min_{k=1,\ldots,m} c_{kj}$.

Example 8.14 (Cost Preprocessing) Applying the first rule we get the first rectangular array shown in Figure 8-11; applying the second rule we get the second rectangular array shown in Figure 8-11.

▷ **Exercise 8.13** Given a general linear program $\min \sum_{j=1}^{n} x_j$ subject to $\sum_{j=1}^{n} a_{ij}x_j = b_i$ for $i = 1, \ldots, m$, $x_j \geq 0$ for $j = 1, \ldots, n$, show that if we replace c_j by $c_j - \sum_{i=1}^{m} \pi_i a_{ij}$ then the corresponding objective values of corresponding feasible solutions differ by a constant.

▷ **Exercise 8.14** Show that if the order of the steps is reversed then we may not get the same reduced array.

▷ **Exercise 8.15** Apply the Least-Remaining-Cost Rule, Vogel's Approximation Method, and Russel's Approximation Method in turn to see whether cost preprocessing gives a better starting solution.

5	0	3	2	3
2	4	3	0	4
0	0	3	3	5
3	4	2	3	

4	0	1	2	3
1	4	1	0	4
0	0	0	3	5
3	4	2	3	

Figure 8-11: Cost Preprocessing

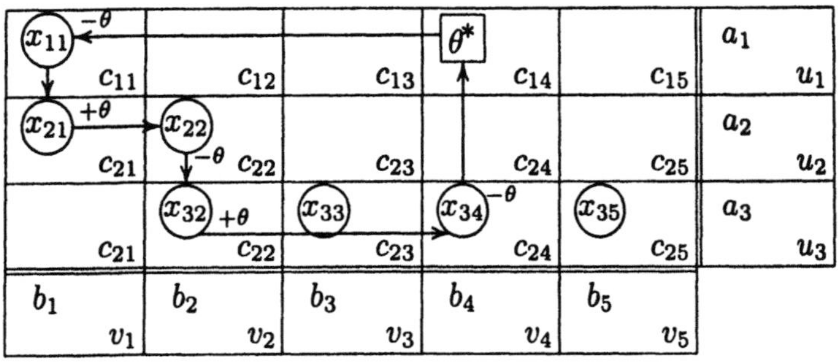

Figure 8-12: Theta-Adjustments for the Standard Transportation Array

8.4 FAST SIMPLEX ALGORITHM FOR THE TRANSPORTATION PROBLEM

Even small general linear programs require a computer; whereas small transportation problems can be easily solved by hand using just a pencil and paper. At each iteration of the Simplex Algorithm applied to the transportation problem, adjustments are made to the x_{ij} in the cells of the rectangular array in Figure 8-3. These adjustments are sometimes referred to as theta-adjustments since a value θ of the incoming variable is added and subtracted from some of the basic variables; see the array in Figure 8-12, where the θ with a star, θ^*, refers to the position of the entering variable.

The steps are the same as those of the Simplex Algorithm except that the calculations of the reduced costs and the changes to the basic variables are simple additions and subtractions.

8.4.1 SIMPLEX MULTIPLIERS, OPTIMALITY, AND THE DUAL

In this section we show how we can take advantage of the structure of the classical transportation problem to easily obtain the simplex multipliers and reduced costs.

SIMPLEX MULTIPLIERS

To distinguish the multipliers corresponding to the rows from those of the columns of the transportation array, let u_i represent the multiplier for the ith row equation, and let v_j represent the multiplier for the jth column equation instead of using π_k for all equations k as we have done earlier.

The values of the u_i and v_j are chosen so that the basic columns price out to zero. However, there appears to be a complication because there are $m+n-1$ equations that are not redundant and $m+n$ unknowns u_i and v_j (including the multiplier on the redundant equation). Recall that any equation in (8.4) may be considered redundant and may be dropped. Dropping is the same as saying that we can assign an arbitrary value, say zero, to the simplex multiplier on the redundant equation and then evaluate the remaining $m+n-1$ multipliers by solving the $m+n-1$ equations corresponding to the $m+n-1$ basic variables. If one id doing hand computations, a good convention is to find a row or column having the greatest number of basic variables and to set its corresponding price to zero. In order for a basic column (i,j) to price out to zero, we must have

$$c_{ij} = u_i + v_j \quad \text{for } x_{ij} \text{ basic,} \tag{8.12}$$

because column (i,j) has exactly two nonzero coefficients: $+1$ corresponding to equation i in the demand equations and $+1$ corresponding to equation j in the supply equations, see (8.3).

As we have seen, the basis is triangular and thus we can determine such prices by scanning the squares corresponding to basic variables until one is found for which either the row price, u_i, or the column price, v_j, has already been evaluated; subtracting either price from c_{ij} determines the other price. The triangularity of the basis guarantees that repeated scanning will result in the evaluation of all u_i and v_j.

For large problems this procedure for scanning the rows and columns is not efficient. More efficient methods are based on the fact that a basis of the transportation problem corresponds to a tree in a graph; the tree structure can be used to determine more efficiently the updated prices and changes in the values of the basic variables. Such techniques are described for general networks in Chapter 9.

REDUCED COSTS AND OPTIMALITY

To determine whether the solution is optimal, multiply the ith row equation of (8.2) by u_i and the jth column equation by v_j and then subtract the resulting equations from the objective function to obtain a modified z-equation,

$$\sum_{i=1}^{m}\sum_{j=1}^{n} \bar{c}_{ij} x_{ij} = z - z_0, \tag{8.13}$$

where the $\bar{c}_{ij}$, which are the reduced costs, are given by

$$\bar{c}_{ij} = c_{ij} - (u_i + v_j) \qquad \text{for } i = 1,\ldots,m,\ j = 1,\ldots,n, \tag{8.14}$$

and

$$z_0 = \sum_{i=1}^{m} a_i u_i + \sum_{j=1}^{n} b_j v_j. \tag{8.15}$$

The $\bar{c}_{ij}$ corresponding to the basic variables are all zero by (8.12). The basic feasible solution is optimal if $\bar{c}_{ij} \geq 0$ for all the nonbasic variables.

▷ **Exercise 8.16** Show that any one of the $m+n$ multipliers may be given an arbitrary value, other than 0, in determining the remaining multiplier. Specifically, show that an arbitrary constant k can be added to all the multipliers u_i and subtracted from all the multipliers v_j, (i.e., the multipliers u_i may be replaced by (u_i+k) and the multipliers v_j may be replaced by (v_j-k)), without affecting the value of $\bar{c}_{ij}$ and z_0 in (8.14) and (8.15).

▷ **Exercise 8.17** Show that if the c_{ij} are replaced by $c_{ij}-u_i-v_j$, where u_i and v_j are arbitrary, that this does not affect the optimal solution x^*_{ij} but does affect the value of the objective.

DUAL OF THE TRANSPORTATION PROBLEM

It is interesting to look at the dual of the transportation problem, i.e.,

$$\begin{array}{ll} \text{Maximize} & \displaystyle\sum_{i=1}^{m} a_i u_i + \sum_{j=1}^{m} b_j v_j = w \\ \text{subject to} & u_i + v_j \le c_{ij} \quad \text{for all } (i,j), \end{array} \tag{8.16}$$

where u_i, v_j are unrestricted in sign for all (i,j). The primal variable x_{ij} gives rise to the dual constraint $u_i+v_j \le c_{ij}$. Thus the slack variable on this dual constraint is

$$y_{ij} = c_{ij} - u_i - v_j, \quad y_{ij} \ge 0, \tag{8.17}$$

and, by complementary slackness, either $x_{ij}=0$ or $y_{ij}=0$. Thus, if x_{ij} is basic we must have $y_{ij}=0$ or $c_{ij}=u_i+v_j$, as we saw in Equation (8.12).

▷ **Exercise 8.18** Prove that any basis of the transportation problem's dual is triangular.

▷ **Exercise 8.19** Prove in general that the transpose of any triangular basis is triangular.

8.4.2 FINDING A BETTER BASIC SOLUTION

As we have seen earlier, in the Simplex Method for a general linear program in standard form, if $\bar{c}_{ij}=\bar{c}_{rs}$ is negative, then the corresponding nonbasic variable x_{rs} is suitable as a candidate for entering into the basic set, replacing a basic variable taht, after it is dropped from the basic set, becomes just another nonbasic variable.

The usual rule for selecting a nonbasic variable to enter the basic set is to choose the one with the most negative reduced cost $\bar{c}_{ij}$, i.e., choose x_{rs} to be the new basic variable if

$$c_{rs} - u_r - v_s = \min_{(i,j)} (c_{ij} - u_i - v_j) < 0. \tag{8.18}$$

The triangularity of the basis makes it very easy to perform the calculations to determine which variable will leave the basis and makes it easy to update the values of the remaining basic variables. This is especially easy to do by hand if the calculations are done directly on the rectangular array itself.

To start, enter the symbol θ^* in the cell (r, s) to indicate that a value, called $\theta = \theta^*$, will be given to the entering nonbasic variable x_{rs}. Next, the basic entries are adjusted so that the row sums and column sums stay unchanged. This requires appending $(+\theta)$ to some entries, $(-\theta)$ to others, and leaving the rest unchanged. Because the basis is triangular, it will always be possible to locate by eye where a single basic entry is to be converted by an unknown amount $+\theta$ or $-\theta$ or remain unchanged. Once the locations where the $+\theta$ or $-\theta$ adjustments have to be done have been determined, θ is replaced by x_{pq}, the largest numerical value that does not make a basic entry negative. That is, at iteration t, θ takes on the value x^t_{pq} of the smallest entry to which the symbol $(-\theta)$ is appended, so that $x^t_{pq} - \theta$ becomes zero. The variable x_{pq} is then dropped from the basic set. Any basic variable whose value is equal to x^t_{pq} and which is appended by $-\theta$ may be chosen to be dropped from the basic set. The value of θ so determined is then used to recompute all the basic entries for the new basic solution x^{t+1}. This completes the iteration and we once again recompute the multipliers and check for optimality and repeat the process if $x = x^{t+1}$ is not optimal.

The eye scanning procedure used above is not efficient for large problems. Instead, the tree structure associated with the basis is exploited to efficiently adjust the basic variables. Such techniques are described for general networks in Chapter 9.

▷ **Exercise 8.20** Show that it is not possible to have $z \to -\infty$.

8.4.3 ILLUSTRATION OF THE SOLUTION PROCESS

In this section we illustrate the application of the algorithm described on two simple transportation problems.

Example 8.15 (Hitchock [1941]) This example is the original example due to Hitchcock. The optimal solution is found in one iteration. The location and values of an initial basic feasible solution are circled in Figure 8-13. The values u_i and v_j are shown in the marginal row and column. The θ variables are introduced into the rectangular array.

We see that x_{34} leaves the basis and x_{32} enters at a value of $\theta^* = 5$.

Example 8.16 (Solution of the Prototype Example) For the purpose of illustrating the simplex algorithm on the prototype Example 8.1, we assume that the Least-Remaining-Cost rule has been applied and that we have obtained the starting basic feasible solution shown in Example 8.11. The steps of the Simplex Algorithm are illustrated in Figure 8-14.

On iteration 1, the value of u_3 was arbitrarily set equal to 0 because it has the highest number of basic variables in a row. Then it is obvious that

$$\begin{aligned} v_1 &= c_{31} - u_3 = 9, \\ v_3 &= c_{33} - u_3 = 4, \\ v_4 &= c_{34} - u_3 = 8. \end{aligned}$$

Iteration 1: $\theta^* = 5$

					u_i
	+2 10	+2 5	+3 6	(25) 7	25 −1
	+1 8	(20) $-\theta$ 2	+5 7	(5) $+\theta$ 6	25 −2
	(15) 9	θ^* −1 3	(30) 4	(5) $-\theta$ 8	50 0
v_j	15 9	20 4	30 4	35 8	$z = 540$

Iteration 2: Optimal

					u_i
	+1 10	+2 5	+2 6	(25) 7	25 0
	0 8	(15) 2	+4 7	(10) 6	25 −1
	(15) 9	(5) 3	(30) 4	+1 8	50 0
v_j	15 9	20 3	30 4	35 7	$z^* = 535$

Figure 8-13: Hitchcock Transportation Problem

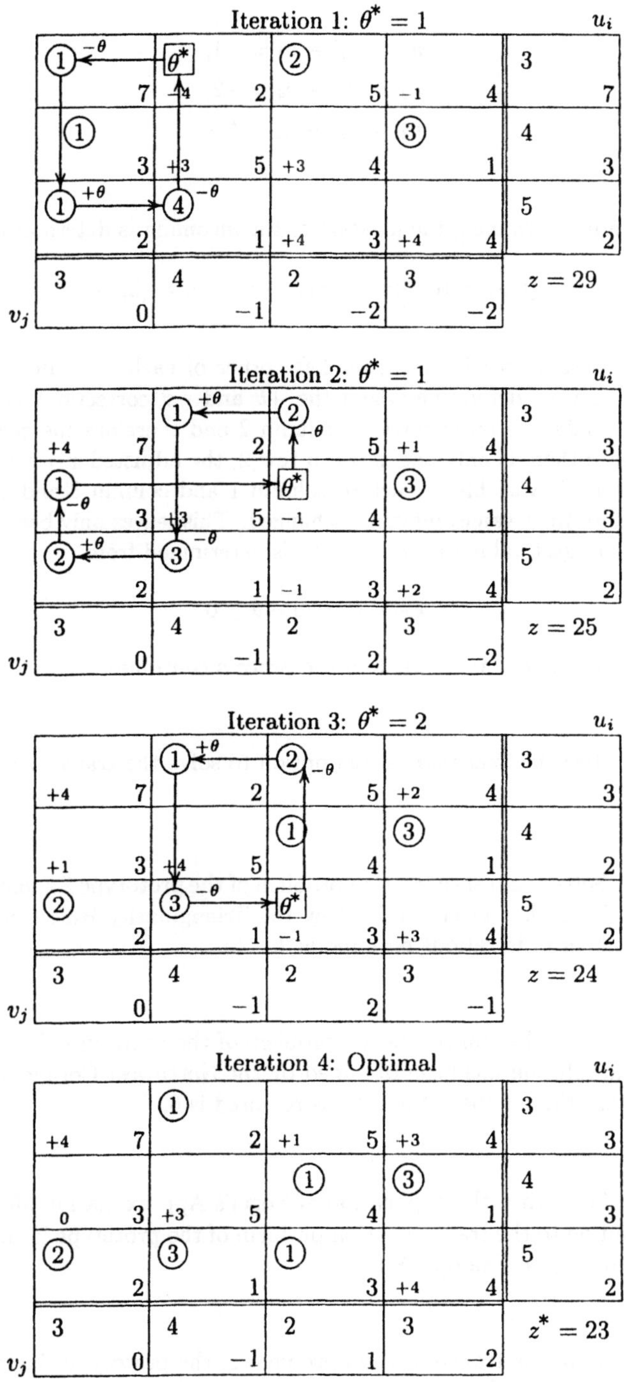

Figure 8-14: Simplex Algorithm on the Prototype Transportation Problem

Using these values we obtain, in turn,

$$\begin{aligned} u_1 &= c_{14} - v_4 = -1, \\ u_2 &= c_{24} - v_4 = -2, \\ v_2 &= c_{22} - u_2 = 4. \end{aligned}$$

Once u_i and v_j are determined, the location of the incoming is determined by variable

$$(r, s) = \operatorname*{argmin}_{(i,j)} \bar{c}_{ij} = c_{ij} - u_i - v_j = (3, 2).$$

Note that the $\bar{c}_{ij}$ are shown in the lower left corner of each cell. Because the basis is triangular, it is easy to determine where the $+\theta$ and $-\theta$ corrections are entered. For example, there is only one basic entry in column 2 and therefore the change in entry is $20 - \theta$. Then, since there is only one entry in row 2, the adjusted x_{42} is $5 + \theta$. Note that there is only one basic variable in each of column 1 and column 3 and also only one in row 1, and therefore their values remain unchanged. This leaves only basic variable x_{34} to be adjusted. The largest value that θ can take is determined from

$$x_{22} - \theta = 20 - \theta \geq 0$$

as $\theta = 20$. The adjustments are made and the process continues.

▷ **Exercise 8.21** Use the `Transportation` option to solve the transportation problem of Example 8.16.

▷ **Exercise 8.22** Solve the transportation problem of the prototype Example 8.1 assuming that the initial solution has been obtained by the Triangularity Rule (see Example 8.9). Show that the number of iterations required is 3.

▷ **Exercise 8.23** Solve the transportation problem of the prototype Example 8.1 assuming that the initial solution has been obtained by the Northwest Corner Rule (see Example 8.10). Show that the number of iterations required is 6.

▷ **Exercise 8.24** Show that the application of Vogel's Approximation Method to find an initial feasible solution to the transportation problem of the prototype Example 8.1 results in an optimal solution (see Example 8.12).

▷ **Exercise 8.25** Solve the transportation problem of the prototype Example 8.1 assuming that the initial solution has been obtained by Russel's Approximation Method (see Example 8.13). Show that the number of iterations required is 2.

8.5 THE ASSIGNMENT PROBLEM

The *assignment problem* is a special case of the classical transportation problem where $m = n$ and $a_i = 1$ for all i and $b_j = 1$ for all j. A typical example is finding the best way to assign n persons to n jobs, assuming that the "desirability" of assigning individual i to job j is d_{ij}. We shall assume that by means of performance tests, the desirability of assigning the ith person to the jth job can in some sense be determined and the objective is to maximize the sum of the d_{ij} for individuals assigned to jobs. The problem can be converted to a minimization problem by using the negative of the d_{ij} and denoting them by c_{ij}, which we will refer to as a cost.

▷ **Exercise 8.26** Show that the minimizing problem is the same as finding a minimizing permutation.

FORMULATION AND PROPERTIES

The assignment problem can be easily solved by reformulating it as a transportation problem with the added requirement that

$$x_{ij} = \begin{cases} 1 \text{ if the } i\text{th individual is assigned to the } j\text{th job,} \\ 0 \text{ otherwise.} \end{cases} \tag{8.19}$$

Because it is assumed that each person can be assigned only one job, we must have

$$\sum_{j=1}^{n} x_{ij} = 1 \quad \text{for } i = 1, \dots, n, \tag{8.20}$$

and because each job is assigned to only one person,

$$\sum_{i=1}^{n} x_{ij} = 1 \quad \text{for } j = 1, \dots, n. \tag{8.21}$$

Finally, the objective of the assignment problem is to choose x_{ij} satisfying (8.19), (8.20), and (8.21) in such a way that the total cost,

$$z = \sum_{i=1}^{n} c_{ip_i} = \sum_{i=1}^{n} \sum_{j=1}^{n} c_{ij} x_{ij} \tag{8.22}$$

is minimized, where $p_1, p_2, \dots, p_n$ is the permutation associated with the assignment.

In general, imposing a condition on a linear program, that the values of the variables be integers cannot be solved by LP techniques. Transportation problems whose constant terms are integers are an exception. To convert (8.19)–(8.22) to a linear program all we need to do is relax the condition $x_{ij} = 0$ or 1 to

$$x_{ij} \geq 0 \quad \text{for } i = 1, \dots, n; \ j = 1, \dots, n. \tag{8.23}$$

x_{11}	x_{12}	x_{13}	x_{14}	$= 1$
x_{21}	x_{22}	x_{23}	x_{24}	$= 1$
x_{31}	x_{32}	x_{33}	x_{34}	$= 1$
x_{41}	x_{42}	x_{43}	x_{44}	$= 1$
$=$ 1	$=$ 1	$=$ 1	$=$ 1	

Figure 8-15: Compact Representation of an Assignment Problem

THEOREM 8.5 (Integral Property of the Assignment Problem) *An optimal solution of the assignment problem with the requirement that $x_{ij} = 0$ or 1 is the same as an optimal solution of the linear programming problem given by* (8.20), (8.21), (8.22), *and* (8.23).

▷ **Exercise 8.27** Use the integral property of the transportation problem to prove Theorem 8.5.

The assignment problem can be represented compactly in an assignment array similar to the transportation problem. See Figure 8-15 for a compact representation of a 4×4 example.

COROLLARY 8.6 (Number of Nonzero Basic Variables Is n) *The integer solution x to the assignment problem has exactly one $x_{ij} = 1$ in every row in the assignment array and the sum n of these x_{ij} counts the number of basic variables that are positive.*

Comment: The linear program equivalent to an assignment problem has the property that every basic solution is degenerate, since there are $2n - 1$ basic variables and exactly n basic variables must receive unit value, the remaining $n - 1$ basic variables must therefore all be zero. Thus the linear program is highly degenerate. To the best of the authors' knowledge it is not known whether a special rule is needed to avoid cycling. The random choice rule can be used to avoid cycling with probability 1.

TWO ILLUSTRATIONS OF THE ASSIGNMENT PROBLEM

Example 8.17 (Prototype Assignment Problem) Suppose that we need to assign four employees to four new tasks at the training costs shown in Figure 8-16. We set up

Task

Employee	1	2	3	4
1	3	7	11	8
2	0	4	4	6
3	0	4	10	9
4	0	0	6	5

Figure 8-16: Training Cost Data for Assigning Employees to Tasks

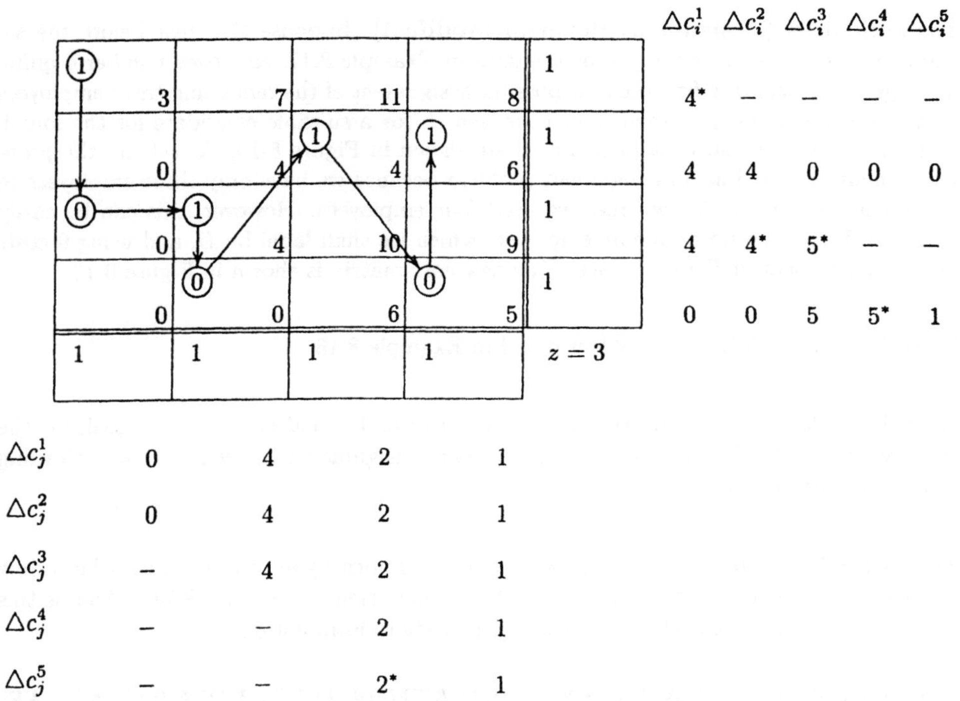

Figure 8-17: Initial Solution to an Assignment Problem by Vogel's Method

Employee	Task 1	Task 2	Task 3	Task 4
1	3	7	11	8
2	0	4	4	6
3	0	4	10	9

Figure 8-18: Training Cost Data for Assigning 3 Employees to 4 Tasks

a 4×4 assignment array and find an initial basic feasible solution by applying Vogel's Method as shown in Figure 8-17.

▷ **Exercise 8.28** Compute the simplex multipliers and reduced costs for the problem in Example 8.17. If not optimal, apply the Transportation Simplex Algorithm to find the optimal solution.

Example 8.18 (Assignment Problem Modified) Suppose that just before the assignments to tasks can be made for the data in Example 8.17, employee number 4 quits. Management decides to find the best possible assignment of the remaining three employees to three of the tasks, and then, over time, search for a suitable candidate for the fourth task. The training cost data in this case are shown in Figure 8-18. As set up, the problem cannot be posed as an assignment problem because we have only three employees to be assigned to four tasks, whereas we need four employees. However, this can be easily corrected by setting up a dummy employee, which we shall label by D, and using 0 costs for the assignment of D to any task. The new cost matrix is shown in Figure 8-19.

▷ **Exercise 8.29** Solve the problem posed in Example 8.18.

▷ **Exercise 8.30** Show that setting an arbitrarily high equal cost to every task for the dummy employee will result in the same optimal assignments but will give a misleading value to the objective.

▷ **Exercise 8.31** Suppose that employee number 2 informs you that he will not be willing to take on Task 3. How would you modify the formulation in Example 8.18 to handle this situation within the context of an assignment problem formulation.

CONVERSION OF A TRANSPORTATION PROBLEM TO AN ASSIGNMENT PROBLEM

We have already seen that the assignment problem, is a special case of the transportation problem. It turns out that the transportation problem is a special case

Employee \ Task	1	2	3	4
1	3	7	11	8
2	0	4	4	6
3	0	4	10	9
D	0	0	0	0

Figure 8-19: Modified Training Cost Data for Assigning 3 Employees to 4 Tasks

of the assignment problem as the next exercise demonstrates.

▷ **Exercise 8.32** Assume that a_i and b_j are integers; if they are rational numbers then these could be replaced by integers through a change of units. If a_i and b_j are irrational numbers, then these could be approximated by rational numbers and then replaced by integers.

1. Show how to convert the transportation problem into an equivalent assignment problem by replacing the ith row by a row set I of a_i equations each summing to 1, replacing the jth column by a column set J of b_j equations each summing to 1, and assigning cost c_{ij} to all new variables that are in the intersection of row set I and column set J.
2. Prove that the resulting optimal basic feasible solution to the resulting assignment problem when aggregated back to the original transportation problem will be an optimal feasible solution but need not be a basic solution to the original transportation problem.
3. Assume that we have solved the transportation problem to obtain an optimal basic feasible solution. Suppose that the solution is unique (if it is not unique, make it unique by adding 1 to the nonbasic costs). Show that the solution to the equivalent assignment problem can be used to generate the unique optimal basic feasible solution to the transportation problem.

8.6 EXCESS AND SHORTAGE

In some applications it may be impossible (or unprofitable) to supply all that is required or to ship all that is available, in which case costs must be given to each unit not supplied and each unit not shipped from the source. In this section we illustrate how to formulate and solve such a problem. This problem can be converted to the

x_{11}	x_{12}	$\cdots$	x_{1n}	$\leq a_1$
x_{21}	x_{22}	$\cdots$	x_{2n}	$\leq a_2$
$\cdots$	$\cdots$	$\cdots$	$\cdots$	$\leq \cdots$
x_{m1}	x_{m2}	$\cdots$	x_{mn}	$\leq a_m$
$\leq$ b_1	$\leq$ b_2	$\leq$ $\cdots$	$\leq$ b_n	

Figure 8-20: Transportation Problem with Inequalities

classical transportation problem by the addition of the slack (excess and shortage) variables. Other variants of this problem are discussed in Exercises 8.38 and 8.39.

8.6.1 MATHEMATICAL STATEMENT

Aside from its objective, a transportation problem with excess and shortage is of the form

$$\begin{aligned} \sum_{j=1}^{n} x_{ij} &\leq a_i, \quad i = 1, \ldots, m, \\ \sum_{i=1}^{m} x_{ij} &\leq b_j, \quad j = 1, \ldots, n, \\ x_{ij} &\geq 0, \quad i = 1, \ldots, m, \; j = 1, \ldots, n. \end{aligned} \tag{8.24}$$

This is shown compactly in Figure 8-20.

Example 8.19 (Variation of Prototype Example) Suppose that in the prototype Example 8.1, which we discussed earlier, we replace the equalities by inequalities of the "$\leq$" type. Then the problem is to find

$$\begin{aligned} \min\ z = {} & 7x_{11} + 2x_{12} + 5x_{13} + 4x_{14} + 3x_{21} + 5x_{22} \\ & + 4x_{23} + 1x_{24} + 2x_{31} + 1x_{32} + 3x_{33} + 4x_{34} \end{aligned} \tag{8.25}$$

subject to

$$\begin{array}{llllllllllllll}
x_{11} & +x_{12} & +x_{13} & +x_{14} & & & & & & & & & \leq & 3 \\
 & & & & x_{21} & +x_{22} & +x_{23} & +x_{24} & & & & & \leq & 4 \\
 & & & & & & & & x_{31} & +x_{32} & +x_{33} & +x_{34} & \leq & 5 \\
x_{11} & & & & +x_{21} & & & & +x_{31} & & & & \leq & 3 \\
 & x_{12} & & & & +x_{22} & & & & +x_{32} & & & \leq & 4 \\
 & & x_{13} & & & & +x_{23} & & & & +x_{33} & & \leq & 2 \\
 & & & x_{14} & & & & +x_{24} & & & & +x_{34} & \leq & 3
\end{array} \tag{8.26}$$

	x_{01}	x_{02}	$\cdots$	x_{0n}	
x_{10}	x_{11}	x_{12}	$\cdots$	x_{1n}	$= a_1$
x_{20}	x_{21}	x_{22}	$\cdots$	x_{2n}	$= a_2$
$\cdots$	$\cdots$	$\cdots$	$\cdots$	$\cdots$	$= \cdots$
x_{m0}	x_{m1}	x_{m2}	$\cdots$	x_{mn}	$= a_m$
	$=$ b_1	$=$ b_2	$=$ $\cdots$	$=$ b_n	

Figure 8-21: Transportation Problem with Row and Column Slacks

and $x_{ij} \geq 0$, for $i = 1, \ldots, 3$, $j = 1, \ldots, 4$.

Introducing slack (excess and shortage) variables x_{i0}, the *excess* at the ith origin, and x_{0j}, the *shortage* at the jth destination, and letting c_{i0} and c_{0j} be the positive penalties for the excesses at the origin and the shortages at the destinations, we have

$$\begin{array}{lrcll} \text{Minimize} & \displaystyle\sum_{i=1}^{m} c_{i0}x_{i0} + \sum_{j=1}^{n} c_{0j}x_{0j} + \sum_{i=1}^{m}\sum_{j=1}^{n} c_{ij}x_{ij} & = & z & \\ \text{subject to} & \displaystyle x_{i0} + \sum_{j=1}^{n} x_{ij} & = & a_i, & i = 1, \ldots, m, \\ & \displaystyle x_{0j} + \sum_{i=1}^{m} x_{ij} & = & b_j, & j = 1, \ldots, n, \\ & x_{i0} \geq 0,\ x_{0j} \geq 0,\ x_{ij} \geq 0, & & & i = 1, \ldots, m,\ j = 1, \ldots, n. \end{array} \tag{8.27}$$

This is shown compactly in Figure 8-21. It should be noted that for problem (8.27) if there is no penalty associated with failure to deliver the required amounts, then there is really no problem at all; simply ship nothing. A meaningful problem exists only when failure to ship means a loss of revenue or goodwill, i.e., when positive cost factors c_{i0} or c_{0j} are assigned to the surplus or shortage.

8.6.2 PROPERTIES OF THE SYSTEM

RANK OF THE SYSTEM

It is straightforward to see that the addition of slack variables to (8.24) now causes all $m + n$ equations to be independent, in contrast to the classical transportation case, in which only $m + n - 1$ are independent. For example, it is easy to see that the slack variables constitute a basic set of $m + n$ variables. In fact,

THEOREM 8.7 (Slack in Basis) *Every basis contains at least one slack variable.*

Proof. Assume on the contrary that some basis has no slack variable. Then this basis would also constitute a basis for the analogous transportation problem without any slack variables. This can only happen if $\sum_{i=1}^{m} a_i = \sum_{j=1}^{n} b_j$. However, in this case, we know that the number of basic variables cannot exceed $m + n - 1$; a contradiction. ∎

▷ **Exercise 8.33** Demonstrate Theorem 8.7 on Example 8.19.

▷ **Exercise 8.34** Suppose that in the classical transportation problem (8.3), all the row equations are replaced by $\geq$ and column equations are replaced by $\leq$ with the condition $\sum_{i=1}^{m} a_i = \sum_{j=1}^{n} b_j$ and arbitrary costs assigned to the surplus and slack variables. Show that this inequality problem is the same as the original problem. Show that the same is true if all the inequalities are reversed.

BASIS TRIANGULARITY

THEOREM 8.8 (Basis Triangularity) *Every basis is triangular.*

▷ **Exercise 8.35** Prove Theorem 8.8.

▷ **Exercise 8.36** Demonstrate Theorem 8.8 on Example 8.19.

8.6.3 CONVERSION TO THE CLASSICAL FORM

The problem (8.27) can be converted to the form of the classical transportation problem, with an additional rank deficiency of 1, by defining

$$x_{00} = \sum_{i=1}^{m} \sum_{j=1}^{n} x_{ij}, \tag{8.28}$$

where $\sum_{i=1}^{m}\sum_{j=1}^{n} x_{ij}$ is the *total* amount shipped from all origins to all destinations. From this definition, it is straightforward to see that

$$\sum_{i=1}^{m} a_i - \sum_{i=1}^{m} x_{i0} = \sum_{j=1}^{n} b_j - \sum_{j=1}^{n} x_{0j} = \sum_{i=1}^{m}\sum_{j=1}^{n} x_{ij} = x_{00}. \tag{8.29}$$

Augmenting the transportation problem (8.24) with the slack variables x_{i0} and x_{0j} and using Equation (8.29), we get the classical transportation problem:

$$\begin{array}{lrcll} \text{Minimize} & \displaystyle\sum_{i=0}^{m}\sum_{j=0}^{n} c_{ij}x_{ij} & = & z, & \text{with } c_{00} = 0 \\ \text{subject to} & \displaystyle\sum_{j=0}^{n} x_{0j} & = & \displaystyle\sum_{j=1}^{n} b_j, & \\ & \displaystyle\sum_{j=0}^{n} x_{ij} & = & a_i, & i = 1,\ldots,m, \\ & \displaystyle\sum_{i=0}^{m} x_{i0} & = & \displaystyle\sum_{i=1}^{m} a_i, & \\ & \displaystyle\sum_{i=0}^{m} x_{ij} & = & b_j, & j = 1,\ldots,n, \\ & x_{ij} & \geq & 0, & i = 0,1,\ldots,m,\ j = 0,1,\ldots,n. \end{array} \tag{8.30}$$

This is shown compactly in Figure 8-22.

▷ **Exercise 8.37** Prove that (8.30) has an additional rank deficiency of 1 over the classical transportation problem, i.e., it has rank $m+n$. Show that adding an arbitrary $k \neq 0$ to $\sum_{i=1}^{m} a_i$ and to $\sum_{j=1}^{n} b_j$ in (8.30) increases the rank by 1 to $m+n+1$. What conditions must be placed on k?

▷ **Exercise 8.38** Write down an equivalent classical transportation problem formulation in equation and array format when there are surpluses only, i.e., when the availabilities exceed the requirements $\left(\text{i.e., } \sum_i a_i > \sum_j b_j\right)$ but requirements must be met exactly.

▷ **Exercise 8.39** Write down an equivalent classical transportation problem formulation in equation and array format when there are shortages only, i.e., when the requirements exceed the availabilities $\left(\text{i.e., } \sum_j b_j > \sum_i a_i\right)$ but all available supplies must be shipped.

▷ **Exercise 8.40** Show that a starting basic feasible solution to (8.30) can be obtained by choosing as the $m+n$ basic variables the slacks x_{i0} for $i = 1,\ldots,m$ and x_{0j} for $j = 1,\ldots,n$. Can you think of a better way to get started?

x_{00}	x_{01}	x_{02}	$\cdots$	x_{0n}	$=\sum_{j=1}^{n} b_j$
x_{10}	x_{11}	x_{12}	$\cdots$	x_{1n}	$= a_1$
x_{20}	x_{21}	x_{22}	$\cdots$	x_{2n}	$= a_2$
$\cdots$	$\cdots$	$\cdots$	$\cdots$	$\cdots$	$= \cdots$
x_{m0}	x_{m1}	x_{m2}	$\cdots$	x_{mn}	$= a_m$
$=$ $\sum_{i=1}^{m} a_i$	$=$ b_1	$=$ b_2	$=$ $\cdots$	$=$ b_n	

Figure 8-22: Equivalent Classical Transportation Model

8.6.4 SIMPLEX MULTIPLIERS AND REDUCED COSTS

As in the classical transportation problem case we let u_i and v_j be the simplex multipliers (or "implicit prices") associated with row i column j. Because the rank of the system is $m+n$, two of the simplex multipliers can be set arbitrarily. Note that we need not define slack multipliers, u_0 or v_0, since there is no equation pertaining to row zero or to column zero. Hence it is convenient to assign fictitious prices to these slack multipliers, i.e., $u_0 = 0$ and $v_0 = 0$. Then the remaining $m + n$ prices u_i and v_j are chosen so that

$$u_i + v_j = c_{ij} \quad \text{if } x_{ij} \text{ is basic.} \tag{8.31}$$

Then the reduced costs are

$$\begin{aligned} \bar{c}_{ij} &= c_{ij} - (u_i + v_j) && \text{for } i \neq 0, j \neq 0, \\ \bar{c}_{0j} &= c_{0j} - v_j && \text{for } j \neq 0, \text{and} \\ \bar{c}_{i0} &= c_{i0} - u_i && \text{for } i \neq 0. \end{aligned} \tag{8.32}$$

If all the reduced costs for the nonbasic variables are nonnegative, the solution is optimal.

▷ **Exercise 8.41** Show that the system of equations in u_i and v_j is triangular.

8.7 PRE-FIXED VALUES AND INADMISSIBLE SQUARES

In practice, it quite often happens that some of the variables must assume predetermined values, either 0 or some other fixed values. For example, there may be no route from an origin i to a destination j in a network; i.e., the variable x_{ij} must be zero. In the problem of assigning people to jobs, certain assignments may be mandatory; for example, assigning a physician to a medical position. Subtracting its predetermined value from the corresponding row and column totals it can be converted to a *zero-restricted* variable. One way to solve the problem with zero-restricted variables is to assign an arbitrarily large cost M to these variables so that if they appear in a feasible optimal basis with a positive value then we know the original zero-restricted variable problem is infeasible. This is the *Big M* method discussed in the Notes & Bibliography section of Chapter 3. For a transportation problem, one way to choose M is to choose it greater than or equal to $\max_{(i,j)} c_{ij} \left(\sum_i a_i\right)$ because

$$\sum_{i=1}^{m}\sum_{j=1}^{n} c_{ij}x_{ij} \leq \max_{(i,j)} c_{ij} \sum_{i=1}^{m}\sum_{j=1}^{n} x_{ij} = \max_{(i,j)} c_{ij} \left(\sum_{i=1}^{m} a_i\right) = \max_{(i,j)} c_{ij} \left(\sum_{j=1}^{n} b_j\right).$$

Another way to solve such a problem is to shade the cells in the array corresponding to the zero-restricted variables. These shaded cells are called *inadmissible* to distinguish them from the regular cells. If only a few cells are inadmissible, the best practical procedure is to attempt to find an initial basic feasible solution by selecting the next basic variable from among the admissible cells, say, by the Least-Remaining-Cost rule discussed in Section 8.3.

In the event that the above procedure fails to furnish enough basic variables, then the zero-restricted variables (or inadmissible cells) can be used to furnish artificial variables for a Phase I procedure, in which a basic feasible solution will be constructed if possible. The Phase I objective of the Simplex Method can be set up as the infeasibility form

$$w = \sum_{i=1}^{m}\sum_{j=1}^{n} d_{ij}x_{ij}, \tag{8.33}$$

where the d_{ij} are defined by

$$d_{ij} = \begin{cases} 1 & \text{if } x_{ij} \text{ is zero-restricted,} \\ 0 & \text{if } x_{ij} \text{ otherwise.} \end{cases} \tag{8.34}$$

If a feasible solution exists, then $\min w$ is zero; otherwise, if the problem is infeasible, $\min w$ will be positive.

Just as in the solution of a linear program with the Simplex Method, it may happen that the problem is feasible, but that some inadmissible variables remain in the basic set at the end of Phase I. Because of the way that the Phase I objective has been defined, we know that these variables will be zero in value, and they must be kept at zero value throughout the remaining procedure. To do this apply the starting Phase II rule of the Simplex Method: drop any nonbasic x_{ij} from further consideration if its relative infeasibility factor $\bar{d}_{ij} = d_{ij} - u_i - v_j$ is positive, where u_i and v_j are the simplex multipliers associated with the infeasibility form w at the end of Phase I.

We demonstrate such a Phase I approach in the context of the capacitated transportation problem, see Example 8.20.

▷ **Exercise 8.42** For the Hitchcock Transportation Problem of Example 8.15, suppose that x_{14} is invalid. Perform a Phase I approach followed by a Phase II to find a new optimal solution.

8.8 THE CAPACITATED TRANSPORTATION PROBLEM

A transportation problem with upper bound on the variables is called a *capacitated transportation problem*:

$$\begin{array}{lrll}
\text{Minimize} & \displaystyle\sum_{i=1}^{m}\sum_{j=1}^{n} c_{ij}x_{ij} & = z & \\
\text{subject to} & \displaystyle\sum_{j=1}^{n} x_{ij} & = a_i, & i = 1,\ldots,m, \\
& \displaystyle\sum_{i=1}^{n} x_{ij} & = b_j, & j = 1,\ldots,n, \\
& 0 \le x_{ij} \le h_{ij}, & i = 1,\ldots,m, \; j = 1,\ldots,n. &
\end{array} \tag{8.35}$$

A rectangular array for recording the data and the primal and dual solution of such a transportation problem is shown in Figure 8-23 for a 3×5 case.

THEOREM 8.9 (Optimality Test) *A feasible solution $x_{ij} = x^*_{ij}$ for the capacitated transportation problem is optimal if there is a set of implicit prices $u_i = u^*_i$ and $v_i = v^*_j$, and relative cost factors $\bar{c}_{ij} = c_{ij} - u^*_i - v^*_j$, such that*

$$\begin{aligned}
0 < x^*_{ij} < h_{ij} &\Longrightarrow \bar{c}_{ij} = 0, \\
x^*_{ij} = 0 &\Longrightarrow \bar{c}_{ij} \ge 0, \\
x^*_{ij} = h_{ij} &\Longrightarrow \bar{c}_{ij} \le 0.
\end{aligned} \tag{8.36}$$

▷ **Exercise 8.43** Prove Theorem 8.9.

x_{11} h_{11} c_{11}	x_{12} h_{12} c_{12}	x_{13} h_{13} c_{13}	x_{14} h_{14} c_{14}	x_{15} h_{15} c_{15}	a_1 u_1
x_{21} h_{21} c_{21}	x_{22} h_{22} c_{22}	x_{23} h_{23} c_{23}	x_{24} h_{24} c_{24}	x_{25} h_{25} c_{25}	a_2 u_2
x_{31} h_{31} c_{21}	x_{32} h_{32} c_{22}	x_{33} h_{33} c_{23}	x_{34} h_{34} c_{24}	x_{35} h_{35} c_{25}	a_3 u_3
b_1 v_1	b_2 v_2	b_3 v_3	b_4 v_4	b_5 v_5	

Figure 8-23: Capacitated Transportation Problem

x_{11} 12 10	x_{12} 13 5	x_{13} 5 6	x_{14} 20 7	25
x_{21} 14 8	x_{22} 20 2	x_{23} 10 7	x_{24} 9 6	25
x_{31} 18 9	x_{32} 4 3	x_{33} 25 4	x_{34} 7 8	50
15	20	30	35	

Figure 8-24: Capacitated Transportation Problem Example

The method for solving a transportation problem with upper bounds is an extension of the method for solving a transportation problem without bounds. While simple rules have been devised for finding an initial solution to an uncapacitated transportation problem, none, to our knowledge, have been found for the capacitated case. If one were able to do this, one could apply it to find a simple solution to the problem of finding an assignment of m men to m jobs where certain men are excluded from certain jobs. In mathematical terms, given an $m \times m$ incidence matrix (elements 0 or 1), pick out a permutation of ones or show that none exists. So far no one has been able to develop a *noniterative procedure* for solving this problem.

We shall illustrate an approach for solving a capacitated transportation problem with the following example.

Example 8.20 (Solution of a Capacitated Transportation Problem) Consider the capacitated transportation problem in Figure 8-24. In trying to find an initial solution for the problem in Figure 8-24, we begin by selecting a box (cell) with the minimum c_{ij}, which in Figure 8-24 is $c_{22} = 2$. Next we assign as high a value as possible to the corresponding variable, in this case x_{22}, without forcing any variable to exceed its upper

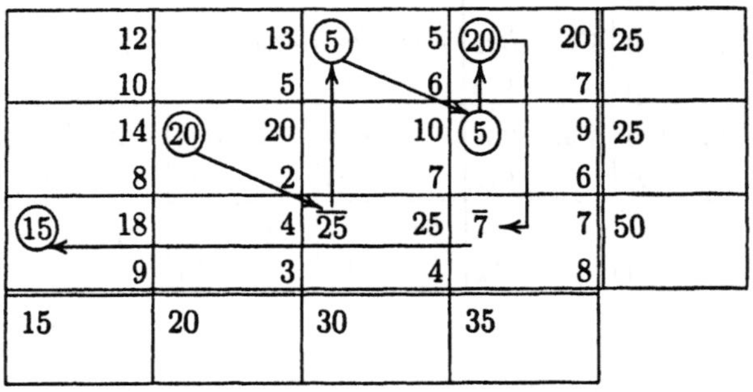

Figure 8-25: Initial Solution of a Capacitated Transportation Problem

bound, that is, we set $x_{22} = 20$. If the value of the variable is limited by a row or column equation rather than its bound, we consider it to be a basic variable and make no more entries in its row or column. If, on the other hand, the value of the variable is limited by its upper bound, then we consider the variable to be nonbasic at its upper bound and indicate this by putting a bar on the entry. If there is a tie between the two types of limitations, we always consider the row or column as limiting and consider the variable basic. Then we repeat the procedure with the remaining boxes.

Application of the above procedure to Figure 8-24 yields, in order, the assignments $x_{22} = 20$ (basic), $x_{33} = 25$ (bounded), $x_{13} = 5$ (basic), $x_{24} = 5$ (basic), $x_{14} = 20$ (basic), $x_{34} = 7$ (bounded), $x_{31} = 15$ (basic) as shown in Figure 8-25. Since the third row and fourth column still have 3 units unassigned, the solution is not feasible. Extra "short" boxes corresponding to an $i = 0$ row and $j = 0$ column are added to the array; the original c_{ij} are replaced by $d_{ij} = 0$, and the shortage boxes have $d_{ij} = 1$ (see iteration 0 of Phase I in Figure 8-26).

Note that $d_{30} = d_{04} = 1$ must equal u_3 and v_4 respectively, since we have previously shown that slack rows and columns can be regarded as having prices u_0 and v_0 equal to zero.

Continuing with Phase I, minimizing the sum $x_{04} + x_{30}$ of the artificial variables, a feasible solution is found in one iteration as shown in Figure 8-26. The original cost factors c_{ij} are next restored, as shown in the array for iteration 1 in Figure 8-26.

This solution is not yet optimal, because x_{34} is a nonbasic variable at its upper bound with relative cost factor $\bar{c}_{34} = \bar{c}_{34} - u_3 - v_4 = 8 - 0 - 7 = +1$. Hence, it pays (except in degenerate cases) to decrease x_{34} from its upper bound value, keeping the other nonbasic variables fixed and adjusting the basic variables. The greatest decrease θ that maintains feasibility is $\theta = 1$ because at this value, x_{24} reaches its upper bound, i.e., $x_{24} = 8 + \theta \leq 9$. The third array in Figure 8-26 is optimal.

The foregoing method illustrated in Example 8.20 implies the following theorem, whose proof for the general bounded transportation problem is left as an exercise.

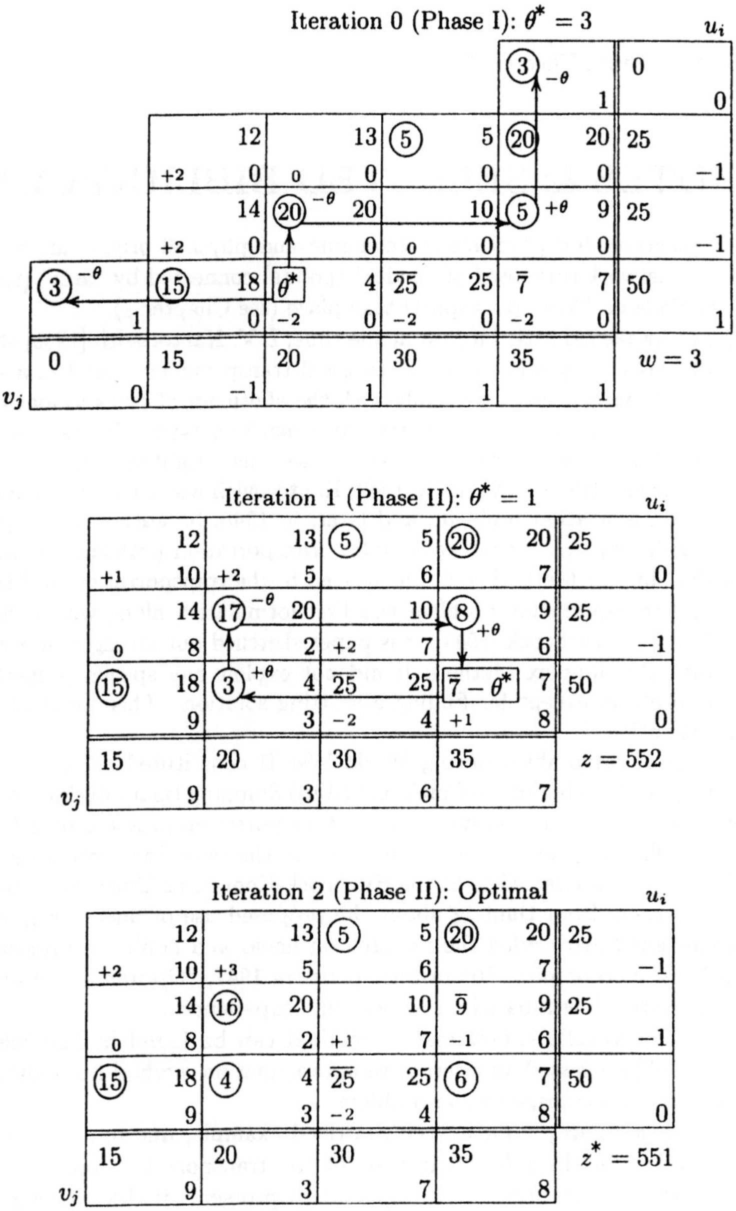

Figure 8-26: Solution of a Capacitated Transportation Problem

THEOREM 8.10 (Integral Property) *In a capacitated transportation problem, if the upper bounds, the availabilities, and the requirements are all integers, then every basic solution will be integral.*

▷ **Exercise 8.44** Prove Theorem 8.10.

8.9 NOTES & SELECTED BIBLIOGRAPHY

Many linear programming problems of economic and physical origin can be abstractly formulated as a network composed of "nodes" (points) connected by "arcs" (routes), over which various kinds of "flow" (transport) take place (see Chapter 9).

Although he awakened little interest at the time, L.V. Kantorovich [1939] showed that a class of problems closely related to the classical transportation case has a remarkable variety of applications concerned typically with the allotment of tasks to machines whose costs and rates of production vary by task and machine type. He gave a useful but incomplete algorithm for solving such problems (see the weighted distribution problem, Chapter 21, in Dantzig [1963]). Again, in 1942, Kantorovich wrote a paper on a *continuous* version of the transportation problem, and later, in 1948, he authored an applicational study, jointly with Gavurin, on the *capacitated* transportation problem. Dantzig [1951b] showed how the Simplex Method could be applied to the transportation problem.

The classical transportation problem was first formulated, along with a constructive solution, by Frank L. Hitchcock [1941]. His paper sketched out an algorithm with points in common with the Simplex Method; it did not exploit any special properties of the transportation problem except for finding a starting solution. This paper also failed to attract much attention.

The shortage of cargo ships during World War II constituted a critical bottleneck. T.C. Koopmans, as a member of the Combined Allied Shipping Board during World War II, used properties of the optimal solutions of the transportation problem to help find ways to reduce overall shipping times. Because of this and the work done earlier by Hitchcock, the classical case is often referred to as the Hitchcock-Koopmans Transportation Problem. In 1947, after learning from Dantzig about the proposed use of linear programming for planning, Koopmans spearheaded the research on linear and nonlinear programs for the study of problems in economics. His historic paper in 1947, "Optimum Utilization of the Transportation System," was based on his wartime experience.

A description of Vogel's approximation method can be found in Reinfield & Vogel [1958]. Russel [1969] suggested an alternative approximation method for finding an initial near-optimal basis for a transportation problem.

See *Linear Programming 2* for a very contrived example, due to L. Johnson, of the occurrence of cycling in the solution process of the transportation problem. It is not known whether cycling can occur in transportation problems if the entering variable is chosen based on the usual rule of picking the one that has the most negative reduced cost. Besides other rules, such as the Random Choice Rule, Bland's Rule, etc., the simple perturbation scheme to avoid degeneracy, discussed in *Linear Programming 2*, can be used with a trivial amount of extra work.

Another work before the era of linear programming was that of mathematician E. Egerváry [1931].. His 1931 paper considered the problem of finding a permutation of ones in a square matrix composed entirely of zero and one elements. Based on this investigation, Kuhn [1955] developed an efficient algorithm (primal-dual), which he called the Hungarian Method, for solving assignment problems (see Nering & Tucker [1993]). Kuhn's approach, in turn, underlies the Ford-Fulkerson Method for solution of the classical transportation problem (see Chapter 20 in Dantzig [1963]). A detailed survey of assignment problem algorithms can be found in Ahuja, Magnanti, & Orlin [1989]. In their 1993 book, Ahuja, Magnanti, & Orlin note that many of the assignment problems share common features; for example, the successive shortest path algorithm described in their book forms the basis for many of these algorithms. Bertsekas [1988] has developed an auction algorithm for solving the assignment problem. Gabow & Tarjan [1989] developed a cost scaling algorithm that runs in $O\big(m^{\frac{1}{2}} n \log(mC)\big)$ time, where C is the largest cost for any assignment, m is the number of nodes, and n is the number of arcs. Orlin & Ahuja [1992] also obtain the same running time of $O\big(m^{\frac{1}{2}} n \log(mC)\big)$ by incorporating scaling within the context of an auction algorithm.

Computational results for the various specialized algorithms for solving the assignment problem can be found in Bertsekas [1988], Kennington & Wang [1991] and Zaki [1990]. Computer code listings in the FORTRAN programming language for some of the algorithms are available in Carpento, Martello, & Toth [1988].

Square arrays of nonnegative numbers x_{ij} with the property that all row and all column sums are unity, frequently appear in statistical theory. They are called *doubly stochastic* matrices, and the x_{ij} are interpreted as probabilities (not necessarily zero or one). When such arrays have all x_{ij} zero or one, they are called *permutation matrices.* Garret Birkhoff [1946] showed that the set of permutation matrices is given by the extreme points of the convex set defined by the conditions for a doubly stochastic matrix with nonnegative entries. Von Neumann [1953] establishes Birkhoff's theorem by reducing an assignment problem to an interesting matrix game. See also Marcus [1960].

The methods we discussed in this chapter to update the values of the basic variables are useful for small problems only. To solve large problems on a computer, a more sophisticated approach for keeping track of and updating basic variables is needed. These techniques are based on graph-theoretical concepts. Discussion of these methods can be found in Ahuja, Magnanti, & Orlin [1993], Chvátal [1983], Glover, Karney, & Klingman [1972, 1973, 1974], Glover, Karney, Klingman, & Napier [1974], and Murty [1983]. Details about one such technique is described in *Linear Programming 2.*

8.10 PROBLEMS

8.1 Consider the transportation array shown below for a problem where the sum of the supplies is equal to the sum of the demands.

5	3	(25) 4	25 −1
8	6	(15) 2	15 −3
(20) 4	(30) 2	(x_{33}) 5	a_3 0
7	6	20 8	20 3
40 4	30 2	b_3 5	

(a) Show that the values $x_{33} = 2$, $a_3 = 52$, and $b_3 = 42$ make the given basic feasible solution optimal.

(b) What values of a_3 and b_3 give a degenerate optimal basic feasible solution?

(c) Why does some $u_i < 0$ and other $u_i > 0$ not contradict dual feasibility?

(d) Assume $a_3 = 60$, $b_3 = 50$, so that $x_{33} = 10$. Is this a unique optimal basic feasible solution? If so explain why; if not find an alternative optimal basic feasible solution.

8.2 Omit the primal basic and dual variables and reorder the rows and columns of Example 8.1 so that the sources are in the order 3, 1, 2, and the destinations are in the order 1, 2, 4, 3. This results in the following transportation array:

2	1	4	3	5
7	2	4	5	3
3	5	1	4	4
3	4	3	2	

(a) Find an initial solution by randomly applying the general Triangularity Rule and then solve the problem by hand by the Simplex Algorithm.

(b) Find an initial solution by the Northwest Corner Rule and then solve the problem by hand by the Simplex Algorithm.

(c) Find an initial solution by the Least-Remaining-Cost Rule and then solve the problem by hand by the Simplex Algorithm.

(d) Find an initial solution by Vogel's Method and then solve the problem by hand by the Simplex Algorithm.

(e) Find an initial solution by Russel's Method and then solve the problem by hand by the Simplex Algorithm.

(f) Compare the above five methods. Comment on the numbers of iterations as compared to those obtained without the reordering.

8.3 Find, an optimal feasible solution to the following transportation problem where the numbers M mean that shipment is not possible.

Week	*Demand*	*Production Limit*	*Production cost/set*	*Storage cost/set*
1	4	10	20.00	1.00
2	6	25	30.00	1.00
3	10	20	25.00	1.00
4	8	4	40.00	N/A

Table 8-1: Data for Dinner Set Production Schedule

M	M	9	4	6
4	M	9	M	16
8	M	3	5	50
1	2	M	M	14
10	14	28	34	

Verify your solution by using the `Transportation` software option.

8.4 Solve the transportation problem Example 1.5 on Page 4 by the Simplex Method discussed in this chapter.

8.5 Your wife has recently taken a ceramics class and discovered that she has a talent for making elegant dinner sets. A specialty store around the corner from the class has recently sold a couple of sets on her behalf. Besides the fact that these sets have been well received, the store's four other suppliers have moved out of town and the store owner has offered your wife the job of supplying dinner sets for the next four weeks to meet the store's demand. With a new baby, it would be difficult for her to meet the demand on her own. As a result she has arranged to hire help over the four weeks. The hired help have different skills and hence different rates. Your wife, on looking over the required demand schedule, availability of firing time at the ceramics class and the cost of inventory storage at the class has realized that the problem is nontrivial. She decides to approach you to see whether your claimed expertise in operations research can help her. The demand, schedule, and costs are displayed in Table 8-1 You immediately realize that it can be set up as a linear program. However, on closer examination you notice that it can be formulated as a transportation problem that can be solved very efficiently.

(a) Formulate this problem as a transportation problem. Hint: Let x_{ij} be the number of dinner sets produced in week i to satisfy demand in week j.

(b) Solve it by hand.

(c) Solve it by the `Transportation` software option to verify your solution.

8.6 (a) The optimal solution displayed in Figure 8-14 has $\bar{c}_{12} = 0$. Demonstrate a different basic feasible optimal solution in which we bring x_{12} into the basis.

(b) How much would the cost of x_{11} have to change in the problem of Figure 8-14 before it becomes a candidate for entering the basis.

8.7 (a) *Alternative Optima.* For the transportation problem show that alternative optima exist if $\bar{c}_{ij} = 0$ for some nonbasic variable x_{ij}. Show how to construct an alternative optimal solution.

(b) *Cost Ranging.* Derive relations for determining ranges on costs for basic variables and nonbasic variables that retain optimality of the current solution of a transportation problem.

(c) *Changes in the Right-Hand Side.* What is the effect on the optimal solution of changing a_i to $a_i + \beta$ for some i and correspondingly changing b_j to $b_j + \beta$ for some j.

8.8 Prove that in the classical transportation problem ($u_m = 0$) the values of the implicit simplex multipliers are always +1 or 0 or −1 if all $c_{ij} = 0$ except that $c_{11} = 1$. What happens if $c_{11} = 100$.

8.9 (a) For the prototype example solved in Figure 8-14 add 10 to the first column of costs and solve it. Show that the optimal solution is the same as that obtained in Figure 8-14.

(b) For the prototype example solved in Figure 8-14 add 5 to the second row of costs and solve it. Show that the optimal solution is the same as that obtained in Figure 8-14.

8.10 Consider the transportation problem shown in the following array:

2	1	4	3	5
7	2	4	5	3
7	6	9	8	4
3	4	3	2	

(a) Solve the problem by hand.

(b) Notice that rows 1 and 3 differ by a constant amount equal to 5. Change the availability of row 1 to be the sum of the availabilities in row 1 and row 3, drop row 3, and re-solve the problem.

(c) Show that the latter solution is equivalent to the original problem; i.e., before the two rows were combined. Show how to obtain the solution to the original problem from this latter solution to the problem with the combined rows.

8.11 (a) Prove for the classic transportation problem that the unit costs c_{iq} of any column q can be replaced by $c_{iq} + \alpha$ without affecting the optimal solution; similarly, for any row r, c_{rj} may be replaced by $c_{rj} + \beta$.

(b) Prove that the classic transportation problem with some (or all) $c_{ij} < 0$ can be replaced by an equivalent problem where all $c_{ij} > 0$.

(c) Suppose corresponding values of c_{ij} in two rows differ by a constant; show that the two rows can be combined into a single row. What happens to the corresponding variables.

8.12 Consider the transportation problem shown in the following array:

4	2	1	5	2
M	7	2	1	7
9	M	6	M	6
5	3	4	3	

where M implies that shipment is not possible (think of it as a prohibitively high cost).

(a) Solve the problem by hand using a Phase I / Phase II approach.

(b) Solve it by hand using the Big-M method of Problem 3.28.

(c) Solve it using the `Transportation` software option to verify your solution.

8.13 *Optimal Allocation of Receivers to Transmitters (Dantzig [1963]).* A certain engine testing facility is fully using four kinds of instruments: 200 thermocouples, 50 pressure gauges, 50 accelerometers, and 40 thrust meters. Each is measuring one type of characteristic and transmitting data about it over a separate communication channel. There are four types of receivers, each capable of recording one channel of information: 200 cameras, 150 oscilloscopes, 256 test instruments called "Idiots," and 50 others called "Hathaways." The setup time per channel varies among the different types and also according to the kind of data to be recorded. The allocation table is shown in Figure 8-27, where the setup time c_{ij} is shown in the lower right corner of the square containing x_{ij}.

(a) Assuming that all data must be recorded, find an allocation of receivers to transmitters that minimizes the total setup time.

(b) When recording instruments are in short supply (or are not of the most suitable types), a decision must be reached as to how much of each kind of data *not to record.* The engineers assign the following unit costs assigned to the shortage row and surplus column.

$$\begin{aligned} &c_{01} = 10, \; c_{02} = 10, \; c_{03} = 4, \; c_{04} = 100, \\ &c_{10} = 0, \; c_{20} = -1, \; c_{30} = 0, \; c_{40} = 0, \text{ and} \\ &c_{00} = 0. \end{aligned}$$

That is, for example, it is 25 times more costly to neglect thrust data ($c_{04} = 100$) than to neglect acceleration data ($c_{03} = 4$). In general, however, it is less costly to *record* data than to *neglect* it.

8.14 Consider the following capacitated transportation problem:

Recording Instrument i	Measuring Instrument j				Total Recording Channels Available
	Temp. *1*	Pressure *2*	Accel. *3*	Thrust *4*	
Cameras *1*	x_{11} 1	x_{12} 3	x_{13} ∞	x_{14} 1	≤ 200
Oscilloscopes *2*	x_{21} .5	x_{22} .5	x_{23} .5	x_{24} .5	≤ 150
"Idiots" *3*	x_{31} 2	x_{32} 2	x_{33} 10	x_{34} 2	≤ 256
"Hathaways" *4*	x_{41} 1.5	x_{42} 1.5	x_{43} 1.5	x_{34} 1.5	≤ 50
Total Channels to be Recorded	200	50	50	50	

Figure 8-27: Allocation of Receivers to Transmitters

x_{11} 8 7	x_{12} 6 4	x_{13} 5 9	x_{14} 5 2	20
x_{21} 15 11	x_{22} 10 3	x_{23} 8 7	x_{24} 10 8	30
x_{31} 4 7	x_{32} 20 5	x_{33} 2 4	x_{34} 15 6	40
20	25	15	30	

(a) Solve the capacitated transportation problem by hand.

(b) Solve it using the `Transportation` software option to verify your solution.

8.15 Solve the following capacitated transportation problem by hand.

x_{11} 5 10	x_{12} 5 5	x_{13} 5 6	x_{14} 9 7	25
x_{21} 14 8	x_{22} 20 2	x_{23} 10 7	x_{24} 9 6	25
x_{31} 18 9	x_{32} 4 3	x_{33} 25 4	x_{34} 7 8	50
15	20	30	35	

Explain how you could have determined in advance that the problem is infeasible.

8.16 (a) Show for the capacitated transportation problem illustrated in Figure 8-23 that no feasible solution exists if there is a row p such that $\sum_{j=1}^{n} h_{pj} < a_p$ or a column q such that $\sum_{i=1}^{m} h_{iq} < b_q$.

(b) Construct an example to show that a feasible solution satisfying a capacitated transportation problem (see Figure 8-23) need not exist even if for *all* $i = 1, \ldots, m$ and $j = 1, \ldots, n$

$$\sum_{j=1}^{n} h_{ij} \geq a_i,$$

$$\sum_{i=1}^{m} h_{ij} \geq b_j.$$

8.17 Consider the assignment problem shown in the following array:

Employee \ *Task*	1	2	3	4
1	6	2	4	1
2	2	5	7	2
3	2	10	M	7
4	8	M	7	5

where M implies that the assignment is not possible.

(a) Solve the above assignment problem by hand.

(b) Solve it using the `Transportation` software option to verify your solution.

8.18 Find, by hand, a feasible solution to the following 5×5 assignment problem where each person can only be assigned to a subset of available jobs (the exclusion is shown by the letter M):

Person \ *Job*	1	2	3	4	4
1	M			M	M
2		M	M		
3		M		M	M
4		M			M
5	M				M

Verify your solution by using the `Transportation` software option.

CHAPTER 9

NETWORK FLOW THEORY

Network theory concerns a class of linear programs having a very special network structure. The combinatorial nature of this structure has resulted in the development of very efficient algorithms that combine ideas on data structures with algorithms from computer science, and mathematics from operations research.

Networks arise in a wide variety of situations. For example, the transportation problem discussed in Chapter 8 is a network representing the shipment of goods from sources to destinations; see Figure 8-1. Networks arise naturally in electrical systems (circuit boards, distribution of electricity) and in communications (local-area communication networks or wide-area communication networks) in which electricity flows from various points in the network to other points. Typically, the analysis of networks requires finding either a maximal-flow solution or a shortest-path solution or a minimal spanning tree solution or a least-cost solution (in the case of the transportation problem), or determining the optimal sequence of tasks. The ability to obtain, under certain conditions, integer-valued solutions has made it possible to extend network analysis to many different areas such as facilities location, project planning (PERT, CPM), resource management, etc.

9.1 TERMINOLOGY

We shall illustrate definitions and concepts of directed networks by referring to Figure 9-1.

NODES AND ARCS OF A NETWORK

In the figure, the circles numbered 1, 2, 3, and 4 are called nodes; the lines joining them are called arcs; and the arrowheads on the arcs show the direction of flow. In all, there are four nodes and six directed arcs.

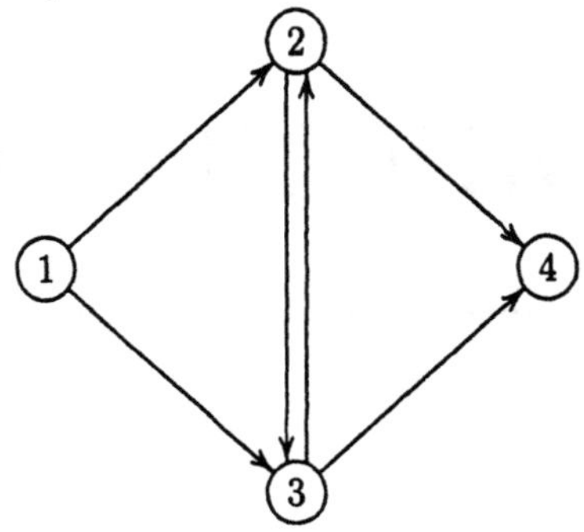

Figure 9-1: A Simple Directed Network

Definition (Graph): A *graph* consists of a finite collection of nodes and a set of node pairs called arcs.

Definition (Directed Network (or Digraph)): A *directed network*, or *digraph*, consists of a set $\mathcal{N}d$ of distinct elements indexed $1, 2, \ldots$ called *nodes* and a set $\mathcal{A}c$ of *ordered pairs* of node indices, $(i, j) \in \mathcal{A}c$, called *directed arcs*. (Thus (j, i) is different from (i, j) if $i \neq j$). In addition we will say that arc (i, j) is *outgoing* from node i (or is *leaving* node i or has a *tail* at node i) and is *incoming* to node j (or is *entering* node j or has a *head* at node j).

We will sometimes denote a directed network by $G = [\mathcal{N}d; \mathcal{A}c]$. Associated with a directed network is an undirected network with the same node set $\mathcal{N}d$ and set of paired nodes $\mathcal{A}c$, except that the pairs are unordered.

Definition (Undirected Network): An *undirected network* consists of a set $\mathcal{N}d$ of elements indexed $1, 2, \ldots$ called *nodes* and a set $\mathcal{A}c$ of *unordered pairs* of these node indices, $(i, j) \in \mathcal{A}c$, called *undirected arcs*. (Thus (j, i) is the same as (i, j)).

We will also say that a directed arc (i, j) connects i **to** j; and that an undirected arc (i, j) *connects* i **and** j.

In the discussion to follow we shall assume that in a directed network, if there is an arc connecting i to j it is unique. Moreover, we have assumed above that there is no arc connecting i to itself, i.e., there are no arcs in $\mathcal{A}c$ of the form (i, i).

Definition (Source, Destination, Intermediate Nodes): A node all of whose arcs are outgoing is called a *source* (or *origin*) node; a node all of whose arcs are incoming is called a *destination* (or *sink* or *terminal*) node. All other nodes are called *intermediate* (or *transshipment*) nodes.

The classical Hitchcock-Koopmans transportation problem of Section 8.1 has m source nodes, n destination nodes, and no intermediate nodes. In Figure 9-1 the node labeled 1 is the place where flow starts; i.e., it is the unique source node. On the other hand, the node labeled 4 is the point where the flows terminate; it is the unique destination node. All other nodes are intermediate nodes.

In addition, we shall use the following notation:

$$Af(k) = \{j \in \mathcal{N}d \mid (k,j) \in \mathcal{A}c\}, \tag{9.1}$$
$$Bf(k) = \{i \in \mathcal{N}d \mid (i,k) \in \mathcal{A}c\}, \tag{9.2}$$

where $Af(k)$ stands for "after" node k and $Bf(k)$ stands for "before" node k.

Definition (Rank or Degree): The *rank*, or *degree*, of a node is the number of arcs that enter and leave it. The *in-degree* of a node is the number of arcs that enter the node; the *out-degree* of a node is the number of arcs that leave the node.

Using $|\mathcal{S}|$ to denote the number of elements in a set $\mathcal{S}$, the in-degree of node k is $|Bf(k)|$ and the out-degree of node k is $|Af(k)|$.

PATHS, CHAINS, CIRCUITS, AND CYCLES

Definition (Path): In a directed network, a *path* (see Figure 9-2(a)) is defined to be a sequence of nodes $i_1, i_2, \ldots, i_m$, $m \geq 2$, such that $(i_k, i_{k+1}) \in \mathcal{A}c$ for $k = 1, \ldots, m-1$. A *simple path* has m distinct nodes, whereas in a *nonsimple path* the nodes can repeat.

Definition (Chain): In a directed network, a *chain* (see Figure 9-2(b)) is defined to be a sequence of nodes $i_1, i_2, \ldots, i_m$, $m \geq 2$, such that either (i_k, i_{k+1}) is an arc or (i_{k+1}, i_k) is an arc for $k = 1, \ldots, m-1$. Thus, a chain is essentially the same as a path except that in "going" from i_1 to i_m we do not require the directed arcs to be traversed all in the same direction. In an undirected network a path and chain are identical because (i_k, i_{k+1}) is the same as (i_{k+1}, i_k).

Definition (Circuit and Cycle): A *circuit* (see Figure 9-2(c)) is a path from some node i_1 to node i_n where $i_n = i_1$, $n \geq 2$; i.e., a circuit is a *closed path.* Similarly, a *cycle* is (see Figure 9-2(d)) a chain from some node i_1 to node i_n where $i_n = i_1$; i.e., a cycle is a *closed chain.*

NODE-ARC INCIDENCE MATRIX

A convenient way to represent the structure of a network mathematically is through the use of a *node-arc incidence matrix.* The nodes correspond to the rows of such a matrix and the arcs correspond to the columns of the matrix. Since each arc connects two nodes, a column corresponding to an arc has exactly two nonzero

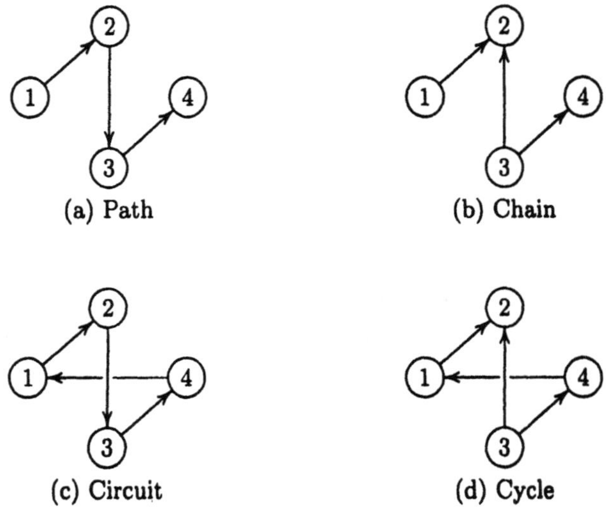

Figure 9-2: A Path, Chain, Circuit, and Cycle

entries; in the directed case, the entries are +1 (corresponding to the tail end of the arc) and −1 (corresponding to the head of the arc), in the undirected case the entries are +1 and +1.

Thus, in the directed case, for a column corresponding to arc (i, j) we use the convention that its entry in row i is +1 and its entry in row j is −1. All other entries in the column corresponding to arc (i, j) are zero. Denoting the node-arc incidence matrix by A, and letting s be the column of the matrix corresponding to arc (i, j),

$$\begin{aligned} A_{is} &= \;\; 1 \\ A_{js} &= -1 \\ A_{ks} &= \;\; 0 \text{ if } k \neq i,\; k \neq j. \end{aligned} \tag{9.3}$$

That is, in matrix notation, $A_{\bullet s} = e_i - e_j$, where e_i is a unit vector with a one in position i and zeros elsewhere, and, similarly e_j is a unit vector with a one in position j and zeros elsewhere.

In the undirected case for a column corresponding to arc (i, j) we use the convention that its entry in row i is +1 and its entry in row j is also +1. All other entries in the column corresponding to arc (i, j) are zero.

Example 9.1 (Node-Arc Incidence Matrix) The node-arc incidence matrix corresponding to Figure 9-1 is shown below, with the row labels being the node numbers and column labels being the arcs.

$$\begin{array}{c} \\ 1 \\ 2 \\ 3 \\ 4 \end{array} \begin{array}{c} \begin{array}{cccccc} (1,2) & (1,3) & (2,3) & (3,2) & (2,4) & (3,4) \end{array} \\ \left(\begin{array}{cccccc} 1 & 1 & 0 & 0 & 0 & 0 \\ -1 & 0 & 1 & -1 & 1 & 0 \\ 0 & -1 & -1 & 1 & 0 & 1 \\ 0 & 0 & 0 & 0 & -1 & -1 \end{array} \right) \end{array}. \tag{9.4}$$

In this example, if we let $x = (x_{12}, x_{13}, x_{23}, x_{32}, x_{24}, x_{34})^T \geq 0$ be a nonnegative flow vector such that component x_{ij} represents the flow from node i to node j, we see that the product of row i of the node-arc incidence matrix and the vector x gives the vector of net flow into and out of node i. If a node net flow is nonnegative, it is the net flow out; if negative, its absolute value is the net flow in. For example, the first component of the product Ax is $1x_{12} + 1x_{13}$, which is the net flow out of node 1.

CONNECTED NETWORKS AND TREES

Definition (Connected Network): Two nodes are said to be *connected* if there is at least one chain that joins the two nodes. A *connected network* is a network in which every pair of nodes is connected. The circuit of Figure 9-2(c) and the cycle of Figure 9-2(d) are examples of connected networks (if there are no other nodes).

Definition (Weakly and Strongly Connected): Sometimes a connected network as defined above is referred to as being *weakly connected*, in which case, a *strongly connected* network is defined as one in which there is a chain connecting every pair of nodes.

Definition (Complete Graph): If each node of a graph connected by an arc to every other node, the graph is called a *complete graph*.

Definition (Tree): A *tree* is a connected network with no cycles.

Definition (Spanning Tree): Given a network with m nodes, a *spanning tree* is a tree that connects all m nodes in the network.

The path of Figure 9-2(a) and the chain of Figure 9-2(b) are examples of spanning trees. A more general example of a tree is shown in Figure 9-3; it is a spanning tree since its arcs connect all the nodes of the network.

▷ **Exercise 9.1** Count the number of nodes and arcs in the tree shown in Figure 9-3.

THEOREM 9.1 (Arcs in a Spanning Tree) *For a network with m nodes, every spanning tree has exactly $m - 1$ arcs.*

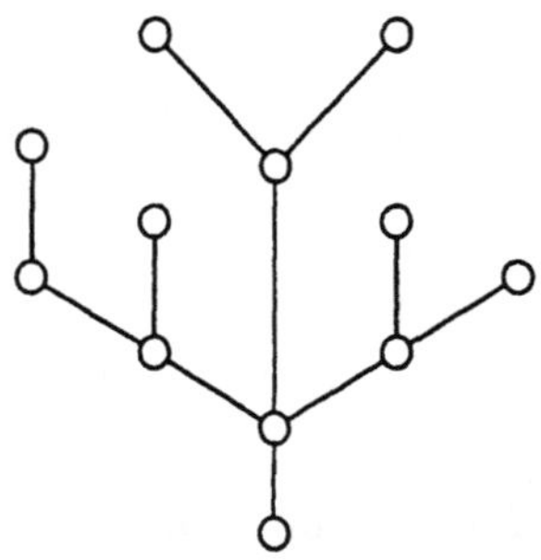

Figure 9-3: A Tree

Another way to state the theorem is that $m-1$ is the minimum number of arcs needed to generate the tree and it is also the maximum number of arcs that can be in the tree without creating undirected cycles.

COROLLARY 9.2 (When a Network Is a Spanning Tree) *A connected network with m nodes, $m-1$ arcs, and no cycles is a spanning tree.*

COROLLARY 9.3 (A Basis Is a Spanning Tree) *Let A be the node-arc incidence matrix of a network. Assuming that the rank of A is $m-1$, the graph associated with any $m-1$ independent columns of A is a spanning tree.*

▷ **Exercise 9.2** Prove Theorem 9.1, Corollary 9.2, Corollary 9.3.

9.2 FLOWS AND ARC-CAPACITIES

We shall sometimes use x to denote a *flow vector*, i.e., a vector consisting of all the arc flows x_{ij} for all $(i,j) \in \mathcal{A}c$ in the network. In some applications $x_{ij} \geq 0$ need not hold. For example, in the maximum flow problem that we will consider in Section 9.3, we require that

$$l_{ij} \leq x_{ij} \leq h_{ij}, \tag{9.5}$$

where the only qualifications on the bounds are that $l_{ij} \leq h_{ij}$ for all (i,j).

> *Definition (Flow Capacity):* If $h_{ij} \geq 0$ and $l_{ij} = 0$ then h_{ij} is referred to as the *flow capacity* of the arc.

In the network in Figure 9-4 each upper bound h_{ij} on the flow is shown by the number on the tail of the arc.

Consider a network with a single source node, $s = 1$, and a single destination node, $t = m$, connected by several intermediate nodes. Except for the nodes 1 and m (the source and destination nodes), the flows into and out of each node

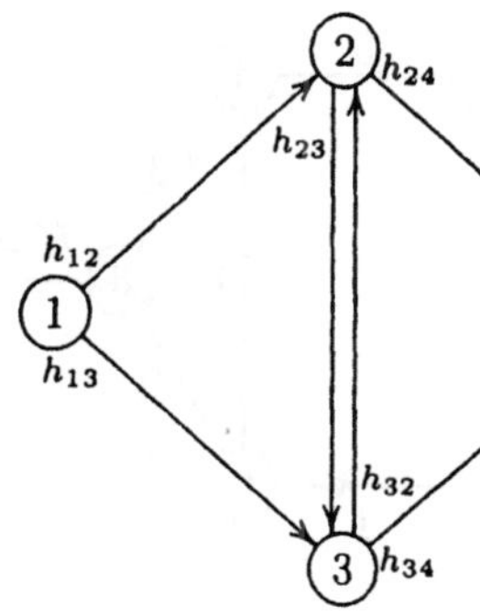

Figure 9-4: A Simple Directed Network with Arc-Capacities

k must balance; such a relation is called *conservation of flows* (in physics, the condition that the quantity of electrons flowing into a point of an electrical network must equal the amount flowing out is referred to as *Kirchoff's Law*). That is, for an intermediate node k,

$$\sum_{i \in Bf(k)} x_{ik} - \sum_{j \in Af(k)} x_{kj} = 0, \qquad \text{for } k = 2, \ldots, m-1, \tag{9.6}$$

where the first summation is over all arcs that have node k as a head node and the second summation is over all arcs that have node k as a tail node. If we denote by F the *exogenous flow* into the source $s = 1$ from *outside* the network, then

$$F - \sum_{j \in Af(1)} x_{1j} = 0, \tag{9.7}$$

because there are no arcs incoming into the source node. If we denote by H the exogenous flow from the destination node t to outside the network, then

$$\sum_{i \in Bf(m)} x_{im} - H = 0. \tag{9.8}$$

If we sum the $m - 2$ relations in (9.6) and (9.7), then each variable x_{ij} appears in exactly two equations with opposite signs (recall the node-arc incidence matrix) and hence cancels, resulting in $F = H$. Therefore

$$\sum_{i \in Bf(m)} x_{im} - F = 0. \tag{9.9}$$

THEOREM 9.4 (Exogenous Flow Balance) *The exogenous flow F into the source node s equals the exogenous flow H out of the destination node t.*

Example 9.2 (Exogenous Flow Balance) Figure 9-5 illustrates an exogenous flow $F = 4$ into node 1, the exogenous flow $H = 4$ out of node 6, and the conservation equations (9.6), (9.7), and (9.9).

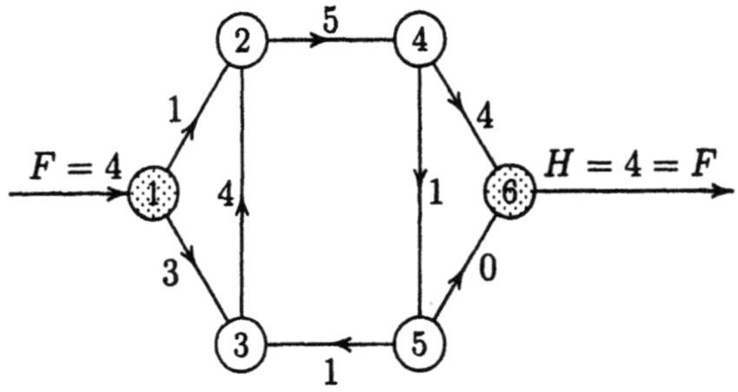

Figure 9-5: Exogenous Flow Balance

▷ **Exercise 9.3** For the network in Figure 9-5, write down the conservation equations in algebraic form and in matrix form using the node-arc incidence matrix.

▷ **Exercise 9.4** Prove that a set of $x_{ij} = x^o_{ij} \geq 0$ satisfying capacity constraints $0 \leq x_{ij} \leq h_{ij}$ and the conservation equations (9.6), (9.7), and (9.9) can be replaced by another set x'_{ij} with the same total flow F in which either x'_{ij} or x'_{ji} is zero by setting

$$\begin{aligned} x'_{ij} &= x^o_{ij} - \min(x^o_{ij}, x^o_{ji}) \text{ and } x'_{ji} = 0 \text{ if } x^o_{ij} \geq x^o_{ji} \\ x'_{ji} &= x^o_{ij} - \min(x^o_{ij}, x^o_{ji}) \text{ and } x'_{ij} = 0 \text{ if } x^o_{ji} \geq x^o_{ij}. \end{aligned} \tag{9.10}$$

Comment: When $l_{ij} = 0$, it follows from Exercise 9.4 that we need only consider flows where either $x_{ij} = 0$ or $x_{ji} = 0$. Thus, for example, if on the directed arc joining nodes i and j the flow $x_{ij} = 4$ appears, it will imply that $x_{ji} = 0$. Hence we could, if we wanted to, replace all the flow variables x_{ij} and x_{ji} by their difference:

$$\bar{x}_{ij} = x_{ij} - x_{ji}, \tag{9.11}$$

in which case

$$\begin{aligned} \bar{x}_{ij} &> 0 \quad \text{corresponds to} \quad \bar{x}_{ij} = x'_{ij} \text{ and } x'_{ji} = 0, \\ \bar{x}_{ij} &< 0 \quad \text{corresponds to} \quad -\bar{x}_{ij} = x'_{ji} \text{ and } x'_{ij} = 0. \end{aligned}$$

▷ **Exercise 9.5** Show that upon making the substitution for $\bar{x}_{ij}$, the arc-capacity constraints (9.5) with $l_{ij} = 0$ and the conservation equations (9.6), (9.7), and (9.9) for a network defined by $\mathcal{N}d$ and $\mathcal{A}c$ become

$$\begin{aligned} -h_{ji} \leq \bar{x}_{ij} &\leq h_{ij}, & (i,j) \in \mathcal{A}c, \\ \sum_i \bar{x}_{ik} &= 0, & k = 2, \ldots, m-1, \\ F + \sum_i \bar{x}_{i1} &= 0, & \\ -F + \sum_i \bar{x}_{im} &= 0. & \end{aligned} \tag{9.12}$$

▷ **Exercise 9.6** Given a feasible solution to (9.12), find a corresponding feasible solution that satisfies the conservation equations (9.6), (9.7), (9.9) and bounds that satisfy (9.5) with $l_{ij} = 0$.

FLOW AUGMENTATION

Suppose that we are given a network $(\mathcal{N}d, \mathcal{A}c)$ with lower and upper bound constraints $l_{ij} \le x_{ij} \le h_{ij}$ on arcs $(i,j) \in \mathcal{A}c$ and a set of feasible flows $l_{ij} \le x^o_{ij} \le h_{ij}$.

Definition (Associated Path): For any chain of arcs connecting node i_p to node i_q, *the associated path* $\mathcal{P}$ is the path formed by reversing, if necessary, the direction of some of the arcs along the chain.

Definition (θ Path Flow): Given any path, a *θ path flow* is a flow with value θ on every arc (i,j) along the path and zero elsewhere.

Definition (θ-Flow Augmentation): Let $\mathcal{C}$ be a chain joining s to t and let $\mathcal{P}$ be the associated path. Let $\mathcal{C}^+$ be the arcs of the chain that are oriented in the direction of the path $\mathcal{P}$ and let $\mathcal{C}^-$, be the remaining arcs of $\mathcal{C}$. The *θ-flow augmentation of a feasible solution* $x = x^o$ is $x_{ij} = x^o_{ij} + \theta$ if $(i,j) \in \mathcal{C}^+$, $x_{ij} = x^o_{ij} - \theta$ if $(i,j) \in \mathcal{C}^-$ and $x_{ij} = x^o_{ij}$ otherwise.

Definition (θ-Path Flow Augmentation is Maximal): A *θ-path flow augmentation is maximal* if θ is the maximum value for which the augmentation remains feasible.

Example 9.3 (Chain flow, θ Path flow, θ-flow Augmentation) A chain flow, θ path flow, and a chain flow after a θ-flow augmentation are illustrated in Figure 9-6. The numbers in brackets at the tail of each arc represent the lower and upper bounds on the flow through the arc. In order for the chain flows to remain feasible after a θ augmentation along the associated path, we need θ to satisfy

$$2 \le 3 + \theta \le 5, \quad 4 \le 5 - \theta \le 6, \quad 1 \le 2 + \theta \le 4.$$

The maximum θ augmentation that maintains feasibility is $\theta = 1$, because the augmentation is blocked by the lower bound on x_{32}.

FLOW DECOMPOSITION

The next theorem shows that a flow can be decomposed into a sum of flows along simple paths and circuits. It is useful because it shows that a solution to a flow problem corresponds to our intuitive notion that items start from nodes of surplus and move from one node to the next without losing their identity along the way.

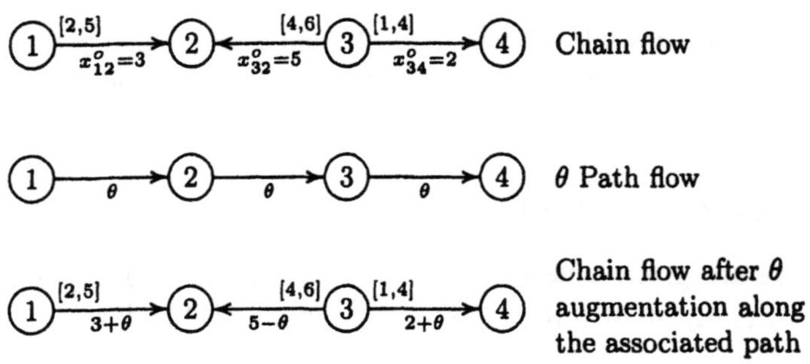

Figure 9-6: Chain Flow, θ Path Flow, θ-Flow Augmentation

THEOREM 9.5 (Flow Decomposition) *Consider a network $(\mathcal{N}d, \mathcal{A}c)$ where the arc-capacity constraints are $0 \le x_{ij} \le h_{ij}$ for $(i,j) \in \mathcal{A}c$. Given an incoming exogenous flow of $F > 0$, a set of flows $x = x^o_{ij}$ that satisfy the capacity constraints and conservation equations* (9.6)–(9.9) *can be decomposed into a vector sum of path flows from source to destination and circuit flows such that the direction of these flows in any common arc (i,j) is the same as that of the directed arc $(i,j) \in \mathcal{A}c$.*

Example 9.4 (Illustration of Flow Decomposition) The Flow Decomposition Theorem 9.5 is illustrated in Figure 9-7. The sum of the two path flows and circuit flow is equal to the total flow shown in Figure 9-5.

9.3 AUGMENTING PATH ALGORITHM FOR MAXIMAL FLOW

The maximal-flow problem for a network is to find the maximum amount that can be transferred from the source to the destination given arc-capacity constraints $l_{ij} \le x_{ij} \le h_{ij}$ and the existence of a feasible flow $x = x^o$. We will also assume that all $l_{ij} = 0$. It is clear that solving the maximal flow problem is the same as solving the linear program

$$
\begin{array}{lrcl}
\text{Maximize} & & & F \\
\text{subject to} & \displaystyle\sum_{j \in Af(1)} x_{1j} & = & -F, \\
& \displaystyle\sum_{i \in Bf(k)} x_{ik} - \sum_{j \in Af(k)} x_{kj} & = & 0, \quad \text{for } k = 2, \ldots, m-1, \\
& \displaystyle\sum_{i \in Bf(m)} x_{im} & = & F, \\
& 0 \le x_{ij} \le h_{ij}, & & \text{for all } (i,j) \in \mathcal{A}c.
\end{array}
\tag{9.13}
$$

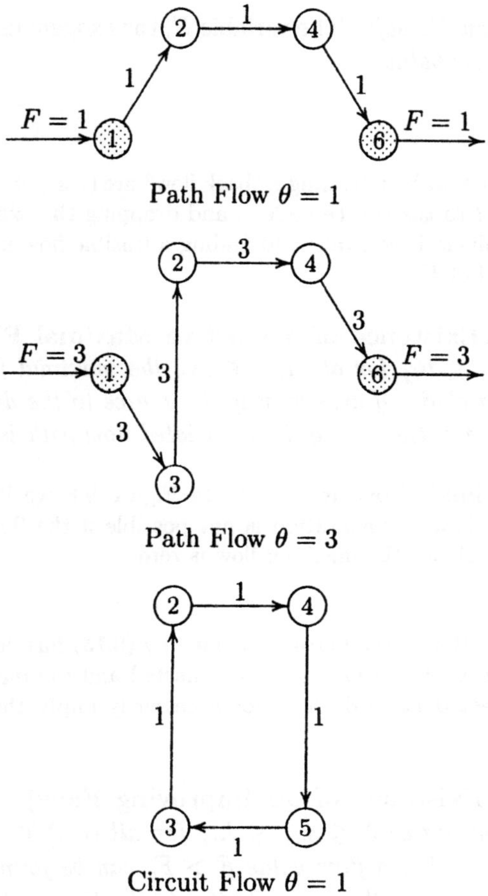

Figure 9-7: Decomposition of Flow in Figure 9-5

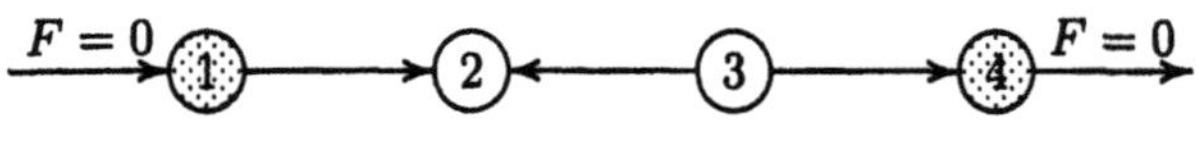

Figure 9-8: Augmentation Not Possible

Definition (Flow Value): The variable F, the exogenous flow into the system, is called the *flow value.*

▷ **Exercise 9.7** Show that by adjoining a "back flow" arc (t, s), of infinite capacity, joining the destination node t to the source node s, and dropping the exogenous flow equations, the maximal flow problem is equivalent to finding a feasible flow in the new network that maximizes x_{ts} instead of F.

THEOREM 9.6 (Existence of a Positive Maximal Flow) *In a network with bounds $0 \leq x_{ij} \leq h_{ij}$ for all $(i,j) \in \mathcal{A}c$, the maximal flow is positive if and only if there exists a chain of arcs joining the source to the destination such that a positive θ-flow augmentation along the associated flow path is possible.*

Example 9.5 (Maximal Flow Is Zero) In Figure 9-8 we illustrate a very simple network in which a θ-flow augmentation is not possible if the flow on the directed arcs must be nonnegative. Hence the maximal flow is zero.

▷ **Exercise 9.8** Show that if the network defined by (9.13) has no chains connecting the source to the destination, the network is disconnected and the maximum flow value is 0. Show that a disconnected network does not necessarily imply that the maximal flow is zero.

THEOREM 9.7 (Existence of an Improving Flow) *Consider a network $(\mathcal{N}d, \mathcal{A}c)$ with arc-capacities $0 \leq x_{ij} \leq h_{ij}$ for all $(i,j) \in \mathcal{A}c$. Given a feasible flow $x = x^o_{ij}$ with $F = F_o$, a flow value $F > F_o$ can be found if and only if there exists a chain of arcs joining the source to the destination such that a positive θ-flow augmentation along its associated flow path is possible.*

Example 9.6 (Illustration of an Improving Flow) Steps to obtain an improving flow are illustrated in Figure 9-9. The numbers in brackets at the tail of each arc represent the lower and upper bounds on the flow through the arc.

COROLLARY 9.8 (Maximal Flow Condition) *The flow value $F = F_o$ is maximal if and only if there is no chain of arcs joining the source to the destination such that a positive θ-flow augmentation along its associated flow path is possible.*

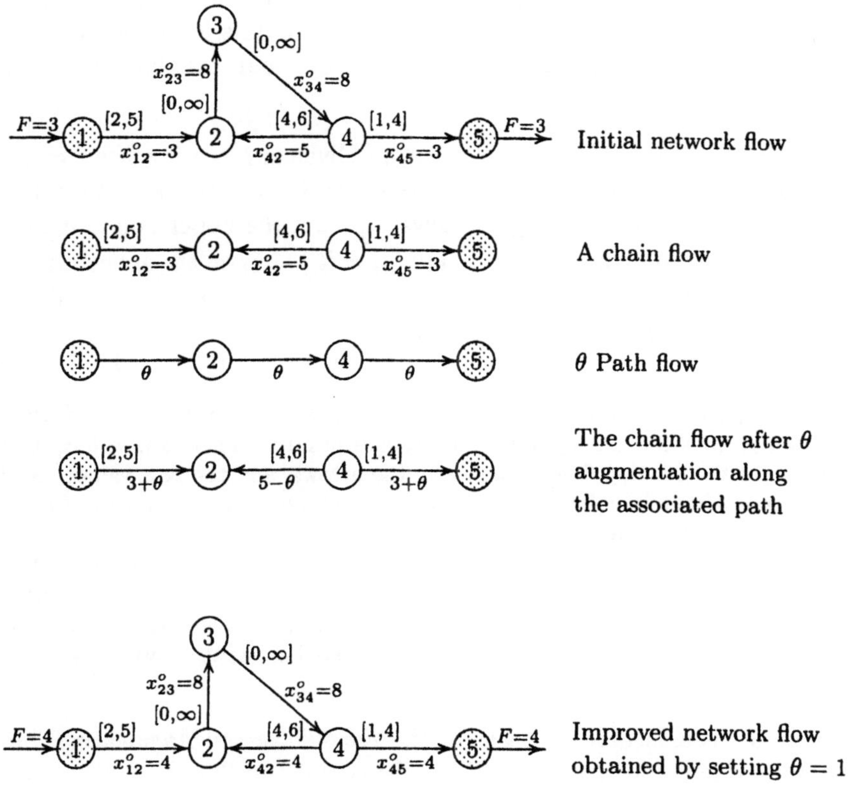

Figure 9-9: Finding an Improved Flow

▷ **Exercise 9.9** Show that the improving flow in Example 9.6 results in a maximal flow for the network.

▷ **Exercise 9.10** Prove Corollary 9.8.

Instead of adjusting flows in a network-flow problem, it is convenient to adjust the arc capacities so that all the existing flows x^o_{ij} are converted to zero and an improving flow is found in this adjusted-capacity network. After finding the improving flow, the arc capacities are again adjusted so that θ improvements are converted to zero, and so on, iteratively. A final step notes the difference between the final adjusted capacity and the original capacity on an arc, and this difference is x^*_{ij}, the maximal flow solution.

> *Definition (Adjusted-Capacity Network)*: Given any existing flow $x = x^0_{ij}$ on a network $(\mathcal{N}d, \mathcal{A}c)$, the ***adjusted-capacity network*** is a network that is computed as follows. Subtract existing arc-flows x^o_{ij} from the upper bound h_{ij} on arc-capacity to obtain a new upper bound $\bar{h}_{ij} = h_{ij} - x^o_{ij}$ on the arc capacity. Add a reverse arc (j, i) with an upper bound $\bar{h}_{ji} = x^o_{ij}$ on arc capacity; if $x^o_{ij} = 0$, the reverse arc (j, i) may be omitted.

Example 9.7 (Construction of an Adjusted-Capacity Network) In Figure 9-10 we illustrate how to obtain an adjusted-capacity network from a network with initial flows. The direction of the existing flow is shown by the arrows, and the flow values are shown as numbers approximately halfway on each arc. It is assumed that each pair of nodes has two arcs connecting it. The capacity on the arc from node i to node j is shown at the tail of the arc; i.e., near node i. In the figure, on the line joining node 1 and node 2, the number 4 near node 1 is the arc capacity for arc $(1, 2)$ and the number 0 near node 2 is the arc capacity for arc $(2, 1)$. In Figure 9-10(a), an initial exogenous flow of $F = 5$ is assumed. In Figure 9-10(b) we have created an adjusted-capacity network by converting the existing flows to zero as discussed above.

Theorem 9.7 is a theoretical basis for the *Ford-Fulkerson Augmenting Path Algorithm* (also due independently to Elias, Feinstein, and Shannon). We shall first illustrate the algorithm through an example.

Example 9.8 (Illustration of the Augmenting Path Algorithm) Consider the network displayed in Figure 9-11(a) with the capacities of the forward and reverse arcs as shown by capacities on both ends of the arcs. Notice that the arc connecting nodes 2 and 3 has a capacity of 1 along the arc, $(2, 3)$ and a capacity of 2 along the arc $(3, 2)$; i.e., the arc joining nodes 2 and 3 is undirected and has been replaced by two arcs $(2, 3)$ and $(3, 2)$. Noting that there are no existing flows, we start the algorithm by setting $F = 0$ and finding an augmenting path such as $(1, 3)$, $(3, 2)$, and $(2, 4)$ with arc capacities 2, 2, and 2 respectively. Thus the improving flow θ through this path is 2, and $F = 0 + 2$. The capacities are then adjusted by subtracting $\theta = 2$ from the tail of the forward arcs along the path and adding $\theta = 2$ to the tail of the reverse arcs along the path to generate the next adjusted-capacity network, shown in Figure 9-11(b).

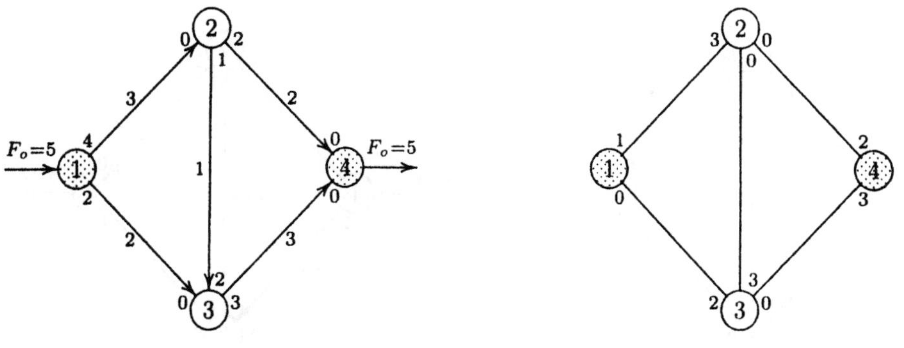

(a) Network with initial nonzero flow

(b) Associated network with zero flows.

Figure 9-10: Construction of an Adjusted-Capacity Network

In the new network there is only one augmenting path, namely the one with arcs $(1,2)$, $(2,3)$, and $(3,4)$ with capacities 4, 3, and 3 respectively. Therefore the improving flow for this iteration is $\theta = 3$, and thus $F = 2 + \theta = 5$. Notice how the flow in arc $(2,3)$ more than counteracts the flow of the previous iteration on arc $(3,2)$. Adjusting the arc capacities on the augmenting path by $\theta = 3$, we get the adjusted-capacity network shown in Figure 9-11(e).

No augmenting path exists, and therefore, by Theorem 9.7, the solution is optimal. From the arc capacities in the adjusted-capacity network of Figure 9-11(e) we compute the difference between original arc-capacities, shown in Figure 9-11(e), and the final arc-capacities shown in Figure 9-11(e) to obtain the final maximal flow solution, shown in Figure 9-11(f).

Algorithm 9.1 (Augmenting Path) In general, we first create an adjusted-capacity network by adding to each arc (i,j) that does not have a reverse arc (j,i) an arc (j,i) with capacity $\bar{h}_{ji} = 0$ (the capacities on the arcs corresponding to the original network are relabeled as $\bar{h}_{ij}$). Set the flow $F = 0$ in this adjusted system. The original network with arc-capacities h_{ij} is set aside until the termination of the algorithm, when its flows are set to the optimal flows.

1. Using any θ-flow augmenting search procedure, try to find an augmenting path for the current adjusted-capacity network. If no such path exists, stop with the updated flow value F being optimal and the optimal arc-flows determined by
$$x^*_{ij} = \bar{h}_{ij} - h_{ij}$$
for each arc in the original network.

2. Determine θ equal to the minimum capacity for the augmenting path in the current adjusted-capacity network:
$$\theta = \min_{(i,j)\in AP} \bar{h}_{ij},$$
where AP is the augmenting path. Set $F \leftarrow F + \theta$.

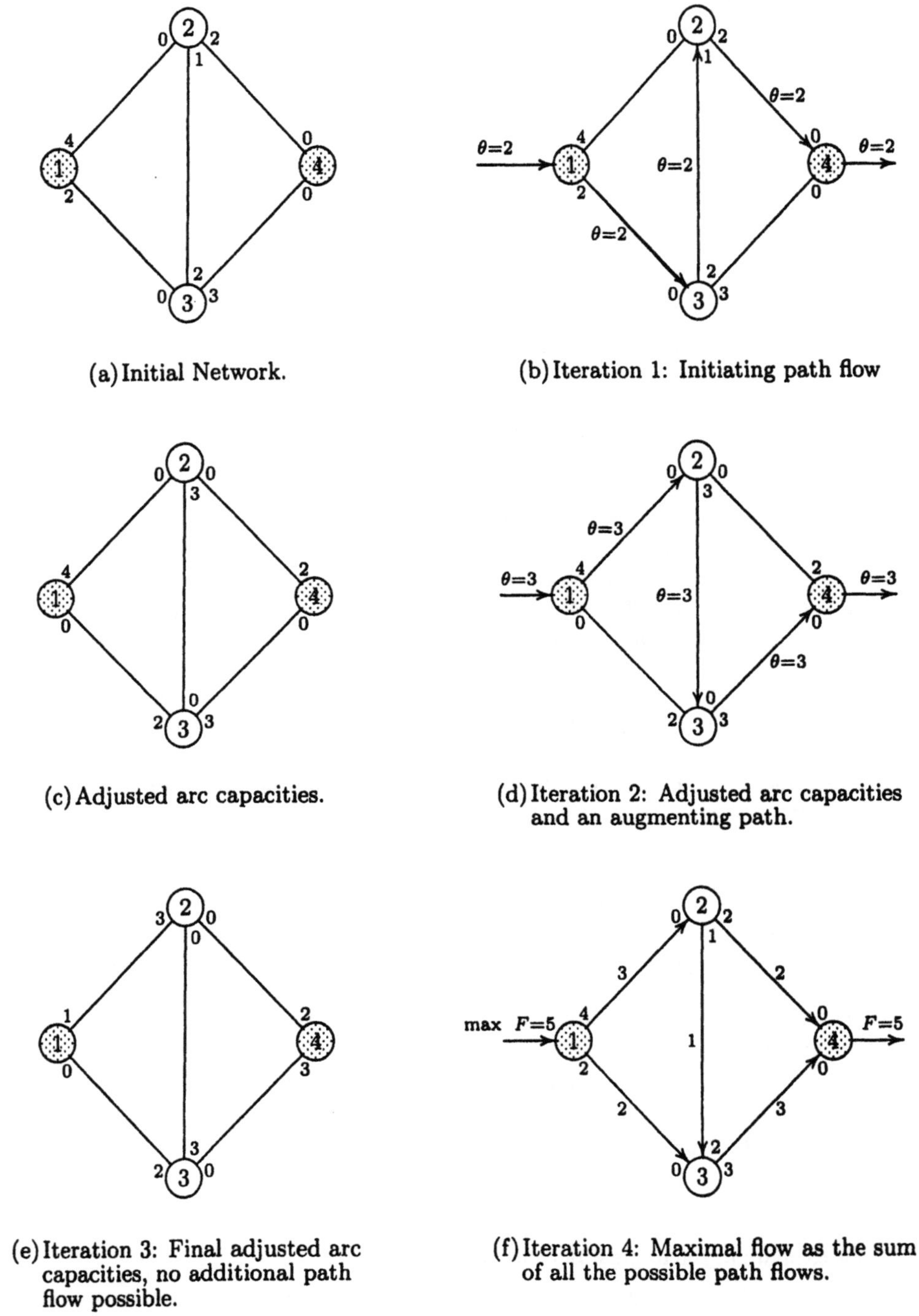

(a) Initial Network.

(b) Iteration 1: Initiating path flow

(c) Adjusted arc capacities.

(d) Iteration 2: Adjusted arc capacities and an augmenting path.

(e) Iteration 3: Final adjusted arc capacities, no additional path flow possible.

(f) Iteration 4: Maximal flow as the sum of all the possible path flows.

Figure 9-11: Illustration of the Augmenting Path Algorithm to Find Maximal Flow

3. For each directed arc (i,j) of the adjusted-capacity network along the augmenting path, adjust the arc-capacity to be

$$\bar{h}_{ij} \leftarrow \bar{h}_{ij} - \theta,$$
$$\bar{h}_{ji} \leftarrow \bar{h}_{ji} + \theta$$

to determine the next adjusted-capacity network.

4. Go to Step 1.

Note that the algorithm only modifies the arc capacities in the adjusted-capacity network and not its arc flows. The optimal arc flows x^*_{ij} for the original network are actually the sums of the individual path flows encountered during the solution process; hence these could have been determined by adjusting the flows x_{ij} at the same time that the arc capacities h_{ij} were adjusted in Step 3. However, a simpler way is to obtain them from the final adjusted-capacity network by comparing the final arc capacities with the initial assigned arc capacities. If for arc (i,j) the final arc capacity $\bar{h}_{ij}$ is less than the initially assigned arc capacity h_{ij}, then arc (i,j) has a flow equal to the difference; i.e., $x_{ij} = h_{ij} - \bar{h}_{ij}$.

▷ **Exercise 9.11** Modify Algorithm 9.1 if the initial feasible flow vector $x = x^o \neq 0$ is given.

▷ **Exercise 9.12** Modify Algorithm 9.1 so that instead of modifying the arc capacities we update the flows directly. In this case we check for augmenting paths such that the arcs in the path can be traversed in the direction of the arc if the flow is less than the upper bound, and in the reverse direction if the flow is greater than the lower bound.

Starting with a feasible solution $x_{ij} = x^o_{ij} = 0$ for all $(i,j) \in \mathcal{A}c$, Algorithm 9.1, regardless of the augmenting path procedure, is guaranteed to solve the maximal flow problem in a finite number of augmenting flow steps provided that the arc capacities are either integers or rational numbers (see Theorem 9.9 and Corollary 9.10). In the event that the flow capacities are not rational numbers, the algorithm may sometimes require an infinite number of steps to converge. However, even if the flow capacities are not rational, the use of a procedure for finding an augmenting path, called the *Breadth-First Unblocked Search* procedure (to be discussed later), is guaranteed by the Edmonds-Karp Theorem 9.11 to solve the maximal flow problem within $mn/2$ flow augmenting steps, where m is the number of nodes and n is the number of arcs.

THEOREM 9.9 (Finite Termination with Integer Capacities) *If the arc capacities are all integers and a finite maximal flow exists, then Algorithm* 9.1 *will construct only a finite number of path flows whose algebraic sum is the maximal flow.*

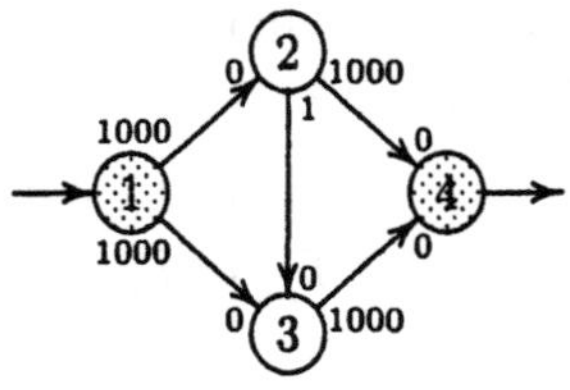

Figure 9-12: Large Number of Iterations of Algorithm 9.1

COROLLARY 9.10 (Finite Termination with Rational Capacities) *If the arc capacities are all rational numbers and a finite maximal flow exists, then Algorithm* 9.1 *will construct only a finite number of path flows whose algebraic sum is the maximal flow.*

The next example demonstrates that it is possible to choose augmenting paths such that the algorithm takes a very large number of iterations to converge.

Example 9.9 (Large Number of Iterations of the Augmenting Path Algorithm) Consider the network in Figure 9-12. It may seem that the greater the number of arcs in the chosen augmenting path at each iteration, the sooner the sum of the augmenting path flows will use up capacity of the arcs and terminate with maximal flow. However if we apply this strategy to the network shown in Figure 9-12, we would choose path $(1,2,3,4)$ on odd iterations and $(1,3,2,4)$ on even iterations and find that Algorithm 9.1 takes 2,000 iterations to converge! If the capacity on each arc with capacity of 1,000 is changed to a capacity of a trillion, the algorithm would require two trillion iterations. If the capacity restrictions on the arcs with capacities of a 1,000 are dropped, the algorithm would never terminate. Notice that if at each iteration we had chosen instead the augmenting path with the fewest number of arcs, the algorithm would have terminated in two iterations!

▷ **Exercise 9.13** Create a network for which the augmenting path algorithm chooses at each iteration an augmenting path with the largest number of arcs, and solves the problem faster than one that chooses an augmenting path with the smallest number of arcs.

The following example demonstrates that if the capacities are irrational numbers then it is possible that the maximal flow Algorithm 9.1 may never terminate even if all the capacities are all finite.

Example 9.10 (Infinitely Many Iterations of the Augmenting Path Algorithm) Consider the network in Figure 9-13 with source node 1, destination node 8, and bounds on arcs

$$\begin{aligned} 0 \le\ & x_{23} \le h, \\ 0 \le\ & x_{47} \le 1, \\ 0 \le\ & x_{57} \le h, \\ 0 \le\ & x_{ij} \le \infty, \end{aligned}$$

where $h = \left(-1+\sqrt{5}\right)/2$ is the golden ratio. It turns out that the algorithm will go through an infinite sequence of iterations τ shown in Table 9-1 if one chooses cyclically for

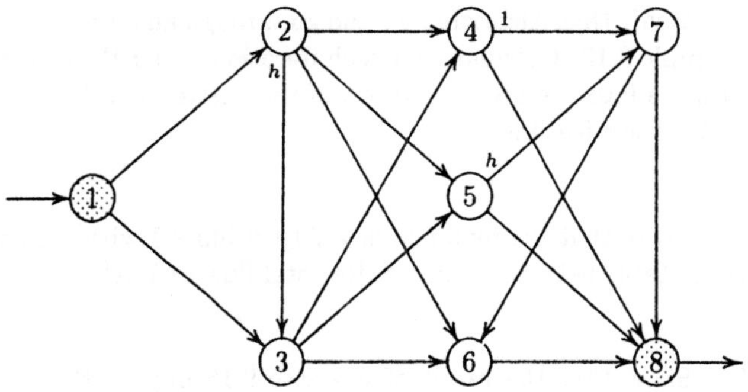

Figure 9-13: Infinite Number of Iterations of Algorithm 9.1

Iteration	Augmenting Path	Path Flow	Resulting Arc Flow (on relevant arcs)			
			x_{23}	x_{47}	x_{57}	x_{76}
1	1,2,3,4,7,8	h	h	$h = 1 - h^2$	0	0
2	1,3,2,4,7,6,8	h^2	$h - h^2$	1	0	h^2
3	1,2,3,5,7,8	h^2	h	1	h^2	h^2
4	1,3,2,5,7,4,8	h^3	$h - h^3$	$1 - h^3$	h	h^2
5	1,2,3,6,7,8	h^3	h	$1 - h^3$	h	h^4
6	1,3,2,6,7,5,8	h^4	$h - h^4$	$1 - h^3$	$h - h^4$	0
⋮	⋮	⋮	⋮	⋮	⋮	⋮
$6\tau + 1$	1,2,3,4,7,8	$h^{3\tau+1}$	h	$1 - h^{3\tau+2}$	$h - h^{3\tau+1}$	0
$6\tau + 2$	1,3,2,4,7,6,8	$h^{3\tau+2}$	$h - h^{3\tau+2}$	1	$h - h^{3\tau+1}$	$h^{3\tau+2}$
$6\tau + 3$	1,2,3,5,7,8	$h^{3\tau+2}$	h	1	$h - h^{3\tau+3}$	$h^{3\tau+2}$
$6\tau + 4$	1,3,2,5,7,4,8	$h^{3\tau+3}$	$h - h^{3\tau+3}$	$1 - h^{3\tau+3}$	h	$h^{3\tau+2}$
$6\tau + 5$	1,2,3,6,7,8	$h^{3\tau+3}$	h	$1 - h^{3\tau+3}$	h	$h^{3\tau+4}$
$6\tau + 6$	1,3,2,6,7,5,8	$h^{3\tau+4}$	$h - h^{3\tau+4}$	$1 - h^{3\tau+3}$	$h - h^{3\tau+4}$	0
⋮	⋮	⋮	⋮	⋮	⋮	⋮

Table 9-1: Infinite Number of Iterations of Algorithm 9.1

the set of successive augmenting paths the ones listed in the table. To see this note that $h^{\tau+2} = h^{\tau} - h^{\tau+1}$ and $h < 1$.

▷ **Exercise 9.14** Verify that Algorithm 9.1 can go through an infinite number of iterations as claimed in Example 9.10. Calculate, for each iteration $6\tau+1$ through $6(\tau+1)+1$, flows on each arc and flows into the origin 1 or out of the destination 8 as a function of τ and show that these flows are feasible.

▷ **Exercise 9.15** Prove that the maximal flow Algorithm 9.1 with the specified augmenting paths shown in Table 9-1 results in a finite total flow of $1/h^2$.

▷ **Exercise 9.16** Show that the result of Exercise 9.15 implies that for the purpose of Example 9.10 the arcs that have infinite upper bounds can be replaced by arcs that have a bound of $1/h^2$.

▷ **Exercise 9.17** Show that the maximal flow through the network of Figure 9-13 is infinite.

BREADTH-FIRST UNBLOCKED SEARCH

A systematic procedure (which is a variation of the shortest path algorithm discussed in Section 9.5) for finding augmenting paths is the *fanning-out* (or *breadth-first search*) procedure. This requires forming a tree of all the nodes j that can be reached from the source s by a flow augmenting path. We first illustrate this by an example and then state the algorithm.

Example 9.11 (Illustration of the Breadth-First Unblocked Search) The steps of the algorithm on the network displayed earlier in Figure 9-12 are displayed in Figure 9-14. We begin with the starting set of nodes $D_o = \{\text{source node } s\} = \{1\}$. Next take the set of nodes j not in D_o that can be reached from nodes i in D_o by some arc (i,j) that has positive remaining capacity in the adjusted capacity network. In Figure 9-14(a) these nodes are 2 and 3. Denote this set by $D_1 = \{2,3\}$ and record the tracing-back node of node 2 as 1 and of node 3 as 1 also. If D_1 contained the destination node 4 we stop the forward search. Since this is not so we continue the forward search by looking for nodes j not in D_o or D_1 that can be reached from nodes i in D_1 by some arc (i,j) that has positive remaining capacity in the adjusted capacity network. In Figure 9-14(a) this is node 4. Denote this set by $D_2 = \{4\}$ and record the tracing-back node of node 4 as 2. Since we have now reached the destination node 4, the θ-augmenting path $(1,2,4)$ is found by tracing back the path in reverse order $(4,2,1)$ with arc capacities on $(2,4)$ of 1000 and on $(1,2)$ also of 1000 so that $\theta = \min(1000, 1000) = 1000$. After adjusting the arc capacities along the path we obtain Figure 9-14(b).

On the next iteration of Algorithm 9.1 applied to Figure 9-14(b), we obtain $D_0 = \{1\}$, $D_1 = \{3\}$ with the tracing-back node 1, and $D_2 = \{4\}$ with the tracing back node 3 as shown in Figure 9-14(c). Since we have now reached the destination node 4, the θ-augmenting path $(1,3,4)$ is found by tracing back the path in reverse order $(4,3,1)$ with arc

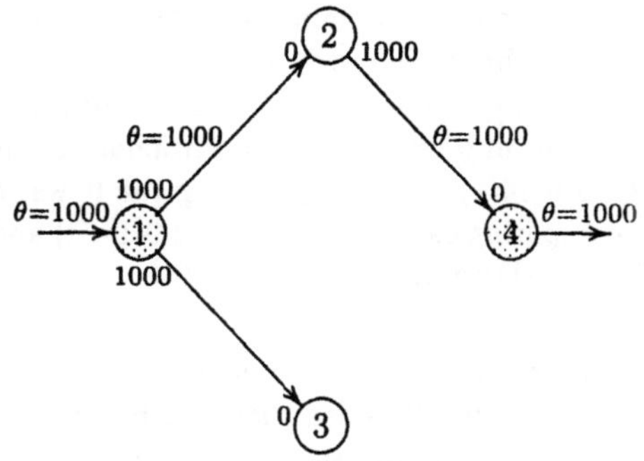

(a) Iteration 1: Breadth-First Search
$D_0 = \{1\},\ D_1 = \{2,3\},\ D_2 = \{4\}$

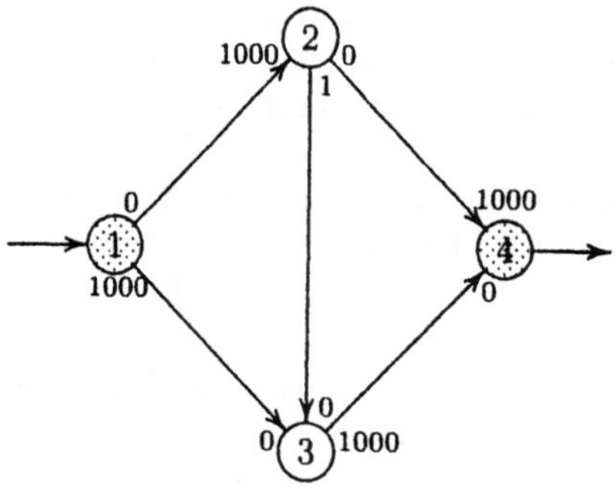

(b) Adjusted arc capacities

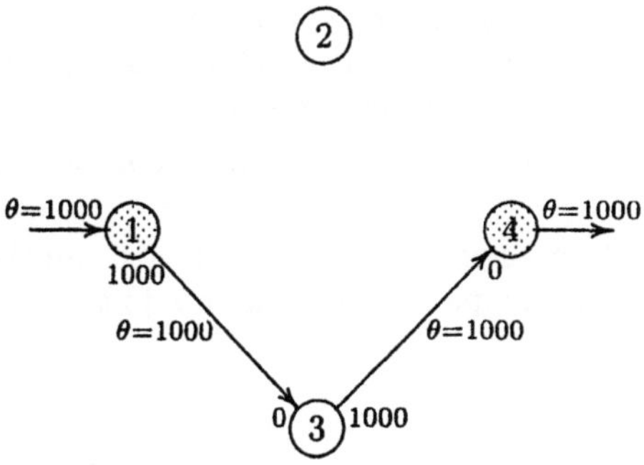

(c) Iteration 2: Breadth-First Search
$D_0 = \{1\},\ D_1 = \{3\},\ D_2 = \{4\}$

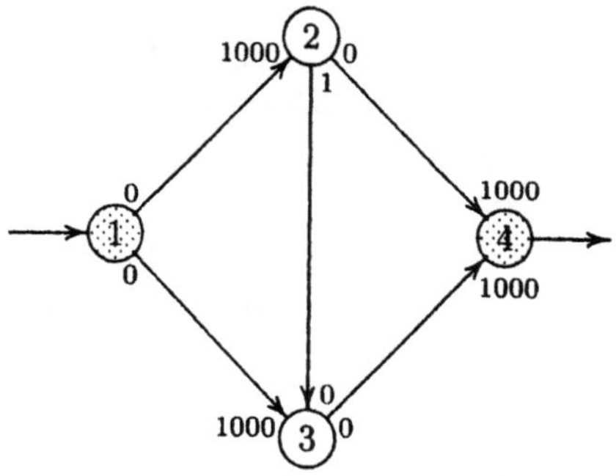

(d) Adjusted arc capacities

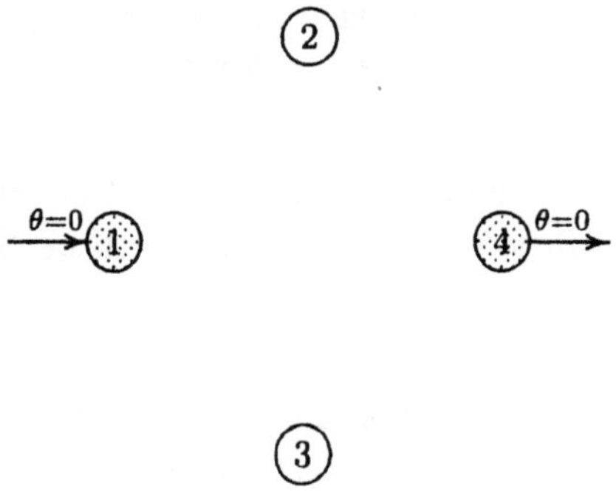

(e) Iteration 3: Breadth-First Search
$D_0 = \{1\},\ D_1 = \emptyset$

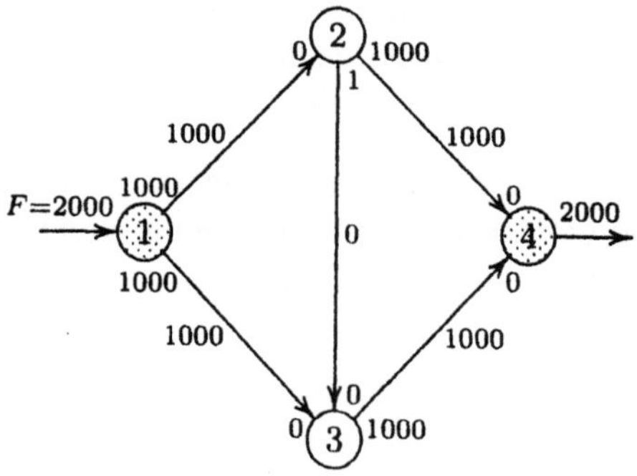

(f) Maximal flow

Figure 9-14: Augmenting Path Algorithm with Breadth-First Unblocked Search

capacities on (3, 4) of 1000 and on (1, 3) of also 1000 so that $\theta = \min(1000, 1000) = 1000$. After adjusting the arc capacities along the path we obtain Figure 9-14(d).

On the next iteration of Algorithm 9.1 applied to Figure 9-14(d), the breadth-first unblocked search algorithm terminates with $D_1 = \emptyset$, implying that no additional flow is possible through the network, as shown in Figure 9-14(e). Either by adding the flows in Figure 9-14(a) and Figure 9-14(c), or better by comparing capacities in Figure 9-14(d) and Figure 9-12, we obtain the maximal flow of 2000, as shown in Figure 9-14(f).

Algorithm 9.2 (Breadth-First Unblocked Search) Let s be the source node, t be the destination node, and let D_ν be the boundary set of all nodes that can be reached from s in exactly ν steps. The nodes in D_ν are said to be a "distance" ν from s.

1. Set $D_0 = \{s\}$ and $\nu = 0$.
2. Set $\nu \leftarrow \nu + 1$.
3. If there exists a directed arc (i, j) along which additional flow is possible (or if there exists a directed arc (j, i) along which decreased flow is possible) such that $i \in D_{\nu-1}$, $j \notin D_h$, $h = 0, 1, \ldots, \nu - 1$, set D_ν to be the set of all such j, and mark each node $j \in D_\nu$ by recording its "predecessor" node $i \in D_{\nu-1}$, which we call its *tracing-back node.* If no such arc (i, j) (or arc (j, i)) exists, set D_ν to be the empty set.
4. If the ν-distance set D_ν is empty, terminate with a statement that no augmenting path exists.
5. If $t \in D_\nu$, terminate with a statement that an augmenting path exists and use the tracing-back nodes (see Step 3) to generate an augmenting path from s to t.
6. Go to Step 2.

▷ **Exercise 9.18** Prove that Algorithm 9.2 terminates with a shortest path in $m(m-1)/2$ comparisons.

THEOREM 9.11 (Edmonds-Karp Max-Flow Theorem) *If a maximal flow exists, the augmenting path Algorithm* 9.1, *when used with the breadth-first unblocked search Algorithm* 9.2 *to find the augmenting paths, will construct at most $mn/2$ path flows whose algebraic sum is the maximal flow, where n is the number of arcs and m is the number of nodes.*

The key idea here is that at each iteration, the flow augmenting path from the source to the destination along which positive flow is possible is chosen to be the one with the fewest number of arcs.

▷ **Exercise 9.19** Verify that Algorithm 9.1 with the breadth-first unblocked search Algorithm 9.2 takes no more than $mn/2$ iterations on the example in Figure 9-14.

▷ **Exercise 9.20** Apply Algorithm 9.1 with the breadth-first unblocked search Algorithm 9.2 to the problem in Example 9.10 and show in this particular case that by selecting at each iteration the augmenting path with the fewest number of arcs, Algorithm 9.1 terminates in a finite number of iterations not exceeding $mn/2$.

9.4 CUTS IN A NETWORK

The search for an augmenting path can be time-consuming, especially in large networks. Thus, it would be nice to be able to recognize optimality without doing an exhaustive search for an augmenting path that may not exist. It turns out that it is sometimes possible to prove that no such path exists by verifying that the conditions of the *Ford-Fulkerson max-flow min-cut* Theorem 9.13 are satisfied. These conditions make use of the notion of a *cut* and its value.

Definition (Cut): A *cut* $\mathcal{Q} = (X, \bar{X})$ in a network is a partition of the node set into two nonempty subsets X and its complement $\bar{X} = \mathcal{N}d \setminus X$. If X contains the source node s and $\bar{X}$ contains the destination node t, the cut is said to ***separate*** node s from node t.

Definition (Cut Value): If $0 \le x_{ij} \le h_{ij}$ for all $(i,j) \in \mathcal{A}c$, then the *cut value* C of a cut $\mathcal{Q} = (X, \bar{X})$ is the sum of the capacities of the arcs that start in set X and end in set $\bar{X}$; i.e.,

$$C = \sum_{i \in X} \sum_{j \in Af(i) \cap \bar{X}} h_{ij}, \tag{9.14}$$

where $Af(i) = \{j \in \mathcal{N}d \mid (i,j) \in \mathcal{A}c\}$. If each such arc (i,j) has a lower bound l_{ij}, not necessarily all zero, on the arc flow x_{ij}, then the *cut value* is

$$C = \sum_{i \in X} \sum_{j \in Af(i) \cap \bar{X}} h_{ij} - \sum_{j \in \bar{X}} \sum_{i \in Bf(j) \cap X} l_{ij}, \tag{9.15}$$

where $Af(i) = \{j \in \mathcal{N}d \mid (i,j) \in \mathcal{A}c\}$ and $Bf(j) = \{i \in \mathcal{N}d \mid (i,j) \in \mathcal{A}c\}$.

Definition (Saturated Arc): An arc is said to be ***saturated*** if it is used to full capacity, i.e., $x_{ij} = h_{ij}$.

▷ **Exercise 9.21** Given $0 \le x_{ij} \le h_{ij}$, show that if flow $F = F_o > 0$ is maximal, then there exists at least one arc that is saturated. Construct a counter example if maximal $F = F_o = 0$.

From Exercise 9.21 it follows that if the saturated arcs associated with a maximal flow are removed from the network by setting their capacities to zero, no flow is possible from the source to the destination over a path of remaining unsaturated arcs, because if such flow were possible, then the path would have positive arc capacity for the adjusted capacity network formed, as discussed in Theorem 9.7, by setting $h'_{ij} = h_{ij} - x^o_{ij}$, implying that the flows $x_{ij} = x^o_{ij}$ can be augmented along the path. Then by Theorem 9.7 the flows $x_{ij} = x^o_{ij}$ would not be maximal.

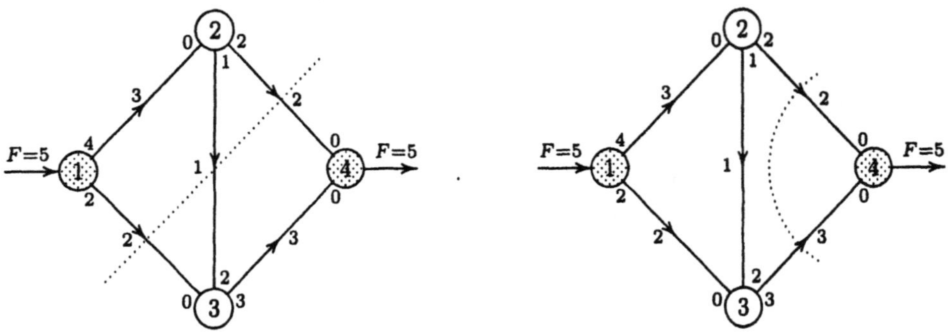

(a) Max flow value = 5
Min cut value = 5

(b) Max flow value = 5
Min cut value = 5

Figure 9-15: Illustration of Min-Cut = Max-Flow

From the above discussion it is clear that the collection of saturated arcs in a maximal solution can be used to define a partition of the nodes into two sets that define a cut that separates the source from the destination.

Example 9.12 (Illustration of Min-Cut = Max-Flow) The network and flows in Figures 9-15(a) and 9-15(b) are identical to those in Figure 9-11(f). The saturated arcs in Figure 9-15(a) are $(1,3)$, $(2,3)$, $(2,4)$, and $(3,4)$ with arc capacities $(2,1,2,3)$ respectively, with total arc capacity of 8. Observe that there are many cuts in the network. In particular, the cut $\mathcal{Q} = (X, \bar{X})$ with $X = \{1,2\}$ and $\bar{X} = \{3,4\}$, depicted by the use of a dotted line that separates X and $\bar{X}$ in Figure 9-15(a), has a cut value of 5. Similarly, the cut depicted in Figure 9-15(b), $\mathcal{Q} = (X, \bar{X})$ with $X = \{1,2,3\}$ and $\bar{X} = \{4\}$ also has a cut value of 5. The cut value 5 by accident happens to be the same in both cuts and also happens to be the same value of that of the maximal flow.

▷ **Exercise 9.22** Find all other cuts in the network of Example 9-15. What is the relationship of the flow value to the cut value of each of these cuts?

The relationship between flow values and cut values is specified in Lemma 9.12 and Theorem 9.13.

LEMMA 9.12 (Flow Value Bounded by Cut Value) *The flow value F of any feasible solution is less than or equal to the value C of any cut separating the source s from the destination t.*

THEOREM 9.13 (Ford-Fulkerson: Min-Cut = Max-Flow) *The max-flow value is equal to the min-cut value.*

▷ **Exercise 9.23 (Duality)** Show that the dual of the maximal flow problem is the min cut problem. Hint: Set up the maximal flow problem as a linear program. Set up the dual by letting u_j be the multipliers corresponding to the nodes and let w_{ij} be the multipliers corresponding to the upper bounds on arc flows, and show that the system is redundant. The redundancy implies that we can set $u_t = 0$, where t is the destination node; show that this implies that $u_s = 1$, where s is the source node. Next show that the remaining variables are all either 0 or 1. Then show that for arc (i, j), we have $w_{ij} = 1$ if and only if $u_i = 1$ and $u_j = 0$. Use this last result to define the cut.

Example 9.13 (Another Example of Max-Flow) Consider the problem of finding the maximal flow in the network illustrated in Figure 9-16(a_1). The capacities for each direction of an arc are stated at the end of the arc where the flow begins. Figure 9-16(b_1) represents a possible tree of positive arc capacities fanning out from the source. The flow can now be increased to $\theta_1 = 1$ along the augmenting path $(1,2)$, $(2,4)$, $(4,6)$, at which point the arcs $(1,2)$ and $(2,4)$ are saturated. The adjusted capacity network is then constructed by adjusting the arc capacities to $h'_{ij} = h_{ij} - \theta_1$ and $h'_{ji} = h_{ji} - \theta_1$ along the augmenting path just considered. The resulting associated network is shown in Figure 9-16(a_2). In Figure 9-16(b_2) is a new possible tree of positive arc capacities fanning out from the source node 1, resulting in the path $(1,3)$, $(3,5)$, $(5,6)$. Arcs $(3,5)$ and $(5,6)$ are saturated by assigning a flow $\theta_2 = 1$. The process is continued and the resulting path and solutions are shown in Figures 9-16(a_3), 9-16(b_3), 9-16(a_4), and 9-16(b_4). Since it is not possible to find an augmenting path in Figure 9-16(b_4), the process is terminated with a maximal flow of $F_o = \theta_1 + \theta_2 + \theta_3 = 3$.

To find the actual flows on each arc we subtract the final arc capacities in Figure 9-16(a_4) from the original arc capacities in Figure 9-16(a_1), interpreting a negative difference as a flow in the opposite direction, to obtain the final solution shown in Figure 9-17. To find the cut with minimum value, choose saturated arcs leading from nodes in X to nodes in $\bar{X}$. The nodes in X can all be reached from the source along unsaturated paths. This set was determined by the nodes in the subtree of positive arc capacities in Figure 9-16(b_4). Hence $X = \{1,3,4,2,5\}$ and $\bar{X} = \{6\}$ forms a cut $Q = (X, \bar{X})$. The set of saturated arcs joining the nodes in X to the nodes in $\bar{X}$ are $(4,6)$ and $(5,6)$, with sum of arc capacities $C_o = 2 + 1 = 3 = F_o$, as shown in Figure 9-17.

▷ **Exercise 9.24** In Figure 9-17 enumerate all the other cuts separating node 1 from node 6. Show that their associated cut values are greater than the maximal flow value.

9.5 SHORTEST ROUTE

The shortest route problem is that of finding the minimum total "distance" along paths in an undirected connected network from the source $s = 1$ to the destination $t = m$. The distance can be actual miles, the cost or time to go between nodes, etc.

A simple method to solve such a problem for *nonnegative distances* (or costs) is a branching out procedure that fans out from the source. Starting from the source, it always picks on the next iteration the closest node i to the source and records its distance. The algorithm is terminated when the shortest distance from the source node to the destination node is recorded. We first illustrate the algorithm and then state it.

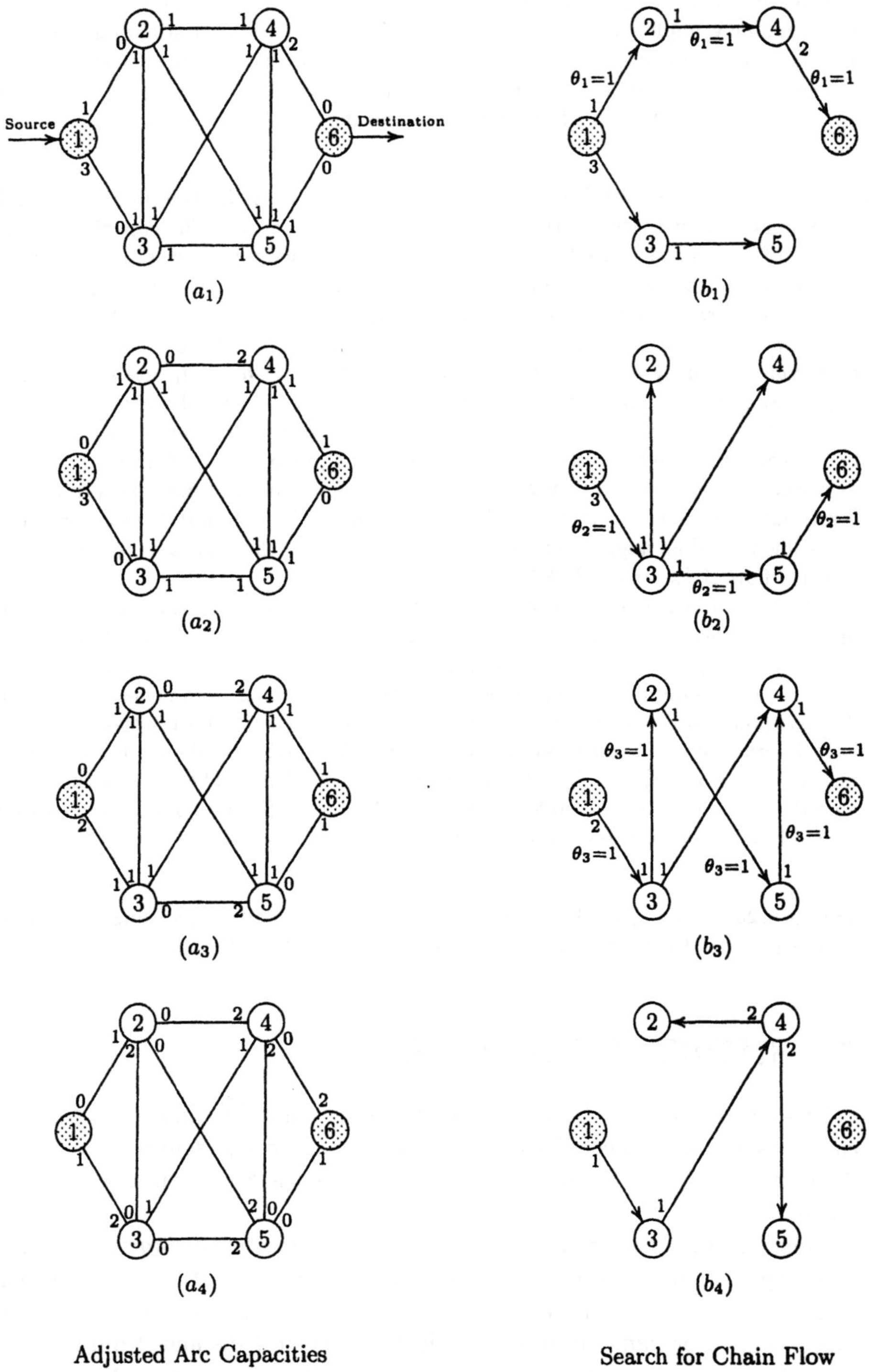

Figure 9-16: Another Example of the Max-Flow Algorithm

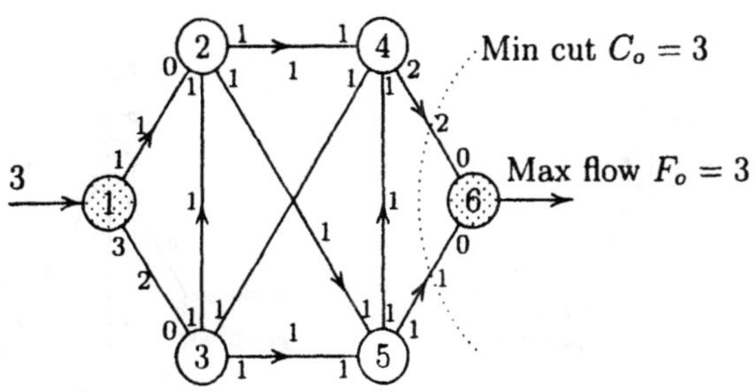

Figure 9-17: Max-Flow Min-Cut Solution

Example 9.14 (Illustration of Dijkstra's Algorithm) The steps of the algorithm are illustrated in Figure 9-18. Let $\mathcal{N}d = \{1, \ldots, 6\}$ be the set of nodes and $\mathcal{A}c$ be the set of arcs in the network. Let $\mathcal{S}$ be the set of nodes to which the shortest distances z_j are known at iteration k. We start by including the source node 1 in the set $\mathcal{S}$, i.e., $\mathcal{S} = \{1\}$, and assign it a distance label $z_1 = 0$ and a fictitious predecessor node index $p_1 = 0$. Note that z_j is called a permanent label if $j \in \mathcal{S}$; this is shown by concentric circles in the figure. Next we assign a temporary label $z_j = d_{ij}$ to each node $j \in \mathcal{N}d \setminus \mathcal{S}$ if an arc joins 1 to j and $z_j = \infty$ if no such arc exists. Thus

$$z_2 = 3, \ z_3 = 8 \ \text{ and } z_4 = \infty, \ z_5 = \infty, \ z_6 = \infty.$$

The corresponding predecessor indices are

$$p_2 = 1, \ p_3 = 1 \ \text{ and } p_4 = 1, \ p_5 = 1, \ p_6 = 1$$

where we interpret $p_4 = p_5 = p_6 = 1$ as meaning that artificial arcs of infinite length join node 1 to nodes 4, 5, and 6 respectively.

On iteration 1 we start by picking the node with the smallest temporary label (with ties broken randomly), i.e.,

$$k^* = \operatorname*{argmin}_{j \in \mathcal{N}d \setminus \mathcal{S}} = \operatorname{argmin}\{ z_2, \ z_3, \ z_4, \ z_5, \ z_6 \} = 2.$$

We now make node 2 a part of set $\mathcal{S}$, i.e.,

$$\mathcal{S} \leftarrow \mathcal{S} \cup \{2\}$$

and call z_2 a permanent label. Next we update the temporary labels $z_j$for $j \in \mathcal{N}d \setminus \mathcal{S}$ if there is an arc from the newly added node 2 to j. To do this define a set

$$\mathcal{Q} = \{ j \mid (2, j) \in \mathcal{A}c, \ \text{ for } j \in \mathcal{N}d \setminus \mathcal{S} \} = \{3, 4, 5\}.$$

Then the temporary labels that need updating are for $j \in \mathcal{Q}$, and they are computed as follows

$$z_j \leftarrow \min\{z_j, \ z_2 + d_{2j}\} \text{ for } j \in \mathcal{Q},$$

or

$$z_3 = 5, \ z_4 = 6, \ z_5 = 4,$$

with predecessors

$$p_3 = 2, \ p_4 = 2, \ p_5 = 2.$$

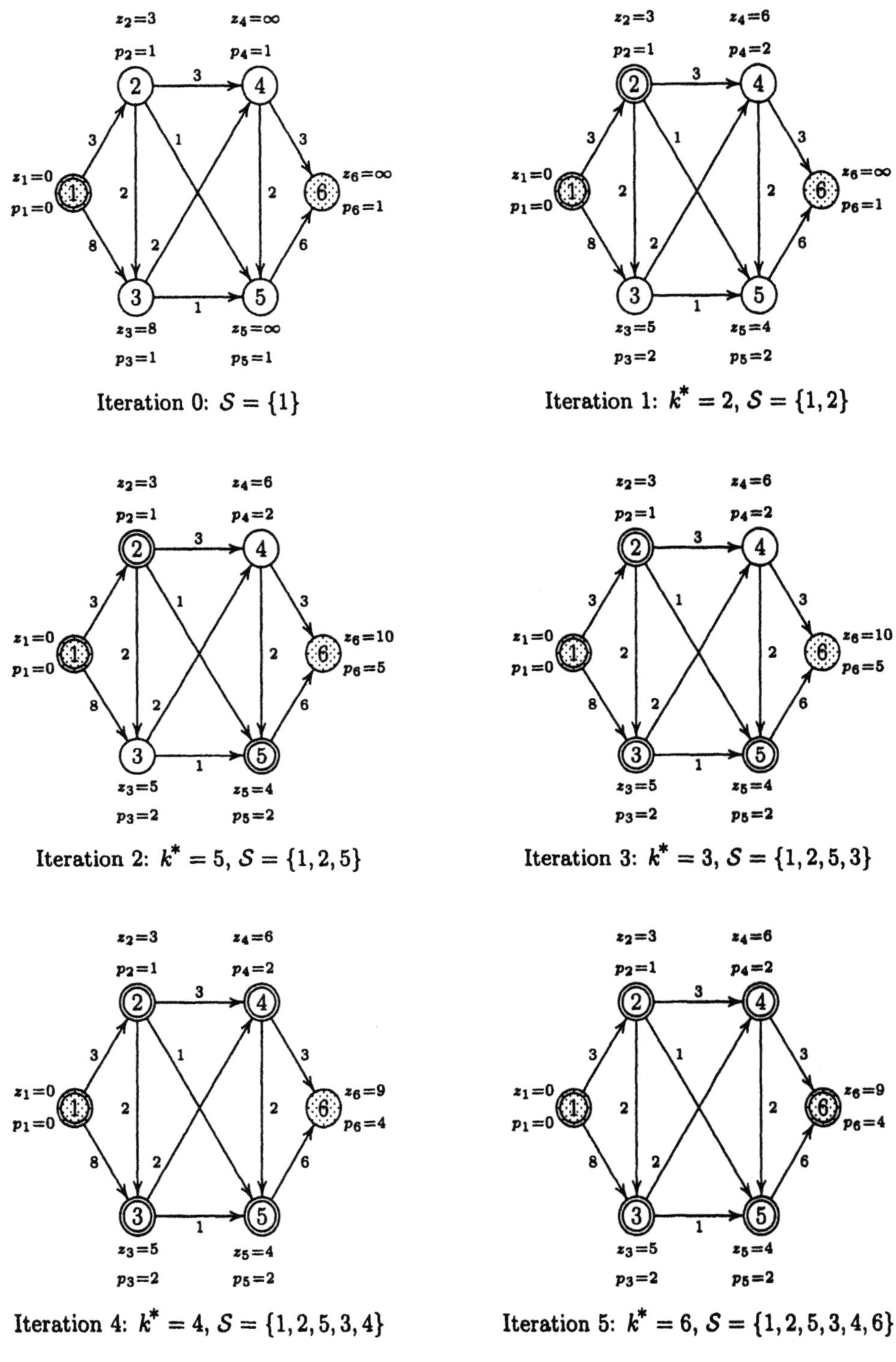

Figure 9-18: Illustration of Dijkstra's Shortest Route Algorithm

We continue with iteration 2 by picking the next node with the smallest temporary label (with ties broken randomly), i.e.,

$$k^* = \operatorname*{argmin}_{j \in \mathcal{N}d \setminus \mathcal{S}} = \operatorname{argmin}\{\, z_3,\ z_4,\ z_5,\ z_6 \,\} = 5.$$

We now make node 5 a part of set $\mathcal{S}$, i.e.,

$$\mathcal{S} \leftarrow \mathcal{S} \cup \{5\},$$

and call z_5 a permanent label. Next we update the temporary labels z_j for $j \in \mathcal{N}d \setminus \mathcal{S}$ if there is an arc from the newly added node 5 to j. To do this define a set

$$\mathcal{Q} = \{\, j \mid (5, j) \in \mathcal{A}c, \quad \text{for } j \in \mathcal{N}d \setminus \mathcal{S} \,\} = \{6\}.$$

Then the temporary labels that need updating are for $j \in \mathcal{Q}$, and they are computed as follows

$$z_j \leftarrow \min\{z_j,\ z_5 + d_{5j}\} \text{ for } j \in \mathcal{Q},$$

or

$$z_6 = 10,$$

with predecessors

$$p_6 = 5.$$

The process then continues, as shown in the figure. Finally, on iteration 5 the shortest distance is determined as $z_6 = 9$. It is straightforward to determine the shortest path through the use of the predecessor nodes.

Algorithm 9.3 (Dijkstra's Shortest Route Algorithm) Given a network $G = (\mathcal{N}d, \mathcal{A}c)$ with $m > 1$ nodes, where $s = 1$ is the source node and $t = m$ is the destination node. It is assumed that the distances between nodes are $d_{ij} \geq 0$. On iteration τ, associated with each node j is (1) a distance z_j from the source to node j along some path P_j, and (2) p_j, which is the predecessor node to node j along the path P_j. Also on iteration τ, the nodes have been partitioned into two sets, $\mathcal{S}$ and $\mathcal{N}d\setminus\mathcal{S}$, where the paths P_j leading to nodes $j \in \mathcal{S}$ are shortest paths. We will refer to z_j as the *label* of node j. If $j \in \mathcal{S}$, z_j will be called its *permanent label*; otherwise it will be called its *temporary* label.

1. Start with the source node $s = 1$ as the initial set $\mathcal{S}$ of permanently labeled nodes; i.e., $\mathcal{S} = \{1\}$. Assign as its permanent label $z_1 = 0$, and predecessor $p_1 = 0$. To each of the nodes $j \in \mathcal{N}d \setminus \mathcal{S}$ assign a temporary label z_j equal to the distance on an arc from the source to node j, and $p_j = 1$; if no such arc exists, the label is set to ∞ and $p_j = 1$. Conceptually this implies that we are adding an artificial arc from the source to every node that is not connected to the source with an arc. Computationally we do not use ∞; instead we use a number greater than the sum of the distances on all the arcs.
2. Set the iteration count: $\tau \leftarrow 1$.
3. Find a node $k^* \in \mathcal{N}d \setminus \mathcal{S}$ such that

$$k^* = \operatorname*{argmin}_{j \in \mathcal{N}d \setminus \mathcal{S}} z_j$$

and set $\mathcal{S} \leftarrow \mathcal{S} \cup \{k^*\}$.

4. If $\mathcal{N}d \setminus \mathcal{S}$ is empty, terminate.
5. Let $\mathcal{Q} = \{j \mid j \in \mathcal{N}d \setminus \mathcal{S},\ (k^*, j) \in \mathcal{A}c\}$ be the set of nodes in set $\mathcal{N}d \setminus \mathcal{S}$ that are attached to node k^* by an arc. For $j \in \mathcal{Q}$,

$$\text{if } z_{k^*} + d_{k^*j} < z_j, \text{ then set } z_j = z_{k^*} + d_{k^*j} \text{ and } p_j = k^*.$$

6. Set $\tau \leftarrow \tau + 1$ and go to Step 3.

Comment: In order to avoid searching all $j \in \mathcal{N}d \setminus \mathcal{S}$ on Step 3, one maintains an ordered list of z_j by updating the list of $j \in \mathcal{Q}$ whose label values z_j have been lowered during the execution of Step 5.

LEMMA 9.14 (Number of Operations in Dijkstra's Algorithm) *Algorithm 9.3 requires $m(m-1)/2$ comparisons in Step 3 and requires at most $m(m-1)/2$ comparisons and additions in Step 5.*

Proof. In iteration τ there are τ nodes in $\mathcal{S}$ and thus $m - \tau$ comparisons are required in Step 3. Hence in a network with m nodes we need

$$1 + 2 + \cdots + (m-1) = m(m-1)/2$$

comparisons. Similarly, we need a maximum of $m(m-1)/2$ comparisons in Step 5. Fewer may be needed, because at any iteration τ, node k^* will typically have fewer than $m - \tau$ nodes in $\mathcal{N}d \setminus \mathcal{S}$ connected to it. ∎

LEMMA 9.15 (Validity of Dijkstra's Algorithm) *Algorithm 9.3, with distances $d_{ij} \geq 0$, finds the shortest path from the source to all nodes in the network.*

▷ **Exercise 9.25** Construct an example to show that Algorithm 9.3 can fail if there are negative arc distances. Construct an example with some negative arc distances but where the sum of distances around every cycle is nonnegative. Demonstrate that in this case Algorithm 9.3 finds the shortest route from the source to the destination.

▷ **Exercise 9.26** Construct an example to show that if the main post office has a collection of shortest routes from it to every other post office, the collection of arcs along the route is not necessarily a tree. Apply Dijkstra's algorithm to your example to find an equivalent tree.

9.6 MINIMAL SPANNING TREE

Consider a connected network $G = (\mathcal{N}d, \mathcal{A}c)$ with m nodes and n arcs; the distances on the arcs are unrestricted in sign. The minimal spanning tree problem is one of finding a tree connecting all m nodes such that the sum of the distances on the arcs of the tree is minimized. The meaning of distance here is very similar

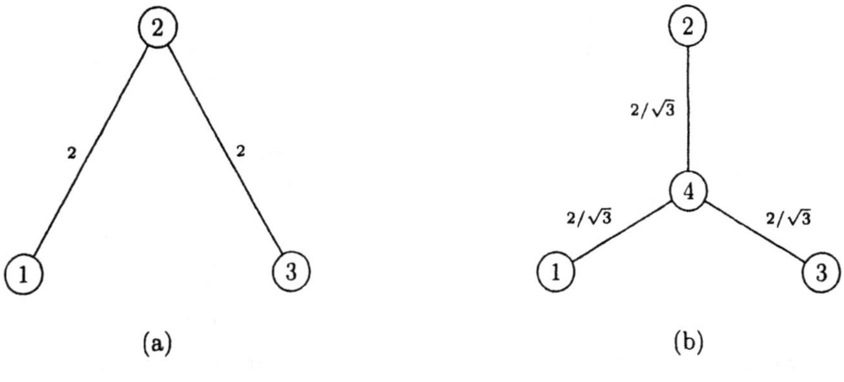

Figure 9-19: Minimal Spanning Tree Example

to that for the shortest route problem in that it can represent cost, time, etc. The minimal spanning tree problem has been applied to design transportation and communication networks; we do not need to have the network already built but just to know what the distance (cost) would be if an arc were added between two nodes.

Example 9.15 (Minimal Spanning Tree and Artificial Node) Suppose that you need to design a new freeway system connecting three equidistant points 1, 2, and 3. The minimal spanning tree has a total distance of 4 miles, as shown in Figure 9-19(a). It is interesting to note that a new freeway system can be built using only $2\sqrt{3} < 4$ miles by building the freeway through an artificial node 4, as shown by the minimal spanning tree in Figure 9-19(b).

▷ **Exercise 9.27** Show that the minimal spanning tree remains the same if all arc distances are changed by a constant value β, i.e., $\bar{d}_{ij} \leftarrow d_{ij} + \beta$.

The solution process is very simple and is one of the rare cases where a *greedy* algorithm (i.e., an algorithm in which one makes a locally best improvement at each iteration) finds a global optimal solution. Since a spanning tree has exactly $m - 1$ arcs, so will the optimal solution. An algorithm to do this chooses the arcs one at a time in increasing order of their lengths, rejecting the selected arc only if it forms a cycle with the arcs already chosen.

Example 9.16 (Illustration of Minimal Spanning Tree Algorithm) The steps of the algorithm are illustrated in Figure 9-20. We start by initializing labels u_j for each node $j = 1, \ldots, 6$ to have the value 0 to imply no label, and we order the arcs by increasing distance. When every node is labeled the same by a nonzero label, the algorithm will terminate.

At iteration 1 we pick arc $(2,5)$ because it has the smallest distance $d_{25} = 1$. The two connecting nodes are relabeled as $u_2 = u_5 = 1$. Next, on iteration 2, from the remaining arcs, we choose arc $(3,4)$ because it has a distance $d_{34} = 2$ (we could equivalently have chosen arc $(4,5)$) and does not create a cycle with the previously chosen arcs. We set

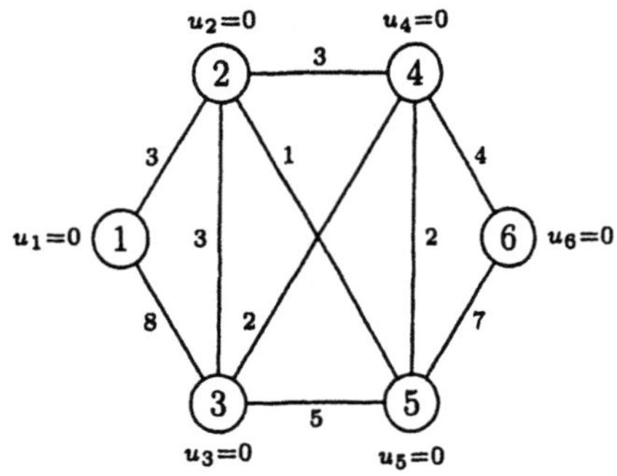

Iteration 0: Initialization

$\tau=\{\emptyset\}$

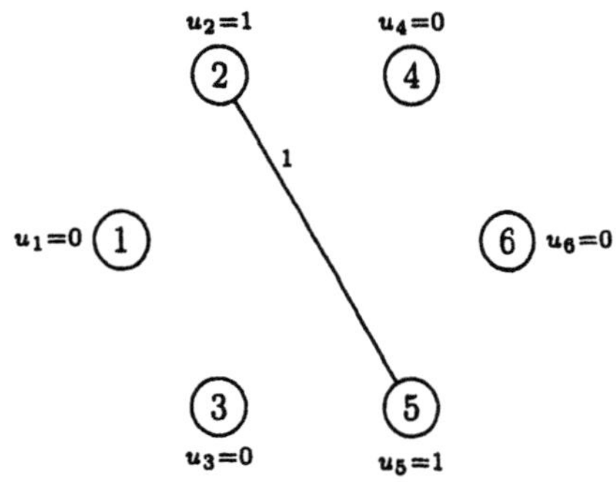

Iteration 1: Arc (2,5) enters

$\tau=\{(2,5)\}$

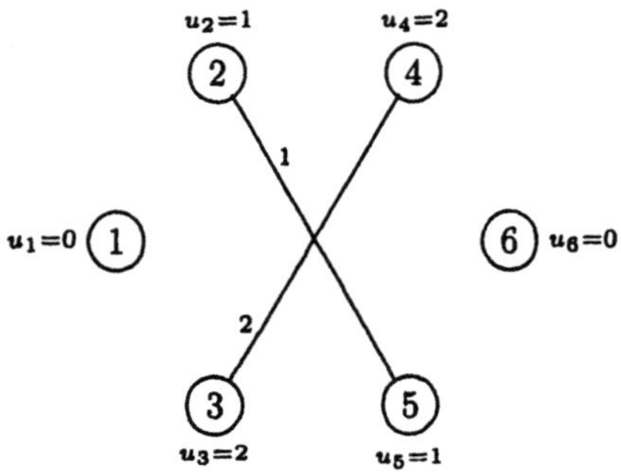

Iteration 2: Arc $(3,4)$ enters

$\tau=\{(2,5),\ (3,4)\}$

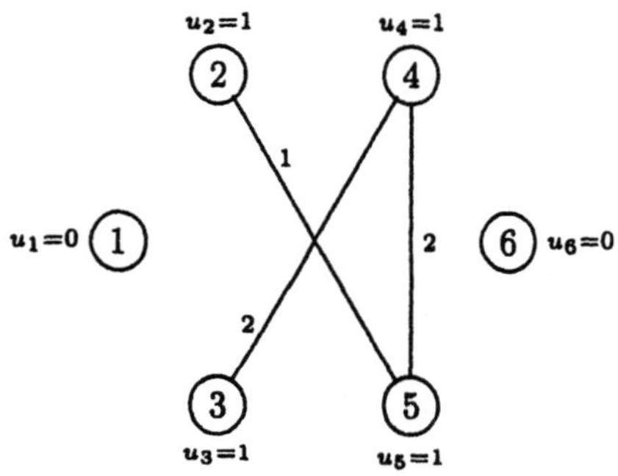

Iteration 3: Arc $(4,5)$ enters

$\tau=\{(2,5),\ (3,4),\ (4,5)\}$

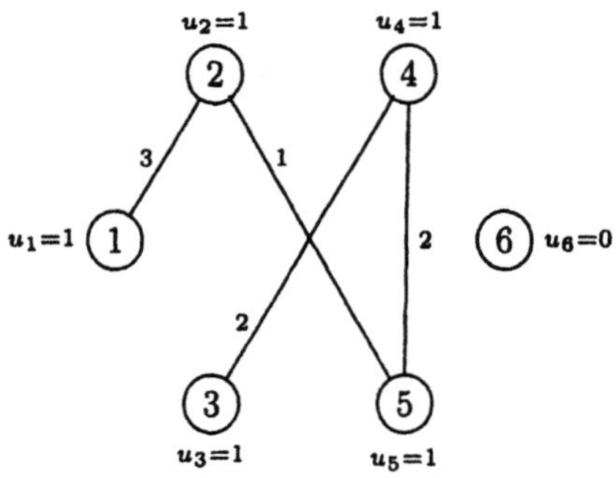

Iteration 4: Arc $(1,2)$ enters

$\tau=\{(2,5),\ (3,4),\ (4,5),\ (1,2)\}$

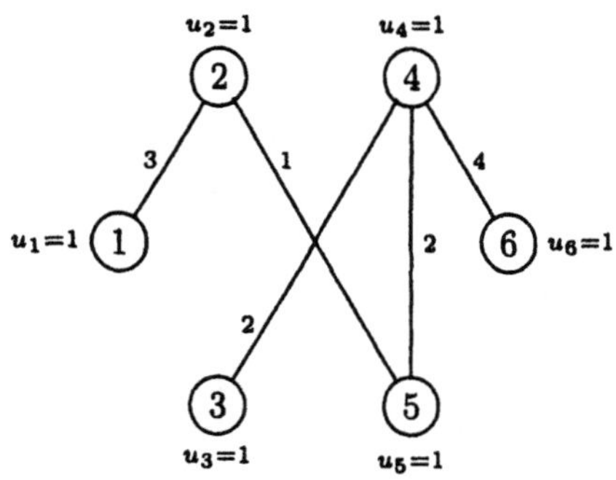

Iteration 5: Arc $(4,6)$ enters

$\tau=\{(2,5),\ (3,4),\ (4,5),\ (1,2),\ (4,6)\}$

Figure 9-20: Illustration of Minimal Spanning Tree Algorithm

the labels for the connecting nodes of this arc to be $u_3 = u_4 = 2$ because the nodes do not intersect any of the other labeled nodes. Continuing, on iteration 3, we pick the next shortest distance arc $(4,5)$, which does not create a cycle with the previously chosen arcs. Now, 4 belongs to a different set of previously labeled nodes than does 5. Hence, we set the labels of the two sets to be equal; we arbitrarily set the labels to be 1 for nodes 2, 3, 4, and 5. We continue in this manner, picking arc $(1,2)$ on iteration 4 and arc $(4,6)$ on the final iteration.

Algorithm 9.4 (Minimal Spanning Tree) The algorithm starts by reindexing the arcs $\alpha = 1, \ldots, m$ so that they are in the order of nondecreasing arc lengths. The nodes will be partitioned into several membership classes, with each node assigned a unique positive membership label $u_j > 0$; upon termination all nodes will belong to only one membership class. Initially all $u_j = 0$, meaning that they do not belong yet to any membership class. Let $\mathcal{T}$ be the set of arcs in the minimal spanning tree, let ν represent the number of arcs assigned so far to the minimal spanning tree, let k represent the last membership class label assigned to a node, and let α be the iteration counter, which is, in this case, the index of the arc being considered for candidacy in the minimal spanning tree.

1. Set $\mathcal{T} \leftarrow \{\emptyset\}$, $\nu \leftarrow 0$, $k \leftarrow 0$, and $\alpha \leftarrow 1$.
2. If $\nu = m - 1$, terminate; we have found a minimal spanning tree.
3. Let arc α connect nodes i and j. We use the node labels to determine whether candidate arc α forms a cycle with the arcs in $\mathcal{T}$. If it forms a cycle, it is rejected; otherwise it is accepted as belonging to the minimal spanning tree.
   (a) *Reject arc α if it will create a cycle with the arcs in $\mathcal{T}$.* If $u_i = u_j \neq 0$, i.e., nodes i and j both belong to the same membership class, then arc α will create a cycle amongst arcs already selected as belonging to the minimal spanning tree, and therefore arc α is rejected. Go to Step 4.
   (b) *Accept arc α if it does not create a cycle with the arcs in $\mathcal{T}$.* In this case set $\mathcal{T} \leftarrow \mathcal{T} \cup \{\alpha\}$, $\nu \leftarrow \nu + 1$, and update the labels of the membership classes by doing one of the following:
      - If $u_i = u_j = 0$, i.e., nodes i and j do not belong to any membership class, then set $k \leftarrow k + 1$ followed by setting $u_i \leftarrow k$ and $u_j \leftarrow k$.
      - If $0 \neq u_i \neq u_j \neq 0$ then the nodes i and j belong to different membership classes. We merge the two membership classes labeled u_i and u_j into one. That is, we change the labels of all nodes p such that $u_p = u_j$ to $u_p \leftarrow u_i$.
      - If $u_i \neq 0$, but $u_j = 0$ set $u_j \leftarrow u_i$.
      - If $u_j \neq 0$, but $u_i = 0$ set $u_i \leftarrow u_j$.
4. Set $\alpha \leftarrow \alpha + 1$. Go to Step 2.

LEMMA 9.16 (Spanning Tree in a Connected Network) *A connected network contains a spanning tree.*

▷ **Exercise 9.28** Prove Lemma 9.16.

$\mathcal{T}$ $\mathcal{T}_1$ $\mathcal{T}_2$

Figure 9-21: Disconnecting an Arc of $\mathcal{T}$ Results in Two Trees: $\mathcal{T}_1$ and $\mathcal{T}_2$

▷ **Exercise 9.29** Prove that Algorithm 9.4 finds a minimal spanning tree for a connected network. How would you modify the algorithm so that it could be used on any network? (If the network is not connected your modified algorithm should determine that it was unconnected.)

▷ **Exercise 9.30** Show how using linked lists can make the operations of merging membership classes in Step 3b of Algorithm 9.4 efficient. Show that you can make it even more efficient by merging the smaller membership class with the larger one.

LEMMA 9.17 (Disconnecting an Arc of a Tree) *Disconnecting an arc (i, j) of a tree $\mathcal{T}$ results in decomposing $\mathcal{T}$ into two trees, $\mathcal{T}_1$ and $\mathcal{T}_2$.*

Example 9.17 (Disconnecting an Arc of a Tree) Figure 9-21 demonstrates how disconnecting an arc of a tree results in two trees.

▷ **Exercise 9.31** Prove Lemma 9.17

LEMMA 9.18 (End Nodes in a Tree) *Any tree has at least two end nodes.*

▷ **Exercise 9.32** Verify Lemma 9.18 for the tree illustrated in Figure 9-21.

▷ **Exercise 9.33** Prove Lemma 9.18.

9.7 MINIMUM COST-FLOW PROBLEM

The minimum cost-flow problem is to find flows x_{ij} through a directed network $G = (\mathcal{N}d, \mathcal{A}c)$ with m nodes indexed $1, \ldots, m$ and n arcs, such that the total cost of the flows is minimized. It can be stated as a standard linear programming problem with very special structure:

$$
\begin{array}{lll}
\text{Minimize} & \displaystyle\sum_{(i,j)\in\mathcal{A}c} c_{ij}x_{ij} = z & \\
\text{subject to} & \displaystyle\sum_{i\in Af(k)} x_{ki} - \sum_{j\in Bf(k)} x_{jk} = b_k & \text{for all } k \in \mathcal{N}d, \\
& l_{ij} \le x_{ij} \le h_{ij} & \text{for all } (i,j) \in \mathcal{A}c,
\end{array}
\tag{9.16}
$$

where c_{ij} is the cost per unit flow on the arc (i,j); b_k is the net flow at node k; l_{ij} is a lower bound on the flow in arc (i,j); and h_{ij} is an upper bound on the flow in arc (i,j). Note that b_k takes on values that depend on the type of node k:

$$
b_k \text{ is } \begin{cases} > 0 & \text{if } k \text{ is a source (supply) node;} \\ < 0 & \text{if } k \text{ is a destination (demand) node;} \\ = 0 & \text{if } k \text{ is a node for transshipment only.} \end{cases}
$$

Definition (Feasible Flow): A set of arc flows x_{ij} that satisfies the conservation of flow constraints and the capacity constraints in (9.16) is called a *feasible* flow.

▷ **Exercise 9.34** Show that a necessary condition for feasibility of the system (9.16) is that $\sum_{i\in\mathcal{N}d} b_i = 0$.

THEOREM 9.19 (Feasibility of a Minimum Cost-Flow Problem) *The minimum cost-flow problem* (9.16) *with* $l_{ij} = 0$, $h_{ij} = \infty$ *for all* $(i,j) \in \mathcal{A}c$ *has a feasible solution if and only if*

$$
\sum_{i\in\mathcal{N}d} b_i = 0
$$

and the cut value of every cut $\mathcal{Q} = (X, \bar{X})$ *satisfies*

$$
C(\mathcal{Q}) \ge \sum_{i\in X} b_i.
$$

▷ **Exercise 9.35** Through a simple example, demonstrate one particular instance of Theorem 9.19. Prove it in general.

The minimum cost-flow formulation has many applications. In fact a number of the problems in this book can be expressed in terms of a minimum cost-flow formulation. Some of these are illustrated through the examples and exercises that follow.

Example 9.18 (Transportation/Assignment Problem) The transportation problem (8.3) is a special case of the minimum cost-flow problem. Here there are *no intermediate nodes*; only source and destination nodes. Furthermore, the upper bounds on the flows are implied by the formulation.

The assignment problem given by (8.20), (8.21), (8.22), and (8.23) is a special case of the transportation problem in which all b_i at supply nodes are $+1$ and all b_i at demand nodes are -1. Thus, it too can be represented as a minimum cost-flow problem.

Example 9.19 (Maximal Flow Problem) The maximal flow problem of Section 9.3 is almost in the minimum cost-flow format. It can be easily converted to a maximal flow problem as shown in Exercise 9.7 on Page 264.

▷ **Exercise 9.36 (Shortest Path Problem)** Show how the shortest path problem of Section 9.5 can be set up as a minimum cost-flow problem. Hint: Reformulate the problem into one of sending exactly one unit of flow from the source to the destination.

9.8 THE NETWORK SIMPLEX METHOD

In this section we show how to simplify the Simplex Method so that it can be applied to solve graphically, by hand, a minimum cost-flow network problem with, say, less than 100 nodes. The methodology can also be applied efficiently to solve, with a computer, larger problems using special data structures. There are three ways in which we can get computational savings:

1. Exploiting the fact that (by deleting a redundant node balance equation or adding an artificial variable) every basis is triangular.
2. Exploiting the fact that all calculations involve additions and subtractions only because the elements of every basis are $+1$, -1, or 0.
3. Exploiting the fact that typically not all the prices need to be updated when a new column corresponding to an arc enters the basis.

We have already seen a specialized version of the Simplex Method applied to the transportation problem in Chapter 8. The network version developed here has similarities and can be used to solve the minimum cost-flow problem very efficiently.

We assume that the minimum cost-flow problem is in, what we call, *standard form*; namely, it is one in which the arc flows are nonnegative with no specified upper bounds:

$$\begin{array}{llll}
\text{Minimize} & \displaystyle\sum_{(i,j)\in\mathcal{A}c} c_{ij}x_{ij} & = z & \\
\text{subject to} & \displaystyle\sum_{i\in Af(k)} x_{ki} - \sum_{j\in Bf(k)} x_{jk} & = b_k & \text{for all } k \in \mathcal{N}d, \\
& 0 \le x_{ij} \le \infty & & \text{for all } (i,j) \in \mathcal{A}c,
\end{array} \tag{9.17}$$

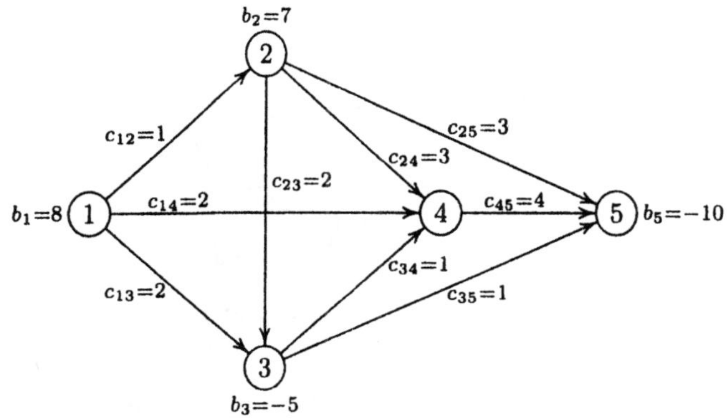

Figure 9-22: Example of a Minimum Cost-Flow Problem

where $\sum_{k \in \mathcal{N}d} b_k = 0$.

Recall from Section 9.1 that the flow through a network is represented by a node-arc incidence matrix where each column corresponding to an arc (i, j) has exactly two nonzero entries: $+1$ in row i and -1 in row j. In matrix notation, each column of A corresponding to an arc $s = (i, j)$ in the network can be written as

$$A_{\bullet s} = e_i - e_j, \tag{9.18}$$

where e_i is a unit-vector with 1 in position i and e_j is a unit-vector with a 1 in position j.

Example 9.20 (Minimum Cost-Flow Problem) An example of a minimum cost-flow problem in standard form is shown in Figure 9-22. The costs c_{ij} and the demand/supply b_i are shown on the network. The resulting coefficient matrix is shown below:

$$A = \begin{array}{c} \\ 1 \\ 2 \\ 3 \\ 4 \\ 5 \end{array} \begin{array}{c} \begin{array}{ccccccccc} (1,2) & (1,3) & (1,4) & (2,3) & (2,4) & (2,5) & (3,4) & (3,5) & (4,5) \end{array} \\ \left(\begin{array}{ccccccccc} 1 & 1 & 1 & 0 & 0 & 0 & 0 & 0 & 0 \\ -1 & 0 & 0 & 1 & 1 & 1 & 0 & 0 & 0 \\ 0 & -1 & 0 & -1 & 0 & 0 & 1 & 1 & 0 \\ 0 & 0 & -1 & 0 & -1 & 0 & -1 & 0 & 1 \\ 0 & 0 & 0 & 0 & 0 & -1 & 0 & -1 & -1 \end{array} \right) \end{array}$$

and

$$b^T = (\,8 \quad 7 \quad -5 \quad 0 \quad -10\,).$$

▷ **Exercise 9.37** Show that an undirected arc (i, j) with cost $c_{ij} \geq 0$, regardless of whether the flow is from i to j or from j to i, can be replaced by two directed arcs (i, j) and (j, i) each with the same cost $c_{ij} = c_{ji}$. Show that this reduction to a linear program is not valid if $c_{ij} < 0$.

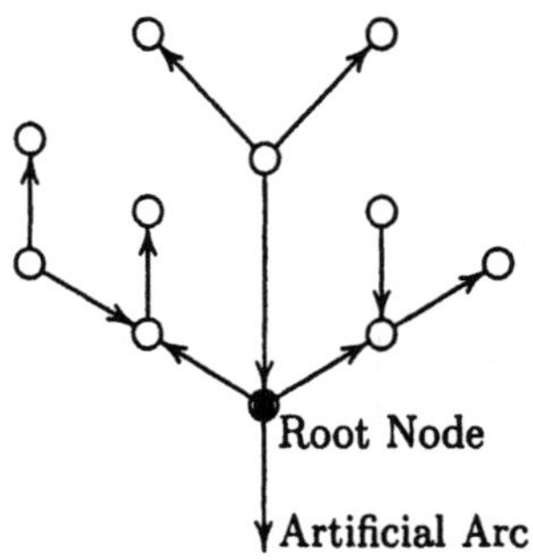

Figure 9-23: Rooted Spanning Tree

BASES, TREES, AND INTEGRAL PROPERTY

Note that the system (9.17) contains at least one redundant equation.

LEMMA 9.20 (Rank of a Connected Network) *The node-arc incidence matrix A of a connected network with $m \geq 2$ nodes has rank $m-1$.*

▷ **Exercise 9.38** Demonstrate that the system shown in Example 9.20 contains exactly one redundant equation.

▷ **Exercise 9.39** Since the rank of any $m-1$ rows of the node-arc incidence matrix of a connected network is rank $m-1$, show that *any* row may be regarded as dependent on the remaining $m-1$ rows and may be dropped as redundant.

▷ **Exercise 9.40** Show that if the network is not connected, there is more than one redundant equation.

We now know that the rank of A is $m-1$ and we know from Exercise 9.39 that we can drop any row of the matrix and solve the resulting problem. Instead of dropping a row, we can augment the columns by an artificial unit-vector column, say e_m, and associate with it an artificial variable $x_{n+1} \geq 0$. This will make the rank equal to m. If we were now to apply the Simplex Method to the resulting linear program, x_{n+1} will be zero in every solution, feasible or otherwise, and therefore it does not matter what cost we assign to the artificial variable.

▷ **Exercise 9.41** Prove this last statement.

Adding such an artificial variable x_{n+1} corresponds to augmenting the directed network by a symbolic arc that starts at node m and has no head node. We will refer to such an arc as an *artificial arc* (see Figure 9-23).

Definition (Root Node): The node associated with the artificial arc is referred to as a *root node.* If the root node corresponds to the last row, then the artificial arc corresponds to the column e_m, the unit vector with 1 in the mth position. The artificial arc will be denoted by (k, ∞) if the root arc is incident at node k.

Definition (Rooted Spanning Tree): A spanning tree in a directed network with a designated root node (and associated artificial arc) is called a *rooted spanning tree.*

THEOREM 9.21 (Basis for the Network) *In a connected network, if we designate a node as the root node and add an artificial arc starting at the root, then the node-arc incidence matrix of any rooted spanning tree is a basis matrix for the network.*

▷ **Exercise 9.42** Prove Theorem 9.21.

Because the basis is triangular, it is easy to efficiently find a solution to $Bx_B = b$ and $B^T\pi = c_B$. Furthermore, because the elements of B are all $+1$, -1, or 0, it is easy to see that if b is integral then so is x_B and if c_B is integral then so is π. This property is formalized through the next two exercises.

▷ **Exercise 9.43** Show that if b_i are integral and the capacity bounds are integral, then the solution to any of the minimum cost-flow problems (9.16) or (9.17) is integral.

▷ **Exercise 9.44** Every element of the inverse of the basis matrix B is $+1$, -1, or 0. Hint: Recall that column j of B^{-1} is the solution to $By = e_j$.

▷ **Exercise 9.45** Show that the representation of nonbasic column$A_{\bullet s}$ in terms of the basis B, namely $B^{-1}A_{\bullet s}$, consists only of $+1$, -1, and 0 components.

SOLVING FOR THE BASIC VARIABLES

Because the system of equations is triangular, it is straightforward to solve the system $Bx_B = b$ for the values of the basic variables x_B. These values can be obtained directly by examining the tree (or subgraph) corresponding to the basis. We start by examining all the end nodes; there are at least two end nodes. Because exactly one arc is incident to each end node, the corresponding equation in the system $Bx_B = b$ contains only one variable, and because their coefficient is $+1$ or -1, it can be easily evaluated. If the arc points out of end node k, the arc flow is set to b_k; if the arc points into the end node k, the arc flow is set to $-b_k$. Once this is done, the nodes are examined to see which have all its incident arcs assigned

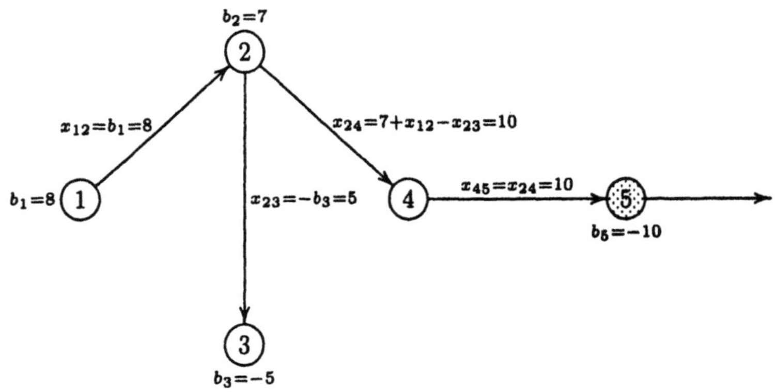

Figure 9-24: Computation of Flow Values from the End Nodes Towards the Root

except for one of the arcs. There must be at least one such node by triangularity of the basis; the conservation of flows around this node can then be used to obtain the arc flow on the unassigned incident arc. The process is continued until all the flows are evaluated.

It turns out that the process of elimination proceeds from the ends of the tree towards the root. This is because the last step evaluates the artificial arc, which will always turn out to be zero since the sum of the b_i's is zero (see Exercise 9.34). The above describes how to solve for the values of the basic variables if the network is small and displayed visually as a graph. An efficient way to carry out this process on a computer requires the use of special data structures that define the tree.

Example 9.21 (Computation of a Basic Feasible Solution) For the minimum cost-flow network of Example 9.20, Figure 9-22, a rooted spanning tree corresponding to a feasible basis is displayed in Figure 9-24; in this figure, node 5 has been arbitrarily made into the root node. The nonbasic x_{ij} are all set to zero. The computation of the flow values x_{ij} for flow on the arcs (except for the artificial arc at the root node) is done as follows. Node 1 is an end node with arc $(1,2)$ leaving it, and hence x_{12} is easily evaluated as

$$x_{12} = b_1 = 8.$$

Next we note that node 3 is an end node with arc $(2,3)$ entering it, and hence

$$x_{23} = -b_3 = 5.$$

At this point, at node 2, except for the flow value on arc $(2,4)$, the flow values on all the arcs incident at the node are known, and hence x_{24} can be easily evaluated because

$$-x_{12} + x_{23} + x_{24} = b_3.$$

Hence we know the value $x_{24} = 10$. Similarly at node 4 we know the value of x_{24}, and hence x_{45} can be evaluated by solving

$$-x_{24} + x_{45} = 0,$$

or $x_{45} = 10$.

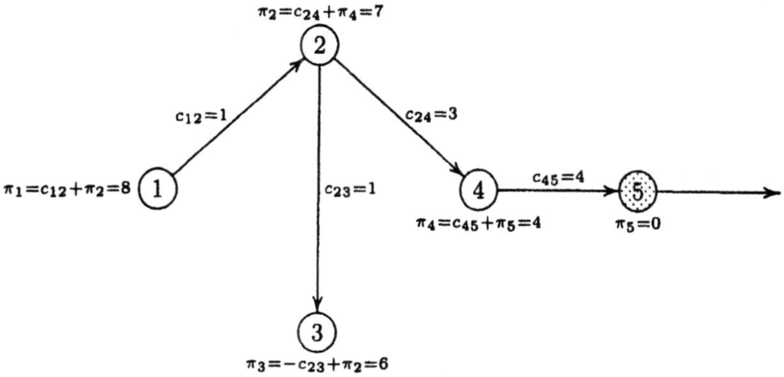

Figure 9-25: Computation of Simplex Multipliers π from the Root

PRICING, OPTIMALITY, AND REDUCED COSTS

The dual of the minimum cost-flow problem (9.17) with (m, ∞) as its root arc is

$$\begin{array}{lrl} \text{Minimize} & \displaystyle\sum_{i \in \mathcal{N}d} b_i \pi_i & = z \\ \text{subject to} & \pi_i - \pi_j & \leq c_{ij} \quad \text{for all } (i,j) \in \mathcal{A}c,\ (i,j) \neq (m,\infty) \\ & \pi_m & \leq 0. \end{array} \tag{9.19}$$

Given any basis B, we can compute the simplex multipliers π, in general, by solving $B^T\pi = c_B$; in the special case of the minimum cost-flow problem this reduces to

$$\pi_i - \pi_j = c_{ij} \qquad \text{for } (i,j) \in \mathcal{A}c,\ (i,j) \neq (m,\infty),\ x_{ij} \text{ basic}, \tag{9.20}$$

and for $(i,j) = (m,\infty)$,

$$\pi_m = 0 \tag{9.21}$$

because the column e_m corresponding to root arc (m, ∞) is always basic. Substituting $\pi_m = 0$ into (9.20) we can solve for the remaining π_i because B triangular implies B^T is triangular. Moreover, since each equation of (9.20) has exactly two variables, only a trivial amount of calculation is needed to solve the system.

The process of computing the remaining dual variables π_i consists in working from the root of the basis tree outwards towards the ends of the tree using the relations (9.20). An efficient way to carry out this process on a computer requires the use of special data structures that define the tree.

Example 9.22 (Computation of the Simplex Multipliers) Continuing with Example 9.20, the rooted spanning tree of Example 9.21 is displayed in Figure 9-25 with the computed values of the simplex multipliers obtained as follows. Proceeding from the root, we start by setting

$$\pi_5 = 0.$$

Next we note that

$$\pi_4 - \pi_5 = c_{45} = 4,$$

implying that $\pi_4 = 4$. For arc $(2,4)$ we note that

$$\pi_2 - \pi_4 = c_{24} = 3,$$

or $\pi_2 = 7$. Similarly, we have

$$\pi_2 - \pi_3 = c_{23} = 1,$$

implying $\pi_3 = 6$, and

$$\pi_1 - \pi_2 = c_{12} = 1,$$

implying $\pi_1 = 8$.

The optimality criterion requires the reduced costs $\bar{c}_N = c_N - N^T\pi \geq 0$; this reduces to the simple criterion

$$\bar{c}_{ij} = c_{ij} - (\pi_i - \pi_j) \geq 0 \qquad \text{for all } (i,j) \in \mathcal{N}, \tag{9.22}$$

where $\mathcal{N}$ is the set of all nonbasic arcs.

Example 9.23 (Computation of the Reduced Costs) For the network of Example 9.20, Figure 9-22, once the multipliers are computed as illustrated in Example 9.22 we compute the reduced costs of the nonbasic variables as follows:

$$\begin{aligned}
\bar{c}_{13} &= c_{13} - \pi_1 + \pi_3 = 2 - 8 + 6 = 0,\\
\bar{c}_{14} &= c_{14} - \pi_1 + \pi_4 = 2 - 8 + 4 = -2,\\
\bar{c}_{25} &= c_{25} - \pi_2 + \pi_5 = 3 - 7 + 0 = -4,\\
\bar{c}_{34} &= c_{34} - \pi_3 + \pi_4 = 1 - 6 + 4 = -1,\\
\bar{c}_{35} &= c_{35} - \pi_3 + \pi_5 = 1 - 6 + 0 = -5.
\end{aligned}$$

INCOMING VARIABLE AND ADJUSTMENTS TO FLOW

In general, if any of the $\bar{c}_{ij} < 0$ for some $(i,j) = (i_s, j_s)$, and if the primal solution is nondegenerate, then the primal solution is not optimal because we can decrease the objective by increasing the flow value of the nonbasic variable $x_{i_s j_s}$ and adjusting the flow values of the basic variables. If the value of $x_{i_s j_s}$ can be increased indefinitely without decreasing the values of any of the basic variables, we have an unbounded solution; otherwise, some basic variable $x_{i_r j_r}$ will be driven to zero. The degenerate case will be discussed in a moment.

The Simplex Method at each iteration adds an arc (corresponding to an incoming variable) to the rooted spanning tree corresponding to the basis, thus creating a cycle that we will denote here by C. The adjustment of the flow around the cycle in the direction from i_s to j_s causes a decrease in the objective value. Because the objective value is decreased by the flow adjusted, such a cycle is referred to as a *negative cycle*. As the flow around the cycle increases, either it can be increased indefinitely (in which case the problem is unbounded), or one of the flows in an arc of the cycle is driven to zero (in which case the arc corresponding to the zero flow is dropped and a new arc corresponding to the incoming basic variable is added to form a new spanning tree).

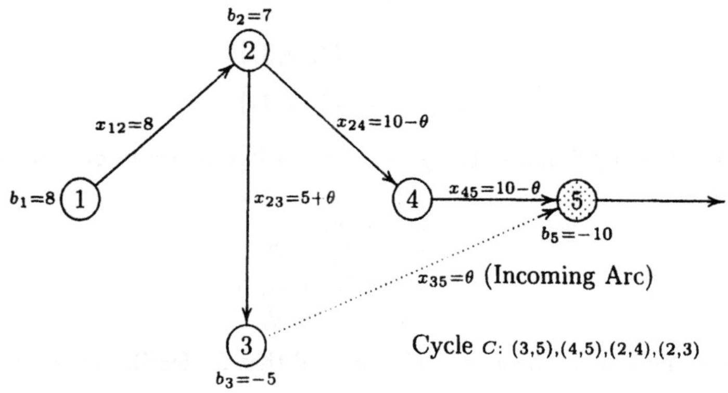

Figure 9-26: Adjustment of Flow Values Around a Cycle

If two or more arc flows are driven to zero simultaneously, the arc dropped from the tree may be chosen arbitrarily in the nondegenerate case. On the other hand, in the degenerate case, the choice is made by some special rule designed to prevent repeating a basis at a subsequent iteration. Since basis repetition in practice is a very rare phenomenon (if it can occur at all when the least cost rule is used for selecting the incoming column), we will assume the tie is "broken" at random (see Section 3.2.3) or by selecting, say, the first arc around the cycle starting with arc (i_s, j_s) and moving around the cycle along the direction of flow along (i_s, j_s).

Let C^+ denote the set of forward arcs in a cycle C, oriented in the direction of arc (i_s, j_s); let C^- denote the set of backward arcs in the cycle C; and let $x_{ij} = x^o_{ij}$ be the current flow values. Then the new flow values are given by

$$x_{ij} = \begin{cases} x^o_{ij} & \text{if } (i,j) \notin C, \\ x^o_{ij} + \theta & \text{if } (i,j) \in C^+, \\ x^o_{ij} - \theta & \text{if } (i,j) \in C^-. \end{cases} \tag{9.23}$$

Then if C^- is empty, the solution is unbounded. Otherwise, let the maximum change to the flow on the incoming arc be

$$x_{i_s j_s} = \theta^o = \min_{(i,j)\in C^-} x^o_{ij}$$

and let the adjustment of x_{ij} be given by (9.23) for $\theta = \theta^o$ and (i,j) basic, and otherwise $x_{ij} = 0$.

Example 9.24 (Adjustment of Flow Values Around a Cycle) For the spanning tree displayed in Figure 9.21 of the network of Example 9.20, the reduced costs are computed as shown in Example 9.23. If we pick the arc with the most negative reduced cost, we find that arc (3,5) would be the incoming arc and would create the cycle shown in Figure 9-26. The cycle C consists of the arcs $(3,5)$, $(4,5)$, $(2,4)$, and $(2,3)$. The sets of forward and backward arcs in C are then

$$C^+ = \{(3,5),(2,3)\},$$
$$C^- = \{(4,5),(2,4)\}.$$

Increasing the flow by θ along the cycle C results in changes to only the flow variables along the cycle:

$$\begin{aligned} x_{35} &= \theta, \\ x_{45} &= 10 - \theta, \\ x_{24} &= 10 - \theta, \\ x_{23} &= 5 + \theta. \end{aligned}$$

Examining the flow along the arcs in C^-, we find that for feasibility, the maximum value of $\theta = \theta^0$ is given by

$$\theta^o = \min_{(i,j)\in C^-} x^o_{ij} = 10.$$

Thus either arc $(2,4)$ or $(4,5)$ is dropped from the spanning tree; for the purposes of this example we assume that a random rule was used and that it selected $(2,4)$ to be dropped.

UPDATING THE PRICES

The simplex multipliers (or prices) π after the change in the basis need not be computed from scratch by solving the updated $B^T\pi = c_B$; instead the prices $\hat{\pi}_i$ from before the addition of the new arc can be updated very efficiently.

Suppose, in general, that arc $s = (p,q)$ enters the basis and $r = (\alpha,\beta)$ leaves the basis. Start by deleting arc $r = (\alpha,\beta)$ from the spanning tree before augmentation by $s = (p,q)$. By Lemma 9.17 on Page 286, this decomposes the tree into two subtrees, $\mathcal{T}_1$ and $\mathcal{T}_2$. Let the root arc $n+1$ belong to tree $\mathcal{T}_1$; in this case the prices π_i corresponding to the tree $\mathcal{T}_1$ will stay unchanged because the prices are computed outward from the root node to all the nodes of $\mathcal{T}_1$ (none of whose arcs have changed). In order to update the prices that correspond to tree $\mathcal{T}_2$, there are two cases to consider: the case that node q belongs to tree $\mathcal{T}_2$ and node p belongs to tree $\mathcal{T}_1$, and the other way around.

Case 1 Notice that (9.20) implies that we can change the values of all the prices π_k by a constant and still satisfy the relations (9.20). Therefore, for the first case, *p belongs to $\mathcal{T}_1$*, once we know the value of π_q we can compute the values of all other prices π_i for nodes $i \in \mathcal{T}_2$ as $\pi_i = \hat{\pi}_i + \pi_q - \hat{\pi}_q$ because $\pi_q - \hat{\pi}_q$ is the amount by which $\hat{\pi}_q$ changes. Now, $\pi_p = \hat{\pi}_p$ is unchanged because p belongs to $\mathcal{T}_1$. Because (p,q) enters the basis, we must have $\pi_p - \pi_q = c_{pq}$. However, we currently have $\bar{c}_{pq} = c_{pq} - \pi_p + \hat{\pi}_q$ and therefore $\pi_q - \hat{\pi}_q = -\bar{c}_{pq}$. Therefore all the prices π_i for $i \in \mathcal{T}_2$ are increased by $-\bar{c}_{pq}$; that is,

$$\pi_i = \hat{\pi}_i - \bar{c}_{pq} \quad \text{for all } i \in \mathcal{T}_2. \tag{9.24}$$

Case 2 If, on the other hand, *p belongs to $\mathcal{T}_2$*, then a similar analysis shows that $\pi_q = \hat{\pi}_q$ is unchanged because q belongs to $\mathcal{T}_1$. Because (p,q) enters the

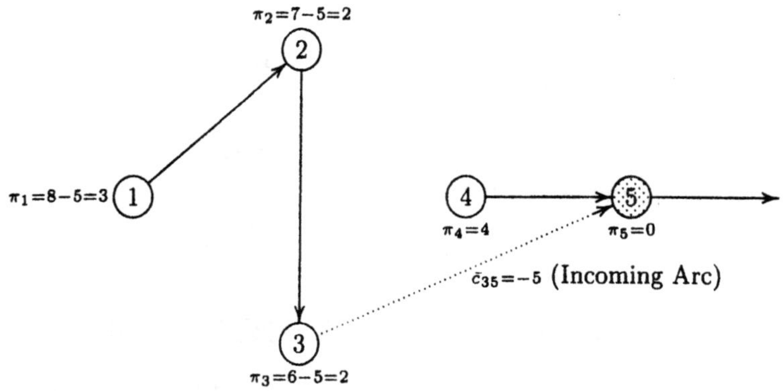

Figure 9-27: Adjustment of Simplex Multipliers π

basis, we must have $\pi_q - \pi_p = c_{pq}$. Finally, all the prices π_i for $i \in \mathcal{T}_2$ are increased by $\bar{c}_{pq}$; that is,

$$\pi_i = \hat{\pi}_i + \bar{c}_{pq} \quad \text{for all } i \in \mathcal{T}_2. \tag{9.25}$$

Example 9.25 (Updating the Simplex Multipliers) Continuing with Example 9.24, we have arc $s = (p, q) = (3, 5)$ being added to the spanning tree and arc $r = (\alpha, \beta) = (2, 4)$ being dropped. The incoming arc is shown by the dotted line in Figure 9-27. We think of this process as dropping arc $(2, 4)$ first so that the tree splits into two subtrees, $\mathcal{T}_1$ (which contains the root node) and $\mathcal{T}_2$, before the addition of arc $(3, 5)$. The subtree $\mathcal{T}_1$ consists of nodes 4 and 5; the subtree $\mathcal{T}_2$ consists of nodes 1, 2, and 3. Let $\hat{\pi}_i$, $i = 1, \ldots, 5$, be the previous prices and let π_i, $i = 1, \ldots, 5$, be the new prices. Clearly, the simplex multipliers on subtree $\mathcal{T}_1$ do not change because $\mathcal{T}_1$ includes the root node. That is,

$$\pi_5 = 0 = \hat{\pi}_5$$

and

$$\pi_4 - \pi_5 = c_{45} = \hat{\pi}_4 - \hat{\pi}_5,$$

implying $\pi_4 = \hat{\pi}_4 = 4$. On the other hand, because incoming arc $(p, q) = (3, 5)$ has node $p = 3$ in $\mathcal{T}_2$, the price on each node in subtree $\mathcal{T}_2$ is increased by $\bar{c}_{35} = -5$ (which means each is decreased by 5) as shown in Figure 9-27. To see this, note that

$$\pi_3 - \pi_5 = c_{35}.$$

We know that $\pi_5 = \hat{\pi}_5$; adding and subtracting $\hat{\pi}_3$, we obtain

$$\pi_3 = \hat{\pi}_3 + c_{35} - (\hat{\pi}_3 - \hat{\pi}_5) = \hat{\pi}_3 + \bar{c}_{35} = 6 - 5 = 2. \tag{9.26}$$

Next, in subtree $\mathcal{T}_2$,

$$\pi_2 - \pi_3 = c_{23} = \hat{\pi}_2 - \hat{\pi}_3.$$

Hence, rearranging and using (9.26),

$$\pi_2 = \hat{\pi}_2 + (\pi_3 - \hat{\pi}_3) = \hat{\pi}_2 + \bar{c}_{35} = 7 - 5 = 2. \tag{9.27}$$

Similarly,

$$\pi_1 - \pi_2 = c_{12} = \hat{\pi}_1 - \hat{\pi}_2.$$

Hence, rearranging and using (9.27),

$$\pi_1 = \hat{\pi}_1 + (\pi_2 - \hat{\pi}_2) = \hat{\pi}_1 + \bar{c}_{35} = 8 - 5 = 3. \tag{9.28}$$

OBTAINING AN INITIAL FEASIBLE SOLUTION

An initial feasible solution can be obtained by adding an artificial node 0 and artificial arcs from 0 to every node of the network as follows: If $b_k > 0$, then an artificial arc $(i,0)$ is added; otherwise an artificial arc $(0,k)$ is added. If we let $\bar{\mathcal{A}c}$ be the set of artificial arcs, the Phase I problem is then

$$\begin{array}{ll} \text{Minimize} & \\ \displaystyle\sum_{(k,0)\in\bar{\mathcal{A}c}} x_{k0} + \sum_{(0,k)\in\bar{\mathcal{A}c}} x_{0k} = z & \\ \text{subject to} & \\ \displaystyle\sum_{i\in Af(k)} x_{ki} - \sum_{j\in Bf(k)} x_{jk} = b_k & \text{for all } k \in \mathcal{N}d \cup \{0\}, \\ 0 \le x_{ij} \le \infty & \text{for all } (i,j) \in \mathcal{A}c \cup \bar{\mathcal{A}c}, \end{array} \tag{9.29}$$

where

$$Af(k) = \{i \in \mathcal{N}d \mid (k,i) \in \mathcal{A}c \cup \bar{\mathcal{A}c}\}; \tag{9.30}$$
$$Bf(k) = \{j \in \mathcal{N}d \mid (j,k) \in \mathcal{A}c \cup \bar{\mathcal{A}c}\}. \tag{9.31}$$

If a feasible solution exists, it will be found with all the flows associated with the artificial arcs at value 0 for the above problem.

▷ **Exercise 9.46**

(a) Suppose that we are given a problem where the total supply at supply nodes is greater than the total demand at demand nodes so that $\sum_{i\in\mathcal{N}d} b_i > 0$. Show how to convert such a problem to one in which the total supply is equal to the total demand.

(b) Show how the feasibility of (9.16) can be determined by solving a related max flow problem with zero lower bounds on the arc flows.

▷ **Exercise 9.47** Extend the methods discussed throughout this section to network problems in which the equations in (9.17) are replaced by inequalities.

9.9 THE BOUNDED VARIABLE PROBLEM

Ideas from the Simplex Method for solving linear programs with bounded variables carry over to network problems.

PROBLEM STATEMENT

The bounded variable minimum cost-flow problem (9.16) is repeated below for convenience:

$$
\begin{array}{lrll}
\text{Minimize} & \displaystyle\sum_{(i,j)\in\mathcal{Ac}} c_{ij}x_{ij} & = z & \\
\text{subject to} & \displaystyle\sum_{j\in Af(k)} x_{kj} - \sum_{i\in Bf(k)} x_{ik} & = b_k & \text{for all } k \in \mathcal{Nd}, \\
& l_{ij} \le x_{ij} \le h_{ij} & & \text{for all } (i,j) \in \mathcal{Ac},
\end{array}
\tag{9.32}
$$

where c_{ij} is the cost per unit flow on the arc (i,j); b_k is the net flow at node k; l_{ij} is a lower bound on the flow in arc (i,j); and h_{ij} is an upper bound on the flow in arc (i,j). Note that b_k takes on values that depend on the type of node k:

$$
b_k \text{ is } \begin{cases} > 0 & \text{if } k \text{ is a source (supply) node;} \\ < 0 & \text{if } k \text{ is a destination (demand) node;} \\ = 0 & \text{if } k \text{ is a node for transshipment only.} \end{cases}
$$

NETWORK SIMPLEX METHOD

We assume that at least one of the bounds is finite and that a full set of artificial variables as described on Page 298 has been added. A starting feasible solution can be obtained by setting each arc flow to one of its bounds and calculating the values for the artificial variables so that the conservation of flow constraints hold. (Of course, the artificial variable will need to be added with appropriate signs so that they are nonnegative.) Another method is to find any basis, say by finding the minimal spanning tree, and create an objective function that *minimizes the sum of infeasibilities*. The latter approach also allows the easy inclusion of arcs that have no lower or upper bound on the flow.

The network Simplex Method for solving problem (9.32) is straightforward and is developed through the following exercises.

▷ **Exercise 9.48** Specify the optimality criterion for the bounded variable minimum cost-flow problem.

▷ **Exercise 9.49** Specify the rules for determining the exiting and entering column (arc) for the bounded variable minimum cost-flow problem.

▷ **Exercise 9.50** Generalize the network Simplex Method described in Section 9.8 to include bounded variables.

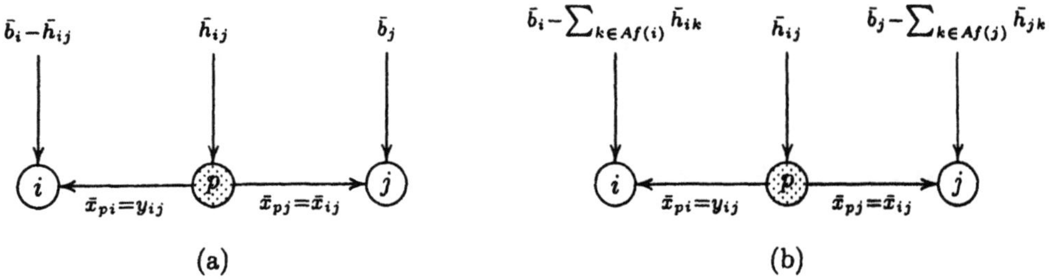

Figure 9-28: Converting Finite Upper Bounds to $+\infty$

TRANSFORMATION TO STANDARD FORM

In the event that all lower bounds are finite, a straightforward translation can be used to replace all the lower bounds on the flows with 0. To do so, substitute $x_{ij} = \bar{x}_{ij} + l_{ij}$ to obtain

$$\begin{array}{lll} \text{Minimize} & \displaystyle\sum_{(i,j)\in\mathcal{A}c} c_{ij}\bar{x}_{ij} = \bar{z} & \\ \text{subject to} & \displaystyle\sum_{j\in Af(k)} \bar{x}_{kj} - \sum_{i\in Bf(k)} \bar{x}_{ik} = \bar{b}_k & \text{for all } k \in \mathcal{N}d, \\ & 0 \le \bar{x}_{ij} \le \bar{h}_{ij} & \text{for all } (i,j) \in \mathcal{A}c, \end{array} \tag{9.33}$$

where

$$\begin{aligned} \bar{h}_{ij} &= h_{ij} - l_{ij}, \\ \bar{b}_k &= b_k - \sum_{j\in Af(k)} l_{kj} + \sum_{i\in B(k)} l_{ik}, \\ \bar{z} &= z - \sum_{(i,j)\in\mathcal{A}c} c_{ij} l_{ij}. \end{aligned}$$

Clearly it is possible to modify the network problem (9.33) to a problem with no upper bounds (nonnegativity constraints only), by introducing slack variables $y_{ij} \ge 0$ so that

$$\bar{x}_{ij} + y_{ij} = \bar{h}_{ij}$$

for each finite $\bar{h}_{ij}$.

What is not so obvious is that this can be done in such a way that it becomes a minimum cost-flow problem with no upper bound on the flows. To do so see Figure 9-28(a). We start by dropping arc (i,j) and then introduce a new node p. Next we draw an arc from p to j and the let the flow on this arc be the same as the flow on the original arc (i,j); thus node j sees no change in the incoming flow $\bar{x}_{ij}$. Because $\bar{x}_{ij}$ (now labeled as $\bar{x}_{pj}$ because this flow originates at node p) is bounded above by $\bar{h}_{ij}$ we set the supply to node p as $b_p = \bar{h}_{ij} > 0$. At the same time, we reduce the supply b_i at node i by $\bar{h}_{ij}$ because that is where we would have to draw from in the original network. However, $\bar{x}_{ij}$ may not necessarily be at its upper bound, i.e., it may not use up all its supply, and hence we need to balance the flows by having an arc out from p to take care of the unused supply (or slack). This unused supply, or slack, must then be returned to node i, and hence we draw the arc from node p to node i with flow y_{ij} (labeled as $\bar{x}_{pi}$ in the figure).

Note that as a result of the transformation all the flows into a node j are unchanged; hence the supply at the node is unchanged, whereas all the flows out of a node i are modified resulting in a reduction in the supply at the node. Figure 9-28(a) illustrates this if we replace only one arc (i, j). When adjustments are made for all the arcs, see Figure 9-28(b), this results in replacing $\bar{b}_k$ by

$$\bar{b}_k - \sum_{q \in Af(k)} \bar{h}_{kq} \quad \text{for all } k \in \mathcal{N}d.$$

These adjustments of course increase the size of the network; in commercial implementations, upper (and lower) bounds are handled directly by the Simplex Method.

Example 9.26 (Illustration of Conversion to Standard Form) Figure 9-29 shows the conversion of a network with upper bounds on the flows to one with no upper bounds. In the figure, the concentric circles are the new nodes that are added for the conversion, and the variables y_{ij} are the slacks added to $x_{ij} \le h_{ij}$ to convert them to equalities.

▷ **Exercise 9.51** Show how to convert a minimum cost-flow problem to a transportation problem. Hint: Apply Figure 9-28 where the new nodes p are supply nodes and the original nodes are destination nodes.

9.10 NOTES & SELECTED BIBLIOGRAPHY

Network optimization theory is a very large field, and our discussion of it has been at a fairly elementary level. For more details on networks and their applications, see, for example, Ahuja, Magnanti, & Orlin [1993], Bertsekas [1991], Ford & Fulkerson [1962], and Lawler [1976]. See also *Linear Programming 2* for more on the theory and details on implementing such algorithms on a computer.

An oft-quoted example of a successful application of network analysis is the award winning study by Klingman, Phillips, Steiger, & Young [1987] and Klingman, Phillips, Steiger, Wirth, & Young [1986] at Citgo Petroleum Corporation. Developed with full top management support, the model takes into account all aspects of the business from production at the refineries to distribution to prices to charge. The study claimed that in the system's initial implementation there was a savings of \$116 million in inventory cost (which translates into a yearly saving in interest) as well as a savings of approximately \$2.5 million as a result of better pricing, transportation, and coordination.

The min-cut max-flow theorem was first established for planar networks at RAND in 1954 and published by Dantzig & Fulkerson [1956a]. Later, Ford & Fulkerson [1956]

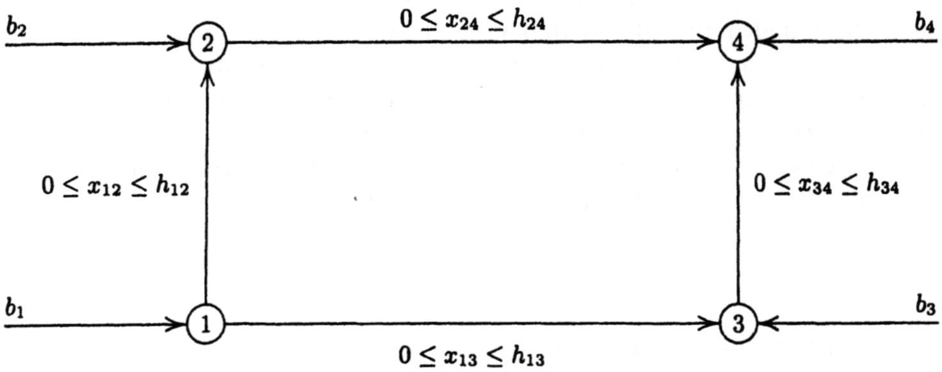

(a) Initial network with upper bounds on flows

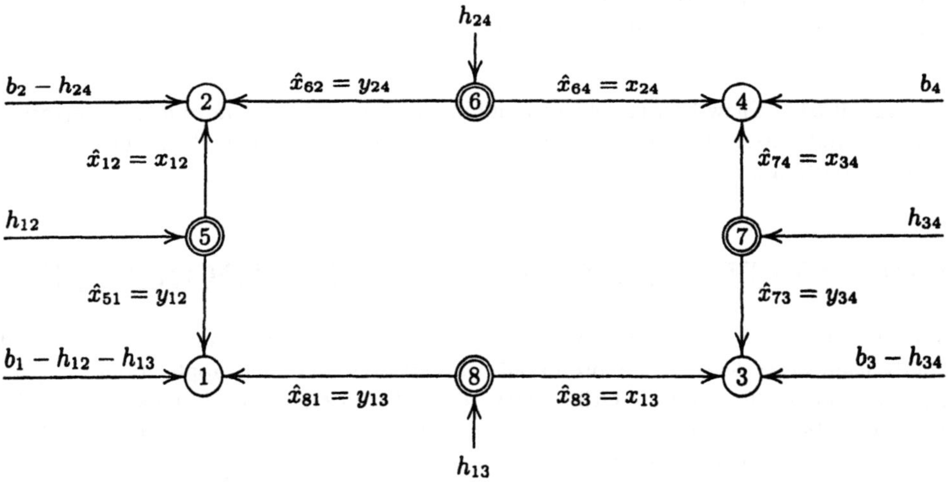

(b) Converted network

Figure 9-29: Converting Finite Upper Bounds to $+\infty$ in a Small Network

established the theorem for general networks. It was also discovered independently by Elias, Feinstein, & Shannon [1956]. A comprehensive treatment of the maximal flow problem and related matters can be found in Ford & Fulkerson [1962].

The classical augmenting path method for finding a maximum flow through a network was developed by Ford & Fulkerson [1957] based on earlier work by Kuhn [1955] and Egerváry [1931]. Fulkerson & Dantzig [1955] and Dantzig & Fulkerson [1956a] developed a tree method for solving maximal flow problems that is also described in Dantzig [1963]. The approach constructs two subtrees, one branching out from the source and the other branching out from the destination so that every intermediate node is reached by just one of the trees. Then a connecting arc between the two trees and an associated path from source to destination is found, and finally the maximum flow along the path is assigned. The network displayed in Example 9.10 due to Chvátal [1983] is a smaller network than that used in the example by Ford & Fulkerson [1962] but based on the same type of recurrence relation. The proofs of Lemma 9.12 and Theorem 9.13 can be found in *Linear Programming 2*.

J. Edmonds & R.M. Karp [1972] showed that an augmenting path method called *first-labeled first-scanned* finds a maximum flow within $mn/2$ iterations, where n is the number of arcs and m is the number of nodes in the network, regardless of what the upper bounds h_{ij} on the arcs are. This method then finds the maximal flow in $O(n^2m)$ operations because it can be shown that each iteration of the augmenting path method takes only $O(n)$ comparisons to find an augmenting path. The proof of Theorem 9.11 can be found in Edmonds & Karp [1972] and in *Linear Programming 2*. Around the same time as Edmonds & Karp's results, Dinic [1970] independently designed a faster algorithm that requires $O(m^2n)$ operations. Later Malhotra, Kumar, & Maheshwari [1978] developed an algorithm that requires $O(m^3)$ operations. For networks that have $n \ll m^2$, an algorithm designed by Galil [1978] takes $O(m^{5/3}n^{2/3})$ operations, and an algorithm designed by Sleator [1980] takes only $O(nm \log m)$ steps.

Shortest path problems come up often in practice and arise as subproblems in many network problems. Dantzig first proposed a method for finding the shortest path from a source node to a destination node in a network, see Dantzig [1960a], based on an earlier RAND research memorandum. At about the same time, Dijkstra [1959] proposed another algorithm for finding the shortest directed paths from a node to all other nodes. The Dijkstra algorithm takes $O(m^2)$ operations. See also Bellman [1958]. Independently, Whiting & Hillier [1960] also developed a shortest route algorithm. Johnson [1977] has shown that this bound can be further reduced to $O(n \log_k m)$ operations, where $k = \max(2, n/m)$. See also Denardo & Fox [1979], Dial [1969], Moore [1959], and Pape [1974]. A summary of various classical algorihtms can be found in Gallo & Pallottino [1988]. Improvements have continued to be made in shortest path algorithms; see, for example, Ahuja, Mehlhorn, Orlin, & Tarjan [1990], Fredman & Willard [1987] Gabow & Tarjan [1989], Goldberg [1993], and Goldberg & Radzik [1993]. Under the assumption that arc lengths are integers between 0 and L where $L \geq 2$, Ahuja, Mehlhorn, Orlin, & Tarjan's algorithm runs in $O(n + m\sqrt{\log L})$. For theory and experimental evaluation of shortest path algorithms see Cherkassky, Goldberg, & Radzik [1996]; in this paper the authors show that some algorithms behave in exactly the same way on two networks, one of which is obtained from the other by replacing the arc lengths by the reduced costs with respect to a potential function; that is, the algorithms are *potential-invariant*. This implies, for example, that a feasible shortest path problem has an equivalent with nonnegative arc lengths.

The minimal spanning tree algorithm described here is due to Kruskal [1956].

For details on the Network Simplex Method, including implementation (in addition to that described in *Linear Programming 2*), see, for example, Ali, Helgason, Kennington, & Lall [1978], Bradley, Brown, & Graves [1977], Chvátal [1983], Cunningham [1979], and Mulvey [1978].

An example of cycling in the Network Simplex Method can be found in Cunningham & Klincewicz [1983]. To the authors' knowledge, cycling as a result of degeneracy, has not been encountered on any practical problem. It is not known whether cycling can occur in minimum cost network flow problems if the entering variable is chosen based on the usual rule of picking the one that has the most negative reduced cost. The interested reader can find strategies used to prevent the possibility of cycling in Bazaraa, Jarvis, & Sherali [1990] and Chvátal [1983], for example.

The Network Simplex Method works very well in practice; in fact, this adaptation of the Simplex Method for networks is typically 200 to 300 times faster than the Simplex Method applied to general linear programs of the same dimensions. However, pathological examples can be constructed in which the Network Simplex Method can take a very large number of iterations. Zadeh [1973] has constructed a sequence of transshipment problems such that the kth problem has only $2k+2$ nodes but if we choose the incoming arc by picking the most negative reduced cost, the Network Simplex Method takes $2^k + 2^{k-2} - 2$ iterations.

The method of minimizing the sum of infeasibilities for Phase I is described in *Linear Programming 2.*

An area that we have not covered at all is that of project planning, and scheduling and coordination of various activities. Methods to do this are called PERT (Program Evaluation and Review Techniques) and CPM (Critical Path Method). Many references exist for such methods; see Hillier & Lieberman [1995]. One such reference relating this to networks is by Elmaghraby [1977].

9.11 PROBLEMS

9.1 Find the maximal flow through the network shown in Figure 9-30. Find a cut that demonstrates the Ford-Fulkerson Min-Cut = Max-Flow Theorem 9.13. Verify your answer by using the `Maximum Flow` software option.

9.2 Find the maximal flow through the network shown in Figure 9-31. Find a cut that demonstrates the Ford-Fulkerson Min-Cut = Max-Flow Theorem 9.13. Verify your answer by using the `Maximum Flow` software option.

9.3 Find, by hand, the maximal flow through the network shown in Figure 9-32; the capacities in each direction are shown on the arcs. Find the cut whose value equals the maximum flow value.

9.4 For the network shown in Figure 9-33 assume that the capacities shown on the arcs are the same in each direction.

(a) Find the maximal flow by hand.

(b) Verify your answer by using the `Maximum Flow` software option.

(c) Find a cut that demonstrates the Ford-Fulkerson Min-Cut = Max-Flow Theorem 9.13.

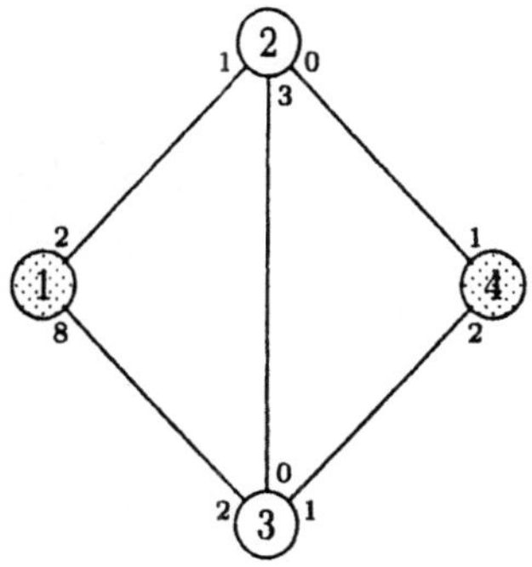

Figure 9-30: Data for a Maximal Flow Problem

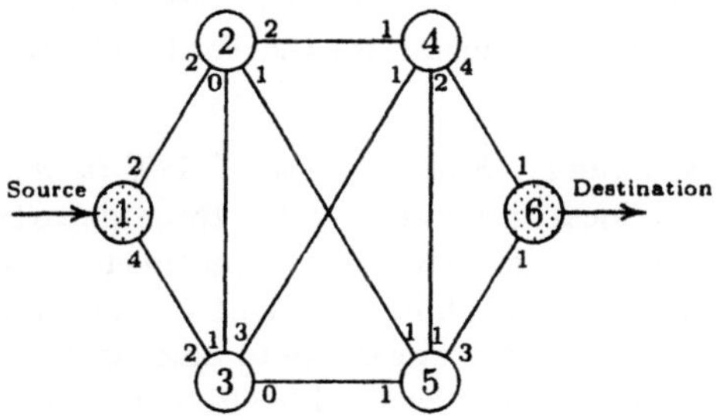

Figure 9-31: Data for a Maximal Flow Problem

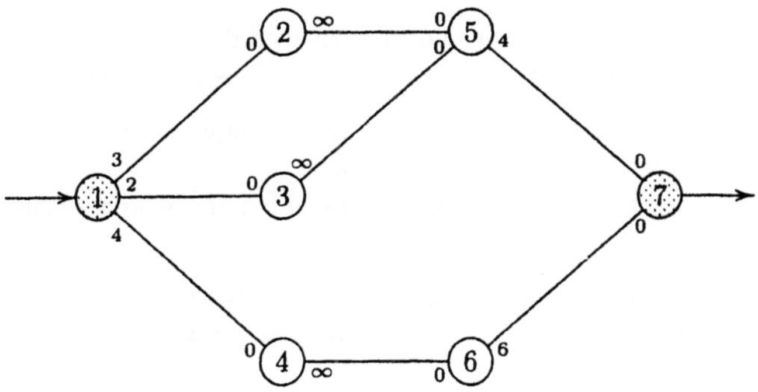

Figure 9-32: Data for a Maximal Flow Problem

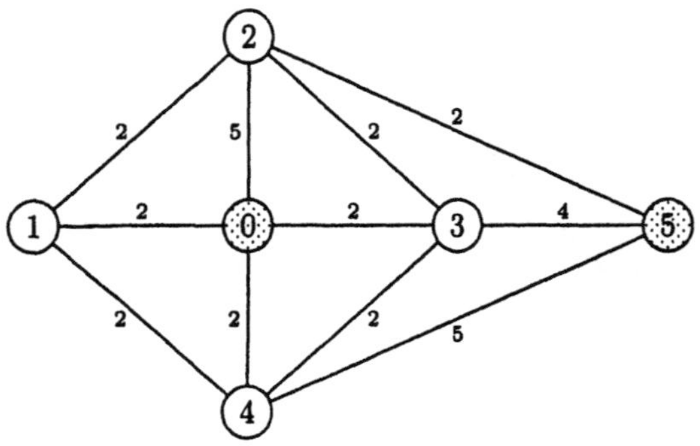

Figure 9-33: Data for a Maximal Flow Problem

9.5 For the directed network shown in Figure 9-34, with capacities shown on the arc, do the following:

(a) Find, by hand, the maximal flow.
(b) Verify your answer by using the **Maximum Flow** software option.
(c) Find a cut that demonstrates the Ford-Fulkerson Min-Cut = Max-Flow Theorem 9.13.

9.6 *Nurse Staff Scheduling (Khan & Lewis [1987] in Ahuja, Magnanti, & Orlin [1993]).* A local hospital operates three departments: Emergency (department 1), neonatal intensive care (department 2), and orthopedics (department 3). It has three work shifts, each with different levels of staffing for nurses. The hospital administrator would like to hold staffing levels as low as possible while providing satisfactory levels of health care. Hence they would like to determine the minimum number of nurses required to meet the following three constraints:

- The hospital must allocate a minimum of 13, 32, and 22 nurses to the three departments respectively.
- The hospital must allocate a minimum of 26, 24, and 19 nurses to each shift.
- The number of nurses allocated to each department in each shift must be between a lower and upper bound as shown in Table 9-2.

Determine the minimum number of nurses required to satisfy the constraints based on a method for maximum flows.

9.7 Consider the proposed network with distances on the arcs as shown in Figure 9-35.

(a) Find the shortest path using Dijkstra's algorithm.
(b) Solve it using the **Dijkstra's Shortest Route** software option to verify your solution.

9.8 *Dantzig [1963].* Find the shortest route from Los Angeles to Boston on one of the routes shown in Figure 9-36.

(a) Solve the problem using Dijkstra's algorithm.

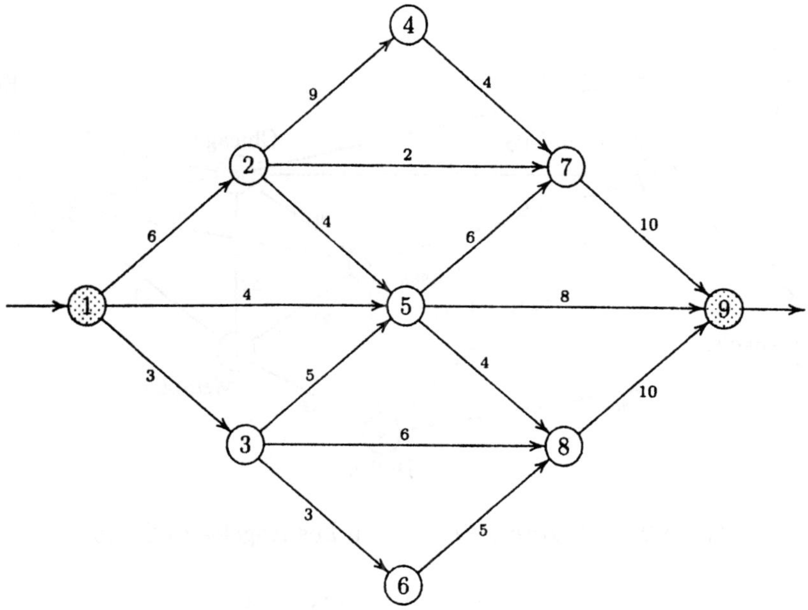

Figure 9-34: Data for a Maximal Flow Problem

		Department 1	Department 2	Department 3
Shift	1	(6,8)	(11,12)	(7,12)
	2	(4,6)	(11,12)	(7,12)
	3	(2,4)	(10,12)	(5,7)

Table 9-2: Bounds on Nurses Allocated to Each Department in Each Shift

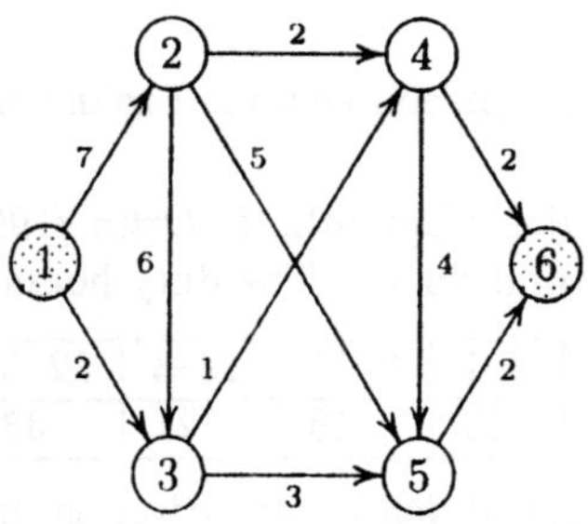

Figure 9-35: Data for a Shortest Route Problem

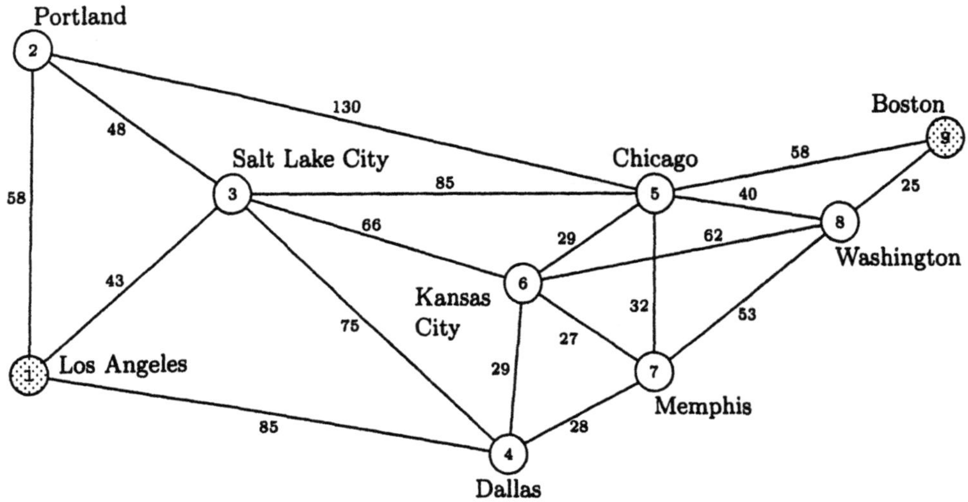

Figure 9-36: Shortest Route from Los Angeles to Boston

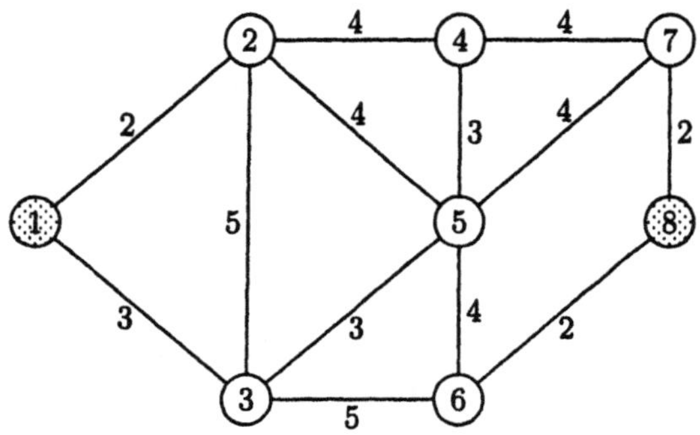

Figure 9-37: Data for a Shortest Route Problem

(b) Solve it using the **Dijkstra's Shortest Route** software option to verify your solution.

9.9 Find the shortest route from source node 1 to destination node 8 for the network shown in Figure 9-37.

9.10 *Crew Scheduling (Ahuja, Magnanti, & Orlin [1993]).* The drivers of a bus company can work several duties. The duty hours and costs are shown below:

Duty Hours	9–1	9–11	12–3	12–5	2–5	1–4	4–5
Cost ($)	30	18	21	38	20	22	9

We wish to ensure that at least one driver is on duty for each hour of the planning period (9am to 5pm). Formulate and solve this problem as a shortest route problem.

9.11 Find the minimal spanning tree for the network shown in Figure 9-38. Verify

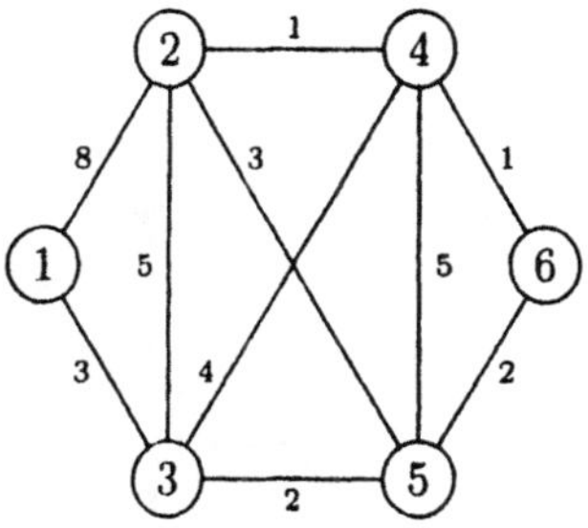

Figure 9-38: Data for a Minimal Spanning Tree Problem

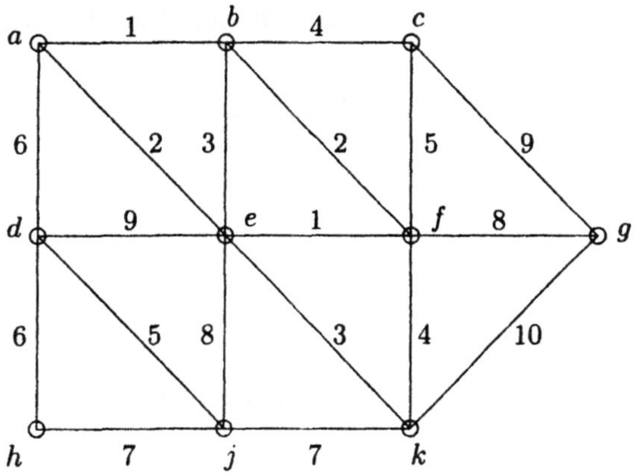

Figure 9-39: Data for a Minimal Spanning Tree Problem

your answer by using the **Minimal Spanning Tree** software option.

9.12 *Ph.D. Comprehensive Exam, September 26, 1987, at Stanford.* Consider the graph and the weights shown in Figure 9-39.

(a) Find the minimal spanning tree of the graph. Briefly outline the algorithm you are using, and give the selection of the arcs of the tree in the order in which they are determined.

(b) Suppose that you are given a graph $G = (V, E)$ and weights as before, but also a particular node $v \in V$. You are asked to find the shortest spanning tree such that v is *not* an end-node. Explain briefly how to modify a minimal spanning tree algorithm to solve the same problem efficiently.

9.13 A minimum cost-flow problem in standard form is shown in Figure 9-40.

(a) Solve it by hand by the Network Simplex Method taking advantage of all efficiencies in computation.

(b) Verify your solution by solving it using the **Network Simplex Method** software option.

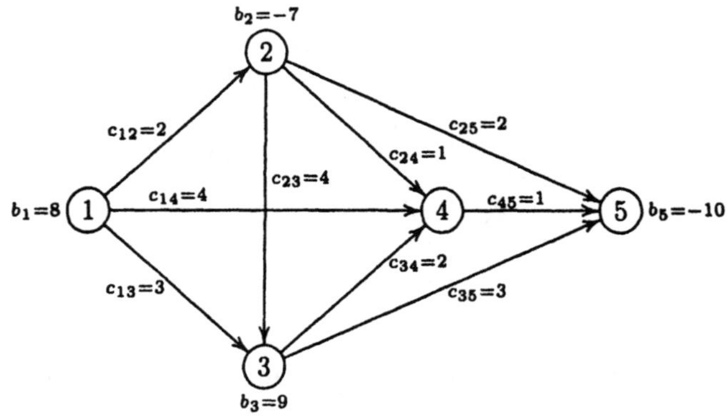

Figure 9-40: Data for a Minimum Cost-Flow Problem

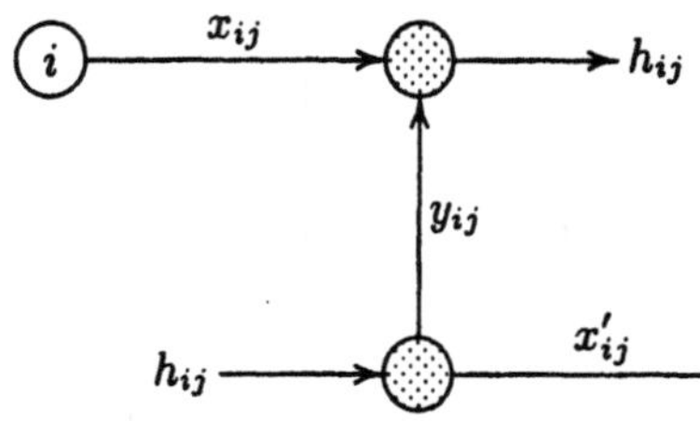

Figure 9-41: Another Method to Convert a Finite Upper Bound to $+\infty$

9.14 Set up the maximal flow problem shown in Figure 9-16 as a minimum cost-flow problem.

(a) Solve it by hand by the Network Simplex Method taking advantage of all efficiencies in computation.

(b) Verify your solution by solving it using the **Network Simplex Method** software option.

9.15 For the minimum cost-flow problem in Figure 9-40, suppose that arc $(1,2)$ has an upper bound of 5, i.e., $0 \le x_{12} \le 5$.

(a) Transform the problem into standard form and solve it by hand using the Network Simplex Method.

(b) Solve the problem by hand using the Network Simplex Method but without performing the transformation to the standard form.

(c) Verify your solution by solving it by the **Network Simplex Method** software option.

9.16 Show that the arc (i,j) with flow value $0 \le x_{ij} \le h_{ij}$ can be replaced by the node and arc representation shown in Figure 9-41. Compare this with that shown in Figure 9-28.

9.17 Solve, by hand, the minimum cost-flow problem in standard form shown in Figure 9-42. Verify your solution by solving it by the **Network Simplex Method** software option.

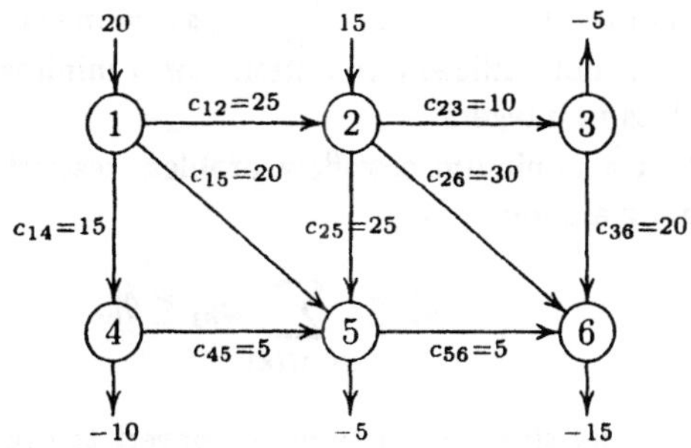

Figure 9-42: Data for a Minimum Cost-Flow Problem

9.18 For the minimum cost-flow problem of Figure 9-42 suppose that arc $(1,2)$ is constrained to be $0 \le x_{12} \le 8$.

(a) Use the technique of Figure 9-28 to convert it to standard form and solve it by hand.

(b) Use the technique of Figure 9-41 to convert it to standard form and solve it by hand.

(c) Verify your solution by solving it by the **Network Simplex Method** software option.

9.19 (a) Using the discussion of the Simplex Algorithm for bounded variables (see Section 3.4) as a guideline, modify the Network Simplex Algorithm to automatically handle lower and upper bounds on the flows.

(b) Suppose that the flows in the network shown in Figure 9-42 have the following lower and upper bounds: $3 \le x_{12} \le 10$, $5 \le x_{14} \le 15$, $0 \le x_{15} \le 20$, $0 \le x_{23} \le 10$, $0 \le x_{25} \le 25$, $5 \le x_{26} \le 30$, $0 \le x_{36} \le 20$, $0 \le x_{45} \le 10$, and $0 \le x_{56} \le 5$. Solve this modified problem, using the variant of the Network Simplex Algorithm that you developed in part (a).

9.20 Consider the following linear program

$$\begin{array}{lrcrcrcrcrcl}
\text{Minimize} & 2x_1 & + & 3x_2 & + & 4x_3 & + & x_4 & + & 2x_5 & = & z \\
\text{subject to} & & & 1x_2 & + & 1x_3 & + & 1x_4 & + & 1x_5 & \ge & 10 \\
 & & & 1x_2 & & & & 1x_4 & + & 1x_5 & \ge & 15 \\
 & 1x_1 & + & 1x_2 & & & & & + & 1x_5 & \ge & 20 \\
 & 1x_1 & + & 1x_2 & & & & & & & \ge & 5.
\end{array}$$

It has a special structure in the sense that each column of the coefficient matrix contains 0's and 1's, with the 1's being consecutive in each column.

(a) Transform this special structure linear program into a minimum cost-flow problem. Hint: Introduce surplus variables and a redundant 0 row to the set of constraints. Subtract row $i+1$ from row i for $i = 4, 3, 2, 1$ in turn.

(b) Draw the network.

(c) Solve the problem by hand using the Network Simplex Method.

(d) Verify your solution by solving it by the **Network Simplex Method** software option.

9.21 Show that conditions $0 \le x_{ij} \le 1$, $\sum_{(i,j)\in\mathcal{A}} x_{ij} = m-1$, $\sum_{(i,j)\in\mathcal{A}} w_{ij}x_{ij} = \min$ are necessary but not sufficient conditions for a minimal spanning tree even if the system solves in integers.

9.22 Suppose that in a minimum cost-flow problem restrictions are placed on the total flow leaving a node k; i.e.,

$$\bar{\theta}_k^l \le \sum_{j\in\mathcal{A}f(k)} x_{kj} \le \hat{\theta}_k^l.$$

Show how to modify these restrictions to convert the problem into a minimum cost-flow problem of the form (9.16).

9.23 A variation of the minimum cost-flow problem (9.16) is to place upper and lower bounds on the conservation of flow constraints at each node; i.e.,

$$\begin{array}{lll} \text{Minimize} & \displaystyle\sum_{(i,j)\in\mathcal{A}c} c_{ij}x_{ij} = z & \\ \text{subject to} & b_k^l \le \displaystyle\sum_{j\in\mathcal{A}f(k)} x_{kj} - \sum_{i\in\mathcal{B}f(k)} x_{ik} \le b_k^u & \text{for all } k \in \mathcal{N}d, \\ & l_{ij} \le x_{ij} \le h_{ij} & \text{for all } (i,j) \in \mathcal{A}c, \end{array}$$

where b_k^l and b_k^u, the lower and upper bounds on the conservation of flow at each node k, are given. Show how to convert this problem into the standard minimum cost-flow form (9.16).

9.24 Pick arbitrary scalars β_i for each node $i \in \mathcal{N}d$ for the minimum cost-flow problem. Show that the optimal flow vectors are unaffected if the arc costs c_{ij} are replaced by

$$\hat{c}_{ij} = c_{ij} + \beta_j - \beta_i.$$

How is

$$\hat{z} = \sum_{(i,j)\in\mathcal{A}c} \hat{c}_{ij}x_{ij}$$

related to

$$z = \sum_{(i,j)\in\mathcal{A}c} C_{ij}x_{ij}?$$

9.25 *The Caterer Problem (Jacobs[1964] in Dantzig[1963]).* A caterer has booked his services for the next T days. He requires r_t fresh napkins on the tth day, $t = 1,\ldots,T$. He sends his soiled napkins to the laundry, which has three speeds of service, $f = 1$, 2, or 3 days. The faster the service, the higher the cost c_f of laundering a napkin. He can also purchase new napkins at a cost c_o. He has an initial stock of s napkins. The caterer wishes to minimize his total outlay.

(a) Formulate as a network problem. Hint: Define the caterer at time t as a "source point" in an abstract network for soiled napkins that are connected to "laundry points" $t+1$, $t+2$, $t+3$. The reverse arc is not possible. The laundry point t is connected to a fresh napkin "destination point t," which in turn is connected to the same type point for $t+1$.

(b) Assign values to the various parameters and solve the problem.

9.26 Suppose that the linear cost function for the minimum cost-flow problem is replaced by a separable piecewise linear function (see Section 6.7.3). Show how to convert this problem to the standard minimum cost-flow problem form (9.16).

APPENDIX A

LINEAR ALGEBRA

In this chapter we briefly review some key concepts of linear algebra. Some of these concepts are elaborated upon in the rest of the book. For details on the concepts discussed here, the reader should refer to any of the standard linear algebra texts.

In this book we have assumed that the reader has working knowledge of linear algebra. However, we start with the basics and briefly cover all the relevant linear algebra required to understand the concepts discussed in this book.

A.1 SCALARS, VECTORS, AND MATRICES

In this section we define scalars, vectors, and matrices.

> *Definition (Scalar)*: A *scalar* is a real or complex number. In this book scalars will often be denoted by lowercase Greek letters in order to distinguish them from vectors and matrices.

> *Definition (Vector)*: A *vector* is an ordered collection of scalars.

Vectors will be represented by lower case letters of the alphabet. A subscript on a vector will denote a particular element of the vector. For example, x_4 will denote the fourth element of the vector x. Superscripts on a vector will denote different vectors. For example, x^k will be a vector different from the vector x^n.

A vector per se is just an ordered collection $(x_1, x_2, \ldots, x_n)$, neither a row nor column vector. A row vector is one whose elements are displayed horizontally, and a column vector is a vector whose elements are displayed in a vertical column.

In this book, a vector, unless otherwise specified, will be understood to be a column vector and have the form

$$x = \begin{pmatrix} x_1 \\ x_2 \\ \vdots \\ x_n \end{pmatrix}. \tag{A.1}$$

To represent a column vector x as a row vector, we will take the *transpose* of x, denoted by x^T ("x-transpose").

Definition (Vector Transpose): The *transpose* of a vector x, denoted by x^T, is defined as the same ordered collection of scalars as in x but with the ordering going from left to right, i.e.,

$$x^T = (x_1, x_2, \ldots, x_n). \tag{A.2}$$

Definition (Matrix): A *matrix* is a rectangular array of numbers. It may be viewed as an ordered collection of column vectors each of which is of the same dimension, or as an ordered collection of row vectors each of which is of the same dimension.

Matrices will be represented by uppercase letters of the alphabet. A double subscript will denote a particular element of the matrix. For example, a_{34} and A_{34} will both denote the element in row 3 and column 4 of the matrix A. Furthermore $A_{\bullet j}$ will denote the jth column of A, and $A_{i\bullet}$ will denote the ith row of A.

The transpose operation is also defined for a matrix.

Definition (Matrix Transpose): The *transpose* of a matrix, denoted by A^T, is a new matrix that is the collection of the transpose of the column vectors in the same order as the columns of the original matrix. Thus, the net effect is to interchange the roles of the row and column indices. For example, if the element in the ith row and jth column of A is A_{ij}, then the element in the jth row and ith column of A^T is A_{ij}.

Example A.1 (Transpose) Consider, for example, the matrix A given by

$$A = \begin{pmatrix} a_{11} & a_{12} & \ldots & a_{1n} \\ a_{21} & a_{22} & \ldots & a_{2n} \\ \vdots & \vdots & \ddots & \vdots \\ a_{m1} & a_{m2} & \ldots & a_{mn} \end{pmatrix}.$$

The transpose of A is given by

$$A^T = \begin{pmatrix} a_{11} & a_{21} & \ldots & a_{m1} \\ a_{12} & a_{22} & \ldots & a_{m2} \\ \vdots & \vdots & \ddots & \vdots \\ a_{1n} & a_{2n} & \ldots & a_{mn} \end{pmatrix}.$$

Definition (Diagonal, Off-diagonal, Super-Diagonal, Sub-Diagonal): The matrix elements on the diagonal, that is, a_{ii}, are called *diagonal* elements. All other elements a_{ij}, $i \neq j$ are called *off-diagonal* elements. The off-diagonal elements are further classified as *super-diagonal* and *sub-diagonal* elements. Super-diagonal elements are the elements above the diagonal, that is, a_{ij} for all $j > i$. Sub-diagonal elements are the elements below the diagonal, that is, a_{ij} for all $i > j$.

A.2 ARITHMETIC OPERATIONS WITH VECTORS AND MATRICES

Arithmetic operations with vectors and matrices are similar to arithmetic operations performed with scalars. The main difference is that the operations are ordered in a special way to take into account the ordering in the definition of vectors and matrices.

Definition (Equal Matrices): Two $m \times n$ matrices A and B are *equal* if and only if each element of A is equal to the corresponding element of B.

That is, A and B are equal if and only if

$$a_{ij} = b_{ij} \quad \text{for } i = 1, \ldots, m,\ j = 1, \ldots, n. \tag{A.3}$$

Similarly two n dimensional vectors x and y are equal if and only if each element of x is equal to the corresponding element of y.

Multiplication of a vector (or matrix) by a scalar amounts to multiplying each element of the vector (or matrix) by the scalar. That is,

$$\alpha A = \begin{pmatrix} \alpha a_{11} & \alpha a_{12} & \ldots & \alpha a_{1n} \\ \alpha a_{21} & \alpha a_{22} & \ldots & \alpha a_{2n} \\ \vdots & \vdots & \ddots & \vdots \\ \alpha a_{m1} & \alpha a_{m2} & \ldots & \alpha a_{mn} \end{pmatrix}. \tag{A.4}$$

The sum of two n-dimensional vectors is another n-dimensional vector whose elements are the sums of the corresponding elements in each of the two vectors. Similarly, the sum of two $m \times n$ matrices is another $m \times n$ matrix formed by summing the corresponding elements in the matrix. The operation of addition is illustrated below for two n-dimensional vectors x and y.

$$x + y = \begin{pmatrix} x_1 \\ x_2 \\ \vdots \\ x_n \end{pmatrix} + \begin{pmatrix} y_1 \\ y_2 \\ \vdots \\ y_n \end{pmatrix} = \begin{pmatrix} x_1 + y_1 \\ x_2 + y_2 \\ \vdots \\ x_n + y_n \end{pmatrix}. \tag{A.5}$$

The rules for matrix (and vector) multiplication are a bit more involved as discussed next.

Definition (Scalar (Dot) Product): The *scalar* (*inner, dot*) product of two vectors is a scalar that is the sum of the products of the corresponding elements of the two vectors, i.e.,

$$\begin{aligned} x^T y = (x_1, x_2, \ldots, x_n) \begin{pmatrix} y_1 \\ y_2 \\ \vdots \\ y_n \end{pmatrix} &= \sum_{j=1}^{n} x_j y_j \\ &= x_1 y_1 + x_2 y_2 + \cdots + x_n y_n. \end{aligned} \tag{A.6}$$

The scalar product of two n-dimensional vectors x and y is denoted by $x^T y$ (or, in some books, by $\langle x, y \rangle$). It is defined in a manner similar to the product between two scalars. The one exception to this is the fact that the product of two nonzero scalars cannot be zero, whereas *the scalar product of two nonzero vectors can be zero* as can be seen by taking the scalar product of the two vectors x and y shown below:

$$x = \begin{pmatrix} 1 \\ 0 \\ 0 \end{pmatrix} \quad \text{and} \quad y = \begin{pmatrix} 0 \\ 1 \\ 0 \end{pmatrix}.$$

Definition (Outer Product): The *outer* product of two vectors, denoted by xy^T, is a matrix as illustrated in (A.7):

$$xy^T = \begin{pmatrix} x_1 \\ x_2 \\ \vdots \\ x_n \end{pmatrix} (y_1, y_2, \ldots, y_n) = \begin{pmatrix} x_1 y_1 & x_1 y_2 & \cdots & x_1 y_n \\ x_2 y_1 & x_2 y_2 & \cdots & x_2 y_n \\ \vdots & \vdots & \ddots & \vdots \\ x_n y_1 & x_n y_2 & \cdots & x_n y_n \end{pmatrix}. \tag{A.7}$$

The matrix xy^T is special and is called a *rank-one matrix* (See Section A.7).

The product of an $m \times n$ matrix A times an n-dimensional vector x is another vector Ax. This product Ax is defined only if the number of columns of A is equal to the dimension of the vector x. The product of a matrix A times a vector x can be thought of in one of two ways. The *first way* is to think of each element of the resultant vector Ax as the scalar product of the corresponding row of the matrix A times the vector x. That is,

$$Ax = \begin{pmatrix} a_{11} & a_{12} & \cdots & a_{1n} \\ a_{21} & a_{22} & \cdots & a_{2n} \\ \vdots & \vdots & \ddots & \vdots \\ a_{m1} & a_{m2} & \cdots & a_{mn} \end{pmatrix} \begin{pmatrix} x_1 \\ x_2 \\ \vdots \\ x_n \end{pmatrix} = \begin{pmatrix} \sum_{j=1}^{n} a_{1j} x_j \\ \sum_{j=1}^{n} a_{2j} x_j \\ \vdots \\ \sum_{j=1}^{n} a_{mj} x_j \end{pmatrix} = \begin{pmatrix} A_{1\bullet} x \\ A_{2\bullet} x \\ \vdots \\ A_{m\bullet} x \end{pmatrix}. \tag{A.8}$$

Before looking at the second way of representing Ax, we need to define *linear combinations* of vectors.

Definition (Linear Combination): A *linear combination* of the n vectors $v^1, v^2, \ldots, v^n$ is another vector, say v, defined by

$$v = \alpha_1 v^1 + \alpha_2 v^2 + \cdots + \alpha_n v^n,$$

where α_i, $i = 1, \ldots, n$, are scalars.

With the definition of a linear combination of vectors, it is straightforward to see that the *second way* to think of Ax is to consider Ax as being the vector formed by taking linear combinations of the columns of the matrix A with the weight on the jth column of A being x_j. This is illustrated below.

$$\begin{aligned} Ax &= \begin{pmatrix} a_{11} \\ a_{21} \\ \vdots \\ a_{m1} \end{pmatrix} x_1 + \begin{pmatrix} a_{12} \\ a_{22} \\ \vdots \\ a_{m2} \end{pmatrix} x_2 + \cdots + \begin{pmatrix} a_{1n} \\ a_{2n} \\ \vdots \\ a_{mn} \end{pmatrix} x_n \\ &= \sum_{j=1}^{n} A_{\bullet j} x_j. \end{aligned} \tag{A.9}$$

The product of an $m \times n$ matrix A times an $n \times p$ matrix B is another matrix $C = AB$, which is of dimension $m \times p$. This product AB is defined only if the number columns of matrix A is equal to the number of rows of matrix B. The product of a matrix A times a matrix B can be thought of in one of two ways. The first way is to think of the ijth element of the resultant matrix $C = AB$ as the scalar product of the ith row of the matrix A times the the jth column of the matrix B. That is, the ijth element of C is given by

$$c_{ij} = \sum_{k=1}^{n} a_{ik} b_{kj} = A_{i\bullet} B_{\bullet j}. \tag{A.10}$$

The second way to think of the matrix product $C = AB$ is to think of each column of C as having been generated by taking linear combinations of the columns of the matrix A. That is, the jth column of the matrix C uses the elements of the jth column of the matrix B as the weights to form the linear combination of the columns of the matrix A. For example,

$$C_{\bullet j} = \sum_{k=1}^{n} A_{\bullet k} b_{kj}.$$

From the above discussion on vector/matrix operations, the reader can easily verify that:

1. Matrix (vector) addition satisfies

associativity:	$A + (B + C)$	$=$	$(A + B) + C$;
commutativity:	$A + B$	$=$	$B + A$.

2. The scalar product for vectors satisfies

$$\begin{array}{lrcl} \textit{commutativity:} & x^Ty & = & y^Tx; \\ \textit{distributivity over vector addition:} & x^T(y+z) & = & x^Ty + x^Tz. \end{array}$$

3. Matrix multiplication satisfies

$$\begin{array}{lrcl} \textit{associativity:} & A(BC) & = & (AB)C; \\ \textit{distributivity over matrix addition:} & A(B+C) & = & AB + AC. \end{array}$$

An additional property that matrix multiplication satisfies is

$$(AB)^T = B^TA^T. \tag{A.11}$$

However, note that matrix multiplication does not satisfy the property of commutativity even if the matrices are square. That is, in general,

$$AB \neq BA. \tag{A.12}$$

▷ **Exercise A.1** Prove (A.12).

A.3 LINEAR INDEPENDENCE

In this section we describe linear independence of vectors.

Definition (Linearly Independent): The vectors $x^1, x^2, \ldots, x^n$ are said to be *linearly independent* if and only if all nontrivial linear combinations of the vectors are nonzero. That is,

$$\alpha_1 x^1 + \alpha_2 x^2 + \cdots + \alpha_n x^n \neq 0 \text{ unless } \alpha_1 = \alpha_2 = \cdots = \alpha_n = 0.$$

Example A.2 (Dependence and Independence) As an illustration of the concept of dependence and linear independence consider the following three 3-dimensional vectors:

$$v^1 = \begin{pmatrix} 2 \\ 0 \\ 4 \end{pmatrix}, \quad v^2 = \begin{pmatrix} 0 \\ 3 \\ -4 \end{pmatrix}, \quad v^3 = \begin{pmatrix} 4 \\ 3 \\ 4 \end{pmatrix}.$$

It is easy to see that $v^3 = 2v^1 + v^2$, or $2v^1 + v_2 - v_3 = 0$. That is, v^3 is a linear combination of v^1 and v^2, or, in other words, v^3 is dependent on v^1 and v^2. On the other hand, any two of the three vectors v^1, v^2, v^3 are linearly independent, since neither of them can be written as a multiple of only one of the other vectors.

A.4 ORTHOGONALITY

The concept of orthogonality plays an important role in the solution of least squares problems and also for the solution of linear programs using interior point methods.

Definition (Orthogonal and Orthonormal): The vectors $q^1, q^2, \ldots, q^n$ are said to be *orthogonal* to each other if and only if

$$(q^i)^T q^j = 0 \text{ if } i \neq j \quad \text{and} \quad (q^i)^T q^j \neq 0 \text{ if } i = j. \tag{A.13}$$

The vectors are said to be *orthonormal* if they are orthogonal and

$$(q^i)^T q^j = 1 \text{ when } i = j. \tag{A.14}$$

THEOREM A.1 (Orthogonality Implies Independence) *A set of orthogonal vectors are linearly independent.*

Proof. In order to prove this, consider any linear combination of the n orthogonal vectors $q^1, q^2, \ldots, q^n$ that equals zero. That is,

$$\alpha_1 q^1 + \alpha_2 q^2 + \cdots + \alpha_n q^n = 0.$$

Take the scalar product of the above vector equation with the vector q^j for arbitrary j. Then, by the definition of orthogonality, all scalar products of the form $(q^j)^T q^i$ must vanish whenever $i \neq j$. Thus, the equation reduces to

$$\alpha_j (q^j)^T q^j = 0.$$

Since we know that $(q^j)^T q^j \neq 0$, α_j must be zero. We showed this for arbitrary α_j, thus $\alpha_i = 0$ for $i = 1, \ldots, n$, and the result follows from the definition of linear independence. ∎

▷ **Exercise A.2** A set of n nonzero vectors $p^1, p^2, \ldots, p^n$ are said to be conjugate, or G-orthogonal, with respect to a matrix G if they satisfy

$$(p^i)^T G p^j = 0 \text{ for all } i \neq j.$$

If G is a positive-definite symmetric $n \times n$ matrix, show that the vectors $p^1, p^2, \ldots, p^n$ are linearly independent.

A.5 NORMS

The norm of a vector or a matrix provides a means for measuring the size of the vector or matrix.

Definition (Vector Norm): The *norm of a vector* is a continuous function with the following properties:

1. $||x|| \geq 0$; with equality holding if and only if $x = 0$.
2. $||\alpha x|| = |\alpha| \, ||x||$ for any scalar α.
3. $||x + y|| \leq ||x|| + ||y||$.

The p-norm of an n-dimensional vector x is denoted by $||x||_p$ and is defined to be

$$||x||_p = \left(\sum_{i=1}^{n} |x_i|^p \right)^{1/p}. \tag{A.15}$$

The most commonly used norms are defined for $p = 1$ (1-*norm*), $p = 2$ (2-*norm*), and $p = \infty$ (∞-*norm*). These are illustrated below:

(a) $||x||_1 = \sum_{i=1}^{n} |x_i|$, (1-norm)

(b) $||x||_2 = \sqrt{x^T x} = \left(\sum_{i=1}^{n} x_i^2\right)^{1/2}$, (2-norm or Euclidean norm)

(c) $||x||_\infty = \max_{i=1,\ldots,n} |x_i|$. ($\infty$-norm)

Unless otherwise stated, the symbol "$||\cdot||$" will always denote the 2-norm.

Example A.3 (Relative Error Using the ∞-Norm) The relative error in the ∞-norm is given by

$$e_R = \frac{||x - \hat{x}||_\infty}{||x||_\infty},$$

where x is the true solution of some problem and $\hat{x}$ is the computed solution of the same problem. It turns out that by using the ∞-norm if $e_R = 10^{-t}$ then we can expect approximately t correct significant digits in the largest component of the computed solution $\hat{x}$. Let $x = (0.5432, 8.789)^T$ and let the computed solution be $\hat{x} = (0.5213, 8.790)^T$. Then $e_R = 0.002492 \approx 10^{-3}$, implying that the largest component of $\hat{x}$ has approximately three correct significant digits, which is clearly true for $\hat{x}_2$ in this example. Nothing is implied about the number of correct digits in other components; in this example, $\hat{x}_1$ has only one correct siginificant digit.

The cosine of angle θ between two vectors x and y in $\Re^n$ is given by

$$\cos\theta = \frac{x^T y}{||x||_2 ||y||_2}. \tag{A.16}$$

Now, the cosine is bounded in absolute value by 1. Thus, this gives us a useful inequality for proving various theoretical results, i.e., the *Cauchy-Schwartz* inequality, which is

$$|x^T y| \leq ||x||_2 \, ||y||_2. \tag{A.17}$$

▷ **Exercise A.3** Show that the Cauchy-Schwartz inequality (A.17) does not hold for the ∞-norm. Does it hold for the 1-norm?

Definition (Matrix Norm): The *norm of a matrix* is a continuous function with the following properties:

1. $||A|| \geq 0$; with equality holding if and only if $A = 0$.
2. $||\alpha A|| = |\alpha| \, ||A||$ for any scalar α.
3. $||A + B|| \leq ||A|| + ||B||$.
4. $||AB|| \leq ||A|| \, ||B||$.

Note: Strictly speaking, a matrix norm need not satisfy the submultiplicative property $||AB|| \leq ||A|| \, ||B||$; however, we have assumed it because all the matrix norms that we work with in this book do satisfy this property.

▷ **Exercise A.4** Define a matrix norm by

$$||A||_\gamma = \max_{i,j} |a_{ij}|.$$

Construct an example to show that it satisfies matrix norm properties 1, 2, and 3, but not the sub-multiplicative property 4.

Definition (Vector Induced Matrix Norm): A matrix norm is said to be *induced* by a vector norm if it is defined as follows:

$$||A|| = \max_{||x|| \neq 0} \frac{||Ax||}{||x||},$$

where the norms on the right-hand side are vector norms. Equivalently, the definition is sometimes stated as

$$||A|| = \max_{||x||=1} ||Ax||.$$

Some specific examples of matrix norms are:

(a) $||A||_1 = \max_{j=1,\ldots,n} \sum_{i=1}^{m} |a_{ij}|,$ (1-norm)

(b) $||A||_2 = \sqrt{\lambda_{\max}(A^T A)},$ (2-norm)

(c) $||A||_\infty = \max_{i=1,\ldots,m} \sum_{j=1}^{n} |a_{ij}|,$ (∞-norm)

(d) $||A||_F = \sqrt{\sum_{i=1}^{m} \sum_{j=1}^{n} a_{ij}^2},$ (Frobenius norm) where $\lambda_{\max}(A^TA)$ is the largest eigenvalue of A^TA (see Section A.12 for a defintion of eigenvalues).

The 1-norm, 2-norm, and ∞-norm are norms that are induced by the corresponding vector norms. The Frobenius norm on the other hand is not a vector induced norm ($\mathcal{V}$-norm) as can be easily seen by considering the identity matrix:

$$||I||_F = \sqrt{n}; \quad \text{whereas} \quad ||I||_{\mathcal{V}} = \max_{||x|| \neq 0} \frac{||Ix||}{||x||} = 1.$$

▷ **Exercise A.5** Show that the 1-norm, 2-norm, and ∞-norm are norms that are induced by the corresponding vector norms.

From the definition of a vector-induced norm, it is straightforward to see that a matrix norm that is vector induced satisfies

$$||Ax|| \leq ||A||\,||x||. \tag{A.18}$$

Definition (Compatible (or Consistent) Norms): In general, any vector norm $||\cdot||$ and any matrix norm $||\cdot||'$ are said to be *compatible*, or *consistent*, if they satisfy the following inequality:

$$||Ax|| \leq ||A||'\,||x||.$$

From (A.18) it follows that every vector norm and its vector-induced matrix norm are compatible.

▷ **Exercise A.6** Show that the Frobenius matrix norm and the Euclidean vector norm are compatible, that is, $||Ax||_2 \leq ||A||_F\,||x||_2$.

▷ **Exercise A.7** Show that given any vector x, for any two vector norms $||\cdot||$ and $||\cdot||'$, there exist two scalars α and β that depend only on the dimension of x such that

$$\alpha||x|| \leq ||x||' \leq \beta||x||.$$

A.6 VECTOR SPACES

Vector spaces play an important role in the development of the theory of active set methods for linear programming.

Definition (Vector Space): A set of vectors is said to form a *vector space* if they satisfy the following two properties:

1. The sum of any two vectors in the space also lies in the space.

2. If we multiply any vector by a scalar, then the multiple also lies in the space.

In other words a vector space is a set of vectors that is *closed under vector addition and closed under scalar multiplication.*

A set of any n linearly independent n-dimensional vectors can be used to define an n-dimensional vector space. This n-dimensional space is denoted by $\Re^n$ (for an n-dimensional real space) or by E^n (for an n-dimensional Euclidean space when the distance between two vectors $x, y \in \Re^n$ is given by $\sqrt{\sum_{i=1}^{n}(x_i - y_i)^2}$).

Definition (Subspace): A subset of a vector space that is a vector space in its own right is called a *subspace.* Thus, a subset of k linearly independent vectors in $\Re^n$ could be used to define a k-dimensional subspace of the vector space $\Re^n$.

Definition (Span): If a vector space V consists of all linear combinations of a particular set of vectors $\{w^1, w^2, \ldots, w^k\}$, then these vectors are said to *span* the vector space V.

Definition (Basis): A spanning set of vectors that are linearly independent are said to form a *basis* for the vector space.

Definition (Dimension of a Vector Space): The *dimension* of a vector space (subspace) is the number of linearly independent vectors that span the vector space (subspace).

Definition (Complementary Subspace, Null Space): For every proper subspace $S \subset \Re^n$, there is a *complementary* subspace $\overline{S} \subset \Re^n$, whose members are defined as follows: if $y \in \overline{S}$ and $x \in S$, then $x^T y = 0$. That is, y is orthogonal to every vector in S. The subspace $\overline{S}$ is termed the *orthogonal complement* of S, and the two subspaces are completely disjoint except for the origin. The vector subspace $\overline{S}$ is also called the *null space* with respect to S, and S is called the null space with respect to $\overline{S}$. If the dimension of S is k, then the dimension of $\overline{S}$ is $n - k$. The vector subspace $S \subset \Re^n$ is therefore defined by k basis vectors, and $\overline{S} \subset \Re^n$ is defined by $n - k$ basis vectors.

▷ **Exercise A.8** Show that if the k basis vectors in S are orthogonal to each of the $n - k$ basis vectors in $\overline{S}$, then the subspaces defined by each of these sets of basis vectors are orthogonal complements of each other.

Definition (Null Space of a Matrix): Let A be an $m \times n$ matrix of rank $r \leq m$ where $m \leq n$. The rows of A belong to $\Re^n$. The orthognal complement to the subspace generated by the rows of A is referred to as the *null space* of A and has dimension $n - r$.

A.7 RANK OF A MATRIX

Analogous to the dimension of a vector space is the rank of a matrix.

Definition (Rank of a Matrix): The *rank* of a matrix is the number of linearly independent rows (or columns). A matrix can only have one rank regardless of the number of rows and columns.

Definition (Column Rank of a Matrix): The *row rank* of a matrix is the dimension of the vector subspace spanned by the rows of A.

Definition (Row Rank of a Matrix): The *column rank* of a matrix is the dimension of the vector subspace spanned by the columns of the matrix.

Definition (Full Rank): An $m \times n$ matrix (with $m \leq n$) is said to be of *full rank* if it has rank m.

Note: The definition of rank implies that row rank is equal to column rank.

A.8 MATRICES WITH SPECIAL STRUCTURE

Matrices with special structure play an important role in optimization and in the solution of systems of equations.

Definition (Diagonal Matrix): A matrix is a *diagonal matrix* if all its off-diagonal elements are zero.

A diagonal matrix is usually denoted by the letter D.

Example A.4 (Diagonal Matrix) An example of an $n \times n$ diagonal matrix is

$$D = \begin{pmatrix} d_{11} & 0 & \dots & 0 \\ 0 & d_{22} & \dots & 0 \\ \vdots & \vdots & \ddots & \vdots \\ 0 & 0 & \dots & d_{nn} \end{pmatrix}.$$

Definition (Identity Matrix): A square diagonal matrix with all diagonal elements equal to 1 is called an *identity matrix*.

An identity matrix is denoted by I or by I_n, where n indicates the dimension of the matrix. The ith column of an identity matrix is often denoted by e_i or e^i.

Example A.5 (Identity Matrix) The following illustrates an n-dimensional identity matrix.

$$I = \begin{pmatrix} 1 & 0 & 0 & \dots & 0 & 0 \\ 0 & 1 & 0 & \dots & 0 & 0 \\ 0 & 0 & 1 & \dots & 0 & 0 \\ \vdots & \vdots & \vdots & \ddots & \vdots & \vdots \\ 0 & 0 & 0 & \dots & 1 & 0 \\ 0 & 0 & 0 & \dots & 0 & 1 \end{pmatrix} = (e_1, e_2, \dots, e_n).$$

Multiplication of a matrix by an identity matrix on the left or on the right does not change the original matrix. This is analogous to the multiplication of a scalar by 1.

> *Definition (Permutation Matrix)*: A *permutation* matrix is a matrix with exactly one 1 in each row and in each column and zeros everywhere else.

Thus, a permutation matrix is a rearrangement of the columns of the identity matrix. A permutation matrix is usually denoted by P.

Example A.6 (Permutation Matrix) An example of a 3×3 permutation matrix is as follows.

$$P = \begin{pmatrix} 0 & 1 & 0 \\ 0 & 0 & 1 \\ 1 & 0 & 0 \end{pmatrix}.$$

Multiplication by a permutation matrix on the left of a matrix A permutes the rows of A whereas multiplication by a permutation matrix on the right of A permutes the columns of A.

Example A.7 (Multiplying by a Permutation Matrix) For example, PA, using the above 3-dimensional matrix P, results in a matrix that consists of the second row of A followed by the third row of A followed by the first row of A. Multiplying by a permutation matrix on the right of A has the effect of permuting the columns of A in a manner similar to that described for the rows of A but in the order 3rd column, 1st column, 2nd column.

> *Definition (Upper (or Right) Triangular Matrix)*: A square matrix is *upper*, or *right, triangular* if all its subdiagonal elements are zero, that is, $A_{ij} = 0$ if $i > j$.

An upper triangular matrix is often denoted by R or U.

Example A.8 (Upper Triangular Matrix) An upper triangular matrix is illustrated below.

$$R = \begin{pmatrix} r_{11} & r_{12} & r_{12} & \dots & r_{1n} \\ 0 & r_{22} & r_{23} & \dots & r_{2n} \\ 0 & 0 & r_{33} & \dots & r_{3n} \\ \vdots & \vdots & \vdots & \ddots & \vdots \\ 0 & 0 & 0 & \dots & r_{nn} \end{pmatrix}.$$

A matrix is said to be *upper trapezoidal* if the definition of upper triangular matrix is applied to a matrix with more columns than rows.

Definition (Lower (or Left) Triangular Matrix): A matrix is *lower triangular* if all the elements above the diagonal are zero, i.e., $A_{ij} = 0$ if $i < j$.

A lower triangular matrix is commonly denoted by L.

Example A.9 (Lower Triangular Matrix) An example of a lower triangular matrix is

$$L = \begin{pmatrix} l_{11} & 0 & 0 & \dots & 0 \\ l_{21} & l_{22} & 0 & \dots & 0 \\ l_{31} & l_{32} & l_{33} & \dots & 0 \\ \vdots & \vdots & \vdots & \ddots & \vdots \\ l_{n1} & l_{n2} & l_{n3} & \dots & l_{nn} \end{pmatrix}.$$

A lower triangular matrix with ones all along the diagonal (that is, $l_{ii} = 1$ for all i) is called a *unit lower triangular* matrix. A matrix is said to be *lower trapezoidal* if the definition of lower triangular matrix is applied to a matrix with more rows than columns.

Definition (Triangular Matrix): We give the following two equivalent definitions of a *triangular* matrix.

1. A square matrix is said to be *triangular* if it satisfies the following properties.
   (a) The matrix contains at least one row having exactly one nonzero element.
   (b) If the row with a single nonzero element and its column are deleted, the resulting matrix will once again have this same property.
2. Equivalently we can define a square matrix to be *triangular*, if its rows and columns can be permuted to be either an upper-triangular or lower-triangular matrix.

▷ **Exercise A.9** Prove that the transpose of a triangular matrix is also triangular.

Definition (Orthogonal (or Orthonormal) Matrix): A square matrix is said to be an *orthogonal* (*orthonormal*) matrix if all its columns are orthogonal (orthonormal).

Usually Q is used to denote an orthonormal matrix. An orthonormal matrix Q satisfies

$$Q^TQ = I, \quad QQ^T = I, \tag{A.19}$$

where I is the identity matrix.

Definition (Elementary Matrix): An *elementary* matrix is a matrix of the form $I + \alpha uv^T$, where the matrix uv^T is called a *rank one matrix*. A *row elementary matrix* is a matrix of the form $I + e_k v^T$ and a *column elementary matrix* is a matrix of the form $I + ue_k{}^T$, where e_k is a vector with a 1 in position k and zeros elsewhere.

Elementary matrices and rank one matrices play an important role in linear programming.

Definition (Symmetric Matrix): A square matrix is said to be a *symmetric* matrix if every element below the diagonal is equal to its mirror image above the diagonal. That is, $A_{ij} = A_{ji}$.

Thus, a symmetric matrix is one whose transpose is equal to itself, that is, $A = A^T$.

A.9 INVERSE OF A MATRIX

The inverse of a matrix is analogous to the reciprocal of a scalar.

Definition (Matrix Inverse): The *inverse* of a square matrix is another square matrix that when multiplying the original matrix on the left or the right results in an identity matrix. The inverse of a matrix A is denoted by A^{-1} and satisfies

$$AA^{-1} = A^{-1}A = I. \tag{A.20}$$

In early implementations of linear equation solvers, the inverse played a critical role since the solution of $Ax = b$ can be written as $x = A^{-1}b$. *However, it can be shown that computing the inverse is inefficient and can be highly unstable.* Thus, current implementations of linear equation solvers use matrix factorization techniques (see *Linear Programming 2*).

THEOREM A.2 (Existence of the Inverse) *The inverse as defined by* (A.20) *only exists for a square matrix of full rank and is unique.*

Proof. It is easy to prove the uniqueness of the inverse. Let B and C be two inverses of A. Then $AB = I$. Multiplying this identity on the left by C and rearranging the multiplication, we get $(CA)B = C$ or $B = C$, thus proving that the two inverses are the same. ∎

Definition (Singular Matrix): A square matrix is said to be *nonsingular* if its inverse exists; otherwise the matrix is said to be *singular*.

It is possible to define one-sided inverses for matrices as follows.

Definition (Left Inverse, Right Inverse): The matrix B is said to be a *left inverse* of a matrix A if B satisfies $BA = I$. The matrix C is said to be a *right inverse* of a matrix A if $AC = I$.

LEMMA A.3 (Left Inverse Equals Right Inverse) *For a square matrix only, if both the left and right inverses exist then they are equal and are equal to the inverse of A.*

Proof. This can be proved by looking at the identity $B(AC) = (BA)C$, which follows from the associativity property of matrix multiplication. Since B is a left inverse and C is a right inverse we get $B = C$. ∎

The inverse operation affects the order of matrix multiplication in exactly the same manner as the transpose operation. That is,

$$(AB)^{-1} = B^{-1}A^{-1}.$$

A.10 INVERSES OF SPECIAL MATRICES

The inverse of an orthonormal matrix Q is given by $Q^{-1} = Q^T$. This follows from the fact that $QQ^T = Q^TQ = I$.

▷ **Exercise A.10** Show that the inverse of a permutation matrix is the same as its transpose.

▷ **Exercise A.11** Show that the inverse of a lower (upper) triangular matrix is a lower (upper) triangular matrix.

▷ **Exercise A.12** The inverse of a nonsingular elementary matrix $I + \sigma uv^T$ (where σ is a scalar) is also an elementary matrix that involves the same two vectors u and v. Show that if $\sigma u^T v \neq -1$, then

$$(I + \sigma uv^T)^{-1} = I + \gamma uv^T, \tag{A.21}$$

where $\gamma = -\sigma/(1 + \sigma u^T v)$.

▷ **Exercise A.13** Let A be an $n \times n$ matrix whose inverse exists. Use the result of Exercise A.12 to determine the form of the inverse of

$$\widehat{A} = A + \alpha uv^T$$

in terms of A^{-1}, the scalar α, and the vectors u and v. Under what conditions does $\widehat{A}^{-1}$ exist? The result obtained is a special case of the Sherman-Morrison-Woodbury formula where u and v are replaced by U and V, which are $n \times k$ matrices; derive this formula.

A.11 DETERMINANTS

We start our discussion by defining a formula for computing the determinant of a square matrix.

Definition (Determinant): The *determinant* of a 1×1 matrix is the value of its element. The *determinant* of a general square $n \times n$ matrix A, with $n > 1$, can be computed by expanding the matrix in the cofactors of its ith row, i.e.,

$$\det A = \sum_{j=1}^{n} a_{ij} \mathcal{A}_{ij}, \tag{A.22}$$

where $\mathcal{A}_{ij}$, the ijth *cofactor*, is the determinant of the *minor* M_{ij} with the appropriate sign:

$$\mathcal{A}_{ij} = (-1)^{i+j} \det M_{ij}. \tag{A.23}$$

The minor M_{ij} is the matrix formed from A by deleting the ith row and jth column of A.

Example A.10 (Determinant) The determinant of a two dimensional matrix

$$A = \begin{pmatrix} a & b \\ c & d \end{pmatrix}$$

can be easily computed as

$$\det A = ad - bc.$$

▷ **Exercise A.14** Show that $\det A = \det A^T$. Use this to show that the determinant of A can be obtained by an expansion of the cofactors along any column j of A.

The determinant possesses several interesting properties. It turns out that every property is a consequence of the following three properties:

1. The determinant of an identity matrix is 1.
2. The determinant is a linear function of the ith row, i.e.,
   (a) Let $\widehat{A}$ be the matrix obtained by replacing the first row $A_{i\bullet}$ by $\alpha A_{i\bullet}$. Then $\det \widehat{A} = \alpha \det A$.
   (b) Let $\widehat{A}$ be the matrix obtained by replacing the first row $A_{i\bullet}$ by a vector $A_{i\bullet} + v^T$. Then $\det \widehat{A} = \det A + \det B$, where B is the matrix formed from A by replacing $A_{i\bullet}$ by v^T.
3. The determinant changes sign if any two rows are interchanged.

▷ **Exercise A.15** Prove the above three properties of a determinant.

▷ **Exercise A.16** Use the above three properties to prove:

1. If the ith row of a matrix is a multiple of the jth row, where $i \neq j$, the determinant is zero.
2. If a multiple of the ith row is subtracted from (or added to) the jth row, where $i \neq j$, the determinant is unchanged.
3. If A has a zero row, the determinant of A is zero.
4. If A is singular, the determinant of A is zero; whereas, if A is nonsingular, the determinant of A is nonzero.
5. The determinant of the product of two square matrices A and B satisfies

$$\det AB = (\det A)(\det B).$$

▷ **Exercise A.17** Show that the determinant of a lower (upper) triangular matrix is the product of the diagonal elements of the matrix.

The inverse of a nonsingular matrix A can be computed using determinants and cofactors. (This has no practical relevance because more efficient and stable methods exist for computing the inverse in practice.) To see this, note that

$$0 = \sum_{j=1}^{n} a_{ij} A_{ij}, \quad \text{for } i \neq k, \tag{A.24}$$

because in effect we are computing the determinant of a matrix formed from A by replacing its ith row by its kth row, and therefore the determinant of this new matrix is zero: it has two identical rows. Using (A.22) and (A.24) we find that

$$A \operatorname{adj} A = (\det A) I, \tag{A.25}$$

where $\operatorname{adj} A$ is the matrix whose ijth component is the cofactor A_{ij}. Therefore

$$A^{-1} = \frac{\operatorname{adj} A}{\det A}. \tag{A.26}$$

For the purpose of solving systems of equations $Ax = b$, this implies that

$$x = A^{-1} b = \frac{(\operatorname{adj} A) b}{\det A}. \tag{A.27}$$

This is equivalent to *Cramer's rule* for solving systems of equations, which is

$$x_j = \frac{\det A^j}{\det A} \quad \text{where } A^j = [A_{\bullet 1}, \ldots, A_{\bullet j-1}, b, A_{\bullet j+1}, \ldots, A_{\bullet n}], \tag{A.28}$$

for $j = 1, \ldots, n$.

▷ **Exercise A.18** Prove that (A.27) and (A.28) are equivalent.

A.12 EIGENVALUES

For this section assume that all matrices are square because it makes no sense to talk about the eigenvalues of a rectangular matrix.

Definition (Eigenvalue, Eigenvector, and Eigensystem): A scalar λ is called an *eigenvalue* of a matrix A and a corresponding nonzero vector x is called an *eigenvector* of A if and only if

$$Ax = \lambda x. \tag{A.29}$$

The set of all eigenvalues and eigenvectors of A is called the *eigensystem* of A.

Rewriting Equation (A.29) it follows that

$$(A - \lambda I)x = 0, \tag{A.30}$$

that is, the vector x must lie in the null space of $A - \lambda I$. Since we are interested in a nonzero vector x, Equation (A.30) implies that λ must be such that $A - \lambda I$ is singular. Therefore, the scalar λ will be an eigenvalue, with corresponding eigenvector x, if and only if

$$\det(A - \lambda I) = 0. \tag{A.31}$$

The roots of Equation (A.31) give the eigenvalues of A. Note that the eigenvector of a matrix is uniquely determined in direction only, because obviously any nonzero scalar multiple of an eigenvector is also an eigenvector for the matrix.

Definition (Characteristic Equation, Characteristic Polynomial): Equation (A.31) is called the *characteristic equation* of the matrix A, and the resulting polynomial is called the *characteristic polynomial*.

▷ **Exercise A.19** Show that for any square matrix there is at least one eigenvalue and one eigenvector.

▷ **Exercise A.20** Show that a lower (upper) triangular matrix has eigenvalues equal to its diagonal entries.

Definition (Similar Matrices): Two matrices are said to be *similar* if they share the same eigenvalues (the eigenvectors can be different).

LEMMA A.4 (Similar Matrices) *If W is any nonsingular matrix then the matrices A and WAW^{-1} are similar.*

Proof. This can be seen by observing that if λ is an eigenvalue with eigenvector x for the matrix A, then λ is also an eigenvalue with eigenvector Wx for the matrix WAW^{-1}, as shown below:

$$Ax = \lambda x \quad \text{and} \quad WAW^{-1}(Wx) = \lambda(Wx).$$

This completes our proof. ∎

If the eigenvalues of A are denoted by $\lambda_i(A)$, the *spectral radius* $\rho(A)$ of the matrix A is defined by:

$$\rho(A) = \max_i |\lambda_i|. \tag{A.32}$$

Definition (Complex Conjugate): The *complex conjugate* of a complex number $c = a + ib$ is $\bar{c} = a - ib$, where a and b are real numbers and $i = \sqrt{-1}$.

Definition (Complex Scalar Product): The *scalar product of the two vectors x and y composed of complex numbers* is defined to be $x^H y$, where x^H is the vector determined from by x by taking the complex conjugate of each element.

Definition (Hermitian Matrix): A *Hermitian matrix* A is defined to be a matrix such that $A = A^H$, where A^H is the *conjugate transpose* of the matrix, i.e., the matrix A such that $a_{ij} = \bar{a}_{ji}$.

▷ **Exercise A.21** Show that $\bar{x}^H A \bar{x}$ is real for any Hermitian matrix A.

▷ **Exercise A.22** Prove that if a matrix A has distinct eigenvalues, then the eigenvectors corresponding to the distinct eigenvalues are linearly independent. Further prove that if A is Hermitian, then the eigenvectors are orthogonal.

LEMMA A.5 (Eigenvalues and Eigenvectors of a Symmetric Matrix) *If A is a real symmetric matrix, then the following two properties hold:*

1. *All eigenvalues of A are real numbers;*
2. *The matrix A has n distinct eigenvectors. Furthermore, the eigenvectors can be made to form an orthonormal set of vectors.*

Proof. Let $\bar{\lambda}$ be an eigenvalue of A and its corresponding eigenvector $\bar{x}$, so that $A\bar{x} = \bar{\lambda}\bar{x}$. Multiplying both sides by $\bar{x}^H$, we get $\bar{x}^H A \bar{x} = \bar{\lambda}\bar{x}^H\bar{x}$. The right-hand side is clearly real, even if $\bar{x}$ is complex. The left-hand side is also real (see Exercise A.21), therefore $\bar{\lambda}$ is real, proving property (1).

Since the eigenvalues are real and A is real, the eigenvectors are also real. To complete the proof, see Exercise A.23. ∎

▷ **Exercise A.23** Show that for any square matrix A, there is an orthonormal matrix U such that $U^{-1}AU$ is upper triangular. Use this to show property (2) of Lemma A.5.

The following lemmas illustrate some useful properties of eigenvalues:

LEMMA A.6 (Product of Eigenvalues Equals Determinant) *The product of the eigenvalues of A is equal to the determinant of A. That is,*

$$\prod_{i=1}^{n} \lambda_i = \det A. \tag{A.33}$$

▷ **Exercise A.24** Prove Lemma A.6 by imagining that the characteristic polynomial is factored into

$$\det(A - \lambda I) = (\lambda_1 - \lambda)(\lambda_2 - \lambda)\cdots(\lambda_n - \lambda), \tag{A.34}$$

and by choosing an appropriate λ.

LEMMA A.7 (Sum of Eigenvalues Equals Trace) *The sum of the diagonal elements of A (i.e., trace of A) is equal to the sum of the eigenvalues of A, that is*

$$\mathit{Trace}(A) = \sum_{i=1}^{n} a_{ii} = \sum_{i=1}^{n} \lambda_i. \tag{A.35}$$

▷ **Exercise A.25** Prove Lemma A.7. Do this by first finding the coefficient of $(-\lambda)^{n-1}$ on the right-hand side of Equation (A.34). Next look for all the terms on the left-hand side of Equation (A.34) that involve $(-\lambda)^{n-1}$ and compare the two. Hint: Show that the terms on the left-hand side of Equation (A.34) involving $(-\lambda)^{n-1}$ all come from the product down the main diagonal.

▷ **Exercise A.26** Show that the n-dimensional matrix $(I + \alpha uv^T)$ has $n-1$ eigenvalues equal to unity. What is its nth eigenvalue?

▷ **Exercise A.27** Obtain an expression for the determinant of $\widehat{A} = A + \alpha uv^T$ in terms of the determinant of A.

LEMMA A.8 (Product of Eigenvalues of a Matrix Product) *The product of the eigenvalues of AB is equal to the product of the eigenvalues of A times the product of the eigenvalues of B,*

$$\prod \lambda(AB) = \prod \lambda(A) \prod \lambda(B). \tag{A.36}$$

▷ **Exercise A.28** Prove Lemma A.8.

A.13 POSITIVE-DEFINITENESS

Positive-definite matrices play an important role in interior point methods for linear programs and in nonlinear optimization.

Definition (Positive-Definite): A symmetric matrix A is said to be *positive-definite* if and only if

$$x^T Ax > 0 \text{ for all } x \neq 0.$$

Definition (Positive Semidefinite): A symmetric matrix A is said to be *positive semidefinite* if and only if

$$x^T Ax \geq 0 \text{ for all } x.$$

Similar definitions hold for *negative definite* matrices and *negative semidefinite* matrices.

▷ **Exercise A.29** Let A be a full rank $m \times n$ matrix with $m < n$. Prove that AA^T is a symmetric positive-definite matrix.

Some important properties of a symmetric positive-definite matrix C are:

1. It is nonsingular.
2. It has all positive eigenvalues.
3. The square root of C exists (that is, there is a matrix A such that $C = AA$; and A is called the square root of C).

▷ **Exercise A.30** Prove the above three properties (a)–(c) of a symmetric positive-definite matrix.

▷ **Exercise A.31** Show that if A is a symmetric positive-definite matrix then so is A^{-1}

A positive-definite matrix C can be used to define a vector C-norm by

$$||x||_C = \sqrt{x^T Cx}. \tag{A.37}$$

The corresponding vector-induced matrix C-norm is

$$||A||_C = \max_{||x||_C \neq 0} \frac{||Ax||_C}{||x||_C}. \tag{A.38}$$

▷ **Exercise A.32** Prove that $||x||_C$ defined by (A.38) is in fact a vector norm.

A.14 NOTES & SELECTED BIBLIOGRAPHY

For an excellent introduction to linear algebra refer to Strang [1988]. Some other references on linear algebra are Dahlquist & Björck [1974], Samelson [1974], and Stewart [1973]. For numerical linear algebra refer to Gill, Murray, & Wright [1991], Golub & Van Loan [1989], Householder [1974], Strang [1986], and the classic work of Wilkinson [1965]. For an extensive discussion on determinants see Muir [1960].

The computation of eigenvalues and eigenvectors is an iterative process because in general, as proved by Galois, there is no algebraic formula for the roots of a quintic polynomial. For an excellent discussion on eigenvalues and eigenvectors and the numerical analysis of such, see the classic work of Wilkinson [1965]. For modern computational techniques for determining the eigenvalues and eigenvectors of a matrix and matrix computations in general, see Golub & Van Loan [1989].

In general, developing software for matrix vector operations is not as easy as it seems, even for the relatively simple operations of vector addition. For details on implementing high-quality linear algebra routines see Dodson & Lewis [1985], Dongarra, DuCroz, Hammarling, & Hanson [1985, 1988a, 1988b, 1988c], Dongara, Gustavson, & Karp [1984], Lawson, Hanson, Kincaid, & Krogh [1979a, 1979b], and Rice [1985].

The multiplication of of two $n \times n$ matrices requires n^3 multiplications and $n^3 - n^2$ additions. However, Strassen [1969] showed that the computations can be reduced by splitting each of the matrices into four equal-dimension blocks. For example, if $n = 2m$, then four $m \times m$ blocks can be constructed; if the Strassen approach is not applied recursively, then the computations require $7m^3$ multiplications and $11m^2$ additions compared to $8m^3$ multiplications and $8(m^3 - m^2)$ additions. If we assume that the matrix size is $n = 2^r$ and recur the idea so that the minimum block size is $q \gg 1$, then the number of additions are roughly the same as the number of multiplications. (We could take q down towards 1, but it turns out that as $q \to 1$, the number of additions becomes high). The number of subdivisions is then $r - k$, where $q = 2^k$. Assuming conventional matrix multiplication for the the final blocks, Strassen's matrix multiplication requires $(2^k)^3 7^{r-k}$ multiplications compared to $(2^r)^3$ multiplications. It can be shown that if the recursion is continued to the 1×1 level, the Strassen procedure requires approximately $n^{2.807}$ multiplications. It turns out that for large matrices this can cut the time down by as much as 40%, see Bailey [1988]. For additional information on fast matrix multiplication, see also Golub & Van Loan [1989] and Pan [1984].

The ∞-norm is sometimes referred to as the *Chebyshev norm* after Chebyshev, the Russian mathematician, who first used it for solving curve fitting problems.

Exercises A.24 and A.25 are adapted from Strang [1986].

A.15 PROBLEMS

A.1 *Putnam Exam, 1969.* Let A be a 3×2 matrix and let B be a 2×3 matrix. Suppose that

$$AB = \begin{pmatrix} 8 & 2 & -2 \\ 2 & 5 & 4 \\ -2 & 4 & -2 \end{pmatrix}.$$

Show that

$$BA = \begin{pmatrix} 9 & 0 \\ 0 & 9 \end{pmatrix}.$$

Hint: Square AB.

A.2 Show that

$$\left\{ \begin{pmatrix} y \\ y \\ z \end{pmatrix} \in \Re^3 : x + y + z = 0 \right\}$$

is a subspace of $\Re^3$, and find a basis for it.

A.3 Let (u_1, u_2, u_3) be a basis of the vector space U. Show that $(u_1 + 2u_2, 3u_2 + u_3, -u_1 + u_3)$ is also a basis of the vector space U.

A.4 Find the dimension of the span in $\Re^3$ of

$$\left\{ \begin{pmatrix} 0 \\ 2 \\ 7 \end{pmatrix}, \begin{pmatrix} -1 \\ -1 \\ -1 \end{pmatrix}, \begin{pmatrix} 1 \\ 2 \\ 3 \end{pmatrix} \right\}.$$

A.5 Show the following:

(a) The vector space of all $m \times n$ matrices has dimension mn.

(b) The set of all diagonal matrices is a subspace of the vector space of all $n \times n$ matrices.

A.6 Is there a 5×3 matrix A and a 3×5 matrix B such that $AB = I_5$?

A.7 Let E_j be a matrix with its jth column different from an identity matrix. That is, $E_j = I + (x - e_j)e_j^T$ for some vector x. What is the inverse of E_j? Under what conditions does E_j^{-1} exist?

A.8 Suppose that a matrix $\bar{A}$ is formed by replacing the jth column of an $m \times m$ matrix A by a vector v. Given A^{-1}, what is the inverse of $\bar{A}$? Under what conditions does $\bar{A}^{-1}$ exist?

A.9 Suppose F is a $n \times n$ matrix and let $||F||$ satisfy the four properties of a matrix norm defined on Page 323.

(a) Show that if $||F|| < 1$, then $I - F$ is nonsingular and

$$(I - F)^{-1} = \sum_{i=0}^{\infty} F^i$$

with

$$||(I - F)^{-1}|| \leq \frac{1}{1 - ||F||}.$$

(b) Show also that

$$||(I - F)^{-1} - I|| \leq \frac{||F||}{1 - ||F||}.$$

A.10 This problem demonstrates how to quantify the change in A^{-1} as a function of changes in A. Show that if A is nonsingular and $||A^{-1}E|| = \gamma < 1$, then $A + E$ is nonsingular and

$$||(A + E)^{-1} - A^{-1}|| \leq \frac{||E||\,||A^{-1}||^2}{(1 - \gamma)}.$$

Hint: Use problem A.9 and also show that

$$B^{-1} = A^{-1} - B^{-1}(B - A)A^{-1},$$

which shows how B^{-1} changes when B changes.

A.11 (a) Show that $(A^T)^{-1} = (A^{-1})^T$.

(b) Without using the above result show that $\det(A^T)^{-1} = \det(A^{-1})^T$.

A.12 *Putnam Exam, 1969.* Let D_n be the determinant of an $n \times n$ matrix of which the element in the ith row and jth column is the absolute value of the difference of i and j. Show that

$$D_n = (-1)^{n-1}(n-1)2^{n-2}.$$

A.13 If the eigenvalues of A are λ_i and corresponding eigenvectors are x^i, show that $A^2 = AA$ (A-squared) has eigenvalues λ_i^2 and the same eigenvectors x^i.

A.14 Let A be an $m \times m$ matrix. Show that there exists an $x \neq 0$ such that $Ax \leq 0$.

A.15 Find the inverse of

$$\begin{pmatrix} 6 & 2 & -1 \\ 3 & 3 & 2 \\ -5 & 0 & 3 \end{pmatrix}.$$

A.16 Consider the system $Ax = b$, $x \geq 0$, where A is $m \times n$. Show that there is a matrix B (perhaps empty) such that $\operatorname{rank}(A, B) = m$ and so that any solution to

$$\begin{aligned} Ax + By &= b \\ x \quad\quad &\geq 0 \\ y &\geq 0 \end{aligned}$$

has $y = 0$.

A.17 *Ph.D. Comprehensive Exam, September 22, 1973, at Stanford.* Let A be an $m \times n$ matrix and b_k an m column vector for $k = 1, \ldots, \infty$. Let $P_k = \{x \mid Ax \leq b_k\}$ for $k = 1, \ldots, \infty$. Suppose that $P_k \neq \phi$ for $k = 1, \ldots, \infty$ and that $b_k \rightarrow b_\infty$ for $k \rightarrow \infty$. Verify that the following formulae are or are not valid.

(a) $\operatorname*{Sup}_{x \in P_k} \operatorname*{Inf}_{y \in P_\infty} ||x - y|| \rightarrow 0$ as $k \rightarrow \infty$.

(b) $\operatorname*{Sup}_{x \in P_\infty} \operatorname*{Inf}_{y \in P_k} ||x - y|| \rightarrow 0$ as $k \rightarrow \infty$.

A.18 *Ph.D. Comprehensive Exam, September 26, 1980, at Stanford.* Consider a set of 6 points A_1, A_2, A_3, A_4, A_5, A_6 situated in 3-dimensional Euclidean space in such a way that:

1. the line segments A_1A_2, A_2A_3, A_3A_4, A_4A_5, A_5A_6, A_6A_1 have equal lengths a;
2. the line segments A_1A_3, A_2A_4, A_3A_5, A_4A_6, A_5A_7, A_6A_2 have equal lengths b.
3. $2a > b$.

That such sets of 6 points exist (even in 2-dimensional space) is shown in Figure A-1.

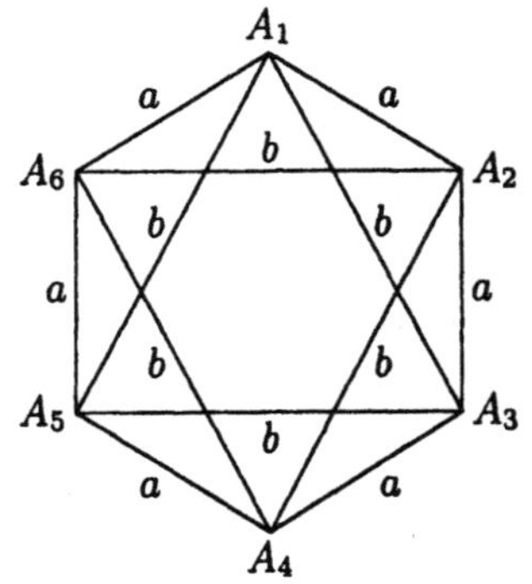

Figure A-1: Six Points in 2-dimensional Space

Given these conditions, the question now arises: What values can the lengths of the line segments A_1A_4 (length x), A_2A_5 (length y), A_3A_6*(lengthz)* have? *Obviously, they cannot be arbitrary, and everyday experience should convince you that among all sets of 6 points* A_1, A_2, A_3, A_4, A_5, A_6 *in 3-space satisfying* (1), (2), *and* (3) *it is possible to achieve various lengths.*

Luckily, the literature contains an answer to this question. Specialized, it is THEOREM [L. Blumenthal, 1970]: Let D denote the matrix

$$\begin{pmatrix} 0 & 1 & 1 & 1 & 1 & 1 & 1 \\ 1 & 0 & a & b & x & b & a \\ 1 & a & 0 & a & b & y & b \\ 1 & b & a & 0 & a & b & z \\ 1 & x & b & a & 0 & a & b \\ 1 & b & y & b & a & 0 & a \\ 1 & a & b & z & b & a & 0 \end{pmatrix}$$

and let D_k denote the submatrix consisting of the first k rows and first k columns of D. Then the realizable lengths of A_1A_4, A_2A_5, A_3A_6 are precisely those for which the matrix D satisfies

$$\begin{array}{lll} \text{(i)} & (-1)^k \det D_k < 0 & \text{for } k = 3, 4, 5, \\ \text{(ii)} & \det D_k = 0 & \text{for } k = 6, 7. \end{array}$$

Now, answer the following questions:

(a) What sort of a matrix is D? *(The answer "nonnegative" is true, but trivial. Say something more profound.)*

(b) Show that under the assumptions (1), (2), and (3) the matrix D_4 has 1 positive eigenvalue and 3 negative eigenvalues.

(c) What does Blumenthal's Theorem say about the eigenvalues of D?

(d) Show that assumptions (1), (2), and (3) imply (i) for $k = 3, 4$.

(e) Assume x has been chosen so that (i) holds; suggest a way to find y and z so that (ii) also holds.

APPENDIX B

LINEAR EQUATIONS

In this chapter we discuss the basic theory behind solving systems of linear equations. In practice, linear equations are typically solved through the use of LU factorizations.

B.1 SOLUTION SETS

Because methods used for solving the linear programming problem depend on familiar methods for solving a system of linear equations, we shall review some elementary concepts.

Example B.1 (Solution of a System of Equations) Consider the solution of the following system of equations:

$$\begin{array}{rcl} x_1 \qquad + 2x_3 + 3x_4 &=& 11 \qquad (a) \\ + x_2 + 4x_3 + 6x_4 &=& 14, \qquad (b) \end{array} \tag{B.1}$$

which in matrix notation can be written as

$$\begin{pmatrix} 1 & 0 & 2 & 3 \\ 0 & 1 & 4 & 6 \end{pmatrix} \begin{pmatrix} x_1 \\ x_2 \\ x_3 \\ x_4 \end{pmatrix} = \begin{pmatrix} 11 \\ 14 \end{pmatrix}. \tag{B.2}$$

The ordered set of values $x_1 = 11, x_2 = 14, x_3 = 0, x_4 = 0$ is said to be a *solution* of the first equation B.1(a) because substitution of these values for x_1, x_2, x_3, x_4 into it produces the identity, $11 = 11$. The solution $(11, 14, 0, 0)$ is said to *satisfy* Equation B.1(a). Substituting the solution $(11, 14, 0, 0)$ in Equation B.1(b) produces the identity $14 = 14$. Thus, the vector $(11, 14, 0, 0)$ is said to be a *solution of the system* of Equations (B.1).

Definition (Solution of a System of Equations): In general, suppose we have a system of m equations in n variables,

$$Ax = b, \tag{B.3}$$

where A is an $m \times n$ matrix. A *solution* of the ith equation is a vector $x' = (x'_1, x'_2, \ldots, x'_n)^T$ such that

$$a_{i1}x'_1 + a_{i2}x'_2 + \cdots + a_{in}x'_n = b_i.$$

The vector x' is said to be a *solution of a system of equations* provided x' is a solution of each equation of the system.

Example B.2 (Solution Set) As we noted earlier, forming the product of a matrix and a vector amounts to taking a linear combination of the columns of the matrix. Since the first two columns of the coefficient matrix in (B.1) are linearly independent, they span the entire space $\Re^2$. Hence, every right-hand side b can be represented as a linear combination of the columns of A and hence a solution to (B.1) exists whatever the right-hand side might be. In other words, every right-hand side belongs to the *range space* (or column space) of A. However, note that there are fewer equations than unknowns, and thus there will be infinitely many solutions, as can be seen by writing x_1 and x_2 in terms of x_3 and x_4 as follows:

$$\begin{aligned} x_1 &= 11 - 2x_3 - 3x_4 \\ x_2 &= 14 - 4x_3 - 6x_4. \end{aligned}$$

The aggregate of solutions of a system is called its *solution set.* The full solution set for (B.1) can be represented by

$$\begin{pmatrix} x_1 \\ x_2 \\ x_3 \\ x_4 \end{pmatrix} = \begin{pmatrix} 11 \\ 14 \\ 0 \\ 0 \end{pmatrix} + \begin{pmatrix} -2 \\ -4 \\ 1 \\ 0 \end{pmatrix} x'_3 + \begin{pmatrix} -3 \\ -6 \\ 0 \\ 1 \end{pmatrix} x'_4$$

by choosing various values for the parameters x'_3 and x'_4. The vectors

$$z_1 = \begin{pmatrix} -2 \\ -4 \\ 1 \\ 0 \end{pmatrix} \quad \text{and} \quad z_2 = \begin{pmatrix} -3 \\ -6 \\ 0 \\ 1 \end{pmatrix}$$

are linearly independent and are orthogonal to the rows of A, as can be easily verified. That is, in other words, $Az_1 = 0$ and $Az_2 = 0$, and the vectors z_1 and z_2 are said to lie in the null space of the rows of A (or simply the null space of A). Note that the row space of A belongs to $\Re^4$. Recalling the definition of vector spaces given in Section A.6, the dimension of the row space of A in our example is 2 and, as expected, the dimension of the null space of A in this example is $(4 - 2)$, which is also 2.

Example B.3 (Incompatible, or Inconsistent, Systems) If the coefficient matrix A in the example above is changed so that we have

$$\begin{pmatrix} 1 & 0 & 2 & 3 \\ 2 & 0 & 4 & 6 \end{pmatrix} \begin{pmatrix} x_1 \\ x_2 \\ x_3 \\ x_4 \end{pmatrix} = \begin{pmatrix} b_1 \\ b_2 \end{pmatrix},$$

the number of independent columns is reduced to 1. Thus, there will no longer be a solution for every right-hand side. For example, the right-hand side $b^T = (1,0)^T$ cannot be represented as a linear combination of the columns of A and thus there is no solution x that satisfies $Ax = b$. In this example, the row space of A has rank or row dimension 1, and thus the dimension of its null space will be $(4-1) = 3$.

Definition (Inconsistent, Unsolvable, or Incompatible): If the right-hand side of a system of equations $Ax = b$ cannot be represented as a linear combination of the columns of A, the system of equations is said to be *inconsistent*, or *unsolvable*, or *incompatible*. A system of equations is said to be *consistent*, or *solvable*, or *compatible*, if the right-hand side b lies in the column space of the matrix A.

Definition (Empty Solution Set): If the system is inconsistent, the solution set is said to be *empty*.

B.2 SYSTEMS OF EQUATIONS WITH THE SAME SOLUTION SETS

We start by illustrating how to generate an equation with the same solution set as the original system of equations.

Example B.4 (Same Solution Set) Given a system of equations such as (B.1) it is easy to construct a new equation from it that has the property that every solution of (B.1) is also a solution of the new equation. The new equation shown in (B.4) below was formed by multiplying Equation B.1(a) by 2 and Equation B.1(b) by 3 and summing.

$$2x_1 + 3x_2 + 16x_3 + 24x_4 = 64. \tag{B.4}$$

Thus the solution $(11, 14, 0, 0)$ of the system of equations (B.1) is also a solution of the new equation.

A scheme for generating a new equation whose solution set includes (among others) all the solutions of a general linear system $Ax = b$, is shown in (B.5). For each equation i an arbitrary number k_i is chosen; the new equation is formed by multiplying the ith equation by k_i and summing. In matrix notation, with $k^T = (k_1, k_2, \ldots, k_m)$, this is

$$k^T Ax = k^T b, \text{ or} \tag{B.5}$$

$$d^T x = \alpha, \tag{B.6}$$

where

$$\begin{aligned} d_j &= k_1 a_{1j} + k_2 a_{2j} + \cdots + k_m a_{mj} \quad \text{for } j = 1, \ldots, n, \\ \alpha &= k_1 b_1 + k_2 b_2 + \cdots + k_m b_m. \end{aligned} \tag{B.7}$$

Definition (Linear Combination): An equation such as $d^T x = \alpha$ that is formed in the above manner is called a *linear combination* of the original equations; and the numbers k_i are called *multipliers*, or *weights*, of the linear combination.

▷ **Exercise B.1** Suppose a linear combination of the *columns* of A equals some other column. Show that the same linear combination applied to $\begin{pmatrix} A \\ d^T \end{pmatrix}$ results in $\begin{pmatrix} b \\ \alpha \end{pmatrix}$, where d^T and α are defined by (B.7).

Whenever a vector $x = (x_1, x_2, \ldots, x_n)^T$ constitutes a solution of (B.3), equation (B.6) becomes, upon substitution, a weighted sum of identities and hence an identity itself. Therefore, *every solution of a linear system is also a solution of any linear combination of the equations of the system.* Such an equation may therefore be inserted into (or adjoined to) a system of equations without affecting the solution set. This fact is used in devising algorithms to solve systems of equations.

Definition (Dependent System, Redundancy, Vacuous Equation): If in a system of equations, an equation is a linear combination of the others, it is said to be *dependent* upon them; the dependent equation is called *redundant.* A *vacuous equation* is an equation of the form

$$0x_1 + 0x_2 + \cdots + 0x_n = 0. \tag{B.8}$$

A system of equations consisting only of vacuous equations is also called redundant.

Definition (Independent System): A system containing no redundancy is called *independent.* In the theorems that follow about a system of equations we assume that the system does not consist of only a single vacuous equation.

A linear system is clearly *unsolvable*, or *inconsistent*, if it is possible to exhibit a linear combination of the equations of the system of the form

$$0x_1 + 0x_2 + \cdots + 0x_n = d \quad \text{with } d \neq 0 \tag{B.9}$$

because any solution of the system would have to satisfy (B.9); but this is impossible no matter what values are assigned to the variables. We shall refer to (B.9) as an *inconsistent equation.*

Example B.5 (Unsolvable System of Equations) The system

$$\begin{aligned} x_1 + x_2 + x_3 &= 3 \\ x_1 + x_2 + x_3 &= 8 \end{aligned} \tag{B.10}$$

is unsolvable because the first equation states that a sum of three numbers is 3, while the second states that this same sum is 8. If we apply multipliers $k_1 = 1$, $k_2 = -1$ to eliminate, say, x_1, we would obtain $0x = -5$, which is clearly a contradiction.

In general, the process of elimination applied to an inconsistent system will *always* lead in due course to an inconsistent equation, as we shall show in the next section.

▷ **Exercise B.2** Show that the only single-equation inconsistent linear system is of the form (B.9).

▷ **Exercise B.3** Show that if a system of 2 or more equations contains a vacuous equation, it is dependent.

B.3 HOW SYSTEMS ARE SOLVED

The usual "elimination" procedure for finding a solution of a system of equations is to *augment* the system by generating new equations by taking linear combinations in such a way that certain coefficients are zero. This is usually followed by the deletion of certain redundant equations.

Example B.6 (Augmentation of a System of Equations) For example, consider,

$$\begin{aligned} x_1 + x_2 + x_3 &= 4 \\ x_1 - x_2 - 2x_3 &= 2. \end{aligned} \tag{B.11}$$

Multiplying the first equation in (B.11) by $k_1 = +1$ and the second equation by $k_2 = -1$ and summing, the coefficient of x_1 vanishes and we obtain

$$2x_2 + 3x_3 = 2.$$

The system of equations (B.11) *augmented* by the above equation clearly has the same solution set as the original system of equations (B.11).

It is interesting to note that the technique of taking linear combinations and augmenting the original system can be used to easily detect whether any equation of the original system is linearly dependent on the others. A second advantage is that it is easy to state the set of all possible solutions, as we have already seen. It turns out that these same basic ideas are used for solving systems of equations on a computer through the use of LU factorizations.

In general, the method of solving a system of equations is one of augmentation by linear combinations until in the enlarged system there is a subsystem whose solution set is readily seen. Moreover, all other equations of the system are linearly dependent on the subsystem. An example of a system whose solution set is readily seen is (B.1) on page 341; it belongs to a class called *canonical.*

> *Definition (Canonical Form)*: A system of m equations in n variables x_j is said to be in *canonical form* with respect to an ordered set of variables $(x_{j_1}, x_{j_2}, \ldots, x_{j_m})$ if and only if x_{j_i} has a unit coefficient in equation i and a zero coefficient in all other equations.

System (B.12) below, with $x_B = (x_1, x_2, \ldots, x_m)^T$ and $x_N = (x_{m+1}, \ldots, x_n)^T$ is canonical because for each i, the variable x_i has a unit coefficient in the ith equation and zero elsewhere:

$$Ix_B + \bar{A}x_N = b. \tag{B.12}$$

▷ **Exercise B.4** Show how by arbitrarily choosing values for $x_{m+1}, \ldots, x_n$, the class of all solutions for (B.12) can be generated. How can (B.12) be used to check easily whether or not another equation $\alpha_1 x_1 + \alpha_2 x_2 + \cdots + \alpha_m x_m = \alpha_0$ is dependent upon it?

If by augmentation we can construct a subsystem that is in canonical form, then it is easy to generate all possible solutions to the original system. When the original system is unsolvable the canonical subsystem does not exist; but in this case the process of generating the canonical form results in an infeasible equation, which tells us that the original system is unsolvable.

Deletion of an equation that is a linear combination of the others is another operation that does not affect the solution set. If after an augmentation, one of the original equations in the system is found to be a linear combination of the others, it may be deleted. In effect the new equation becomes a "substitute" for one of the original equations. Although the limited capacity of electronic computers to store information keeps growing year by year still there is likely always to be a practical limit, and so this ability to throw away equations will remain important in the future.

> *Definition (Equivalent Systems)*: Two systems are called *equivalent* if one system may be derived from the other by inserting and deleting a redundant equation or if one system may be derived from the other through a chain of systems each linked to its predecessor by such insertions and deletions.

THEOREM B.1 (Same Solution Sets) *Equivalent systems have the same solution set.*

▷ **Exercise B.5** Prove Theorem B.1.

B.4 ELEMENTARY OPERATIONS

There are two simple but important types of linear combinations, called *elementary operations*, that may be used to obtain equivalent systems.

Type I. Replacing any equation E_t by the equation kE_t with $k \neq 0$, and leaving all remaining equations unchanged.

Type II. Replacing any equation E_t by the equation $E_t + kE_i$, where E_i is any other equation of the system, and leaving all remaining equations unchanged.

To prove that an elementary operation of Type I results in an equivalent system, insert kE_t as a new equation after E_t, then delete E_t. Note that E_t is a redundant equation, for it can be formed from kE_t by $(1/k)kE_t$ if $k \neq 0$. Similarly, to prove that an elementary operation of Type II results in an equivalent system, insert $E_t + kE_i$ after E_t and then delete E_t. Note that E_t is a redundant equation, for it is given by $(E_t + kE_i) - kE_i$.

Therefore one way to transform a system of equations into an equivalent system is by a sequence of elementary operations. For example, system (B.14) is equivalent to system (B.13) because it can be obtained from (B.13) by replacing (B.13b) by subtracting from it 2 times (B.14a). Furthermore, system (B.15) is equivalent to (B.14) because (B.15b″) can be obtained from (B.14b′) by dividing it by 3.

$$\begin{aligned} x_1 - x_2 + 2x_3 &= 4 && \text{(a)} \\ 2x_1 + x_2 - 2x_3 &= 2 && \text{(b)} \end{aligned} \tag{B.13}$$

$$\begin{aligned} x_1 - x_2 + 2x_3 &= 4 && \text{(a}'\text{)} = \text{(a)} \\ 3x_2 - 6x_3 &= -6 && \text{(b}'\text{)} = \text{(b)} - 2\text{(a)} \end{aligned} \tag{B.14}$$

$$\begin{aligned} x_1 - x_2 + 2x_3 &= 4 && \text{(a}''\text{)} = \text{(a}'\text{)} \\ x_2 - 2x_3 &= -2 && \text{(b}''\text{)} = \text{(b}'\text{)}/3 \end{aligned} \tag{B.15}$$

THEOREM B.2 (Inverse of a Sequence of Elementary Operations) *Corresponding to a sequence of elementary operations is an inverse sequence of elementary operations of the same type by which a given system can be obtained from the derived system.*

▷ **Exercise B.6** Prove Theorem B.2 by proving the following:

1. Show that the inverse operation of Type I is itself of Type I and is formed by replacing equation E_t by $(1/k)E_t$ with $k \neq 0$.
2. Show that the inverse operation of Type II is itself of Type II and is formed by replacing equation E_t by $E_t - kE_i$.

▷ **Exercise B.7** Perform the inverse operations on (B.15) and show that the result is the original system (B.13).

We can also see that if a system can be derived from a given system by a sequence of elementary operations, this implies that it is possible to obtain each row of the derived system in detached coefficient form directly by a linear combination of the rows of the given system. Conversely, each row of the given system is some linear combination of the rows of the derived system.

THEOREM B.3 (Obtaining Rows of an Equivalent System) *The rows of two equivalent systems in detached coefficient form can be obtained one from the other by linear combinations.*

THEOREM B.4 (Obtaining Equivalent Systems) *If the rth equation of a given system is replaced by a linear combination of the equations i with multipliers k_i where $k_i \neq 0$, an equivalent system is obtained.*

▷ **Exercise B.8** Prove Theorems B.3 and B.4. What is the inverse operation of a linear combination performed in Theorem B.4.

▷ **Exercise B.9** Show that if a second system of equations can be obtained from the original system of equations by rearranging the order of equations, the two systems are equivalent, i.e., one can be obtained from the other by elementary operations.

The most important property of a system derived by elementary operations is that it has the *same solution set* as the system it was derived from.

An interesting converse question now arises. Are all systems of linear equations with the same solution set obtained by a sequence of inserting and deleting of redundant equations? As we will show in Section B.5, the answer is *yes* if the solution set is *nonempty* but may be *no* if the solution set is *empty*, as can be easily seen by comparing the two systems

$$\{0x = 1\} \quad \text{and} \quad \left\{ \begin{array}{c} 0x = 1 \\ 1x = 1 \end{array} \right\};$$

both have *empty*, hence identical, solution sets. It is obvious that if these two systems were equivalent, some multiple (linear combination) of the equation $0x = 1$ of the first system would yield the equation $1x = 1$ of the second system, but this is clearly impossible.

THEOREM B.5 (Reversal of a Sequence of Elementary Operations) *Let S be a system of equations $E_1, E_2, \ldots, E_m$ and let E_0 be formed from S by a linear combination:*

$$\lambda_1 E_1 + \lambda_2 E_2 + \cdots + \lambda_k E_k = E_0,$$

where λ_i are not all zero. Let $\bar{S}$ be a system of equations $\bar{E}_1, \bar{E}_2, \ldots, \bar{E}_m$ obtained from S by a sequence of elementary operations. Then there exists $\bar{\lambda}_i$ such that

$$\bar{\lambda}_1 \bar{E}_1 + \bar{\lambda}_2 \bar{E}_2 + \cdots + \bar{\lambda}_k \bar{E}_k = E_0,$$

where $\bar{\lambda}_i$ are not all zero.

Proof. If we can show that the theorem is true for a single elementary operation of Type I and a single elementary operation of Type II, then by induction, it will be true for a sequence of such operations.

Suppose that E_1 is replaced by $E_1' = kE_1$ where $k_1 \neq 0$ (an elementary operation of Type I). Then it is easy to verify that

$$\frac{\lambda_1}{k}E_1' + \lambda_2 E_2 + \cdots + \lambda_k E_k = E_0,$$

where $\lambda_1/k, \lambda_2, \ldots, \lambda_k$ are not all zero.

Next suppose that E_1 is replaced by $E_1' = E_1 + kE_2$ (an elementary operation of Type II), then $E_1 = E_1' - kE_2$ and it is easy to verify that

$$\lambda_1 E_1' + (\lambda_2 - \lambda_1 k)E_2 + \cdots + \lambda_k E_k = E_0.$$

By induction the theorem holds for any number of elementary operations. ∎

B.5 CANONICAL FORMS, PIVOTING, AND SOLUTIONS

Suppose that we have a system of m equations in n variables with $m \leq n$,

$$Ax = b, \tag{B.16}$$

where A is an $m \times n$ matrix. We are interested in ways of replacing this system, if possible, by an equivalent *canonical* system (see Section B.3 for a definition of canonical):

$$Ix_B + \bar{A}x_N = \bar{b}, \tag{B.17}$$

where $\bar{A}$ is $m \times (n-m)$. For square systems, it is clear that the canonical system is simply of the form $Ix = \bar{b}$. In this form the solution set is evident and it is easy to check whether or not any other system is equivalent to it.

REDUCTION TO A CANONICAL FORM

The standard procedure for reducing (if possible) a general system (B.16) to an equivalent canonical form (B.17) will now be discussed. The principles are best illustrated with an example and then generalized.

Example B.7 (Reduction to Canonical Form) Consider the 2×4 system

$$\begin{array}{rcrcrcrcl} & + & \mathbf{2x_2} & + & 2x_3 & + & 2x_4 & = & 10 \\ x_1 & - & 2x_2 & - & x_3 & + & x_4 & = & 2. \end{array} \tag{B.18}$$

Choose as "pivot element" any term with nonzero coefficient such as the **boldfaced** term in the first equation. Next perform a Type I elementary operation by dividing its equation (the first one) by the coefficient of the **boldfaced** term, and then eliminating its corresponding variable (x_2) from the other equation(s) by means of elementary operation(s) of Type II:

$$\begin{aligned} x_2 + x_3 + x_4 &= 5 \\ + \mathbf{1}x_1 \qquad + x_3 + 3x_4 &= 12. \end{aligned} \tag{B.19}$$

Next choose as pivot any term in the remaining equations (such as the boldfaced term in the second equation above). Eliminate the corresponding variable (in this case x_1) from all the other equations. (Because x_1 happens to have a zero coefficient in the first equation, no further work is required to perform the eliminations.) From this canonical system with the ordered set of basic variables x_2, x_1, it is evident, by setting $x_3 = x_4 = 0$, that one solution is $x_1 = 12$, $x_2 = 5$, $x_3 = x_4 = 0$.

▷ **Exercise B.10** Show that system (B.19) is in canonical form with respect to basic variables (x_2, x_1). Show that it is not possible to get system (B.18) into canonical form with respect to variables (x_1, x_2) by first choosing a pivot element in the first equation and then the second, etc., but that it is clearly possible by rearrangement of the order of the equations. What are the elementary operations that would rearrange the order of the equations?

PIVOTING

Pivoting forms the basis for the operations to reduce a system of equations to canonical form and to maintain it in such form.

> *Definition (Pivot Operation)*: A *pivot operation* consists of m elementary operations that replace a system by an equivalent system in which a specified variable has a coefficient of unity in one equation and zero elsewhere.

The detailed steps for pivoting on a term $a_{rs}x_s$, called the *pivot term*, where $a_{rs} \neq 0$ are as follows:

1. Perform a Type I elementary operation by replacing the rth equation by the rth equation multiplied by $(1/a_{rs})$.
2. For each $i = 1, \ldots, m$ except $i = r$, perform a Type II elementary operation by replacing the ith equation by the sum of the ith equation and the replaced rth equation multiplied by $(-a_{is})$.

Pivoting can be easily explained in matrix notation as follows. Let (B.20a) and (B.20b) denote the systems before and after pivoting.

$$\begin{aligned} (a)\ Ax &= b \\ (b)\ \tilde{A}x &= \tilde{b}. \end{aligned} \tag{B.20}$$

The pivoting operations described above are the same as multiplying (B.20a) by the elementary matrix M:

$$\begin{aligned} (a)\ M \quad &= I + \frac{1}{a_{rs}}(e_r - A_{\bullet s})e_r^T, \\ (b)\ M^{-1} &= I + (A_{\bullet s} - e_r)e_r^T, \end{aligned} \tag{B.21}$$

where e_r is the unit vector with 1 in row r and zero elsewhere.

▷ **Exercise B.11** Show that (B.21b) is the inverse of M. Prove that multiplying (B.20b) by (B.21b) restores the original problem (B.20a). Show by way of an example that multiplying (B.20b) by (B.21b) is not necessarily the same as performing a pivot on (B.20b).

▷ **Exercise B.12** Show that $\bar{A}_{\bullet s}$ is the same as e_r. How may this be applied to show the second part of Exercise B.11.

In general the reduction to a canonical form can be accomplished by a sequence of pivot operations. For the first pivot term select any term $a_{rs}x_s$ such that $a_{rs} \neq 0$. After the first pivoting, the second pivot term is selected using a nonzero term from any equation r' of the modified system except r. After pivoting, the third pivot term is selected in the resulting modified m-equation system from any equation r'' except r and r'. In general, repeat the pivoting operation, always choosing the pivot term from modified equations other than one corresponding to those previously selected. Continue in this manner, terminating either when m pivots have been performed or when, after selecting $k < m$ pivot terms, it is not possible to find a nonzero pivot term in any equation except those corresponding to equations previously selected because all coefficients in the remaining equations are zero.

Let the successive pivoting be, for example, on variables $x_1, x_2, \ldots, x_r$ in the corresponding equations $i = 1, \ldots, r$, where $r \leq m$; then the original system (B.16) has been reduced to an equivalent system of form (B.22) below, which we will refer to as the *reduced system* with respect to the basic variables $x_1, x_2, \ldots, x_r$. We shall also refer to a system as reduced with respect to r basic variables if by changing the order of the variables and equations, it can be put into form (B.22), where $x_B = (x_1, x_2, \ldots, x_r)^T$ and $x_N = (x_{r+1}, \ldots, x_n)^T$:

$$\begin{aligned} Ix_B + \bar{A}x_N &= \bar{b}_I \\ 0x_B + 0x_N &= \bar{b}_{II}, \end{aligned} \tag{B.22}$$

where $\bar{A}$ is $r \times (n-r)$, $\bar{b}_I$ is of length r, and $\bar{b}_{II}$ is of length $m - r$.

Since (B.22) was obtained from (B.16) by a sequence of pivoting operations each of which consists of m elementary operations, it follows that the reduced system is (a) formed from *linear combinations* of the original system, and (b) *equivalent* to the original system.

The original system (B.16) is solvable if and only if its reduced system (B.22) is solvable, and (B.22) is solvable if and only if $\bar{b}_{II} = 0$, i.e.,

$$\bar{b}_{r+1} = \bar{b}_{r+2} = \cdots = \bar{b}_m = 0. \tag{B.23}$$

If (B.23) holds, the solution set is immediately evident because any values of the (independent) variables $x_{r+1}, \ldots, x_n$ determine corresponding values for the (dependent) variables $x_1, \ldots, x_r$. On the other hand, if $\bar{b}_{r+i} \neq 0$ for some i, the solution set is *empty* because the $(r+i)$th equation is clearly inconsistent; it states that $0 = \bar{b}_{r+i}$. In this case, the original system (B.16) and the reduced system (B.23) are both inconsistent (unsolvable).

CANONICAL SYSTEM

If the original system is consistent, the system formed by dropping the vacuous equations from the reduced system is called its *canonical equivalent* with the pivotal variables as basic:

$$Ix_B + \bar{A}x_N = \bar{b}_I, \tag{B.24}$$

where $\bar{A}$ is $r \times (n-r)$ and $\bar{b}_I$ is of length r.

UNIQUENESS OF A CANONICAL EQUIVALENT

The fundamental property of a canonical system resulting from the reduction process of a consistent system of equations is that for any other system with the same solution set, a reduction can be effected using the same pivotal variables, and the resulting canonical system will be *identical* if the equations are reordered (permuted) so that the unit coefficient of the corresponding basic variables is on the same row in both systems (in detached coefficient form).

Example B.8 (Canonical Equivalent) The following two canonical systems are clearly equivalent if in the second system, the second row is exchanged with the first row.

$$\begin{array}{rcrcl} x_1 & & + 2x_3 & = & 5 \\ & x_2 & - 3x_3 & = & 2 \end{array} \quad \text{and} \quad \begin{array}{rcrcl} & x_2 & - 3x_3 & = & 2 \\ x_1 & & + 2x_3 & = & 5. \end{array}$$

THEOREM B.6 (Uniqueness of a Canonical Equivalent) *There is at most one equivalent canonical system with respect to a fixed ordered set of basic variables.*

Proof. Let there be two equivalent canonical systems relative to the ordered set of basic variables $x_1, x_2, \ldots, x_r$. We wish to show that the two canonical systems are identical. Substituting $x_{r+1} = x_{r+2} = \cdots = x_n = 0$ into the first system, we get $x_1 = \bar{b}_1, x_2 = \bar{b}_2, \cdots, x_r = \bar{b}_r$. Because of equivalence, substitution into the second system should yield the same values; this will only be true if their respective constant terms $\bar{b}_i$ are the same. Similarly, substituting the values for independent variables of $x_{r+1} = x_{r+2} = \cdots = x_n = 0$ except $x_s = 1$ will show (after equating constant terms) that their corresponding coefficients $\bar{a}_{is}$ of x_s are the same. It follows that the corresponding coefficients are the same for each $s = r+1, r+2, \ldots, n$. ∎

The above theorem can also be established by applying

LEMMA B.7 (Generating a System from its Canonical Equivalent) *If a system of equations S is equivalent to a canonical system, it can be generated from the canonical system by a unique linear combination of the equations of the canonical system, the weights being the coefficients of the basic variables of the equation of the system S being generated.*

▷ **Exercise B.13** Prove Lemma B.7. Use Lemma B.7 to prove Theorem B.6.

▷ **Exercise B.14** Apply Lemma B.7 to test whether a system is equivalent to a canonical system.

BASIC SOLUTIONS AND DEGENERATE SOLUTIONS

Basic solutions play a critical role in the solution of linear programs.

> *Definition (Basic Solution)*: The special solution obtained by setting the independent variables equal to zero and solving for the dependent variables is called a *basic solution.*

Thus if (B.17) is the canonical system of (B.16) with basic variables $x_1, x_2, \ldots, x_m$, the corresponding basic solution is $x_B = \bar{b}$ and $x_N = 0$, i.e.,

$$x_1 = \bar{b}_1, x_2 = \bar{b}_2, \ldots, x_m = \bar{b}_m; \quad x_{m+1} = x_{m+2} = \cdots = x_n = 0. \tag{B.25}$$

> *Definition (Degenerate Solution)*: A basic solution is *degenerate* if the value of one or more of the dependent (basic) variables is zero. In particular, the basic solution (B.25) is degenerate if $\bar{b}_i = 0$ for at least one i.

BASIS

A set of columns (of a system of equations in detached coefficient form) is said to be a *basis* if they are linearly independent and each of the other columns can be generated from them by a linear combination.

> *Definition (Basic Columns (Activities))*: In accordance with the special usage in linear programming, the term *basis* refers to the *ordered set* of columns of the original *independent* system (in detached coefficient form) corresponding to the ordered set of basic variables of the canonical equivalent. The columns of the basis are called *basic columns* (or *basic activities*).

Example B.9 (Basis) For example, the canonical equivalent of system (B.18) is (B.19) with respect to variables (x_2, x_1). Columns corresponding to this ordered set of variables in (B.18) are

$$\begin{pmatrix} 2 & 0 \\ -2 & 1 \end{pmatrix}.$$

It is easy to verify that these columns are linearly independent and any other column can be generated from them by a linear combination and hence they form a basis. For instance, $A_{\bullet 4} = 1.A_{\bullet 2} + 3.A_{\bullet 1}$ in the original system of equations (B.18) and the same linear combination also works in the canonical system (B.19)—see Exercise B.9. In the canonical system it is easy to find what linear combination of basic columns will generate column j; namely, the weights are the same as the coefficients in column j. Thus the weights $(1, 3)$ are the same as $\bar{A}_{\bullet 4}$ in the canonical system (B.19).

B.6 PIVOT THEORY

The purpose of this section is to extend the discussion regarding properties of pivot operations and other characteristics of pivot operations. The *first* important property of the pivot operation is its *irreversibility* except in certain situations.

THEOREM B.8 (Canonical Form After a Pivot Operation) *If a system is in canonical form before a pivot operation, then it is in canonical form after the pivot operation.*

Proof. Suppose that the system is in canonical form with respect to the variables $x_1, x_2, \ldots, x_m$, i.e.,

$$Ix_B + \bar{A}x_N = \bar{b}.$$

Pivoting on $\bar{a}_{rs}x_s$, where $s > m$, is the same as multiplying on the left by M given by Equation (B.21), that is,

$$Mx_B + M\bar{A}x_N = M\bar{b}. \tag{B.26}$$

Column $s - m$ of $M\bar{A}$ is e_r, as we have seen earlier. This together with columns $1, \ldots, r-1, r+1, \ldots, m$ of M shows that the system is in canonical form with respect to $1, \ldots, r-1, s, r+1, \ldots, m$. ∎

COROLLARY B.9 (Inverse of a Pivot Operation) *If a system is in canonical form, the inverse of a pivot operation is a pivot operation.*

COROLLARY B.10 (Pivoting on a Canonical Subsystem of a General System) *Let S be a subsystem of equations $S_1, S_2, \ldots, S_k$ that is in canonical form and let E be the remaining equations $E_1, E_2, \ldots, E_l$ such that there are coefficients of zero for basic variables in the noncanonical equations. That is,*

$$\begin{array}{lcl} Ix_B + \bar{A}x_N = \bar{b}_I & & : S \\ 0x_B + Tx_N = \bar{b}_{II} & & : E \end{array}$$

If the pivot term is selected in an equation of S, then the corresponding subsystem after a pivot operation is in canonical form; the inverse of the pivot operation is also a pivot operation.

COROLLARY B.11 (Inverse of a Pivot Operation on a Subsystem) *Consider the systems S and E defined in Corollary B.10. If the pivot term is selected in an equation E not in S, then the subsystem corresponding to $\{S, E\}$ after a pivot operation is in canonical form; the inverse of the pivot operation is not a pivot operation on $\{S, E\}$ unless $\{S, E\}$ was in canonical form initially.*

▷ **Exercise B.15** Prove Corollaries B.9, B.10, and B.11.

The *second* important property of the pivot operation is that there is a one-to-one correspondence between the original system of equations and the system of equations derived by pivoting. Furthermore, easily defined subsets of the original and the derived systems are equivalent.

> *Definition (Pivotal Subsystem): The pivotal subsystem P* is that set of equations of the *original system* corresponding to those selected for pivot terms in a sequence of pivot operations.

It is clear that the number of equations in the pivotal subsystem increases or remains the same during a sequence of pivot operations, depending on whether or not the successive pivot terms are selected from among equations corresponding to the pivotal system or from among the remainder.

THEOREM B.12 (Equivalence and Canonical Form) *Let S' be the system obtained after a sequence of pivot operations on S using rows corresponding to P of S, and let P' be the set of equations of S' corresponding to the pivotal subsystem P of S. The system S' is equivalent to S; in particular, P' is equivalent to P and moreover, P' is in canonical form.*

THEOREM B.13 (Pivotal Subsystem Properties) *The pivotal subsystem P is independent and solvable.*

▷ **Exercise B.16** Prove Theorems B.12 and B.13.

THEOREM B.14 (Redundancy and Inconsistency Tracing Theorem) *If an equation E'_t of a reduced system is vacuous (or inconsistent), then in the original system, E_t is either redundant with respect to the pivotal system P or a linear combination of the equations of P and E_t can form an inconsistent equation.*

Proof. Let S be the equations of the original system, P be the pivotal subset, and R be the remainder. Let S', P', and R' be the corresponding equations after pivoting. If equation E_t in R is not originally vacuous but becomes so after pivoting, then

$$\lambda_1 E_1 + \lambda_2 E_2 + \cdots + \lambda_k E_k + E_t = E'_t, \tag{B.27}$$

where $(E_1, \ldots, E_k)$ are the pivotal equations (not necessarily the first k) and E'_t is alternatively either a vacuous or an inconsistent equation. Hence, if E'_t is vacuous, E_t is dependent on the others. ∎

The *third* important property of the pivoting operation is that it provides a way to show whether or not two systems have *the same solution set* by trying to reduce them simultaneously by pivoting step by step, each step using the *same* pivotal variables in both systems. The same process will test the *equivalence* of two systems.

Definition (Homogeneous System): It is convenient at this point to recall the definition of a *homogeneous system.* A system $Ax = b$ is homogeneous if the right-hand side vector is $b = 0$.

Definition (Homogeneous Solution): A solution $x = x^h$ is called a *homogeneous solution* associated with $Ax = b$, $b = 0$ or $b \neq 0$, if $Ax^h = 0$. A fundamental property of homogeneous solutions is that any scalar multiple of a homogeneous solution is a homogeneous solution. A homogeneous solution is called nontrivial if $x^h \neq 0$.

THEOREM B.15 (Equivalence of Homogeneous Systems) *Two homogeneous systems $Ax = 0$ and $\hat{A}x = 0$ are equivalent if and only if it is possible to pivot on both systems with respect to the same set of variables and after pivoting, the two systems are identical or can be made so after row permutations.*

Proof. Let us suppose first that it is possible to reduce two homogeneous systems using the same set of pivot variables. We assume that the equations of the canonical parts are reordered so that both subsystems are canonical with the same set of basic variables. If the two systems are to be *equivalent*, it is necessary that their canonical parts be identical (see proof of Theorem B.6).

Next let us suppose that the two homogeneous systems are equivalent. By Theorem B.6 there is at most one equivalent canonical system with respect to a fixed ordered set of variables for each system. Thus, by the definition of equivalence, if it is possible to pivot on one system with respect to a fixed set of variables, it is also possible to pivot on the second system with respect to the same fixed set of variables. ∎

COROLLARY B.16 (Equivalence of Linear Systems) *Two systems $Ax = b$ and $\hat{A}x = \hat{b}$ are equivalent if and only if $Ax - bx_0 = 0$ and $\hat{A}x - \hat{b}x_0 = 0$ are equivalent.*

Proof. It is clear that if a linear combination of $Ax = b$ results in $\hat{A}x = \hat{b}$ then the same linear combination when applied to $Ax - bx_0 = 0$ also results in $\hat{A}x - \hat{b}x_0 = 0$. The converse also is clearly true. ∎

THEOREM B.17 (Equivalent Systems Have Same Solution Sets) *Two solvable systems have the same solution sets if and only if they are equivalent.*

▷ **Exercise B.17** Prove Theorem B.17.

The *fourth* important property of pivoting is that it provides a way to prove a number of interesting theorems concerning the number of independent and dependent equations of a system.

THEOREM B.18 (Number of Equations in Equivalent Systems) *Two equivalent, independent, and consistent systems have the same number of equations.*

Proof. Since the two systems are equivalent, independent, and consistent, it is possible simultaneously to reduce by pivoting the two systems to canonical forms that are identical. No vacuous equations can happen during the process because pivoting is actually a sequence of elementary operations, so that by Theorem B.5, the appearance of such equations would imply the presence of redundant equations in the original systems, contrary to the assumption of independence. Therefore, the identical canonical equivalents have the same number of equations as their respective original systems. ∎

The following three theorems are consequences of the above.

THEOREM B.19 (Equations in Equivalent Canonical Systems) *Two equivalent canonical systems have the same number of equations.*

THEOREM B.20 (Number of Equations in the Canonical Equivalent) *If a system can be partitioned into r independent equations and a set of equations dependent on it, then the canonical equivalent of the original system has r equations.*

THEOREM B.21 (Independent Set Can Generate the Others) *If a system has a canonical equivalent with r equations, then any r independent equations of the system can generate the remainder by linear combinations.*

▷ **Exercise B.18** Prove Theorems B.19, B.20, and B.21. Show that r in Theorem B.20 is the rank of the system.

B.7 NOTES & SELECTED BIBLIOGRAPHY

For additional discussions on canonical forms and pivoting see Dantzig [1963]. For a discussion on LU factorizations and the numerical solution of systems of linear equations see *Linear Programming 2*.

B.8 PROBLEMS

B.1 Find a basis for the solution sets of:

(a) $$\begin{aligned} x_1 + 2x_2 + 3x_3 &= 0 \\ 4x_1 + 5x_2 + 6x_3 &= 0 \\ 7x_1 + 8x_2 + 9x_3 &= 0. \end{aligned}$$

(b) $$\begin{aligned} x_1 \qquad\quad - x_3 &= 0 \\ 2x_1 + 10x_2 - 2x_3 &= 0 \\ x_1 + x_2 + x_3 &= 0. \end{aligned}$$

(c) $$\begin{aligned} x_1 \qquad + x_3 &= 0 \\ x_1 + 5x_2 \qquad &= 0. \end{aligned}$$

(d) $$\begin{aligned} x_1 + x_2 - 4x_3 + 2x_4 &= 0 \\ 3x_1 \qquad + x_3 - 3x_4 &= 0 \\ 6x_1 + 3x_2 - 11x_3 + 3x_4 &= 0 \\ 2x_1 - x_2 + 5x_3 - 5x_4 &= 0. \end{aligned}$$

B.2 Determine whether or not there exist solutions to the following systems of equations

(a) $$\begin{aligned} x_1 + 2x_2 + 3x_3 &= 0 \\ 4x_1 + 5x_2 + 6x_3 &= 0 \\ 7x_1 + 8x_2 + 9x_3 &= 1. \end{aligned}$$

(b) $$\begin{aligned} x_1 + x_2 - 4x_3 &= 2 \\ 3x_1 \qquad + x_3 &= -3 \\ 6x_1 + 3x_2 - 11x_3 &= 3 \\ 2x_1 - x_2 + 5x_3 &= -5. \end{aligned}$$

(c) $$\begin{aligned} 3x_1 \qquad + 2x_3 &= 2 \\ -2x_1 + x_2 \qquad &= 0 \\ 6x_2 - 3x_3 &= 4. \end{aligned}$$

(d) $$\begin{aligned} x_1 + 2x_2 + 3x_3 + 4x_4 &= 5 \\ 6x_1 + 7x_2 + 8x_3 + 9x_4 &= 10 \\ 3x_1 + 3x_2 + 3x_3 + 3x_4 &= 3 \\ 2x_1 + x_2 \qquad - x_4 &= -2. \end{aligned}$$

B.3 Give the values of λ for which the following system of linear equations

$$\begin{aligned} 2x_1 + 2x_2 + 7x_3 &= 7 \\ 3x_1 + (\lambda+5)x_2 + (\lambda+11)x_3 &= 10 \\ 3x_1 + (\lambda+3)x_2 + 11x_3 &= 10 \end{aligned}$$

has (a) no solutions, (b) exactly one solution, and (c) infinitely many solutions.

B.4 Reduce the following system $Ax = b$ to canonical form

$$\begin{aligned} x_1 + x_2 + 2x_3 + x_4 + x_5 &= 1 \\ 7x_1 + 2x_2 + 9x_3 + 9x_4 \qquad &= 8 \\ 9x_1 + 4x_2 + 13x_3 + 11x_4 + 2x_5 &= 10. \end{aligned}$$

(a) What is the rank of A? What is the rank of $[A\ b]$?

(b) Change the right-hand side of the third equation from a 10 to an 11. Show that the system is infeasible by finding the infeasibility multipliers for this system of equations. Now what is the rank of A? What is the rank of $[A\ b]$?

B.5 Show by a sequence of pivots and row permutations that the following canonical systems are equivalent:

$$\begin{aligned} x_1 \qquad\qquad\qquad + x_6 &= 1 \\ x_2 \quad + x_4 \quad + x_6 &= 2 \\ x_3 + x_4 + x_5 + x_6 &= 3 \end{aligned}$$

and

$$\begin{aligned} -x_1 + x_2 \qquad + x_4 \qquad &= 1 \\ -x_2 + x_3 \qquad + x_5 &= 1 \\ x_1 \qquad\qquad\qquad + x_6 &= 1. \end{aligned}$$

Use the result just obtained to determine the inverse of the following matrix:

$$\begin{pmatrix} 0 & 0 & 1 \\ 1 & 0 & 1 \\ 1 & 1 & 1 \end{pmatrix},$$

which consists of some of the coefficients of the system on the left. Explain briefly where you got your answer. Change one of the coefficients in the second system and show that the two systems are not equivalent in general.

B.6 Given the system

$$Ix + Bu = 0, \tag{B.28}$$

where I is an $m \times m$ matrix and B is square and nonsingular, prove that in m pivots one can convert (B.28) to an equivalent canonical form

$$B^{-1}x + Iu = 0. \tag{B.29}$$

Discuss how this can be viewed as an algorithm for finding the inverse of B.

B.7 Let A be an $m \times n$ matrix and suppose that we pivot on element (r, s) of A in the canonical form $[I_m \mid A]$ and in the resulting tableau exchange columns r and $m + s$ to obtain $[I_m \mid \bar{A}]$. Show that if we pivot on element (s, r) of $-A^T$ in the canonical form $[I_n \mid -A^T]$ and in the resulting tableau exchange columns s and $m + r$, then we obtain $[I_n \mid -\bar{A}^T]$.

REFERENCES

A

Abadie, J. (1965). Verbal Communication to Dantzig.

Abrahamson, P.G. (1983). "A Nested Decomposition Approach for Solving Staircase Linear Programs," Technical Report SOL 83-4, Department of Operations Research, Stanford University, Stanford, CA.

Ackoff, R.L. and Rivett, P. (1963). *A Manager's Guide to Operations Research*, Wiley, New York.

Adler, I. (1976). "Equivalent Linear Programs," Department of IE & OR, University of California, Berkeley, California.

Adler, I. and Ülkücü, A. (1973). "On the Number of Iterations in the Dantzig-Wolfe Decomposition," in D. M. Himmelblau (ed.), *Decomposition of Large Scale Problems*, North-Holland, Amsterdam, the Netherlands, 181–187.

Adler, I., Karmarkar, N., Resende, M.G.C. and Veiga, G. (1990). "Data Strucutres and Programming Techniques for the Implementation of Karmarkar's Algorithm," *ORSA Journal on Computing* **1**, 84–106.

Adler, I., Resende, M.G.C., Veiga, G., and Karmarkar, N. (1989). "An Implementation of Karmarkar's Algorithm for Linear Programming," *Mathematical Programming* **44**, 297–336.

Ahuja, R.K., Magnanti, T.L., and Orlin, J.B. (1991). "Network Flows," in G.L. Nemhauser, A.H.G. Rinnooy Kan, and M.J. Todd (eds.), *Handbooks in Operations Research and Management Science, Vol 1: Optimization*, North-Holland, Amsterdam, the Netherlands, 211–369.

Ahuja, R.K., Magnanti, T.L., and Orlin, J.B. (1993). *Network Flows: Theory, Algorithms, and Applications*, Prentice-Hall, Inc., Englewood Cliffs, New Jersey.

Ahuja, R.K., Mehlhorn, K., Orlin, J.B., and Tarjan, R.E. (1990). "Faster Algorithms for the Shortest Path Problem," *Journal of the Association for Computing Machinery*, **37**, 211–369.

Ali, A.I., Helgason, R.V., Kennington, J.L., and Lall, H.S. (1978). "Primal Simplex Network Codes: State of the Art Implementation Technology," *Networks*, **8**, 315–339.

Andersen, E.D. and Ye, Y. (1995). "Combining Interior-Point and Pivoting Algorithms for Linear Programming," Working Paper, Department Sciences, The University of Iowa, Iowa City, IA 52242; to appear in Management Science.

Anstreicher, K.M. (1989). "A Worst-Case Step in Karmarkar's Algorithm," *Mathematics of Operations Research* **14**, 294–302.

Anstreicher, K.M. (1990). "On Long Step Path Following and SUMT for Linear and Quadratic Programming," Technical Report, Yale School of Management, Yale University, New Haven, CT.

Anstreicher, K.M. and Watteyne, P. (1993). "A Family of Search Directions for Karmarkar's Algorithm," *Operations Research* **41**, 759–767.

Arid, T.J., and Lynch, J.E. (1975). "Computable Accurate Upper and Lower Error Bounds for Approximate Solutions of Linear Algebra Systems," *ACM Transactions on Mathematical Software,* **1**, 217–231.

Arntzen, B.C., Brown, G.G., Harrison, T.P., and Trafton, L.L. (1995). *Interfaces* **25:1**, 69–93.

Avi-Itzhak, H. (1994). "High Accuracy Correlation Based Pattern Recognition," Ph.D. thesis, Electrical Engineering, Stanford University, Stanford, CA.

Avriel, M. (1976). *Nonlinear Programming: Analysis and Methods*, Prentice-Hall, Inc., Englewood Cliffs, New Jersey.

Axelsson, O. and Munksgaard, N. (1983). "Analysis of Incomplete Factorizations with Fixed Storage Allocation," in D. Evans (ed.), *Preconditioning Methods Theory and Applications*, Gordon and Breach, New York, 219–241.

B

Bahn, O., Goffin, J.L., Vial, J.-Ph., and Merle, O.D. (1991). "Implementation and Behavior of an Interior Point Cutting Plane Algorithm for Convex Programming: An Application to Geometric Programming," Working Paper, University of Geneva, Geneva, Switzerland to appear in *Discrete Applied Mathematics* .

Bailey, D. (1988). "Extra High Speed Matrix Multiplication on the Cray-2," *SIAM J. on Scientific and Statistical Computing* **9**, 603–607.

El-Bakry, A., Tapia, R., and Zhang, Y. (1991). "Numerical Comparisons of Local Convergence Strategies for Interior-Point Methods in Linear Programming," Technical Report TR 91-18, Department of Computational and Applied Mathematics, Rice University, Houston, TX.

Balanski, M.L. and Gomory, R.E. (1963). "A Mutual Primal-Dual Simplex Method," in Graves, R.L. and Wolfe, P. (eds.), *Recent Advances in Mathematical Programming*, McGraw-Hill, New York.

Barnes, E.R. (1986). "A Variation on Karmarkar's Algorithm for Linear Programming," *Mathematical Programming* **36**, 174–182.

Bartels, R.H. (1971). "A Stabilization of the Simplex Method," *Numerische Mathematik* **16**, 414–434.

Bartels, R.H. and Golub, G.H. (1969). "The Simplex Method of Linear Programming Using the LU Decomposition," *Communications of the Association for Computing Machinery* **12**, 266–268.

Bartels, R.H., Stoer, J. and Zenger, Ch. (1971). "A Realization of the Simplex Method Based on Triangular Decompositions," in J.H. Wilkinson and C. Reinsch (eds.), *Contributions I/II in Handbook for Automatic Computation, Volume II: Linear Algebra*, Springer-Verlag, Berlin and New York.

Bastian, M. (1984). "Implicit Representation of Generalized Variable Upper Bounds Using the Elimination Form of the Inverse on Secondary Storage," *Mathematical Programming* **30**, 357–361.

Bazaraa, M.S., Jarvis, J.J., and Sherali, H.D., 1990, *Linear Programming and Network Flows*, John Wiley and Sons, New York.

Beale, E.M.L. (1954a). "Linear Programming by the Method of Leading Variables," *Report of the Conference on Linear Programming*, (May), arranged by Ferranti Ltd., London.

Beale, E.M.L. (1954b). "An Alternative Method for Linear Programming," *Proceedings of the Cambridge Philosophical Society* **50**, 513–523.

Beale, E.M.L. (1955a). "On Minimizing a Convex Function Subject to Linear Inequalities," *J. Roy. Society*, **17b**, 173–184.

Beale, E.M.L. (1955b). "Cycling in the dual simplex algorithm," *Naval Research Logistics Quarterly*, **2**, 269–275.

Beale, E.M.L. (1970). "Matrix Generators and Output Analyzers," in H.W. Kuhn (ed.), *Proceedings of the Princeton Symposium on Mathematical Programming*, Princeton University Press, Princeton, New Jersey, 25–36.

Beale, E.M.L., Hughes, P.A.B., and Small, R. (1965). "Experiences in Using a Decomposition Program," *Computer Journal* **8**, 13–18.

Bell, C.G. and Newell, A. (1971). *Computer Structures: Readings and Examples*, McGraw-Hill, New York.

Bellman, R.E. (1958). "On a Routing Problem," *Quart. Appl. Math.* **1958**, 87–90.

Bellman, R.E. and Dreyfus, S. (1962). *Applied Dynamic Programming*, Princeton University Press, Princeton, New Jersey.

Ben-Dov, Y., Hayre, L., and Pica, V. (1992). "Mortgage Valuation Models at Prudential Securities," *Interfaces* **22**:1, 55–71.

Benichou, M., Gauthier, J.M., Hentges, G., and Ribiére, G. (1977). "The Efficient Solution of Large-Scale Linear Programming Problems—Some Algorithmic Techniques and Computational Results," *Mathematical Programming* **13**, 280–322.

Benders, J.F., (1962). "Partitioning Procedures for Solving Mixed-Variables Programming Problems," *Numerische Mathematik* **4**, 238–252.

Bennett (1963). "An Approach to Some Structured Linear Programming Problems," Bassar Computing Department, School of Physics, University of Sydney, Australia.

Bertsekas, D.P. (1988). "The Auction Algorithm: A Distributed Relaxation Method for the Assignment Problem," *Annals of Operations Research* **14**, 105–123.

Bertsekas, D.P. (1991). *Linear Network Optimization: Algorithms and Codes*, The MIT Press, Cambridge, Massachusetts.

Berge, C. (1963). *Topological Spaces*, Oliver & Boyd Ltd., Edinburgh.

Birge, J.R. (1980). "Solution Methods for Stochastic Dynamic Linear Programs," Technical Report SOL 80-29, Department of Operations Research, Stanford University, Stanford, CA.

Birge, J.R. (1984). "Aggregration in Stochastic Linear Programming," *Mathematical Programming* **31**, 25–41.

Birge, J.R. (1985). "Decomposition and Partitioning Methods Methods for Multi-Stage Stochastic Linear Programming," *Operations Research* **33**, 989–1007.

Birge, J.R., Dempster, M.A., Gassman, H.I., Gunn, E.A., King, A.J., and Wallace, S. W. (1987). "A Standard Input Format for Multi-Period Stochastic Linear Programs," *Mathematical Programming Society COAL Newsletter* **17**.

Birge, J.R. and Wallace, S. W. (1988). "A Separable Piecewise Linear Upper Bound for Stochastic Linear Programs," *SIAM J. on Control and Optimization* **26** .

Birge, J.R. and Wets, R. J. (1986). "Designing Approximate Schemes for Stochastic Optimization Problems, in Particular for Stochastic Programs with Recourse," *Mathematical Programming Study* **27**, 54–102.

Birge, J.R. and Wets, R. J. (1989). "Sublinear Upper Bounds for Stochastic Programming Problems with Recourse," *Mathematical Programming* **43**, 131–149.

Birkhoff, G. (1946). "Three Observations on Linear Algebra," *Rev. Univ. Nac. Tucumán, Ser.* **A.,** *Vol. 5*, 147–151.

Bisschop, J. and Meeraus, A. (1981). "Towards Successful Modeling Applications in a Strategic Planning Environment," in Dantzig, G.B., Dempster, M.A.H., and Kallio, M.J. (eds.), *Large-Scale Linear Programming,* Vol. 2, CP-81-51, IIASA Collaborative Proceedings Series, Laxenberg, Austria, 712–745.

Bisschop, J. and Meeraus, A. (1982). "On the Development of a General Algebraic Modeling System," *Mathematical Programming Study* **20**, 1–29.

Bixby, R.E. (1981). "Hidden Structure in Linear Programs," in H. Greenberg and J. Maybee (eds.), *Computer Assisted Analysis and Model Simplification*, Academic Press, London and New York, 327–360.

Bixby, R.E. (1984). "Recent Algorithms for Two Versions of Graph Realization and Remarks on Applications to Linear Programming," in W.R. Pulleybank (ed.), *Progress in Combinatorial Optimization*, Academic Press, Canada, 39–67.

Bixby, R.E. and Cunningham, W.H. (1980). "Converting Linear Programs to Network Problems," *Mathematics of Operations Research* **5**, 321–357.

Bixby, R.E. and Fourer, R. (1988). "Finding Embedded Network Rows in Linear Programs I. Extraction Heuristics," *Management Science* **34**, 342–376.

Bixby, R.E., Gregory, J.W., Lustig, I.J., Marsten, R.E., and Shanno, D.F. (1992). "Very Large Scale Linear Programming: A Case Study in Combining Interior Point and Simplex Methods," *Operations Research* **40**, 885-897.

Bixby, R.E. and Saltzman, M.J. (1992). "Recovering an Optimal LP Basis from the Interior Point Solution," Technical Report 607, Department of Mathematical Sciences, Clemson University, Clemson, SC.

Bixby, R.E. and Wagner, D.K. (1987). "A Note on Detecting Simple Redundancies in Linear Systems," *Operations Research Letters* **6**, 15–17.

Bixby, R.E. and Wagner, D.K. (1988). "An Almost Linear-Time Algorithm for Graph Realization," *Mathematics of Operations Research* **13**, 99–123.

Björck, Å. (1967). "Solving Linear Least Squares Problems by Gram-Schmidt Orthogonalization," *BIT* **7**, 1–21.

Björck, Å., Plemmons, R.J., and Schneider, H. (1981). *Large-Scale Matrix Problems*, North-Holland, Amsterdam, the Netherlands.

Bland, R.G. (1977). "New Finite Pivoting Methods for the Simplex Method," *Mathematics of Operations Research* **2**, 103–107.

Blue, J.L. (1975). "Automatic Numerical Quadrature—DQUAD," *Bell Laboratories Report*, Murray Hill, N.J.

Borel, E. (1921). "La Théorie du Jeu et les Équations Íntegrales à Noyau Symétrique," *Compt. Rend. Acad. Sci. U.S.S.R.* **173**, 1304–1307. Translated by Leonard J. Savage in *Econometrica,* **21**, 97–100, January 1953.

Borel, E. (1924). "Sur les Jeux où Interviennent le Hasard et l'habileté des Joueurs," *Théorie des Probabilités*, Librairie scientifique, Hermann, Paris, 204–224. Translated by Leonard J. Savage in *Econometrica,* **21**, 101–115, January 1953.

Borel, E. (1927). "Sur les Systèmes de Formes Linéaires à Déterminant Symétrique Gauches et la Théorie Générale du Jeu," from "Algèbre et Calcul des Probabilités," *Compt. Rend. Acad. Sci. U.S.S.R.* **184**, 52–53. Translated by Leonard J. Savage in *Econometrica,* **21**, 101–115, January 1953.

Borgwardt, K.H. (1982a). "The Average Number of Pivot Rules Required by the Simplex Method is Polynomial," *Zeitschrift für Operations Research* **26**, 157–177.

Borgwardt, K.H. (1982b). "Some Distribution Independent Results About the Asymptotic Order of the Average Number of Pivot Steps in the Simplex Method," *Mathematics of Operations Research* **7**, 441–462.

Borgwardt, K.H. (1987a). *The Simplex Method: A Probabilistic Analysis*, Springer-Verlag, Berlin and New York.

Borgwardt, K.H. (1987b). "Probabilistic Analysis of the Simplex Method," in *Operations Research Proceedings 16th Dgor Meeting*, 564–576.

Bradley, G.H., Brown, G.G., and Graves, G.W. (1977). "Design and Implementation of Large Scale Primal Transshipment Algorithms," *Management Science* **24**, 1–34.

Bradley, S.P., Hax, A.C., and Magnanti, T.L. (1977). *Applied Mathematical Programming*, Addison-Wesley Publishing Company, Reading, Massachusetts.

Brearley, A.L., Mitra, G., and Williams, H.P. (1975). "Analysis of Mathematical Programs Prior to Applying the Simplex Algorithm," *Mathematical Programming* **8**, 54–83.

Brent, R. P. (1973a). "Hashing," *Communications of the Association for Computing Machinery* **16,2**, 105–109.

Brent, R.P., (1973b). *Algorithms for Finding Zeros and Extrema of Functions Without Calculating Derivatives*, Prentice-Hall, Inc., Englewood Cliffs, New Jersey.

Brickman, L. (1989). *Mathematical Introduction to Linear Programming and Game Theory*, Springer-Verlag, Berlin and New York.

Brooke, A., Kendrick, D., and Meeraus, A. (1988). *GAMS: A User's Guide*, Scientific Press, South San Francisco.

Brown, G.G. and McBride, R.D. (1984). "Solving Generalized Networks," *Management Science* **30**, 1497–1523.

Brown, G.G. and Thomen, D.S. (1980). "Automatic Identification of Generalized Upper Bounds in Large-Scale Optimization Models," *Management Science* **26**, 1166–1184.

Brown, G.G. and Wright, W. (1981). "Automatic Identification of Embedded Structure in Large-Scale Optimization Models," in H. Greenberg and J. Maybee (eds.), *Computer Assisted Analysis and Model Simplification*, Academic Press, London and New York, 369–388.

Brown, G.G. and Wright, W. (1984). "Automatic Identification of Embedded Network Rows in Large-Scale Optimization Models," *Mathematical Programming* **29**, 41–56.

Broyden, C.G. (1973). "Some Condition Number Bounds for the Gaussian Elimination Process," *J. Institute of Mathematics and its Applications* **12**, 273–286.

Burks, A.W., Goldstine, H.H., and von Neumann, J. (1946). "Planning and Coding of Problems for an Electronic Computing Instrument," Parts 1 and 2, Institute for Advanced Study, Princeton, New Jersey, USA. Reprinted in *Datamation* **8**, 9, 24–31, September, 1962 and *Datamation* **8**, 10, 36–41, October, 1962.

Bunch, J.R. and Parlett, B.N. (1971). "Direct Methods for Solving Symmetric Indefinite Systems of Linear Equations," *SIAM J. on Numerical Analysis* **8**, 639–655.

Bunch, J.R. and Kaufman, L.C. (1977). "Some Stable Methods for Calculating Inertia and Solving Symmetric Linear Equations," *Linear Algebra and its Applications* **34**, 341–370.

Businger, P.A. (1968). "Matrices Which can be Optimally Scaled," *Numerische Mathematik* **12**, 346–348.

Buzbee, B.L. (1986). "A Strategy for Vectorization," *Parallel Computing* **3**, 187–192.

C

Cariño, D.R., Kent, T., Myers, D.H., Stacy, C., Sylvanus, M., Turner, A.L., Watanabe, K., and Ziemba, W.T. (1994). "The Russel-Yasuda Kasai Model: An Asset/Liability Model for a Japanese Insurance Company Using Multistage Stochastic Programming," *Interfaces* **14**:1, 29–49.

Carpento, G., Martello, S., and Toth, P. (1988). "Algorithms and Codes for the Assignment Problem," in B. Simeone, P. Toth, G. Gallo, F. Maffioli, and S. Pallotino (eds.), *FORTRAN Codes for Network Optimization, Annals of Operations Research* **13**, 193-224.

Carpenter, T.J., Lustig, I.J., Mulvey, J.M., and Shanno, D.F. (1993). "Higher Order Predictor-Corrector Interior Point Methods with Applications to Quadratic Objectives," *SIAM Journal on Optimization* **3**, 696–725.

Chambers, J.M. (1977). *Computational Methods for Data Analysis*, John Wiley and Sons, New York.

Chan, T.F. (1985). "On the Existence and Computation of LU Factorizations with Small Pivots," *Mathematics of Computation* **42**, 535–548.

Charnes, A. (1952). "Optimality and Degeneracy in Linear Programming," *Econometrica,* **20**, 160–170.

Charnes, A., Cooper, W.W., and Mellon, B. (1952). "Blending Aviation Gasoline—A Study of Programming Interdependent Activities in an Integrated Oil Company," *Econometrica,* **20**.

Charnes, A. and Kortanek, K.O. (1963). "An Opposite Sign Algorithm for Purification to an Extreme Point Solution," Memorandum No. 89, Office of Naval Research.

Charnes, A. and Lemke, C. (1960). "Multi-Copy Generalized Networks and Multi-Page Programs," R.P.I. Math Report No. 41, Rensselaer Polytechnic Institute; Troy, New York.

Cheng, M.C. (1985). "Generalized Theorems for Permanent Basic and Nonbasic Variables," *Mathematical Programming* **31**, 229–234.

Cheng, M.C. (1987). "General Criteria for Redundant and Nonredundant Linear Inequalities," *Journal of Optimization Theory and Applications* **53**, 37–42.

Cherkassky, B.V., Goldberg, A.V., and Radzik, T. (1996). "Shortest Path Algorithms: Theory and Experimental Evaluation," *Mathematical Programming* **73**, 129-174.

Chvátal, V. (1983). *Linear Programming,* W. H. Freeman and Company, New York.

Choi, I.C., Monma, C.L., and Shanno, D.F. (1990). "Further Developments of a Primal-Dual Interior Point Method," *Operations Research* **40**, 885-897.

Clements, D.W. and Reid, R.A. (1994). "Analytical MS/OR Tools Applied to a Plant Closure," *Interfaces* **24**:2, 1–12.

Cline, A.K., Conn, A.R., and van Loan, C. (1982). "Generalizing the LINPACK Condition Estimator," in Hennart, J.P. (ed.), *Numerical Analysis,* Lecture Notes in Mathematics No. 909, Springer-Verlag, Berlin and New York.

Cline, A.K., Moler, C.B., Stewart, G.W., and Wilkinson, J.H., (1979). "An Estimate for the Condition Number of a Matrix," *SIAM J. on Numerical Analysis* **16**, 368–375.

Cline, A.K. and Rew, R.K. (1983). "A Set of Counter Examples to Three Condition Number Estimators," *SIAM J. on Scientific and Statistical Computing* **4**, 602–611.

Cody, W.J. (1974). "The Construction of Numerical Subroutine Libraries," *SIAM Review,* **16**, 36–46.

Cohen, A.M. (1974). "A Note on Pivot Size in Gaussian Elimination," *Linear Algebra and its Applications* **8**, 361–368.

Coleman, T.F. and More, J.J. (1983). "Estimation of Sparse Jacobian Matrices and Graph Coloring Problems," *SIAM J. on Numerical Analysis* **20**, 187–209.

Cornfield, J. (1940). Verbal Communication to G.B. Dantzig.

Cope, J.E. and Rust, B.W. (1979). "Bounds on Solutions of Systems with Accurate Data," *SIAM J. on Numerical Analysis* **16**, 950–963.

Cottle, R.W. (1974). "Manifestations of the Schur Complement," *Linear Algebra and its Applications* **8**, 189-211.

Courant, R. (1943). "Variational Methods for the Solution of Problems of Equilibrium and Vibrations," *Bull. Amer. Math. Society* **49**, 1–23.

Cowell, W. R. (ed.), (1983). *Sources and Development of Mathematical Software*, Prentice-Hall, Inc., Englewood Cliffs, New Jersey.

Cox, L.A., Kuehner, Parrish, S.H., and Qiu, Y. (1993). "Optimal Expansion of Fiber-Optic Telecommunications Networks in Metropolitan Areas," *Interfaces* **23**:2, 35–48.

Cryer, C.W. (1968). "Pivot Size in Gaussian Elimination," *Numerische Mathematik* **12**, 335–345.

Cunningham, W.H. (1979). "Theoretical Properties of the Network Simplex Method," *Mathematics of Operations Research* **4**, 196–208.

Cunningham, W.H. and Klincewicz, J.G. (1983). "On Cycling in the Network Simplex Method," *Mathematical Programming* **26**, 182–189.

Curtis, A.R., Powell, M.J.D. and Reid, J.K. (1974). "On the Estimation of Sparse Jacobian Matrices," *J. Institute of Mathematics and its Applications* **13**, 117–119.

Curtis, A.R. and Reid, J.K. (1971). "The Solution of Large Sparse Unsymmetric Systems of Linear Equations," *J. Institute of Mathematics and its Applications* **8**, 344–353.

D

Dahl, O., Dijkstra, E., and Hoare, C.A.R. (1972). *Structured Programming*, Academic Press, London and New York

Dahlquist, G. and Björk, Å. (1974). *Numerical Methods*, Prentice-Hall, Inc., Englewood Cliffs, New Jersey.

Dantzig, G.B. (1939). "On a Class of Distributions that Approach the Normal Distribution Function," *Annals of Math. Stat.* **10**, September, 247–253.

Dantzig, G.B. (1940). "On the Non-Existence of Tests of Students' Hypothesis Involving Power Functions Independent of Sigma," *Annals of Math. Stat.* **11**, June, 186–192.

Dantzig, G.B. (1948, January 5). "A Theorem on Linear Inequalities," unpublished report.

Dantzig, G.B. (1949). "Programming in a Linear Structure," Report of the September 9, 1948 meeting in Madison, *Econometrica*, **17**, 73–74.

Dantzig, G.B. (1951a). "Maximization of a Linear Function of Variables Subject to Linear Inequalities," in T.C. Koopmans (ed.), *Activity Analysis of Production and Allocation*, July–October 1949, Cowles Commission Monograph 13, Proceedings of Linear Programming Conference, June 20–24, 1949, John Wiley and Sons, New York, 339–347.

Dantzig, G.B. (1951b). "Application of the Simplex Method to the Transportation Problem," in T.C. Koopmans (ed.), *Activity Analysis of Production and Allocation*, July–October 1949, Cowles Commission Monograph 13, Proceedings of Linear Programming Conference, June 20–24, 1949, John Wiley and Sons, New York, 359–373.

Dantzig, G.B. (1951c). "Programming of Interdependent Activities, II: Mathematical Model," *Econometrica*, **17**, 200–211; also in T.C. Koopmans (ed.), *Activity Analysis of Production and Allocation*, July–October 1949, Cowles Commission Monograph 13, Proceedings of Linear Programming Conference, June 20–24, 1949, John Wiley and Sons, New York, 19–32.

Dantzig, G.B. (1951d). "A Proof of the Equivalence of the Programming Problem and the Game Problem, in T.C. Koopmans (ed.), *Activity Analysis of Production and Allocation*, July–October 1949, Cowles Commission Monograph 13, Proceedings of Linear Programming Conference, June 20–24, 1949, John Wiley and Sons, New York, 330–335.

Dantzig, G.B. (1951e). "Linear Programming," in Problems for the Numerical Analysis of the Future, Proceedings of Symposium on Modern Calculating Machinery and Numerical Methods, UCLA, July 29–31, 1948, Appl. Math. Series 15, National Bureau of Standards, June 1951, 18–21.

Dantzig, G.B. (1954a). "The Dual Simplex Algorithm (Notes on Linear Programming: Part VII)," *RM*-**1270**, The RAND Corporation, July.

Dantzig, G.B. (1954b). "Upper Bounds, Secondary Constraints, and Block Triangularity in Linear Programming (Notes on Linear Programming: Part VIII, IX, X)," *RM*-**1367**, The RAND Corporation, October.

Dantzig, G.B. (1954c). "Notes on Linear Programming: Part XI, Composite Simplex-Dual Simplex Algorithm—I," *RM*-**1274**, The RAND Corporation, April.

Dantzig, G.B. (1954d). "A Comment on Eddie's 'Traffic Delays at Toll Booths'," *Journal of Operations Research Society of America* **2**, 339–341.

Dantzig, G.B. (1955a). "Linear Programming Under Uncertainty," *Management Science* **1**, 197–206. Also in A.F. Veinott (ed.), *Mathematical Studies in Management Science*, The Macmillian Co., New York, 1965, 330-339.

Dantzig, G.B. (1955b). "Optimal Solution of a Dynamic Leontief Model with Substitution (Notes on Linear Programming: Part XIII)," *RM*-**1281-1**, The RAND Corporation, April. Also *Econometrica,* **23**, 151–176.

Dantzig, G.B. (1955c). "Developments in Linear Programming," Proceedings Second Symposium on Linear Programming, National Bureau of Standards and Comptroller, U.S.A.F. Headquarters, January, 667–685.

Dantzig, G.B. (1956a). "Constructive Proof of the Min-Max Theorem," *Pacific Journal of Mathematics* **6**, 25–33.

Dantzig, G.B. (1956b). "Recent Advances in Linear Programming," *Management Science* **2**, 131–144.

Dantzig, G.B. (1956c). "Formulating a Linear Programming Model," *P*-**893**, The RAND Corporation, July. Also in Linear Programming and Extensions, Princeton University Press, Princeton, N.J., 1963, Chapter 3.

Dantzig, G.B. (1956d). "Note on Klein's 'Direct Use of Extremal Principles in Solving Certain Problems Involving Inequalities'," *Operations Research* **4**, 247–249.

Dantzig, G.B. (1957a). "Thoughts on Linear Programming and Automation," *Management Science* **3**, 131–139. Also: *P*-**824**, The RAND Corporation, March 2, 1956.

Dantzig, G.B. (1957b). "Concepts, Origins and Uses of Linear Programming," in Davis, Eddison, Page (eds.), *Proceedings of First International Conference on Operations Research*, Operations Research Society of America, Baltimore, December, 100–108.

Dantzig, G.B. (1957c). "Discrete Variable Extremum Problems," *Operations Research* **5**, 226-277.

Dantzig, G.B. (1957d). "On The Status of Multi-Stage Linear Programs," *RM*-**1028**, The RAND Corporation, February. Also in A.F. Veinott (ed.), *Mathematical Studies in Management Science*, The Macmillian Co., New York, 1965, Chapter 6, Section III: Topics in Linear Programming, 303–320. Also in *Management Science* , 1959, 71–90.

Dantzig, G.B. (1958a). "Solving Two-Move Games with Perfect Information," *P*-**1459**, The RAND Corporation, August.

Dantzig, G.B. (1958b). "On Integer and Partial Integer Linear Programming Problems," *P*-**1410**, The RAND Corporation, June.

Dantzig, G.B. (1958c). "Chemical Equilibrium in Complex Mixtures," *Journal of Chemical Physics* **28**, 751–755.

Dantzig, G.B. (1959). "Note on Solving Linear Programs in Integers," *Naval Research Logistics Quarterly,* **6**, 75–76.

Dantzig, G.B. (1960a). "Inductive Proof of the Simplex Method," *IBM J. Res. Develop.*, **4**, 505–506. Also: *P*-**1851**, The RAND Corporation, December 28, 1959.

Dantzig, G.B. (1960b). "On the Shortest Route Through a Network," *Management Science* **6**, 187–190. Also in Fulkerson, D. R. (eds.), *Some Topics in Graph Theory,* MAA Studies, No. 11, 1975, 89–93.

Dantzig, G.B. (1960c). "On the Significance of Solving Linear Programming Problems with Some Integer Variables," *Econometrica,* **28**, 30–44.

Dantzig, G.B. (1960d). "General Convex Objective Forms," in Arrow, K., Karlin, S., and Suppes, P. (eds.), *Mathematical Methods in the Social Sciences*, Stanford University Press, Stanford, California, 151–158.

Dantzig, G.B. (1960e). "A Machine-Job Scheduling Model," *Management Science* **6**, 191–196.

Dantzig, G.B. (1961a). "Future Developments of Operations Research," Operations Research Center, University of California, Berkeley, Summary Report 1, 1961 Also in Proceedings Thirteenth Annual Industrial Engineering Institute, University of California, February 1961, 25–28, Special Topic; and in Decentralized Planning, J. Operations Research Society, Switzerland, 1962, 53–55.

Dantzig, G.B. (1961b). "Quadratic Programming—-A Variant of the Wolfe-Markowitz Algorithms," Technical Report RR-2, Operations Research Center, University of California, Berkeley. Also in Linear Programming and Extensions, Princeton University Press, Princeton, N.J., 1963, Chapter 24-4.

Dantzig, G.B. (1962a). "A Proof of a Property of Leontief and Markov Matrices," Technical Report RR-25, Operations Research Center, University of California, Berkeley.

Dantzig, G.B. (1962a). "Compact Basis Triangularization for the Simplex Method," Technical Report RR-28, Operations Research Center, University of California, Berkeley. Also in Graves, R.L. and Wolfe, P. (eds.), *Recent Advances in Mathematical Programming*, McGraw-Hill, New York, 125–133.

Dantzig, G.B. (1963). *Linear Programming and Extensions*, Princeton University Press, Princeton, New Jersey, August. Revised edition Fall 1966; fourth printing, 1968, 621 pages. [Japanese translation, Tutte-Mori, Inc., Tokyo, 1983.]

Dantzig, G.B. (1964a). "New Mathematical Methods in the Life Sciences," *The American Mathematical Monthly* **71**, 4–15. Also in Stacy and Waxman (eds.), *Computers and Biomedical Research 1*, Academic Press, New York, March 1966.

Dantzig, G.B. (1964b). "Research Problems," *Bull. Amer. Math. Society* **70**, 499–501.

Dantzig, G.B. (1964c). "Linear Control Processes and Mathematical Programming," Technical Report 64-13, Operations Research Center, University of California, Berkeley. Also in *SIAM J. on Control and Optimization* **4**, 1966, 56–60. Also in in Dantzig, G.B. and Veinott, A.F., Jr. (eds.), *Mathematics of the Decision Sciences Part 2*, the American Mathematical Society Summer Seminar, Providence, RI, 1968, 31–36. Also in J. Abadie (ed.), *Nonlinear Programming*, North-Holland, Amsterdam, the Netherlands, 1967, 283–286. Also in Aziz, A. K. (ed.), *Lecture Series in Differential Equations 1*, Van Nostrand Reinhold Publishing Co., 1969, 1–7.

Dantzig, G.B. (1965a). "Operations Research in the World of Today and Tomorrow," Technical Report 65-7, Operations Research Center, University of California, Berkeley. Also in Technion Yearbook; also in commemorative volume for Professor Keller, Institut fur Okonometri; presidential address, TIMS 1966, entitled "Management Science in the World of Today and Tomorrow" in *Management Science* **13**, February 1967, pages C-107–C-111.

Dantzig, G.B. (1965b). "Large-Scale System Optimization: A Review," Technical Report 65-9, Operations Research Center, University of California, Berkeley.

Dantzig, G.B. (1965c). "Optimization in Operations Research," Technical Report 65-10, Operations Research Center, University of California, Berkeley. Also in Wayne Kalenich (ed.), *Proceedings, International Federation for Information Processing Congress, 1965*, May.

Dantzig, G.B. (1965c). "The Simplex Method," in Machol, R. (ed.), *Engineering Handbook*, McGraw-Hill, New York, June 1965, Chapter 25, 10 pages.

Dantzig, G.B. (1966a). "Linear Programming and its Progeny," Naval Research Reviews XIX (6), June 1966, 1; also in Beale, E.M.L. (ed.), *Application of Mathematical Programming Techniques*, English Universities Press, Ltd., London, 1970, 3–16.

Dantzig, G.B. (1966b). "All Shortest Routes in a Graph," Technical Report 66-3, Department of Operations Research, Stanford University, Stanford, CA, November. Also in Théorie des Graphes, International Symposium, Rome, Italy, July 1966, 91–92, published by DUNOD, Paris.

Dantzig, G.B. (1966c). "On Positive Principal Minors," Technical Report 67-1, Department of Operations Research, Stanford University, Stanford, CA, January.

Dantzig, G.B. (1967). "Large-Scale Linear Programming," Technical Report 67-8, Department of Operations Research, Stanford University, Stanford, CA, November. Also in "Large-Scale Systems and the Computer Revolution," in H.W. Kuhn (ed.), *Proceedings of the Princeton Symposium on Mathematical Programming*, Princeton University Press, Princeton, New Jersey, August 1967, 51–72. Also in Dantzig, G.B. and Veinott, A.F., Jr. (eds.), *Mathematics of the Decision Sciences*, the American Mathematical Society Summer Seminar, Providence, RI, 1968, 77–92.

Dantzig, G.B. (1969a). "Complementary Spanning Trees," Technical Report CS 126, Department of Computer Science, Stanford University, Stanford, CA, March. Also in J. Abadie (ed.), *Integer and Nonlinear Programming*, North-Holland, Amsterdam, the Netherlands, 1970, 499–505.

Dantzig, G.B. (1969b). "A Hospital Admission Problem," Technical Report 69-15, Department of Operations Research, Stanford University, Stanford, CA, December.

Dantzig, G.B. (1970a). "On a Model for Computing Round-Off Error of a Sum," Technical Report STAN-CS-70-156, Department of Computer Science, Stanford University, Stanford, CA.

Dantzig, G.B. (1970b). "A Control Problem of Bellman," Technical Report 70-15, Department of Operations Research, Stanford University, Stanford, CA, September. Also in Management Science: Theory 16, May 1971, 542–546.

Dantzig, G.B. (1972a). "Health Care in Future Cities," Technical Report 72-22, Department of Operations Research, Stanford University, Stanford, CA, September.

Dantzig, G.B. (1972b). "On the Relation of Operations Research to Mathematics," Technical Report 72-23, Department of Operations Research, Stanford University, Stanford, CA, October.

Dantzig, G.B. (1972c). "The Price Lecture on Compact City," Price Lecture Series University of Pittsburgh.

Dantzig, G.B. (1973a). "The ORSA New Orleans Address on Compact City," *Management Science* **19**, 1151-1161.

Dantzig, G.B. (1973b). "Solving Staircase Linear Programs by a Nested Block-Angular Method," Technical Report 73-1, Department of Operations Research, Stanford University, Stanford, CA, January.

Dantzig, G.B. (1973c). "Drews'Institutionalized Divvy Economy," Technical Report 73-7, Department of Operations Research, Stanford University, Stanford, CA, September. Revised, International Institute for Applied Systems Analysis, December 1973. Revised 1974, as Technical Report TR 74-14. Also in "An Institutionalized Divvy Economy," Technical Report SOL 75-17 (revision of TR 74-14), Department of Operations Research, Stanford University, Stanford, CA. Also in *Journal of Economic Theory* **11, 1975**, 372–384.

Dantzig, G.B. (1974a). "On a Convex Programming Problem of Rozanov," *Applied Mathematics and Optimization* **1**, 189-192. Also entitled "A Generalized Programming Solution to a Convex Programming Problem with a Homogeneous Objective," IIASA Research Report RR-73-21, December 1973. Also in Symposia Mathematica, Mongraf, Italy, Vol. XIX, Academic Press, 1976, 209–214.

Dantzig, G.B. (1974b). "On the Reduction of an Integrated Energy and Interindustry Model to a Smaller Linear Program," Technical Report SOL 74-20, Department of Operations Research, Stanford University, Stanford, CA, December. Also in *Review of Econ. and Statist.* **LVIII**, May 1976, 248–250.

Dantzig, G.B. (1974c). "Formulating a PILOT Model for Energy in Relation to the National Economy," Technical Report SOL 75-10, Department of Operations Research, Stanford University, Stanford, CA, April.

Dantzig, G.B. (1975). "Note on the Objective Function for the PILOT Model," Technical Report SOL 75-20, Department of Operations Research, Stanford University, Stanford, CA, August. Also in Prekopa, A. (ed.), *Survey of Mathematical Programming*, Proceedings, IX International Symposium on Mathematical Programming, Publishing House of the Hungarian Academy of Sciences, Budapest, 1980, 325–328.

Dantzig, G.B. (1976). "Linear Programming: Its Past and Its Future," in Salkovitz, Edward I. (ed.), *Science Technology, and the Modern Navy, Thirtieth Anniversary*, ONR-37, Office of Naval Research, 84–95.

Dantzig, G.B. (1977). "Large-Scale Systems Optimization with Application to Energy," Technical Report SOL 77-3, Department of Operations Research, Stanford University, Stanford, CA, April. Also in Proceedings of Symposia in Applied Mathematics, American Mathematical Society, Providence, RI.

Dantzig, G.B. (1978). "Are Dual Variables Prices? If Not, How to Make them More So," Technical Report SOL 78-6, Department of Operations Research, Stanford University, Stanford, CA, March. Also in Franco Angeli (ed.), *Mathematical Programming and its Economics Applications*, Milano, Italy, 1981, 135–148.

Dantzig, G.B. (1979a). "The Role of Models in Determining Policy for Transition to a More Resilient Technological Society," IIASA Distinguished Lecture Series /1, Vienna, June 12, International Institute for Applied Systems Analysis, Laxenburg, 1979.

Dantzig, G.B. (1979b). "Comments on Khachian's Algorithms for Linear Programming," Technical Report SOL 79-22, Department of Operations Research, Stanford University, Stanford, CA, November. Also in *SiamNews* **13**, October 1980.

Dantzig, G.B. (1980a). "Expected Number of Steps of the Simplex Method for a Linear Program with a Convexity Constraint," Technical Report SOL 80-3, Department of Operations Research, Stanford University, Stanford, CA, March (revised October 1980).

Dantzig, G.B. (1980b). "Time-Staged Methods in Linear Programming; Comments and Early History, Technical Report SOL 80-18, Department of Operations Research, Stanford University, Stanford, CA, June. Also in Dantzig, G.B., Dempster, M.A.H., and Kallio, M.J. (eds.), *Large-Scale Linear Programming*, Vol. 2, CP-81-51, IIASA Collaborative Proceedings Series, Laxenberg, Austria, 1981, 3–16. Also in "Large Scale Systems.," in Haims, Y.Y. (ed.), *Studies in Management Science and Systems, Vol. 7*, North-Holland Publishing Company, Amsterdam, 1982, 19–30.

Dantzig, G.B. (1980c). "Time-Staged Linear Programs," Technical Report SOL 80-28, Department of Operations Research, Stanford University, Stanford, CA, October.

Dantzig, G.B. (1981a). "Reminiscences About the Origins of Linear Programming," Technical Report SOL 81-5, Department of Operations Research, Stanford University, Stanford, CA, April. Also in Contemporary Mathematics, American Mathematical Society. Also in Cottle, R.W., Kelmanson, M.L., Korte, B. (eds.), *Mathematical Programming*, Proceedings of the International Congress on Mathematical Programming, Rio de Janeiro, Brazil, April 6–8, 1981, North-Holland Publishing Co., Amsterdam, 1984, 105–112. Also in *Operations Research Letters* **1**, 43–48, April 1982. Also in in A. Bachem, M. Grotschel, and B. Korte (eds.), *Mathematical Programming: The State of the Art, Bonn 1982*, Springer-Verlag, Berlin and New York, 78–86. Also in *Australian Society for OR Bulletin*, 1986.

Dantzig, G.B. (1981b). "Concerns About Large-Scale Models," Technical Report SOL 81-27, Department of Operations Research, Stanford University, Stanford, CA, December. Also in *Large-Scale Energy Models*, AAAS Selected Symposia Series 73, Westview Press, Inc. Boulder, CO, for the Amer. Assoc. for the Adv. of Sc., Washington, D.C., 15–20, 1983.

Dantzig, G.B. (1981b). "The PILOT Energy-Economic Model for Policy Planning," Technical Report SOL 81-26, Department of Operations Research, Stanford University, Stanford, CA, December. Also in Veziroglu, T. Nejat (ed.), *Energy Programs Policy Economics*, Proceedings of the 4th Miami International Conference on Alternative Energy Sources, Volume 8, Ann Arbor Science Publishers, Ann Arbor Michigan, 1982, 409–415.

Dantzig, G.B. (1981c). "Contributions of Mathematics to Planning During and Immediately After World War II," Working Paper, November. Also in *History of Mathematics in World War II*, MAA Series.

Dantzig, G.B. (1982a). "Time-staged Methods in Linear Programs," in Y.Y. Haims (ed.), *Studies in Management Science*, Vol. 7: *Large-Scale Systems*, North-Holland, Amsterdam, the Netherlands, 19–30.

Dantzig, G.B. (1982b). "Mathematical Programming and Decision Making in a Technological Society," Technical Report SOL 82-11, Department of Operations Research, Stanford University, Stanford, CA, August. Also in *Information Processing Society of Japan* **24**, May 1983 (in Japanese).

Dantzig, G.B. (1983). "Can Leontief and P-matrices be Rescaled Positive Definite," Technical Report SOL 83-23, Department of Operations Research, Stanford University, Stanford, CA, November.

Dantzig, G.B. (1984). "Economic Growth and Dynamic Equilibrium," Technical Report SOL 84-8, Department of Operations Research, Stanford University, Stanford, CA, October.

Dantzig, G.B. (1985a). "Deriving a Utility Function for the Economy," Technical Report SOL 85-6, Department of Operations Research, Stanford University, Stanford, CA, June; revised April 1986.

Dantzig, G.B. (1985b). "Impact of Linear Programming on Computer Development," Technical Report SOL 85-7, Department of Operations Research, Stanford University, Stanford, CA. Revised version, Stanford University, July 1986, in Chudnovsky, D.V. and Jenks, R.D. (eds.), *Proceedings Computers in Mathematics*, Lecture Notes in Pure and Applied Mathematics, Marcel Dekker, Inc., 1990, 233–240. Also in *ORMS Today* **14**, August 1988, 12–17.

Dantzig, G.B. (1986). "Need to do Planning Under Uncertainty and the Possibility of Using Parallel Processors for this Purpose," Technical Report SOL 86-11, Department of Operations Research, Stanford University, Stanford, CA, April.

Dantzig, G.B. (1987a). "Simplex Method for Solving Linear Programs," *The New Palgrave: A Dictionary of Economic Theory and Doctrine*, The Macmillian Press, Ltd., London.

Dantzig, G.B. (1987b). "Linear Programming," *The New Palgrave: A Dictionary of Economic Theory and Doctrine*, The Macmillian Press, Ltd., London, to appear.

Dantzig, G.B. (1987c). "Planning Under Uncertainty Using Parallel Computing," Technical Report SOL 87-1, Department of Operations Research, Stanford University, Stanford, CA, January. Also in *Annals of Operations Research* **14**, 1988, 1–16.

Dantzig, G.B. (1987d). "Origins of the Simplex Method," Technical Report SOL 87-5, Department of Operations Research, Stanford University, Stanford, CA, May. Also in Nash, S.G. (ed.), *Proceedings of the ACM Conference on a History of Scientific Computing*, ACM Press, Addison-Wesley Publishing Company, 1990, 141–151.

Dantzig, G.B. (1988a). "Dikin's Interior Method for Solving LP," manuscript, Department of Operations Research, Stanford University, Stanford, CA.

Dantzig, G.B. (1988b). "Making Progress During a Stall in the Simplex Algorithm," Technical Report SOL 88-5, Department of Operations Research, Stanford University, Stanford, CA, February. Also in *Linear Algebra and its Applications* **114/115**, 1989, 251–259.

Dantzig, G.B. (1989). "Decomposition Techniques for Large-Scale Electric Power Systems Planning Under Uncertainty," in Sharda, R., *et al.* (ed.), *Impact on Recent Computer Advances on Operations Research* North Holland, 3–20.

Dantzig, G.B. (1990). "The Diet Problem," *Interfaces* **20**:4, July/Aug, 43–47.

Dantzig, G.B. (1991). "Converting a Converging Algorithm into a Polynomial Bounded Algorithm," Technical Report SOL 91-5, Department of Operations Research, Stanford University, Stanford, CA, March.

Dantzig, G.B. (1992a). "An ϵ-Precise Feasible Solution to a Linear Program with a Convexity Constraint in $1/\epsilon^2$ Iterations Independent of Problem Size," Technical Report SOL 92-5, Department of Operations Research, Stanford University, Stanford, CA, October.

Dantzig, G.B. (1992b). "Bracketing to Speed Convergence Illustrated on the von Neumann Algorithm for Finding a Feasible Solution to a Linear Program with a Convexity Constraint," Technical Report SOL 92-6, Department of Operations Research, Stanford University, Stanford, CA, October.

Dantzig, G.B. (1995). Working Paper.

Dantzig, G.B., *et al.* (1972). "On The Need for a Systems Optimization Laboratory," Technical Report 72-11, Department of Operations Research, Stanford University, Stanford, CA, September. Also in Hu, T.C. and Robinson, S.M. (ed.), *Mathematical Programming*, Proceedings of an Advanced Seminar Conducted by the Mathematics Research Center, University of Wisconsin, September 1972, Academic Press, London and New York, 1–31.

Dantzig, G.B. and Adler, I. (1971). "Maximum Diameter of Abstract Polytopes," Technical Report 71-12, Department of Operations Research, Stanford University, Stanford, CA, August. Also in Harry Williams (ed.), *IDA Economics Papers*, January 1972. Also in *Mathematical Programming Study* **1**, 1974, 20–40.

Dantzig, G.B., Adler, I., and Murty, K. (1970). "Existence of x-Paths in Abstract Polytopes," Technical Report 70-1, Department of Operations Research, Stanford University, Stanford, CA, March.

Dantzig, G.B., Adler, I., and Murty, K. (1974). "Existence of A-Avoiding Paths in Abstract Polytopes," *Mathematical Programming Study* **1**, 41–42.

Dantzig, G.B., Avi-Itzhak, B., Connolly, T.J., McAllister, P.H., and Winkler, W.D. (1982a). "A Dynamic Equilibrium Model for Energy-Economic Planning," Technical Report SOL 82-2, Department of Operations Research, Stanford University, Stanford, CA, March.

Dantzig, G.B., Avi-Itzhak, B., Connolly, T.J., McAllister, P.H., and Winkler, W.D. (1982b). "Mathematical Appendix: A Dynamic Equilibrium Model for Energy-Economic Planning," Technical Report SOL 82-3, Department of Operations Research, Stanford University, Stanford, CA, March.

Dantzig, G.B., Avi-Itzhak, and Iusem, A. (1983). "The Consumers Energy Services Model of the PILOT System," in Lev, Benjamin (ed.), *Energy Models and Studies*, Studies in Management Science and Systems 9, North-Holland Publishing Co., Amsterdam, 195–220.

Dantzig, G.B., Beale, E.M.L., Watson, R.D. (1986). "A First Order Approach to a Class of Multi-Time-Period Stochastic Programming Problems," *Mathematical Programming Study* **27**, 103–117.

Dantzig, G.B. and Avriel, M. (1976). "Determining Prices and Monetary Flows of the PILOT Energy Model," Technical Report SOL 76-28, Department of Operations Research, Stanford University, Stanford, CA, October.

Dantzig, G.B., Bigelow, J., Golub, G., Gordon, R., Montalbano, M., Pinsky, P., Sahberwal, F., Wirth, N., and Witzgall, C. (1967). "Mathematical Programming Language," Technical Report 67-4, Department of Operations Research, Stanford University, Stanford, CA, June,

Dantzig, G.B., Bigelow, J., Golub, G., Gordon, R., Montalbano, M., Pinsky, P., Sahberwal, F., Wirth, N., and Witzgall, C. (1968). "Mathematical Programming Language," Technical Report CS 119, Department of Computer Science, Stanford University, Stanford, CA.

Dantzig, G.B. and Blattner, W.O., and Rao, M.R. (1966). "Finding a Cycle in a Graph with Minimum Cost to Time Ratio with Application to a Ship Routing Problem," Technical Report 66-1, Department of Operations Research, Stanford University, Stanford, CA, November. Also in Theorie Des Graphes, International Symposium, Rome, Italy, July 1966, 77–84, published by DUNOD, Paris.

Dantzig, G.B. and Blattner, W.O., and Rao, M.R. (1966). "All Shortest Routes from a Fixed Origin in a Graph," Technical Report 66-2, Department of Operations Research, Stanford University, Stanford, CA, November. Also in Theorie Des Graphes, International Symposium, Rome, Italy, July 1966, 85–90, published by DUNOD, Paris.

Dantzig, G.B., Connolly, T.J., and Parikh, S.C. (1977). "Stanford PILOT Energy/Economic Model," Technical Report SOL 77-19, Department of Operations Research, Stanford University, Stanford, CA, July. Also in *Advances in the Economics of Energy and Resources, Volume 1 – The Structure of Energy Markets*, JAI Press, 1979, 77–103. Also in El Mallakh, R. and El Mallakh, D.H. (eds.), *Proceedings of the 4th International Conference on Energy Options and Conservation*, October 17–19, 1977, The International Research Center for Energy and Economic Development, Boulder, Colorado, 1978, 87-119. *Policy Analysis and Information Systems* **2**, 1978, 23–51.

Dantzig, G.B., Connolly, T.J., and Parikh, S.C. (1978). "Stanford PILOT Energy/Economic Model," EA-626, Volumes 1 and 2, Interim Report, Electric Power Research Institute, Palo Alto, California, May.

Dantzig, G.B., Connolly, T.J., Parikh, S.C., Riddel, J.M. (1978). "A Description and Demonstration of the Stanford PILOT Energy/Economic Model," *Stanford PILOT Energy/Economic Model*, EA-626, Volumes 1 & 2, Electric Power Research Institute, May, 1–40. Also in Proceedings of Second US-USSR Symposium on Econometric Modeling, Skyland, Virginia, 1978, University of Maryland, College Park, Maryland 1980.

Dantzig, G.B. and Cottle, R.W. (1963). "Positive (Semi-) Definite Matrices and Mathematical Programming," Technical Report RR-35, Operations Research Center, University of California, Berkeley. Also in "Positive (Semi-) Definite Programming," in in J. Abadie (ed.), *Nonlinear Programming*, North-Holland, Amsterdam, the Netherlands, 55–73.

Dantzig, G.B. and Cottle, R.W. (1967). "Complementary Pivot Theory of Mathematical Programming," Technical Report 67-2, Department of Operations Research, Stanford University, Stanford, CA, April. Also in Dantzig, G.B. and Veinott, A.F., Jr. (eds.), *Mathematics of the Decision Sciences*, the American Mathematical Society Summer Seminar, Providence, RI, 1968. Also in *Linear Algebra and its Applications* **1**, 103–125. Also in Dantzig, G.B. and Eaves, B.C. (eds.), *Studies in Optimization,* MAA Studies in Mathematics, Vol. 10, Mathematical Association of America, 1974, 27–51.

Dantzig, G.B. and Cottle, R.W. (1968). "A Generalization of the Linear Complementarity Problem," Technical Report 68-9, Department of Operations Research, Stanford University, Stanford, CA. Also in *Journal of Combinatorial Theory* **8**, January 1970, 79–90.

Dantzig, G.B. and Cottle, R.W. (1974). "Optimization, Mathematical Theory of (Linear and Nonlinear Programming)," Encyclopaedia Britannica, Vol. 13, 628–632.

Dantzig, G.B., Collen, M.F., *et al.* (1964). "Automated Multiphasic Screening and Diagnosis,", *American Journal of Public Health* **54**, 741–750.

Dantzig, G.B., DeHaven, J., and Sams, C.F. (1960). "A Mathematical Model of the Respiratory System," in Proceedings, Fourth Air Pollution Medical Research Conference, San Francisco, December, 72–95. Also in *P*-**2048**, The RAND Corporation, July 1960.

Dantzig, G.B., DeHaven, J., and Sams, C.F. (1961a). "A Mathematical Model of the Chemistry of the External Respiratory System," in J. Neyman (ed.), *Proceedings 4th Berkeley Symposium on Mathematical Statistics and Probability, 1961*, University of California Press, Berkeley, California, 181–196.

Dantzig, G.B., DeHaven, J., and Sams, C.F. (1961b). "A Mathematical Model of the Human External Respiratory System," Perspectives of Biology and Medicine IV (3), Spring 1961, 324–376.

Dantzig, G.B. and DeHaven, J. (1961c). "On The Reduction of Certain Multiplicative Chemical Equilibrium Systems to Mathematically Equivalent Additive Systems," *P*-**2419**, The RAND Corporation, August. Also *Journal of Chemical Physics* **36**, May, 1962, 2620–2627.

Dantzig, G.B., Dempster, M.A.H., and Kallio, M.J. (eds.), (1981). *Large-Scale Linear Programming,* Vol. 1, CP-81-51, IIASA Collaborative Proceedings Series, Laxenberg, Austria.

Dantzig, G.B., Dempster, M.A.H., and Kallio, M.J., editors (1981). *Large-Scale Linear Programming,* Vol. 2, CP-81-51, IIASA Collaborative Proceedings Series, Laxenberg, Austria.

Dantzig, G.B. and Eaves, B.C. (1972). "Fourier-Motzkin Elimination and Its Dual," Technical Report 72-18, Department of Operations Research, Stanford University, Stanford, CA, June 1972. Also in *Journal of Combinatorial Theory* **14**, May 1973, 288–297. Also in Roy, B. (ed.), *Combinatorial Programming: Methods and Applications,* D. Reidel Publishing Co., Boston, 1975, 93–102.

Dantzig, G.B. and Eaves, B.C., editors (1974). *Studies in Optimization, MAA Studies in Mathematics,* Vol. 10, Mathematical Association of America.

Dantzig, G.B., Eaves, B.C., and Gale, D. (1976). "An Algorithm for a Piecewise Linear Model of Trade and Production with Negative Prices and Bankruptcy," Technical Report SOL 76-19, Department of Operations Research, Stanford University, Stanford, CA. Also in *Mathematical Programming* **16**, 1979, 190–209.

Dantzig, G.B., Eaves, B.C., and Rothblum, U. (1983). "A Decomposition and Scaling-inequality for Line-sum-symmetric Nonnegative Matrices," Technical Report SOL 83-21, Department of Operations Research, Stanford University, Stanford, CA, December. Also in *SIAM J. on Alg. Disc. Meth.,* **6**, April 1985.

Dantzig, G.B., Eisenberg, E., and Cottle, R.W. (1962). "Symmetric Dual Nonlinear Programs," Technical Report RR-35, Operations Research Center, University of California, Berkeley. Also in *Pacific Journal of Mathematics* **15**, 1965, 809–812.

Dantzig, G.B., Eisenstat, S., Magnanti, T.L., Maier, S., McGrath, M., Nicholson, V., and Reidl, C. (1970). "MPL: Mathematical Programming Language Specification Manual for Committee Review," Technical Report STAN-CS-70-187, Department of Computer Science, Stanford University, Stanford, CA.

Dantzig, G.B., Kawaratani, T.K., and Ullman, R.J. (1960). "Computing Tetraethyl-Lead Requirements in a Linear-Programming Format," *Operations Research* **8**, 24–29.

Dantzig, G.B., Friel, J., Golightly, R., Harvey, R.P., and McKnight, R.D. (1975). "Solution of a Large-Scale Air Force Ordnance Planning Problem by Mathematical Programming," Proceedings of the Fire Support Requirements Methodology Workshop, Ketron, Inc., August.

Dantzig, G.B., Folkman, J., and Shapiro, M. (1965). "On the Continuity of the Minimum Set of a Continuous Function," The RAND Corporation, June. Also in *Journal of Math. Anal. and Appl.* **17**, March 1967, 519–548.

Dantzig, G.B., Ford, L.R., and, Fulkerson, D.R. (1956). "A Primal-Dual Algorithm for Linear Programs," in H.W. Kuhn and A.W. Tucker (eds.), *Linear Inequalities and Related Systems,* Annals of Mathematics Study No. 38, Princeton University Press, Princeton, New Jersey, 171–181; also, *RM*-**1709**, The RAND Corporation, May.

Dantzig, G.B. and Fulkerson, D.R. (1955). "Computation of Maximal Flows in Networks," *Naval Research Logistics Quarterly,* **2**, 277–283.

Dantzig, G.B. and Fulkerson, D.R. (1956a). "On the Max-Flow Min-Cut Theorems of Networks," in H.W. Kuhn and A.W. Tucker (eds.), *Linear Inequalities and Related Systems,* Annals of Mathematics Study No. 38, Princeton University Press, Princeton, New Jersey, 215–221.

Dantzig, G.B. and Fulkerson, D.R. (1956b). "Minimizing the Number of Tankers to Meet a Fixed Schedule," *Naval Research Logistics Quarterly,* **1**, 217–222.

Dantzig, G.B., Fulkerson, D.R., and Johnson, S.M. (1954). "Solution for a Large-Scale Traveling Salesman Problem," *Journal of Operations Research Society of America* **2**, 393–410.

Dantzig, G.B., Fulkerson, D.R., and Johnson, S.M. (1959). "On a Linear Programming Combinatorial Approach to the Traveling Salesman Problem," *Operations Research* **7**, 58–66.

Dantzig, G.B. and Glynn, P.W. (1990). "Parallel Processors for Planning Under Uncertainty," *Annals of Operations Research* **22**, 1–21. Also in Technical Report SOL 88-8, Department of Operations Research, Stanford University, Stanford, CA, June.

Dantzig, G.B., Glynn, P.W., Avriel, M., Stone, J.C., Entriken, R., and Nakayama, M. (1989). "Decomposition Techniques for Multi-Area Transmission Planning Under Uncertainty," Final Report of EPRI Project RP 2940-1, prepared by Systems Optimization Laboratory, Department of Operations Research, Stanford University, Stanford, CA.

Dantzig, G.B., Hax, R, Pomeroy, R., Sanderson, R., and van Slyke, R. [with contributions by G. Buerk, I. Durrer, B. Laurent, S. Mukerjee] (1970). "Natural Gas Transmission System Optimization," American Gas Association, Inc., April.

Dantzig, G.B. and Harvey, R.P., *et al.* (1969). "Sparse Matrix Techniques in Two Mathematical Programming Codes,", Technical Report 69-1, Department of Operations Research, Stanford University, Stanford, CA. Also in D.J. Rose and R.A. Willoughby (eds.), *Sparse Matrices and their Applications*, Plenum Press, New York, in Sparse Matrices and Their Applications, Proceedings of the Symposium on Sparse Matrices and Their Applications IBM RA-1, IBM Watson Research Center, September 1968), March 1969, 85–100.

Dantzig, G.B., Harvey, R.P., Lansdowne, Z.F., and Muth, R. (1979). "Framework for a System of Transportation Spatial Form Research Tools," Report DOT-TSC-RSPA-79-12, Final Report to U.S. Department of Transportation, Washington, D.C., April.

Dantzig, G.B., Harvey, R.P., Lansdowne, Z.F., Maier, S.F., and Robinson, D.W. (1977a). "Computational Experience with a Continuous Network Design Code," GSBA Working Paper 236, Graduate School of Business Administration, Duke University, Durham, N.C., December.

Dantzig, G.B., Harvey, R.P., Lansdowne, Z.F., Maier, S.F., and Robinson, D.W. (1977b). "Formulating and Solving the Network Design Problem by Decomposition," GSBA Working Paper 215, Graduate School of Business Administration, Duke University, Durham, N.C., January 1977. Also in "A Convex Network Design Model Based on a Decomposition Procedure," *Transportation Research* **B, 13B**, 1979, 5–17,

Dantzig, G.B., Harvey, R.P., Lapin, L.L., and Uren, J. (1966). "An Integer Branching Algorithm for Locating Warehouses," Standard Oil Company of California Report, Operations Research Division, October (revised December 1968).

Dantzig, G.B., Harvey, R.P., and McKnight, R. (1964). "Updating the Product Form of the Inverse for the Revised Simplex Method," Technical Report 64-33, Operations Research Center, University of California, Berkeley. Also in A.C.M. Proceedings, August 1965; and Abstract in J.A.C.M., October 1965.

Dantzig, G.B. and Hirsch, W. (1954). "The Fixed Charge Problem," *P*-**648**, The RAND Corporation; also in *Naval Research Logistics Quarterly,* **15**, 413–424.

Dantzig, G.B., Ho, J.K., and Infanger, G. (1991). "Solving Stochastic Linear Programs on a Hypercube Multicomputer," Technical Report SOL 91-10, Department of Operations Research, Stanford University, Stanford, CA, August.

Dantzig, G.B. and Hoffman, A. (1956). "Dilworth's Theorem on Partially Ordered Sets," in H.W. Kuhn and A.W. Tucker (eds.), *Linear Inequalities and Related Systems*, Annals of Mathematics Study No. 38, Princeton University Press, Princeton, New Jersey, 207–213. Also in "New Directions in Mathematical Programming," RAND Symposium on Mathematical Programming, March 16-20, 1959, RAND R-351, page 1.

Dantzig, G.B. Hoffman, A.J., and Hu, T.C. (1983). "Triangulations (Tilings) and Certain Block Triangular Matrices," Technical Report SOL 83-17, Department of Operations Research, Stanford University, Stanford, CA, September. Also in *Mathematical Programming* **31**, 1985, 1–14.

Dantzig, G.B. and Holling, C.S. (1974). "Determining Optimal Policies for Ecosystems," Technical Report 74-11, Department of Operations Research, Stanford University, Stanford, CA, August.

Dantzig, G.B., Holling, C.S., Baskerville, C., Jones, D.D., and Clark, W. C. (1975). "A Case Study of Forest Ecosystem/Pest Management," Prepared for Proceedings International Canadian Conference on Applied Systems Analysis, 1975, WP-75-60, International Institute for Applied Systems Analysis, Laxenburg, Austria, June 1975.

Dantzig, G.B., Holling, C.S., Clark, W.C., Jones, D.D., Baskerville, G., and Peterman, R.M. (1976). "Quantitative Evaluation of Pest Management Options: The Spruce Budworm Case Study," in Wood, D.L. (ed.), *Proceedings of the XVth International Congress of Entomology*, August. Also in in Waters, W.E. (ed.), *Current Topics in Forest Entomology*, U.S. Government Printing Office, Washington, D.C., February 1979, 82–102.

Dantzig, G.B., Holling, C.S., and Winkler, C. (1986). "Determining Optimal Policies for Ecosystems," *TIMS Studies in the Management Sciences* **21**, 1986, 453-473.

Dantzig, G.B. and Infanger, G. (1991). "Large-Scale Stochastic Linear Programs: Importance Sampling and Benders Decomposition," Technical Report SOL 91-4, Department of Operations Research, Stanford University, Stanford, CA, March. Also in *Proceedings of the 13th IMACS World Congress on Computation and Applied Mathematics*, Dublin, Ireland, July 22–26, 1991.

Dantzig, G.B. and Infanger, G. (1992). "Approaches to Stochastic Programming with Applications to Electric Power Systems," *Proceedings of tutorial on Optimization in Planning and Operation of Electric Power Systems*, October 15-16, 1992, Thun, Switzerland, 141–157.

Dantzig, G.B. and Infanger, G. (1993). "Multi-Stage Stochastic Linear Programs for Portfolio Optimization," *Annals of Operations Research* **45**, 59–76. Also in Technical Report SOL 91-11, Department of Operations Research, Stanford University, Stanford, CA, September. Also in *Proceedings of the Annual Symposium of RAMP (Research Association on Mathematical Programming)*, Tokyo, 1991.

Dantzig, G.B. and Iusem, A. (1981). "Analyzing Labor Productivity Growth with the PILOT Model," Technical Report SOL 81-4, Department of Operations Research, Stanford University, Stanford, CA, March. Also in Energy, Productivity and Economic Growth, A Workshop sponsored by the Electric Power Research Institute, Oelgeschlager, Gunn & Hain, Cambridge, Mass., 1983, 347–366.

Dantzig, G.B. and Jackson, P. (1979). "Pricing Underemployed Capacity in a Linear Economic Model," Technical Report SOL 79-2, Department of Operations Research, Stanford University, Stanford, CA, February. Also in Cottle, R. W., Giannessi, F., and Lions, J. L. (eds.), *Variational Inequalities and Complementarity Problems: Theory and Applications*, John Wiley and Sons, Ltd., London, 1980, 127–134.

Dantzig, G.B. and Johnson, D.L. (1963). "Maximum Payloads per Unit Time Delivered Through an Air Network," Report D1-82-0265, Boeing Scientific Research Laboratories, June. Also in *Operations Research* **12**, 230–236.

Dantzig, G.B. and Johnson, S. (1955). "A Production Smoothing Problem," Proceedings, Second Symposium on Linear Programming, National Bureau of Standards and Comptroller, U.S.A.F. Headquarters, January, 151–176.

Dantzig, G.B., Johnson, S., and Wayne, W. (1958). "A Linear Programming Approach to the Chemical Equilibrium Problem," *Management Science* **5**, 38–43.

Dantzig, G.B., Leichner, S.A., and Davis, J.W. (1992a). "A Strictly Improving Phase I Algorithm Using Least-Squares Subproblems," Technical Report SOL 92-1, Department of Operations Research, Stanford University, Stanford, CA, April.

Dantzig, G.B., Leichner, S.A., and Davis, J.W. (1992b). "A Strictly Improving Linear Programming Phase I Algorithm," Technical Report SOL 92-2, Department of Operations Research, Stanford University, Stanford, CA, April. To appear in Annals of Operations Research.

Dantzig, G.B., Levin, S., and Bigelow, J. (1965). "On Steady-State Intercompartmental Flows," Technical Report 65-26, Operations Research Center, University of California, Berkeley. Also in *J. Colloid and Interface Sci.* **23**, April 1967, 572–576.

Dantzig, G.B. and Madansky, A. (1961). "On the Solution of Two-Staged Linear Programs Under Uncertainty," in J. Neyman (ed.), *Proceedings 4th Berkeley Symposium on Mathematical Statistics and Probability, 1961*, University of California Press, Berkeley, California, 165–176. Also in *P*-**2039**, The RAND Corporation.

Dantzig, G.B., Magnanti, T.L., and Maier, S. (1972). "The User's Guide to MPL/T.1 (Revised)," GSBA Working Paper 76, Graduate School of Business Administration, Duke University, Durham, N.C., December 1972.

Dantzig, G.B., Maier, S.F., and Lansdowne, Z.F. (1976). "The Application of Decomposition to Transportation Network Analysis,", DOT Report, Control Analysis Corporation, Palo Alto, California, March.

Dantzig, G.B. and Manne, A. (1974). "A Complementarity Algorithm for an Optimal Capital Path with Invariant Proportions," International Institute for Applied Systems Analysis, Laxenburg, Austria. Also in Technical Report 74-1, Department of Operations Research, Stanford University, Stanford, CA, March. Also in *Journal of Economic Theory* **9**, November, 312–323.

Dantzig, G.B., McAllister, P.H., and Stone, J.C. (1985). "Changes Made for the PILOT-1983 Model," Technical Report SOL 85-12, Department of Operations Research, Stanford University, Stanford, CA, July.

Dantzig, G.B., McAllister, P.H., and Stone, J.C. (1988a). "Deriving a Utility Function for the U.S. Economy," Technical Report SOL 88-6, Department of Operations Research, Stanford University, Stanford, CA, April. Parts I, II, III, in *Journal for Policy Modeling* **11**, 1989, 391–424 and Parts IV, V in *Journal for Policy Modeling* **11**, 1989, 569–592.

Dantzig, G.B., McAllister, P.H., and Stone, J.C. (1988b). "Analyzing the Effects of Technological Change: A Computational General Equilibrium Approach," Technical Report SOL 88-12, Department of Operations Research, Stanford University, Stanford, CA, July.

Dantzig, G.B., McAllister, P.H., and Stone, J.C. (1990). "An Interactive Model Management System: User Interface and System Design," Technical Report SOL 90-3, Department of Operations Research, Stanford University, Stanford, CA, January.

Dantzig, G.B. and Orchard-Hays, W. (1953). "Alternate Algorithm for the Revised Simplex Method Using Product Form for the Inverse (Notes on Linear Programming: Part V)," *RM*-**1268**, The RAND Corporation, November.

Dantzig, G.B. and Orchard-Hays, W. (1954). "The Product Form for the Inverse in the Simplex Method," Mathematical Tables and Other Aids to Computation VIII, April, 64–67.

Dantzig, G.B. and Orden, A. (1952). "A Duality Theorem Based on the Simplex Method," Symposium on Linear Inequalities and Programming, Report 10, Project SCOOP, Planning Research Division, Director of Management Analysis Service, Comptroller, U.S.A.F. Headquarters, April, 51–55.

Dantzig, G.B. and Orden, A. (1953). "Duality Theorems (Notes on Linear Programming: Part II)," *RM*-**1526**, The RAND Corporation.

Dantzig, G.B., Orden, A., and, Wolfe, P. (1955). "The Generalized Simplex Method for Minimizing a Linear Form Under Linear Inequality Constraints (Notes on Linear Programming: Part I)," *Pacific Journal of Mathematics* **5**, 183–195. See also *RM*-**1264**, The RAND Corporation, April 5, 1954.

Dantzig, G.B. and Pace, N. (1963a). "Molecular-Sized Channels and Flows Against the Gradient," Technical Report 63-14, Operations Research Center, University of California, Berkeley.

Dantzig, G.B. and Pace, N. (1963b). "A Model for Sodium-Potassium Transport in Red Cells," Technical Report 63-26, Operations Research Center, University of California, Berkeley.

Dantzig, G.B. and Parikh, S.C. (1975). "On a PILOT Linear Programming Model for Assessing Physical Impact on the Economy of a Changing Energy Picture," Technical Report SOL 75-14, Department of Operations Research, Stanford University, Stanford, CA, June 1975 (revised SOL 75-14R, August 1975). Also in Roberts, F.S. (ed.), *Energy: Mathematics and Models*, Proceedings of a SIMS Conference on Energy, held at Alta, Utah, July 7–11, 1975, SIAM, 1976, 1–23. Also in IIASA Conference '76, May 10–13 1976, 183–200. Also in Proceedings of Symposia in Applied Mathematics, Vol. 21, American Mathematical Society, 1977, pp. 93–106.

Dantzig, G.B. and Parikh, S.C. (1976). "Energy Models and Large-Scale Systems Optimization," Technical Report SOL 76-23, Department of Operations Research, Stanford University, Stanford, CA, November. Also in White, W.W. (ed.), *Computers and Mathematical Programming*, Proceedings of the Bicentennial Conference on Mathematical Programming, November 1976. Also in NBS Special Publication 502, February 1978, 4–10.

Dantzig, G.B. and Parikh, S.C. (1977). "At the Interface of Modeling and Algorithms Research" Technical Report SOL 77-29, Department of Operations Research, Stanford University, Stanford, CA, October. Proceedings of Nonlinear Programming Symposium 3, University of Wisconsin, July 1977, Academic Press, 1978, 283–302.

Dantzig, G.B. and Parikh, S.C. (1978). "PILOT Model for Assessing Energy-Economic Options," in Bagiotti, T. and Franco, G. (eds.), *Pioneering Economics*, Edizioni Cedam - Padova, Italy, 271–276.

Dantzig, G.B. and Pereira, M.V.F., *et al.* (1988). "Mathematical Decomposition Techniques for Power System Expansion Planning," EPRI EL-5299, Volumes 1–5, February 1988, Electric Power Research Institute, Palo Alto, CA.

Dantzig, G.B. and Perold, A.F. (1978). "A Basic Factorization Method for Block Triangular Linear Programs," SOL78-7, April. Also in I.S. Duff and G.W. Stewart (eds.), *Sparse Matrix Proceedings*, SIAM, Philadelphia, 1979, 283–312.

Dantzig, G.B. and Ramser, J.H. (1959a). "Optimum Routing of Gasoline Delivery Trucks," Proceedings, World Petroleum Congress, Session VIII, Paper 19, 1959.

Dantzig, G.B. and Ramser, J.H. (1959b). "The Truck Dispatching Problem," *Management Science* **6**, 80–91.

Dantzig, G.B. and Reynolds, G.H. (1966). "Optimal Assignment of Computer Storage by Chain Decomposition of Partially Ordered Sets," Technical Report 66-6, Operations Research Center, University of California, Berkeley.

Dantzig, G.B. and Saaty, T.L. (1973). *Compact City*, Freeman, San Francisco.

Dantzig, G.B. and Sethi, S.P. (1981). "Linear Optimal Control Problems and Generalized Linear Programming," *Journal of the Operational Research Society* **32**, 467–476.

Dantzig, G.B. and Shapiro, M. (1960). "Solving the Chemical Equilibrium Problem Using the Decomposition Principle," *P*-**2056**, The RAND Corporation, August.

Dantzig, G.B., Stone, J.C., and McAllister, P.H. (1986). "Using the PILOT Model to Study the Effects of Technological Change," Technical Report SOL 86-16, Department of Operations Research, Stanford University, Stanford, CA, December. Also in Lev, B., Bloom, J., Gleit, A., Murphy, F., and Shoemaker, C. (eds.), *Strategic Planning in Energy and Natural Resources*, Studies in Management Science and Systems, Vol. 15, Proceedings of the 2nd Symposium on Analytic Techniques for Energy, Natural Resources and Environmental Planning April 1986; North-Holland, Amsterdam, 1987, 31–42.

Dantzig, G.B., Stone, J.C., and McAllister, P.H. (1988). "Formulating an Objective for an Economy," Proceedings of the Martin Beale Memorial Symposium, Mathematical Programming 42, (Series B), 11–32.

Dantzig, G.B. and Tomlin, J.A. (1987). "E.M.L. Beale, FRS: Friend and Colleague," Technical Report SOL 87-2, Department of Operations Research, Stanford University, Stanford, CA, January 1987. Also in *Mathematical Programming* **38**, 117–131.

Dantzig, G.B. and Veinott, A.F. (eds.), (1967). *Mathematics of the Decision Sciences*, Proceedings of the American Mathematical Society Summer Seminar, Providence, RI.

Dantzig, G.B. and Veinott, A.F., editors (1968a). *Lectures in Applied Mathematics*, Vol. 11, American Mathematics Society, Providence, RI.

Dantzig, G.B. and Veinott, A.F. (1968b). "Integral Extreme Points," Technical Report 67-7, Department of Operations Research, Stanford University, Stanford, CA, November. Also in *SIAM Review*, **10**, 371–372.

Dantzig, G.B. and Veinott, A.F. (1977). "Discovering Hidden Totally Leontief Substitution Systems," Technical Report SOL 77-17, Department of Operations Research, Stanford University, Stanford, CA, June. Revised for *Mathematics of Operations Research* **3**, May 1978, 102–103.

Dantzig, G.B. and Wald, A. (1951). "On the Fundamental Lemma of Neyman and Pearson," *Annals of Math. Stat.* **22**, 87–93.

Dantzig, G.B. and Wolfe, P., (1960). "Decomposition Principle for Linear Programs," *Operations Research* **8**, 101–111. Also in Wolfe, P. (ed.), *RAND Symposium on Mathematical Programming*, March 1959, RAND R-351, page 5. Also in Dantzig, G.B. and Eaves, B.C. (eds.), *Studies in Optimization*, MAA Studies in Mathematics, Vol. 10, Mathematical Association of America, 1974.

Dantzig, G.B. and Wolfe, P. (1961). "The Decomposition Algorithm for Linear Programming," *Econometrica,* **29**, 767–778.

Dantzig, G.B. and Wolfe, P. (1962). "Linear Programming in a Markov Chain," *Operations Research* **10**, 702–710. Also *RM-2957-PR*, The RAND Corporation, April.

Dantzig, G.B. and Wood, M.K. (1951). "Programming of Interdependent Activities, I: General Discussion," *Econometrica,* **17**, 193–199; also in T.C. Koopmans (ed.), *Activity Analysis of Production and Allocation,* July–October 1949, Cowles Commission Monograph 13, Proceedings of Linear Programming Conference, June 20–24, 1949, John Wiley and Sons, New York, 15–18.

Dantzig, G.B. and Van Slyke, R.M. (1964a). "Generalized Upper Bounded Techniques for Linear Programming — I," Technical Report 64-17, Operations Research Center, University of California, Berkeley. Also in Proceedings IBM Scientific Computing Symposium, Combinatorial Problems, March 16–18, 1964, 249–261.

Dantzig, G.B. and Van Slyke, R.M. (1964b). "Generalized Upper Bounded Techniques for Linear Programming — II," Technical Report 64-18, Operations Research Center, University of California, Berkeley.

Dantzig, G.B. and Van Slyke, R.M. (1967). "Generalized Upper Bounding Techniques," *Journal of Computer and System Science* **1**, 213–226.

Dantzig, G.B. and Van Slyke, R.M. (1971). "Generalized Linear Programming," in David Wismer (ed.), *Optimization Methods and Applications for Large Systems*, McGraw-Hill, New York, 75–120.

Dantzig, G.B. and Ye, Y. (1990). "A Build-Up Interior Method for Linear Programming: Affine Scaling Form," Technical Report SOL 90-4, Department of Operations Research, Stanford University, Stanford, CA, March.

Davidon, W.C. (1979). "Variable Metric Methods for Optimization," *A. E. C. Research and Development Report ANL-5990*, Argonne National Laboratory, Argonne, Illinois.

Davis, K.D. and McKeown, P.G. (1981). *Quantitative Models for Management*, Kent Publishing Company, Boston, Massachusetts, A Division of Wadswirth Inc., Belmont, California.

Davis, P.J. and Rabinowitz, P. (1984). *Methods of Numerical Integration*, Academic Press, London and New York.

Day, J. and Peterson, B. (1988). "Growth in Gaussian Elimination," *The American Mathematical Monthly* **95**, 489–513.

de Boor, C.W. (1971). "CADRE: An Algorithm for Numerical Quadrature," in J. Rice (ed.), *Mathematical Software*, Academic Press, London and New York, 417–450, 315–330.

Deák, I. (1988). "Multidimensional Integration and Stochastic Programming," in Y. Ermoliev and R.J.B. Wets (eds.), *Numerical Techniques for Stochastic Optimization*, Springer-Verlag, Berlin and New York, 187–200.

Dembo, R.S. and Steihaug, T. (1983). "Truncated Newton Methods for Large Scale Unconstrained Optimization," *Mathematical Programming* **26**, 190–212.

Demmel, J.W. (1987). "On the Distance to the Nearest Ill-Posed Problem," *Numerische Mathematik* **51**, 251–289.

Denardo, E.V. and Fox, B.L. (1979). "Shortest-route Methods: 1. Reaching, Pruning, and Buckets," *Operations Research* **27**, 161–186.

Dennis, J.E., Jr. (1977) "Nonlinear least squares," in D. Jacobs (ed.), *The State of the Art in Numerical Analysis*, Academic Press, London and New York, 269–312.

Dennis, J.E. and More, J.J. (1977). "Quasi-Newton Methods, Motivation and Theory," *SIAM Review,* **19**, 46–89.

Dennis, J.E. and Moré, J.J. (1974). "A Characterization of Superlinear Convergence and its Application to Quasi-Newton Methods," *Mathematics of Computation* **28**, 549–560.

Dial, R.B. (1969). "Algorithm 360: Shortest Path Forest with Topological Ordering," *Communications of the Association for Computing Machinery* **12**, 632–633.

Dijkstra, E. (1959). "A Note on Two Problems in Connection with Graphs," *Numerische Mathematik* **1**, 269–271.

Dilworth, R.P. (1950). "A Decompositon Theorem for Partially Ordered Sets," *Annals of Mathematics* **51**, 161–166.

Dinic, E.A. (1970). "Algorithm for Solution of a Problem of Maximum Flow in a Network with Power Estimation," *Soviet Mathematics Doklady* **11**, 1277–1280.

Dikin, I.I. (1967). "Iterative Solution of Problems of Linear and Quadratic Programming," *Doklady Akademiia Nauk USSR* **174**, 747–748, Translated in *Soviet Mathematics Doklady* **8**, 674–675.

Dikin, I.I. (1974). "On the Convergence of an Iterative Process," *Upravlyaemye Sistemi* **12**, 54–60.

Dikin, I.I. (1990). "The Convergence of Dual Variables," Technical Report, Siberian Energy Institute, Irkutsk, Russia.

Dikin, I.I. (1992). "Determination of an Interior Point of one System of Linear Inequalities," *Kibernetika and System Analysis* **1**, 74–96.

Dikin, I.I. and Zorkaltsev, V.I. (1980). *Iterative Solution of Mathematical Programing Problems: Algorithms for the Method of Interior Points*, Nauka, Novosibirsk, USSR.

Dixon, L.C.W. (1972a). "Quasi-Newton Methods Generate Identical Points," *Mathematical Programming* **2** 383–387.

Dixon, L.C.W. (1972b). "Quasi-Newton Methods Generate Identical Points. II. The Proof of Four New Theorems," *Mathematical Programming* **3** 345–358.

Dodson, D.S. and Lewis, J.G. (1985). "Proposed Sparse Extensions to the Basic Linear Algebra Subprograms," *SIGNUM Newsletter* **20**, 22–25.

Doig, A.G. and Belz, M.H. (1956). "Report on Trim Problems for May, 1956." Department of Statistics, University of Melbourne, Australia. The report is addressed to Australia Paper Manufacturers, Melbourne, Australia, July 31, 1956.

Doig, A.G. and Land, A.H. (1960). "An Automatic Method of Solving Discrete Programming Problems," *Econometrica,* **28**, 497–520.

Dongarra, J.J., Bunch, J.R., Moler, C.B., and Stewart, G.W. (1979). "LINPACK Users Guide," SIAM, Philadelphia.

Dongarra, J.J., Duff, I.S., Sorensen, D.C., and van der Vorst, H.A. (1991). *Solving Linear Systems on Vector and Shared Memory Computers*, SIAM, Philadelphia.

Dongarra, J.J. and Eisenstat, S. (1984). "Squeezing the Most out of an Algorithm in Cray Fortran," *ACM Transactions on Mathematical Software*, **10**, 221–230.

Dongarra, J.J. and Hinds, A. (1979). "Unrolling Loops in Fortran," *Software Practice and Experience*, **9**, 219–229.

Dongarra, J.J., DuCroz, J., Hammarling, S. and Hanson, R.J. (1985). "A Proposal for an Extended Set of Fortran Basic Linear Algebra Subprograms," *SIGNUM Newsletter* **20**, 2–18.

Dongarra, J.J., DuCroz, J., Hammarling, S. and Hanson, R.J. (1988a). "An Extended Set of Fortran Basic Linear Algebra Subprograms," *ACM Transactions on Mathematical Software*, **14**, 1–17.

Dongarra, J.J., DuCroz, J., Hammarling, S. and Hanson, R.J. (1988b). "Algorithm 656 An Extended Set of Fortran Basic Linear Algebra Subprograms: Model Implementation and Test Programs," *ACM Transactions on Mathematical Software*, **14**, 18–32.

Dongarra, J.J., DuCroz, J., Hammarling, S. and Hanson, R.J. (1988c). "A Set of Level 3 Basic Linear Algebra Subprograms," Report ANL-MCS-TM-88, Argonne National Laboratory, Argonne, Illinois.

Dongarra, J.J., Gustavson, F.G., and Karp, A. (1984). "Implementing Linear Algebra Algorithms for Dense Vectors on a Vector Pipeline Machine," *SIAM Review*, **26**, 91–112.

Duff, I.S., Erisman, A.M., and Reid, J.K. (1986). *Direct Methods for Sparse Matrices*, Oxford University Press, Oxford and New York.

Duff, I.S. (1976). "A Survey of Sparse Matrix Research," *Report AERE CSS 28*, Atomic Energy Research Establishment, Harwell, England.

Duff, I.S. (1977). "MA28—A Set of Fortran Subroutines for Sparse Unsymmetric Linear Equations," *Report AERE AERE R8730*, Atomic Energy Research Establishment, Harwell, England.

Duff, I.S. (1981). "An Algorithm for Obtaining a Maximum Transversal," *ACM Transactions on Mathematical Software*, **7**, 315–330.

Duff, I.S. and Reid, J.K., (1978). "An Implementation of Tarjan's Algorithm for the Block Triangularization of a Matrix," *ACM Transactions on Mathematical Software*, **4**, 137–147.

Duff, I.S. and Stewart, G.W., eds. (1979). *Sparse Matrix Proceeedings*, SIAM, Philadelphia.

Duffin, R.J. (1974). "On Fourier's Analysis of Linear Inequality Systems," *Mathematical Programming Study* **1**, 71–95.

Dupačová, J. (1990). "Stability and Sensitivity Analysis for Stochastic Programming," *Annals of Operations Research* **27**, 115–142.

E

Eaves, B.C., (1979). "A View of Complementary Pivot Theory (or Solving Equations with Homotopies," in: *Constructive Approaches to Mathematical Models*, Academic Press, London and New York, 153–170.

Edmonds, J. and Karp, R.M. (1972). "Theoretical Improvements in Algorithmic Efficiency for Network Flow Algorithms," *Journal of the Association for Computing Machinery*, **19**, 248–264.

Edmondson, J.H. (1951). "Proof of the Existence Theorem of an Epsilon Transformation," class exercise dated March 28, 1951, for Department of Agriculture Graduate School course in linear programming given by George Dantzig.

Egerváry, E. (1931). "Matrixok Kombinatorius Tulajfonságairól," *Matematikai és Fizikai Lapok*, No. 38, 16–28. "On Combinatorial Properties of Matrices," translated by H.W. Kuhn, Paper No. 4, George Washington University Logistics Research Project. Published in *Logistics Papers*, Issue No. 11, Appendix I to Quarterly Progress Report No. 21, (November 16, 1954 to February 15, 1955).

Elias, P., Feinstein, A., and Shannon, C.E. (1956). "Note on Maximum FLow Through a Network," *IRE Transactions on Information Theory*, **IT-2**, 117–119.

Elmaghraby, S.E. (1977). *Activity Networks: Project Planning and Control by Network Models*, Wiley, New York.

Entriken, R. (1989). "The Parallel Decomposition of Linear Programs," Ph.D. thesis, Department of Operations Research, Stanford University, Stanford, CA.

Erisman, A.M., Grimes, R.G., Lewis, J.G., and Poole, W.G., Jr. (1985). "A Structurally Stable Modification of Hellerman-Rarick's P^4 Algorithm for Reordering Unsymmetric Sparse Matrices," *SIAM J. on Numerical Analysis* **22**, 369–385.

Erisman, A.M. and Reid, J.K. (1974). "Monitoring the Stability of the Triangular Factorization of a Sparse Matrix," *Numerische Mathematik* **22**, 183–186.

Ermoliev, Y. (1983). "Stochastic Quasi-Newton Methods and Their Applications to Systems Optimization," *Stochastics*, **9**, 1–36.

Ermoliev, Y. (1988). "Stochastic Quasi-Gradient Methods," in Y. Ermoliev and R.J.B. Wets (eds.), *Numerical Techniques for Stochastic Optimization*, Springer-Verlag, Berlin and New York, 141–186.

Ermoliev, Y. and Wets, R.J.B., eds. (1988). *Numerical Techniques for Stochastic Optimization*, Springer-Verlag, Berlin and New York.

Eisemann, K. (1957). "The Trim Problem," *Management Science* **3**, 279–284.

F

Fabian, T. (1954). "Process Type Analysis of the Iron and Steel Industry, Part IV—Programming Open Hearth Steel Production," Discussion Paper No. 46, Management Sciences Research Project, University of California, Los Angeles, California, 12 pages.

Fabian, T. (1955). "Process Analysis of the Iron and Steel Industry: A Model," *Econometrica,* **23**, 347–348 (abstract). Also Research Report No. 47, Management Sciences Research Project, University of California, Los Angeles, California.

Fabian, T. (1958). "A Linear Programming Model of Integrated Iron and Steel Production," *Management Science* **4**, 425–449.

Farkas, J. (1902). "Über die Theorie der einfachen Ungleichungen," *J. Reine Angew. Math.,* **124**, 1–24.

Faddeev, D.K., and Faddeeva, V.N. (1963). *Computational Methods in Linear Algebra,* Freeman, San Francisco. Translated from the Russian version.

Faulkner, J.C. (1988). "Bus crew scheduling and the set partitioning model," Ph.D. thesis, Department of Theoretical and Applied Mechanics, University of Auckland, Auckland, New Zealand.

Feinstein, C.D. and Thapa, M.N. (1993). "A Reformulation of a Mean-Absolute Deviation Portfolio Optimization Model," *Management Science* **39**, 1552–1553.

Feller, W. (1957). *An Introduction to Probability Theory and its Applications,* Volume I, John Wiley and Sons, New York.

Feller, W. (1969). *An Introduction to Probability Theory and its Applications,* Volume II, John Wiley and Sons, New York.

Fenner, T. and Loizou, G. (1974). "Some New Bounds on the Condition Numbers of Optimally Scaled Matrices," *Journal of the Association for Computing Machinery,* **21**, 514–524.

Ferguson, A.R. and Dantzig, G.B., (1955). "Notes on Linear Programming: Part XVI—The Problem of Routing Aircraft—a Mathematical Solution," *Aeronautical Engineering Review* **14**, 51–55. Also: *RM*-**1369**, The RAND Corporation, September 1, 1954, and *P*-**561**, The RAND Corporation, 1954.

Ferguson, A.R. and Dantzig, G.B., (1956). The Allocation of Aircraft to Routes—An Example of Linear Programming under Uncertain Demand," *Management Science* **3** 1, 45–73. Also: *P*-**727**, The RAND Corporation, December 7, 1956. Also in Bowman and Fetter (eds.), *Analysis of Industrial Operations,* Richard D. Irwin, Inc., Homewood, Illinois, 1959.

Fisher, W.D. and Schruben, L.W. (1953). "Linear Programming Applied to Feed-Mixing Under Different Price Conditions," *J. Farm Econ.* **35**, 471–483.

Fiacco, A.V. and McCormick, G.P. (1968). *Nonlinear Programming: Sequential Unconstrained Minimization Techniques,* John Wiley and Sons, New York.

Fletcher, R. (1985). "Degeneracy in the Presence of Round-Off Errors," Technical Report NA/89, Department of Mathematical Sciences, University of Dundee, Dundee.

Fletcher, R. (1987). "Recent Developments in Linear and Quadratic Programming", in Iserles, A. and Powell, M.J.D. (eds.), *The State and Art in Numerical Analysis,* Oxford University Press, Oxford and New York, 213–243.

Fletcher, R., and Mathews, S.P.J. (1984). "Stable Modification of Explicit LU Factors for Simplex Updates," *Mathematical Programming* **30**, 267–284.

Ford, L.R., Jr., and Fulkerson, D.R. (1956). "Maximal Flow Through a Network," *Canadian Journal of Mathematics* **8**, 399–404. This appeared first as the RAND Corporation Research Memorandum RM-1400, November 19, 1954.

Ford, L.R., Jr., and Fulkerson, D.R. (1957). "A Simple Algorithm for Finding Maximal Network Flows and an Application to the Hitchcock Problem," *Canadian Journal of Mathematics* **9**, 210–218.

Ford, L.R., Jr., and Fulkerson, D.R. (1958a). "Constructing Maximal Dynamic Flows from Static Flows," *Operations Research* **6**, 419–433.

Ford, L.R., Jr., and Fulkerson, D.R. (1958b). "Suggested Computation for Maximal Multi-Commodity Network Flows," *Management Science* **5**, 97–101.

Ford, L.R., Jr., and Fulkerson, D.R. (1962). *Flows in Networks*, Princeton University Press, Princeton, New Jersey.

Forrest, J.J.H. and Goldfarb, D. (1992). "Steepest-edge Simplex Algorithms for Linear Programming," *Mathematical Programming* **57**, 341–376.

Forrest, J.J.H. and Tomlin, J. A. (1972). "Updating Triangular Factors of the Basis to Maintain Sparsity in the Product Form Simplex Method," *Mathematical Programming* **2**, 263–278.

Forsgren, A.L. and Murray, W. (1990). "Newton Methods for Large-Scale Linearly Constrained Minimization," Technical Report SOL 90-6, Department of Operations Research, Stanford University, Stanford, CA.

Forsythe, G.E. (1960). "Crout with Pivoting," *Communications of the Association for Computing Machinery* **3**, 507–508.

Forsythe, G.E. (1970). "Pitfalls in Computation, or why a Math Book isn't Enough," *The American Mathematical Monthly* **9**, 931–956.

Forsythe, G.E., Malcolm, M.A., and Moler, C.B. (1977). *Computer Methods for Mathematical Computations*, Prentice-Hall, Inc., Englewood Cliffs, New Jersey.

Forsythe, G.E. and Moler, C.B. (1967). *Computer Solution of Linear Algebraic Systems*, Prentice-Hall, Inc., Englewood Cliffs, New Jersey.

Forsythe, G.E. and Wastow, W.R. (1960). *Finite Difference Methods for Partial Differential Equations*, John Wiley and Sons, New York.

Foster, L.V. (1986). "Rank and Null Space Calculations Using Matrix Decomposition Without Column Interchanges," *Linear Algebra and its Applications* .

Fourer, R. (1979). "Sparse Gaussian Elimination of Staircase Linear Systems," Technical Report SOL 79-17, Department of Operations Research, Stanford University, Stanford, CA.

Fourer, R. (1982). "Solving Staircase Linear Programs by the Simplex Method, 1: Inversion," *Mathematical Programming* **23**, 274–313.

Fourer, R. (1983a). "Solving Staircase Linear Programs by the Simplex Method, 2: Pricing," *Mathematical Programming* **25**, 251–292.

Fourer, R. (1983b). "Modeling Languages versus Matrix Generators for Linear Programming," *ACM Transactions on Mathematical Software,* **9**, 143–183.

Fourer, R. (1984). "Staircase Systems and Matrices," *SIAM Review,* **26**, 1–70.

Fourer, R. (1985). "A Simplex Algorithm for Piecewise-Linear Programming I: Derivation and Proof," *Mathematical Programming* **33** 204–233.

Fourer, R., Gay, D.M., and Kernighan, B.W. (1992). *AMPL: A Modeling Language for Mathematical Programming,* Scientific Press, South San Francisco.

Fourer, R. and Mehrotra, S. (1991). "Performance of an Augmented System Approach for Solving Least-Squares Problems in an Interior-Point Method for Linear Programming," Department of Industrial Engineering and Management Sciences, Northwestern University, Evanston, IL.

Fourer, R. and Mehrotra, S. (1992). "Solving Symmetric Indefinite Systems in an Interior-Point Method for Linear Programming," Technical Report 92-01, Department of Industrial Engineering and Management Sciences, Northwestern University, Evanston, IL.

Fourier, Jean Baptiste Joseph (1826). "Solution d'une question particulière du calcul des inégalités," 1826, and extracts from "Histoire de l'Académie," 1823, 1824, *Oeuvres* II, 317–328.

Fourier, Jean Baptiste Joseph (1890). "Second Extrait," in G. Darboux (ed.), *Oeuvres*, Gauthiers-Villars, Paris, 325–328. English Translation by Kohler [1973].

Fox, L. (1964). *Introduction to Numerical Linear Algebra*, Oxford University Press, Oxford and New York.

Fraley, C. and Vial, J.-Ph. (1989). "Numerical Study of Projective Methods for Linear Programming," in S. Dolecki (ed.), *Optimization Proceedings of the 5th French-German Conference in Castel-Novel, Varetz, France, October 1988*, 25–38.

Frauendorfer, K. (1988). "Solving SLP Recourse Problems with Arbitrary Multivariate Distributions — The Dependent Case," *Mathematics of Operations Research* **13**, *3*, 377–394.

Frauendorfer, K. (1992). *Stochastic Two-Stage Programming*, Lecture Notes in Economics and Mathematical Systems 392, Springer-Verlag, Berlin and New York.

Frauendorfer, K., and Kall, P. (1988). "Solving SLP Recourse Problems with Arbitrary Multivariate Distributions — The Independent Case," *Problems of Control and Information Theory* **17** *(4)*, 177–205.

Fredman, M.L. and Willard, D.E. (1994). "Trans-dichotomous Algorithms for Minimum Spanning Trees and Shortest Paths," *J. Comp. and Syst. Sci.* **48**, 533–551.

Freund, R. (1991). "Theoretical Efficiency of a Shifted Barrier Function Algorithm for Linear Programming," *Linear Algebra and its Applications* **152**, 19–41.

Freund, R. (1991). "Polynomial-Time Algorithms for Linear Programming Based only on Primal Scaling and Projected Gradients of a Potential Function," *Mathematical Programming* **51**, 203–222.

Freund, R. (1988). "Projective Transformation for Interior Point Methods, Part I: Basic Theory and Linear Programming," Working Paper OR 179-88, Operations Research Center, Massachusetts Institute of Technology.

Freund, R. (1987). "An Analog of Karmarkar's Algorithm for Inequality Constrained Linear Programs, with a 'New' Class of Projective Transformations for Centering a Polytope," Working Paper OR 1921-87, Operations Research Center, Massachusetts Institute of Technology.

Frisch, K.R. (1957). *Linear dependencies and a mechanized form of the multiplex method linear programming*, Memorandum of September, 1957, University Institute of Economics, Oslo, Norway.

Frisch, K.R. (1955). *The logarithmic potential method of convex programs*, Memorandum of May 13, 1955, UIEOSLO.

Fulkerson, D.R., and Dantzig, G.B. (1955). "Computations of Maximal Flows in Networks," *Naval Research Logistics Quarterly*, **2**, 277–283.

G

Gabow, H.N. and Tarjan, R.E. (1989). "Faster Scaling Algorithms for Network Problems," *SIAM J. on Computing*, **18**, 1013–1036.

Gainen, L. (1955). "Linear Programming in Bid Evaluations," in H.A. Antosiewicz (ed.), *Proceedings of the Second Symposium in Linear Programming*, Vol. 2, National Bureau of Standards and Directorate of Management Analysis, DCS/Comptroller, USAF, Washington, D.C., 29–38.

Gaivoronski, A. (1988). "Implementation of Stochastic Quasi-Gradient Methods," in Y. Ermoliev and R.J.B. Wets (eds.), *Numerical Techniques for Stochastic Optimization*, Springer-Verlag, Berlin and New York, 313–352.

Gale, D.H., (1956). "Polyhedral Convex Cones and Linear Inequalities," in T.C. Koopmans (ed.), *Activity Analysis of Production and Allocation*, July–October 1949, Cowles Commission Monograph 13, Proceedings of Linear Programming Conference, June 20–24, 1949, John Wiley and Sons, New York,

Gale, D.H., Kuhn, H.W., and Tucker, A.W. (1951). "Linear Programming and the Theory of Games," in T.C. Koopmans (ed.), *Activity Analysis of Production and Allocation*, July–October 1949, Cowles Commission Monograph 13, Proceedings of Linear Programming Conference, June 20–24, 1949, John Wiley and Sons, New York, Chapter 19

Galil, Z. (1978). "A New Algorithm for the Maximal Flow Problem," *Proceedings of the Nineteenth Annual Symposium on Foundations of Computer Science, IEEE*, 231–245.

Gallivan, K., Jalby, W., Meier, U., and Sameh, A.H. (1988). "Impact of Hierarchical Memory Systems on Linear Algebra Algorithm Design," International Journal of Supercomputer Applications **2**, 12–48.

Gallo, G. and Pallottino, S. (1988). "Shortest Path Algorithms," *Annals of Operations Research* **13**, 3–79.

Garbow, B.S., Boyle, J.M., Dongarra, J.J., and Moler, C.B. (1977). *Matrix Eigensystem routines—EISPACK Guide Extensions, Lecture Notes in Computer Science* **51**, Springer-Verlag, Berlin and New York.

Garvin, W.W., Crandall H.W., John, J.B., and Spellman, R.A. (1957). "Applications of Linear Programming in the Oil Industry," *Management Science* **3**, 407–430.

Gass, S.I. (1991). "Model World: Models at the OK Corral," *Interfaces* **21**:6, 80–86.

Gass, S.I. (1985). *Linear Programming: Methods and Applications*, McGraw-Hill, New York.

Gass, S.I., and Saaty, T.L. (1955a). "The Computational Algorithm for the Parametric Objective Function," *Naval Research Logistics Quarterly,* **2**, 39–45.

Gass, S.I., and Saaty, T.L. (1955b). "The Parametric Objective Function, Part 1," *Operations Research* **2**, 316–319.

Gass, S.I., and Saaty, T.L. (1955c). "The Parametric Objective Function, (Part 2)—Generalization," *Operations Research* **3**, 395–401.

Gassner, B. J. (1964). "Cycling in the Transportation Problem," *Naval Research Logistics Quarterly,* **11**, 43-58.

Gay, D.M. (1978). "On Combining the Schemes of Reid and Saunders for Sparse LP Bases," in I.S. Duff and G.W. Stewart (eds.), *Sparse Matrix Proceedings*, SIAM, Philadelphia, 313–334.

Gay, D.M. (1985). "Electronic Mail Distribution of Linear Programming Test Problems," *Mathematical Programming Society COAL Newsletter* **13**, 10–12.

Gay, D.M. (1987). "A Variant of Karmarkar's Linear Programming Algorithm for Problems in Standard Form," *Mathematical Programming* **37**, 81–90.

Gentleman, M. (1973). "Least Squares Computations by Givens Transformations Without Square Roots," *J. Institute of Mathematics and its Applications* **12**, 329–336.

Geoffrion, A.M. (1974). "Elements of Large-Scale Mathematical Programming," *Management Science* **16** .

George, A. and Liu, J. (1981). *Computer Solution of Large Sparse Positive Definite Systems*, Prentice-Hall, Inc., Englewood Cliffs, New Jersey.

George, A. and Ng, E., (1984). "Symbolic Factorization for Sparse Gaussian Elimination with Partial Pivoting," Technical Report CS-84-43, Department of Computer Science, University of Waterloo, Waterloo, Ontario, Canada.

de Ghellinck, G. and Vial, J.-Ph. (1986). "A Polynomial Newton Method for Linear Programming," Special issue of *Algorithmica* **1**, 425–453.

Gill, P.E., Golub, G.H., Murray, W. and Saunders, M.A. (1974). "Methods for Modifying Matrix Factorizations," *Mathematics of Computation* **28**, 505–535.

Gill, P. E., and Murray, W. (1974a). *Safeguarded Steplength Algorithms for Optimization Using Descent Methods*, Report NAC 37, National Physical Laboratory, England.

Gill, P.E. and Murray, W. (1974b). "Newton Type Methods for Unconstrained and Linearly Constrained Optimization," *Mathematical Programming* **7**, 311–350.

Gill, P.E. and Murray, W. (1974c). "Quasi-Newton Methods for Linearly Constrained Optimization," in P.E. Gill and W. Murray (eds.), *Numerical Methods for Unconstrained Optimization*, Academic Press, London and New York, 67–92.

Gill, P.E. and Murray, W. (1974d). *Numerical Methods for Constrained Optimization*, Academic Press, London and New York.

Gill, P.E. and Murray, W. (1977). "Linearly Constrained Problems Including Linear and Quadratic Programming," in D. Jacobs (ed.), *The State of the Art in Numerical Analysis*, Academic Press, London and New York, 313–363.

Gill, P.E. and Murray, W. (1978). "Algorithms for the Solution of the Nonlinear Least Squares Problem," *SIAM J. on Numerical Analysis* **15**, 977–992.

Gill, P.E. and Murray, W. (1979). *Conjugate-Gradient methods for Large Scale Nonlinear Optimization*, Technical Report SOL 79-15, Department of Operations Research, Stanford University, Stanford, CA.

Gill, P.E., Murray, W. and Picken, S.M. (1972). "The Implementation of Two Modified Newton Algorithms for Unconstrained Optimization," Report NAC 11, National Physical Laboratory, England.

Gill, P.E., Murray, W. and Wright, M.H. (1981). *Practical Optimization*, Academic Press, London and New York.

Gill, P.E., Murray, W., and Saunders, M.A. (1988). "Interior-Point Methods for Linear Programming: A Challenge to the Simplex Method," Technical Report SOL 88-14, Department of Operations Research, Stanford University, Stanford, CA.

Gill, P.E., Murray, W., Saunders, M.A., Tomlin, J.A. and Wright, M.H. (1986). "On Projected Newton Barrier Methods for Linear Programming and an Equivalence to Karmarkar's Projective Method," *Mathematical Programming* **36**, 183–209.

Gill, P.E., Murray, W., Saunders, M.A. and Wright, M.H. (1981a). "A Procedure for Computing Forward Finite-Difference Intervals for Numerical Optimization," Technical Report SOL 81-25, Department of Operations Research, Stanford University, Stanford, CA.

Gill, P.E., Murray, W., Saunders, M.A., and Wright, M.H. (1981b). "QP-based Methods for Large Scale Nonlinearly Constrained Optimization," in O.L. Mangasarian, R.R. Meyer, and S.M. Robinson (eds.), *Nonlinear Programming 4*, Academic Press, London and New York, .

Gill, P.E., Murray, W. Saunders, M.A. and Wright, M.H. (1983). "On the Representation of a Basis for the Null Space," Technical Report SOL 83-19, Department of Operations Research, Stanford University, Stanford, CA.

Gill, P.E., Murray, W., Saunders, M.A. and Wright, M.H. (1984a). "Sparse Matrix Methods in Optimizations," *SIAM J. on Scientific and Statistical Computing* **5**, 562–589.

Gill, P.E., Murray, W., Saunders, M.A., and Wright, M.H. (1984b). "Software and its Relationship to Methods," Technical Report SOL 84-10, Department of Operations Research, Stanford University, Stanford, CA.

Gill, P.E., Murray, W., Saunders, M.A. and Wright, M.H. (1984c). "Model Building and Practical Implementation Aspects in Nonlinear Programming," Presented at the NATO Advanced Study Institute on *Computational Mathematical Programming* Bad Windsheim, July 23–August 2, 1984.

Gill, P.E., Murray, W., Saunders, M.A. and Wright, M.H., (1985). "LUSOL User's Guide," Technical Report SOL , Department of Operations Research, Stanford University, Stanford, CA.

Gill, P.E., Murray, W., Saunders, M.A. and Wright, M.H. (1987). "Maintaining LU Factors of a General Sparse Matrix," *Linear Algebra and its Applications* **88/99** 239–270.

Gill, P. E., Murray, W., Saunders, M. A. and Wright, M. H. (1989). "A Practical Anti-Cycling Procedure for Linearly Constrained Optimization," *Mathematical Programming* **45**, 437–474.

Gill, P.E., Murray, W., Picken, S.M., and Wright, M.H. (1979). "The Design and Structure of a FORTRAN Program Library for Optimization," *ACM Transactions on Mathematical Software,* **5**, 259–283.

Gill, P.E., Murray, W., Poncelön, D.B., and Saunders, M.A. (1991a). Primal-Dual Methods for Linear Programming," Technical Report SOL 91-3, Department of Operations Research, Stanford University, Stanford, CA.

Gill, P.E., Murray, W., Poncelön, D.B., and Saunders, M.A. (1991b). Solving Reduced KKT Systems in Barrier Methods for Linear and Quadratic Programming," Technical Report SOL 91-7, Department of Operations Research, Stanford University, Stanford, CA.

Gill, P.E., Murray, W., and Wright, M.H. (1981). *Practical Optimization*, Academic Press, London and New York.

Gill, P.E., Murray, W., and Wright, M.H. (1991). *Numerical Linear Algebra and Optimization*, Addison-Wesley Publishing Company, Reading, Massachusetts.

Gille, P. and Loute, E. (1982). "Updating the LU Gaussian Decomposition for Rank-One Corrections; Application to Linear Programming Basis Partitioning Techniques," *Cahier No. 8201, Séminaire de Mathématiques Appliquées aux Sciences Humaines, Facultés Universitaires Saint-Louis, Brussels, Belgium.*

Givens, W. (1954). "Numerical Computation of the Characteristic Values of a Real Symmetric Matrix," *Technical Report ORNL-1574*, Oak Ridge National Laboratory, Oak Ridge, Tennesse.

Glassey, C.R., (1971). "Dynamic LP's for Production Scheduling," *Operations Research* **19**, 45–56.

Glover, F., Karney, D., and Klingman, D. (1972). "The Augmented Predecessor Index Method for Locating Stepping Stone Paths and Assigning Dual Prices in Distribution Problems," *Transportation Science* **6**, 171–180.

Glover, F., Karney, D., and Klingman, D. (1973). "A Note on Computational Studies for Solving Transportation Problems," *Proceedings of the ACM* , 180–187.

Glover, F., Karney, D., and Klingman, D. (1974). "Implementation and Computational Comparisons of Primal, Dual, and Primal-Dual Computer Codes for Minimum Cost Flow Network Problems," *Networks*, **4**, 191–212.

Glover, F., Karney, D., Klingman, D., and Napier, A. (1974). "A Computational Study on Start Procedures, Basis Change Criteria and Solution Algorithms for Transportation Problems," *Management Science* **20**, 793–813.

Glynn, P.W. and Iglehart, D.L. (1989). "Importance Sampling for Stochastic Simulation," *Management Science* **35**, 1367–1392.

Goffin, J.L. and Vial, J.-Ph. (1990). "Cutting Plane and Column Generation Techniques with the Projective Algorithm," *Journal of Optimization Theory and Applications* **65**, 409–429.

Goldberg, A.V. (1993). "Scaling Algorithms for the Shortest Path Problems," in: *Proceedings 4th ACM-SIAM Symposium on Discrete Algorithms*, 222–231.

Goldberg, A.V. and Radzik, T. (1993). "A Heuristic Improvement of the Bellman-Ford Algorithm," *Applied Math. Let.* **6**, 3–6.

Goldfarb, D. and Reid, J.K., (1977). "A Practical Steepest-Edge Simplex Algorithm," *Mathematical Programming* **12**, 361–371.

Goldman, A.J. (1956). "Resolution and Separation Theorems for Polyhedral Convex Sets," in H.W. Kuhn and A.W. Tucker (eds.), *Linear Inequalities and Related Systems*, Annals of Mathematics Study No. 38, Princeton University Press, Princeton, New Jersey, 41–51.

Goldman, A.J. and Tucker, A.W. (1956a). "Polyhedral Convex Cones," in H.W. Kuhn and A.W. Tucker (eds.), *Linear Inequalities and Related Systems*, Annals of Mathematics Study No. 38, Princeton University Press, Princeton, New Jersey, 19–39.

Goldman, A.J. and Tucker, A.W. (1956b). "Theory of Linear Programming," in H.W. Kuhn and A.W. Tucker (eds.), *Linear Inequalities and Related Systems*, Annals of Mathematics Study No. 38, Princeton University Press, Princeton, New Jersey, 53–97.

Goldstein, L. (1952). "Problem of Contract Awards," in Alex Orden and Leon Goldstein (eds.), *Symposium on Linear Inequalities and Programming*, Project SCOOP No. 10, Planning Research Division, Director of Management Analysis Service, Comptroller, USAF, Washington, D.C., April, 147–154.

Golub, G.H. and Van Loan, C.F. (1989). *Matrix Computations*, The John Hopkins University Press, Baltimore, Maryland, USA.

Gonzaga, C.C. (1989). "Conical Projection Algorithms for Linear Programming," *Mathematical Programming* **43**, 151–173.

Gonzaga, C.C. (1991). "Large Steps Path-Following Methods for Linear Programming, Part II: Potential Reduction Method," *SIAM Journal on Optimization* **1**, 280–292.

Gonzaga, C.C. (1992). "Path Following Methods for Linear Programming," *SIAM Review*, **34**, 167–227.

Goodman, S.E. and Hedetniemi, S.T. (1985). *Introduction to the Design and Analysis of Algorithms*, McGraw-Hill, San Francisco.

Gordan, P., (1873). "Über die Auflösung linearer Gleichungen mit reelen Koeffizienten," *Math. Ann.* **6**, 23–28.

Gould, N. (1991). News Clip in *SiamNews*, the bimonthly news journal of the Society for Industrial and Applied Mathematics, January, 9.

Grandzol, J.R. and Traaen, T. (1995). "Using Mathematical Programming to Help Supervisors Balance Workloads," *Interfaces* **25**:4, 92–103.

Graves, G.W. (1965). "A Complete Constructive Algorithm for the General Mixed Linear Programming Problem," *Naval Research Logistics Quarterly*, **12**, 1–34.

Grcar, J.F. (1990). "Matrix Stretching for Linear Equations," Report SAND90-8723, Sandia National Laboratories, Albuquerque, NM.

Greenberg, H.J. (1978c). "Pivot selection tactics," in H.J. Greenberg (ed.), *Design and Implementation of Optimization Software*, Sijthoff and Noordhoff, Alpen aan den Rijn, 143–178.

Greenberg, H.J. (ed.), (1978a). *Design and Implementation of Optimization Software*, Sijthoff and Noordhoff, Alpen aan den Rijn.

Greenberg, H.J. (1978b). "A Tutorial on Matricial Packing," in H.J. Greenberg (ed.), *Design and Implementation of Optimization Software*, Sijthoff and Noordhoff, Alpen aan den Rijn, 109–142.

Greenberg, H.J. and Kalan, J., (1975). "An Exact Update for Harris' TREAD," *Mathematical Programming Study* **4**, 26–29.

Greenstadt, J.L. (1967). "On the Relative Efficiencies of Gradient Methods," *Mathematics of Computation* **24**, 145–166.

Grimes, R.G. and Lewis, J.G. (1981). "Condition Number Estimation for Sparse Matrices," *SIAM J. on Scientific and Statistical Computing* **2**, 384–388.

Grunbaum, B., (1967). *Convex Polytopes*, John Wiley and Sons, New York.

Gunderman, R.E. (1973). "A Glimpse into Program Maintenance," *Datamation* **19** 99–101.

H

Hadley, G. (1972). *Linear Programming*, Addison-Wesley Publishing Company, Reading, Massachusetts.

Hall, P. (1982). "Rates of Convergence in the Central-Limit Theorem," *Research Notes in Mathematics,* **62** .

Hall, L.A. and Vanderbei, R.J. (1993). "Two-thirds is Sharp for Affine Scaling," *Operations Research Letters* **13**, 197–201.

Harris, P.M.J., (1975). "Pivot Selection Codes in the Devex LP Code," *Mathematical Programming Study* **4**, 30–57.

Hammersly, J.M. and Handscomb (1964). *Monte Carlo Methods*, Matheun, London.

Hellerman, E. and Rarick, D., (1971). "Reinversion in the Preassigned Pivot Procedure," *Mathematical Programming* **1**, 195–216.

Hellerman, E. and Rarick, D., (1972). "The Partitioned Preassigned Pivot Procedure," in D.J. Rose and R.A. Willoughby (eds.), *Sparse Matrices and their Applications*, Plenum Press, New York, 67–76.

Henderson, A. and Schlaifer, R. (1954). "Mathematical Programming," *Harvard Business Review,* **32,** May-June, 73–100.

Hestenes, M.R. and Stiefel, E. (1952). "Methods of Conjugate Gradients for Solving Linear Systems," *J. Res. National Bureau of Standards* **49**, 409–436.

Hertog, D.D. and Roos, C. (1991). "A Survey of Search Directions in Interior Point Methods for Linear Programming," *Mathematical Programming* **52**, 481–509.

Higham, N.J. (1985). "Nearness Problems in Numerical Algebra," Ph.D. thesis, University of Manchester, Manchester, England.

Higham, N.J. (1987). "A Survey of Condition Number Estimation for Triangular Matrices," *SIAM Review,* **29**, 575–596.

Higham, N.J. and Higham, D.J. (1988). "Large Growth Factors in Gaussian Elimination with Pivoting," *SIAM J. on Matrix Analysis and Applications* .

Higle, J.L. and Sen, S. (1991). "Stochastic Decomposition: An Algorithm for Two Stage Linear Programs with Recourse," *Mathematics of Operations Research* **16/3**, 650–669.

Hillier, F.S., and Lieberman, G.J. (1995). *Introduction to Operations Research*, McGraw-Hill, San Francisco.

Hitchcock, Frank, L. (1941). "The Distribution of a Product from Several Sources to Numerous Localities," *J. Math. Phys.* **20**, 224–230.

Hirsch, W.M. (1957). "Hirsch Conjecture," verbal communication to Dantzig.

Ho, J.K., (1974). "Nested Decomposition for Large-Scale Programs with the Staircase Structure," Technical Report SOL 74-4, Department of Operations Research, Stanford University, Stanford, CA.

Ho, J.K., (1984). "Convergence Behaviour of Decomposition Algorithms for Linear Programs," *Operations Research Letters* **3**, 91–94.

Ho, J.K. and Loute, E. (1981). "An Advanced Implementation of the Dantzig-Wolfe Decomposition Algorithm for Linear Programming," *Mathematical Programming* **20**, 303–326.

Ho, J.K. and Manne, A., (1974). "Nested decomposition for dynamic models," *Mathematical Programming* **6**, 121–140.

Hoffman, A.J., (1953). "Cycling in the Simplex Algorithm," *National Bureau of Standards Report No. 2974*.

Hoffman, A.J., Mannos, M., Sokolowsky, D., and Wiegmann, N. (1953). "Computational Experience in Solving Linear Programs," *J. Society for Industrial and Applied Mathematics,* **1**, 17–33.

Hoffman, A.J. and Hirsch, W. (1961). "Extreme Varieties, Concave Functions and the Fixed Charge Problem," *Communications in Pure and Applied Mathematics* **14**, 355–369.

Hockney, R.W. and Jesshope, C.R. (1988). *Parallel Computers 2*, Adam Hilger, Bristol and Philadelphia.

Holmes, D. (1994). "A Collection of Stochastic Programming Problems," Technical Report 94-11, Department of Industrial Engineering and Operations Research, University of Michigan, Ann Arbor, Michigan.

Hooker, J.N. (1986). "Karmarkar's Linear Programming Algorithm," *Interfaces* **16:4**, 75–90.

Householder, A.S. (1974). *The Theory of Matrices in Numerical Analysis*, Dover Publications, New York.

Huang, C.C., Ziemba, W.T., and Ben-Tal, A. (1977). "Bounds on the Expectation of a Convex Function with a Random Variable with Applications to Stochastic Programming," *Operations Research* **25**, 315–325.

Huard, P. (1970). "A Method of Centers by Upper-Bounding Functions with Applications," in J.B. Rosen, O.L. Mangasarian and K. Ritter (eds.), *Nonlinear Programming: Proceedings of a Symposium held at the University of Wisconsin, Madison, May 1970*, Academic Press, London and New York, 1–30.

I

Ignizio, J.P. (1976). *Goal Programming and Extensions*, Heath, Lexington, Massachusetts.

Infanger, G. (1991). "Monte Carlo (Importance) Sampling within a Benders Decomposition Algorithm for Stochastic Linear Programs Extended Version: Including Results of Large-Scale Problems," Technical Report SOL 91-6, Department of Operations Research, Stanford University, Stanford, CA. Shorter version in *Annals of Operations Research* **39**, 69–95.

Infanger, G. (1994). *Planning Under Uncertainty: Solving Large-Scale Stochastic Linear Programs*, Boyd and Fraser Publishing Company, Massachusetts.

International Business Machines Corporation (1978). *Mathematical Programming System Extended and Generalized Upper Bounding (GUB)*, IBM Manual SH20-0968-1, White Plains, New York.

Isaacson, E. and Keller, H.B., (1966). *Analysis of Numerical Methods*, John Wiley and Sons, New York.

J

Jack, C., Kai, S., and Shulman, A. (1992). "NETCAP—An Interactive Optimization System for GTE Telephone Network Planning," *Interfaces* **22**:1, 72–89.

Jackson, J.R. (1957). "Simulation Research on Job Shop Production," *Naval Research Logistics Quarterly,* **1**, 287.

Jacobs, W.W. (1954). "The Caterer Problem," *Naval Research Logistics Quarterly,* **1**, 154–165.

Jankowski, M. and Wozniakowski, M. (1977). "Iterative Refinement Implies Numerical Stability," *BIT* **17**, 303–311.

Jewell, W.S. (1958). "Optimal Flows Through Networks," Interim Technical Report No. 8, on Fundamental Investigations in Methods of Operations Research, Massachusetts Institute of Technology, Cambridge, Massachusetts.

Johnson, D.B. (1977). "Efficient Algorithms for Shortest Paths in Sparse Networks," *Journal of the Association for Computing Machinery,* **24**, 1–13.

Jones, M.T. and Plassmann, P.E. (1995). "An Improved Incomplete Cholesky Factorization," *ACM Transactions on Mathematical Software,* **21**, 5–17.

Judin, D.B., and Nemirovskii, A.S. (1976a). "Estimation of the Informational Complexity of Mathematical Programming Problems," *Ekonomika i Matematicheskie Metody,* **12**, 128–142. (English Translation in *Matekon: Translation of Russian and East European Mathematical Economics,* **13**, 3–25, 1977.)

Judin, D.B., and Nemirovskii, A.S. (1976b). "Informational Complexity and Effective Methods for the Solution of Convex Extremal Problems," *Ekonomika i Matematicheskie Metody,* **12**, 357–369. (English Translation in *Matekon: Translation of Russian and East European Mathematical Economics,* **13**, 25–45, 1977.)

Judin, D.B., and Nemirovskii, A.S. (1976c). "Informational Complexity of Strict Convex Programming," *Ekonomika i Matematicheskie Metody,* **12**, 550–559.

K

Kahan, W. (1966). "Numerical Linear Algebra," *Canadian Math. Bull.* **9**, 757–801.

Kahaner, D., Moler, C.B., and Nash, S. (1988). *Numerical Methods and Software*, Prentice-Hall, Inc., Englewood Cliffs, New Jersey.

Kalaba, R.E. and Juncosa, M.L. (1956). "Optimal Design and Utilization of Communications Networks," *Management Science* **3**, 33–44.

Kalan, J.E. (1971). "Aspects of Large-Scale In-Core Linear Programming," *Proceedings of the ACM Annual Conference*, 304–313.

Kalan, J.E. (1976). "Machine Inspired Enhancements of the Simplex Algorithm," Technical Report CS75001-R, Computer Science Department, Virginia Polytechnical University, Blacksburg, Virginia.

Kall, P. (1974). "Approximations to Stochastic Programs with Complete Fixed Recourse," *Numerische Mathematik* **22**, 333-339.

Kall, P. (1979). "Computational Methods for Two Stage Stochastic Linear Programming Problems," *Z. angew. Math. Phys.* **30**, 261–271.

Kall, P. and Stoyan, D. (1982). "Solving Stochastic Programming Problems with Recourse Including Error Bounds," *Mathematische Operationsforschung und Statistik, Series Optimization* **13**, 431–447.

Kall, P. and Wallace, S.W. (1993). *Stochastic Programming*, John Wiley and Sons, New York.

Kantorivich, L.V. (1939). "Mathematical Methods in the Organization and Planning of Production," Publication House of the Leningrad State University. Translated in *Management Science* **6**, 1960, 366–422.

Kantorivich, L. V. (1942). "On the Translocation of Masses," *Compt. Rend. Acad. Sci. U.S.S.R.* **37**, 199–201.

Kantorivich, L.V. and Gavurin, M.K. (1949). "The Application of Mathematical Methods to Freight Flow Analysis," (translation), *Akademii Nauk SSSR* .

Karmarkar N. (1984). "A New Polynomial-Time Algorithm for Linear Programming," *Combinatorica* **4**, 373-395.

Katz, P., Sadrian, A., and Tendick, P. (1994). "Telephone Companies Analyze Price Quotations with Bellcore's PDSS Software," *Interfaces* **24**:1, 50–63.

Kaul, R.N. (1965). "An Extension of Generalized Upper Bounding Techniques for Linear Programming," Technical Report ORC 65-27, Operations Research Center, University of California, Berkeley.

Kennedy, W.J., Jr. and Gentle, J.E. (1980). *Statistical Computing*, Marcel Dekker, Inc. New York and Basel.

Kennington, J.L. and Wang, Z. (1991). "An Empirical Analysis of the Dense Assignment Problem: Sequential and Parallel Implementations," *ORSA Journal on Computing* **3**, 299-306.

Kernighan, B.W. and Plauger, P.J. (1970). *The Elements of Programming Style*, McGraw-Hill, New York.

Khachian, L.G. (1979). "A Polynomial Algorithm for Linear Programming," *Doklady Akademiia Nauk USSR* **244**, 1093–1096. [English translation: *Soviet Mathematics Doklady* **20**, 191–194].

Khan, M.R. and Lewis, D.A. (1987). "A Network Model for Nurse Staff Scheduling," *Zeitschrift für Operations Research* **31**, B161–B171.

Kotiah, T.C.T. and Steinberg, D.I. (1977). "Occurrences in Cycling and Other Phenomena Arising in a Class of Linear Programming Models," *Communications of the Association for Computing Machinery* **20**, 107–112.

Kotiah, T.C.T. and Steinberg, D.I. (1978). "On the Possibility of Cycling with the Simplex Method," *Operations Research* **26**, 374–375.

Kim, K. and Nazareth, J.L. (1994). "A Primal Null-Space Affine Scaling Method," *ACM Transactions on Mathematical Software*, .

Klee, V.L. and Minty, G.J. (1972). "How Good is the Simplex Algorithm?" in O. Shisha (ed.), *Inequalities III*, Academic Press, London and New York, 159–175.

Klee, V.L. and Walkup, D.W. (1967). "The *d*-Step Conjecture for Polyhedra of Dimension $d < 6$," *Acta Mathematica* **117**, 53–78.

Klingman, D., Phillips, N., Steiger, D., and Young, W. (1987). "The Successful Deployment of Management Science throughout Citgo Petroleum Corporation," *Interfaces* **17**:, 4–25.

Klingman, D., Phillips, N., Steiger, D., Wirth, R., and Young, W. (1986). "The Challenges and Success Factors in Implementing an Integrated Products Planning System for Citgo," *Interfaces* **16**:, 1–19.

Klotz, E.S. (1988). "Dynamic Pricing Criteria in Linear Programming," Technical Report SOL 88-15, Department of Operations Research, Stanford University, Stanford, CA.

Kohler, D.A. (1973). "Translation of a Report by Fourier on his Work on Linear Inequalities," *OPSEARCH* **10**, 38–42.

Kojima, M., Megiddo, N., and Mizuno, S. (1993). "A Primal-Dual Infeasible Interior Point Algorithm for Linear Programming," *Mathematical Programming* **61**, 263–280.

Kojima, M., Mizuno, S., and Yoshise, A. (1989a). "A Primal-Dual Interior Point Method for Linear Programming," in N. Megiddo (ed.), *Progress in Mathematical Programming: Interior Point and Related Methods* Springer-Verlag, Berlin and New York, 29–48.

Kojima, M., Mizuno, S., and Yoshise, A. (1989b). "A Polynomial-Time Algorithm for a Class of Linear Complementarity Problems," *Mathematical Programming* **44**, 1–26.

Koopmans, T.C. (1947). "Optimum Utilization of the Transportation System," *Proceedings of the International Statistical Conference*, Washington, D.C. Volume 5 was reprinted as a supplement to *Econometrica*, **17**, 1949.

Koopmans, T.C. (ed.), (1951).*Activity Analysis of Production and Allocation*, John-Wiley and Sons, New York.

Koopmans, T.C. and Reiter, S. (1951). "A Model of Transportation," in T.C. Koopmans (ed.), *Activity Analysis of Production and Allocation*, July–October 1949, Cowles Commission Monograph 13, Proceedings of Linear Programming Conference, June 20–24, 1949, John Wiley and Sons, New York, 33–97.

Kortanek, K.O. and Zhu, J. (1988). "New Purification Algorithms for Linear Programming," *Naval Research Logistics Quarterly,* **35**, 571–583.

Knuth, D.E. (1973a). *Fundamental Algorithms, The Art of Computer Programming, Vol. 1,* Addison-Wesley Publishing Company, Reading, Massachusetts. Originally published in 1968.

Knuth, D.E. (1973b). *Sorting and Searching, The Art of Computer Programming, Vol. 3,* Addison-Wesley Publishing Company, Reading, Massachusetts.

Knuth, D.E. (1981). *Seminumerical Algorithms, The Art of Computer Programming, Vol. 2,* Addison-Wesley Publishing Company, Reading, Massachusetts. Originally published in 1969.

Kranich, E. (1991) "Interior Point Methods for Mathematical Programming: A Bibliography Discussion Paper 171, Institute of Economy and Operations Research, Fern Universität Hagen, P.O. Box 940, D-5800 Hagen 1, West Germany.

Krishna, A.S. (1989). Unpublished Manuscript.

Krishna, A.S. (1993). "Enhanced Algorithms for Stochastic Programming," Technical Report SOL 93-8, Department of Operations Research, Stanford University, Stanford, CA.

Kruskal, J.B., Jr., (1956). "On the Shortest Spanning Subtree of a Graph and the Traveling Salesman Problem," *Proceedings of the American Mathematical Society* **7**, 48–50.

Kuhn, H.W. (1955). "The Hungarian Method for the Assignment Problem," *Naval Research Logistics Quarterly,* **2**, 1 and 2, 83–97.

Kuhn, H.W. (1956). "Solvability and Consistency for Linear Equations and Inequalities," *The American Mathematical Monthly* **63**, 217–232.

Kuhn, H.W. and Tucker, A.W. (1950). "Nonlinear Programming," in J. Neyman (ed.), *Proceedings 2nd Berkeley Symposium on Mathematical Statistics and Probability, 1950,* University of California Press, Berkeley, California, 481–492.

Kuhn, H.W. and Tucker, A.W. (1958). "John von Neumann's Work in the Theory of Games and Mathematical Economics," *Bull. Amer. Math. Society* **64**, 100–122.

Kusy, M.I. and Ziemba, W.T. (1986). "A Bank Asset and Liability Management Model," *Operations Research* **34**, 356–378.

Draft Paper, University of British Columbia, Vancouver, B.C., Canada. To appear in Operations Research.

L

Laderman, J. (1947).

Lansdowne, Z.F., (1979). "Survey of Research on Model Simplification," Technical Report SOL 79-26, Department of Operations Research, Stanford University, Stanford, CA.

Lasdon, L.S., (1970). *Optimization Theory for Large Systems,* Macmillan, London.

Lawler, E.L. (1976). *Combinatorial Optimization: Networks and Matroids,* Holt, Rinehart and Winston, New York.

Lawson, C.L., Hanson R.J., Kincaird D.R., and Krogh, F.T. (1979a). "Basic Linear Algebra Subprograms for Fortran Usage," *ACM Transactions on Mathematical Software*, **5**, 308–323.

Lawson, C.L., Hanson R.J., Kincaird D.R., and Krogh, F.T. (1979b). "Algorithm 539, Basic Linear Algebra Subprograms for Fortran Usage," *ACM Transactions on Mathematical Software*, **5**, 324–325.

Lawson, C.L., Hanson, R.J., (1974). *Solving Least Squares Problems*, Prentice-Hall, Inc., Englewood Cliffs, New Jersey.

Lee, S.M. (1972). *Goal Programming for Decision Analysis*, Auerbach, Philadelphia.

Lemke, C.E. (1954). "The Dual Method of Solving the Linear Programming Problem," *Naval Research Logistics Quarterly*, **1**, 36–47.

Lenstra, J.K., Rinnooy Kan, A.H. G., and Schrijver, A. (eds.), (1991). *History of Mathematical Programming*, Elsevier Science Publishing Company, New York.

Levin, A.Ju. (1965). "On an Algorithm for the Minimization of Convex Functions," *Soviet Mathematics Doklady* **6**, 286–290.

Lewis, R.E. (1955). "Top Management Looks at Linear Programming and Inventory Management," *Proceedings of the Linear Programming and Inventory Management Seminar*, Methods Engineering Council, Pittsburgh, B-1 to B-8.

Litty, C.J. (1994). "Optimal Lease Structuring at GE Capital," *Interfaces* **24**:3, 34–45.

Liu, J. (1985). "Modification of the Minimum-Degree Algorithm by Multiple Elimination," *ACM Transactions on Mathematical Software*, **11**, 141–153.

Luenberger, D.G. (1973). *Introduction to Linear and Non-Linear Programming*, Addison-Wesley Publishing Company, Reading, Massachusetts.

Lustig, I.J. (1987). "An Analysis of an Available Set of Linear Programming Test Problems," Technical Report SOL 87-11, Department of Operations Research, Stanford University, Stanford, CA.

Lustig, I.J. (1991). "Feasibility Issues in an Interior Point Method for Linear Programming," *Mathematical Programming* **49**, 145–162.

Lustig, I.J., Marsten, R.E., and Shanno, D.F. (1994). "Interior Point Methods for Linear Programming: Computational State of the Art," *ORSA Journal on Computing* **6**, 1–14.

Lustig, I.J., Marsten, R.E., and Shanno, D.F. (1992a). "Computational Experience with a Globally Convergent Primal-Dual Predictor-Corrector Algorithm for Linear Programming," Technical Report SOR 92-10, School of Engineering and Applied Science, Department of Civil Engineering and Operations Research, Princeton University, Princeton, New Jersey.

Lustig, I.J., Marsten, R.E., and Shanno, D.F. (1992b). "The Interaction of Algorithms and Architectures for Interior Point Methods, in P.M. Pardolos (ed.), *Advances in Optimization and Parallel Computing, North-Holland,* the Netherlands, 190–205.

Lustig, I.J., Marsten, R.E., and Shanno, D.F. (1991a). "Computational Experience with a Primal-Dual Interior Point for Linear Programming," *Linear Algebra and its Applications* **152**, 191–222.

Lustig, I.J., Marsten, R.E., and Shanno, D.F. (1991b). "Interior Method vs Simplex Method: Beyond NETLIB," *Mathematical Programming Society COAL Newsletter* **19**, 41–44.

Lustig, I.J., Marsten, R.E., and Shanno, D.F. (1992c). "On Implementing Mehrotra's Predictor-Corrector Interior Point Method for Linear Programming," *SIAM Journal on Optimization* **2**, 435–449.

Lustig, I.J., Marsten, R.E., and Shanno, D.F. (1990). "The Primal-Dual Interior Point Method on the Cray Supercomputer, in T.F. Coleman and Y. Li (eds.), *Large-Scale Numerical Optimization*, papers from the workshop held at Cornell University, Ithaca, NY, October 1989, vol. 46 of SIAM Proceedings in Applied Mathematics, Society of Industrial and Applied Mathematics (SIAM), Philadelphia, PA, 70–80.

Lustig, I.J., Mulvey, J.M., and Carpenter, T.J. (1990). "Formulating Programs for Interior Point Methods," Technical Report SOR 89-16, School of Engineering and Applied Science, Department of Civil Engineering and Operations Research, Princeton University, Princeton, New Jersey.

M

Madansky, A. (1959). "Bounds on the Expectation of a Convex Function of a Multivariate Random Variable," *Ann. Math. Stat.* **30**, 743–746.

Makuch, W.M., Dodge, J.L., Ecker, J.G., Granfors, D.C., and Hahn, G.J. (1992). "Managing Consumer Credit Delinquency in the US Economy: A Multi-Billion Dollar Management Science Application," *Interfaces* **22**:1, 90–109.

Malhotra, V.M., Kumar, M.P., and Maheshwari, S.N. (1978). "An $O(|V|^3)$ Algorithm for Finding Maximum Flows in Networks," *Information Processing Letters,* **7**, 277–278.

Manley, B.R. and Threadgill, J.A. (1991). "LP Used for Valuation and Planning of New Zealand Plantation Forests," *Interfaces* **21**:6, 66–79.

Manne, A.S. (1956). *Scheduling of Petroleum Refinery Operations*, The Harvard University Press, Cambridge, Massachusetts.

Marcus, M. (1960). "Some Properties of Doubly Stochastic Matrices," *The American Mathematical Monthly* **67**, 215–221.

Markowitz, H.M. (1952). "Portfolio Selection," *The Journal of Finance* **7** .

Markowitz, H.M. (1957). "The Elimination Form of the Inverse and its Applications to Linear Prorgamming," *Management Science* **3**, 255–269.

Markowitz, H.M. (1959). *Portfolio Selection: Efficient Diversification of Investments*, John Wiley and Sons, New York.

Maros, I.G., (1986). "A General Phase-I Method in Linear Programming," *European Journal of Operations Research* **23**, 64–77.

Marshall, K.T. and Suurballe, J.W. (1969). "A Note on Cycling in the Simplex Method," *Naval Research Logistics Quarterly,* **16**, 121–137.

Marsten, R.E. (1981). "The Design of the XMP Linear Programming Library," *ACM Transactions on Mathematical Software,* **7**, 481–497.

Marsten, R.E., Saltzman, M.J., Shanno, D.F., Ballinton, J.F., and Pierce, G.S. (1989). "Implementation of a Dual Affine Interior Point Algorithm for Linear Programming," *ORSA Journal on Computing* **1**, 287–297.

Mascarenhas, W.F. (1993). "The Affine Scaling Algorithm Fails for $\lambda = 0.999$," Technical Report, Universidade Estadual de Campinas, Campinas S. P., Brazil.

Massé, P. and Gibrat, R. (1957). "Applications of Linear Programming to Investments in the Electric Power Industry," *Management Science* **3**, 149–166.

Maynard, H.B. (1955). "Putting New Management Tools to Work," *Proceedings of the Linear Programming and Inventory Management Seminar*, Methods Engineering Council, Pittsburgh.

McCarthy, C. and Strang, G. (1973). "Optimal Conditioning of Matrices," *SIAM J. on Numerical Analysis* **10**, 370–388.

McCormick, S.T. (1983). "Optimal Approximation of Sparse Hessians and its Equivalence to a Graph Coloring Problem," *Mathematical Programming* **26**, 153–171.

McDiarmid, C. (1990). "On the Improvement per Iteration in Karmarkar's Algorithm for Linear Programming," *Mathematical Programming* **46**, 299-320.

McKeeman, W.M. (1962). "Crout with Equilibration and Iteration," *Communications of the Association for Computing Machinery* **5**, 553–555.

McGowan, C.L. and Kelly, J.R. (1975). *Top-Down Structured Programming Techniques*, Petrocelli/Charter, New York.

McShane, K.A., Monma, C.L., and Shanno, D.F. (1989). "An Implementation of a Primal-Dual Interior Point Method for Linear Programming," *ORSA Journal on Computing* **1**, 70–83.

Megiddo, N. (1986). "Introduction: New Approaches to Linear Programming," Special issue of *Algorithmica* **1**, 387–394, Springer-Verlag, Berlin and New York.

Megiddo, N. (1988). "Pathways to the Optimal Set in Linear Programming," in N. Megiddo (ed.), *Progress in Mathematical Programming: Interior Point and Related Methods* Springer-Verlag, Berlin and New York, 131–158.

Megiddo, N. (1991). "On Finding Primal- and Dual- Optimal Bases," *ORSA Journal on Computing* **3**, 63–65.

Mehrotra, S. (1992a). "On the Implementation of a Primal-Dual Interior Point Method," *SIAM Journal on Optimization* **2**, 575–601.

Mehrotra, S. (1992b). "Implementation of Affine Scaling Methods: Approximate Solutions of Systems of Linear Equations Using Preconditioned Conjugate Gradient Methods," *ORSA Journal on Computing* **4**, 103–118.

Mehrotra, S. (1993). "Quadratic Convergence in a Primal-Dual Method," *Mathematics of Operations Research* **18**, 741–751.

Meijerink, J. van der Vorst, H.A. (1981). "Guidelines for the Usage of Incomplete Decomposition in Solving Sets of Linear Equations as they Occur in Practical Problems," *Journal of Comput. Physics* **44**, 134–155.

Mitchell, J.E. and Borchers, B. (1992). "A Primal-Dual Interior Point Cutting Plane Method for the Linear Ordering Problem," *COAL Bulletin* **21**, 13–18.

Mitchell, J.E. and Todd, M.J. (1992). "Solving Combinatorial Optimization Problems Using Karmarkar's Algorithm," *Mathematical Programming* **56**, 245–284.

Millham, C.B. (1976). "Fast Feasibility Methods for Linear Programming," *OPSEARCH* **13**, 198–204.

Minkowski, H. (1896). *Geometrie der Zahlen*, B.G. Teubner, Leipzig and Berlin, 1910. First printing, 1896; also reprinted by Chelsea Publishing Company, New York, 1953.

Mizuno, S. (1992). "Polynomiality of the Kojima-Megiddo-Mizuno Infeasible Interior Point Algorithm for Linear Programming," Technical Report 1006, School of Operations Research and Industrial College of Engineering, Cornell University, Ithaca, NY 14853.

Mizuno, S., Kojima, M., and Todd, M. (1995). "Infeasible-Interior-Point-Primal-Dual Potential-Reduction Algorithms for Linear Programming," *SIAM Journal on Optimization* **5**, 52–67.

Monma, C.L. and Morton, A.J. (1987) "Computational Experiments with a Dual Affine Variant of Karmarkar's Method for Linear Programming," *Operations Research Letters* **6**, 261–267.

Monteiro, R.C. and Adler, I. (1989a) "Interior Path Following Primal-Dual Algorithms—Part I: Linear Programming," *Mathematical Programming* **44**, 27–42.

Monteiro, R.C. and Adler, I. (1989b) "Interior Path Following Primal-Dual Algorithms—Part II: Convex Quadratic Programming," *Mathematical Programming* **44**, 43–66.

Monteiro, R.C., Tsuchiya, T., and Wang, Y. (1993). "A Simplified Global Convergence Proof of the Affine Scaling Algorithm," *Annals of Operations Research* **47**, 443–482.

Mood, A.M., Graybill, F.A., and Boes, D.C. (1974). *Introduction to the Theory of Statistics*, McGraw-Hill, New York.

Mooney, J.W. (1975). "Organized Program Maintenance," *Datamation* **21**, 63–64.

Moore, E.F. (1959). "The Shortest Path Through a Maze," in *Proceedings of the International Symposium on the Theory of Switching*, The Harvard University Press, Cambridge, Massachusetts, 285–292.

More, J.J. and Sorenson, D.C. (1981). "Computing a Trust Region step," Report ANL-81-83, Argonne National Laboratory, Argonne, Illinois.

Morin, G. (1955). "More Effective Production Planning with Linear Programming," Paper F from *Proceedings of the Linear Programming and Inventory Management Seminar*, Methods Engineering Council, Pittsburgh.

Morton, D.P. (1993). "Algorithmic Advances in Multi-Stage Stochastic Programming," Ph.D. thesis, Department of Operations Research, Stanford University, Stanford, CA.

Morton, D.P. (1995). "Stochastic Network Interdiction," Working Paper presented at a Workshop in Stochastic Programming at UC Davis, March, 1995.

Morris, W.T. (1967). "On the Art of Modeling," *Management Science* **13**, 707–717.

Motzkin, T.S., (1936). "Beiträge zur Theorie der linearen Ungleichungen," doctoral thesis, University of Zurich.

Muir, T., (1960). *Determinants*, Dover Publications, New York.

Mulvey, J.M., (1978). "Pivot Strategies for Primal-Simplex Network Codes," *Journal of the Association for Computing Machinery*, **25**, 266–270.

Munksgaard, N., (1980). "Solving Sparse Symmetric Sets of Linear Equations by Preconditioned Conjugate Gradient," *ACM Transactions on Mathematical Software*, **6**, 206–219.

Murray, W. (1976). "Constrained Optimization" in L.C.W. Dixon (ed.), *Optimization in Action*, Academic Press, London and New York, 217–251.

Murray, W. and Wright, M.H. (1978). *Projected Lagrangian Methods Based on Trajectories of Penalty and Barrier Functions*, Technical Report SOL 78-23, Department of Operations Research, Stanford University, Stanford, CA.

Murtagh, B.A. and Saunders, M.A., (1978). "Large-Scale Linearly Constrained Optimization," *Mathematical Programming* **14**, 41–72.

Murtagh, B.A. and Saunders, M.A. (1995). "MINOS 5.4 User's Guide," Technical Report SOL 83-20R, Department of Operations Research, Stanford University, Stanford, CA.

Murty, K.G. (1980). "Computational Complexity of Parametric Linear Programming," *Mathematical Programming* **19**, 213–219.

Murty, K.G. (1983). *Linear Programming*, John Wiley and Sons, New York.

Musalem, J. (1980). "Converting Linear Models to Network Models," Ph.D. thesis, University of California, Los Angeles, California.

Myers, D.C. and Shih, W. (1988). "A Constraint Selection Technique for a Class of Linear Programs. *Operations Research Letters* **7**, 191–194.

N

Nazareth, J.L. (1984). "Numerical Behavior of LP Algorithms Based Upon the Decomposition Principle," *Linear Algebra and its Applications* **57**, 181–189.

Nazareth, J.L. (1986). "Implementation Aids for the Optimization Algorithms that Solve Sequences of Linear Programs," *ACM Transactions on Mathematical Software*, **12**, 307–323.

Nazareth, J.L. (1987). *Computer Solution of Linear Programs*, Monographs on Numerical Analysis, Oxford University Press, Oxford and New York.

Nering, E.D. and Tucker, A.W. (1993). *Linear Programs and Related Problems*, Academic Press, London and New York.

Ng E. (1988) "On the Solution of Sparse Linear Least-Squares Problems," Presentation at Stanford, Mathematical Science Section, Oak Ridge National Laboratory, Oak Ridge, Tennesse.

Nishiya, T. (1983). "A Basis Factorization Method for Multi-Stage Linear Programming with an Application to Optimal Operation of an Energy Plant," Draft Report.

O

Oettli, W., and Prager, W., (1964). "Compatability of Approximate Solutions of Linear Equations With Given Error Bounds For Coefficients and Right-Hand Sides," *Numerische Mathematik* **6**, 405–409.

O'Leary, D.P. (1980). "Estimating Matrix Condition Numbers," *SIAM J. on Scientific and Statistical Computing* **1**, 205–209.

Orchard-Hays, W. (1954). *A Composite Simplex Algorithm—II, RM*-**1275**, The RAND Corporation, May.

Orchard-Hays, W. (1956). "Evolution of Computer Codes for Linear Programming," The RAND Corporation, Paper P-810, (March 14), 22–24.

Orchard-Hays, W. (1968). *Advanced Linear-Programming Computing Techniques*, McGraw-Hill, New York.

Orchard-Hays, W., (1978). "History of Mathematical Programming Systems," in H.J. Greenberg (ed.), *Design and Implementation of Optimization Software*, Sijthoff and Noordhoff, Alpen aan den Rijn, 1–26.

Orchard-Hays, W., Cutler, L., and Judd, H. (1956). "Manual for the RAND IBM Code for Linear Programming on the 704," The RAND Corporation, Paper P-842, (May 16), 24–26.

Orden, A. (1956). "The Transshipment Problem," *Management Science* **2**, 276–285.

Orden, A. (1993). "LP from the 40s to the 90s," *Interfaces* **23**:5, 2–12.

Orlin, J.B. and Ahuja, R.K. (1992). "New Scaling Algorithms for the Assignment and Minimum Cycle Mean Problems," *Mathematical Programming* **54**, 41–56.

Ortega, J.M. (1973). *Numerical Analysis: A Second Course*, Academic Press, London and New York.

Ortega, J.M. (1988). "The *ijk* Forms of Factorization Methods I: Vector Computers," *Parallel Computing* **7**, 135–147.

Ortega, J.M. and Rheinboldt, W.C., (1970). *Iterative Solutions of Nonlinear Equations in Several Variables*, Academic Press, London and New York.

Osterby, O. and Zlatev Z. (1983). "Direct Methods for Sparse Matrices," *Lecture notes in Computer Science*, Springer-Verlag, Berlin and New York

P

Pan, V. (1984). "How Can We Speed Up Matrix Multiplication," *SIAM Review,* **26**, 393–416.

Papadimitriou, C.H. and Stieglitz, K. (1982). *Combinatorial Optimization, Algorithms and Complexity,* Prentice-Hall, Inc., Englewood Cliffs, New Jersey.

Pape, U. (1974). "Implementation and Efficiency of Moore Algorithms for the Shortest Root Problem," *Mathematical Programming* **7**, 212–222.

Parlett, B.N., and Wang, Y. (1975). "The Influence of the Compiler on the Cost of Mathematical Software—in Particular on the Cost of Triangular Factorization," *ACM Transactions on Mathematical Software,* **1**, 35–46.

Paull, A. E. and Walter, J.R. (1955). "The Trim Problem: An Application of Linear Programming to the Manufacture of Newsprint Paper," *Econometrica*, **23**, 336 (abstract).

Peters, G. and Wilkinson, J.H. (1970). "The least-squares problem and pseudo-inverses," *Computer Journal* **13**, 309–316.

Pissanetzky, Sergio (1984). *Sparse Matrix Technology*, Academic Press, London and New York.

Pereira, M.V. and Pinto, L.M.V.G. (1989). "Stochastic Dual Dynamic Programming," Technical Note, DEE-PUC/RJ-Catholic University of Rio de Janeiro, Caixa Postal 38063 Gávea, Rio de Janeiro, RJ CEP, Brazil.

Pereira, M.V., Pinto, L.M.V.G., Oliveira, G.C., and Cunha, S.H.F. (1989). "A Technique for Solving LP-Problems with Stochastic Right-Hand Sides," CEPEL, Centro del Pesquisas de Energia Electria, Rio de Janeiro, Brazil.

Powell, M.J.D. (1970). A Hybrid Method for Nonlinear Equations, in P. Rabinowitz (ed.), *Numerical Methods for Nonlinear Algebraic Equations*, Academic Press, London and New York, 29–55.

Perold, A. F. (1984). "Large Scale Portfolio Optimizations," *Management Science* **30**, 1143-1160.

Powell, M.J.D. (1971). "On the Convergence of the Variable Metric Algorithm," *J. Institute of Mathematics and its Applications* **7**, 21–36.

Powell, M.J.D. (1976). "Some Global Convergence Properties of a Variable Metric Algorithm Without Exact Line Searches," in R.C. Cottle and C.E. Lemke (eds.), *SIAM-AMS proceedings, Volume IX, Mathematical Programming*, American Mathematical Society, Providence, Rhode Island, 53–72.

Powell, M.J.D. (1991). "On the Number of Iterations of Karmarkar's Algorithm for Linear Programming for Linear Programming," Technical Report DAMTP 1991/NA23, Department of Applied Mathemathics and Theoretical Physics, University of Cambridge, Cambridge, UK.

Powell, M.J.D. and Toint, Ph.L. (1979). "On the Estimation of Sparse Hessian Matrices," *SIAM J. on Numerical Analysis* **16**, 1060–1073.

Prékopa, A. (1989). "Sharp Bounds on Probabilities Using Linear Programming," *Operations Research.*

R

Reid, J.K. (1971). "A Note on the Stability of Gaussian Elimination," *J. Institute of Mathematics and its Applications* **8**, 374–375.

Reid, J.K. (1976). "Fortran Subroutines for Handling Sparse Linear Programming Bases," *Report AERE R8269*, Atomic Energy Research Establishment, Harwell, England.

Reid, J.K. (1982). "A Sparsity-Exploiting Variant of the Bartels-Golub Decomposition for Linear Programming Bases," *Mathematical Programming* **24**, 55–69.

Reinfield, N.V., and Vogel, W.R. (1958). *Mathematical Programming*, Prentice-Hall, Inc., Englewood Cliffs, New Jersey.

Renegar, J. (1988). "A Polynomial-Time Algorithm, Based on Newton's Method, for Linear Programming," *Mathematical Programming* **40**, 59–93.

Rice, J.R. (1966a). "Experiments with Gram-Schmidt orthogonalization," *Mathematics of Computation* **20**, 325–328.

Rice, J.R. (1966b). "A Theory of Condition," *SIAM J. on Numerical Analysis* **3**, 287–310.

Rice, J.R. (1976a). "Parallel Algorithm for Adaptive Quadrature III: Program Correctness," *ACM Transactions on Mathematical Software,* **2**, 1–30.

Rice, J.R. (1976b). "The Algorithm Selection Problem," in M. Rubicoff and M. Yovits (eds.), *Advances in Computers,* Vol. 15, Academic Press, London and New York, 65–118.

Rice, J.R. (1985). *Matrix Computations and Mathematical Software,* McGraw-Hill, San Francisco.

Rigby, B., Lasdon, L.L., and Waren, A.D. (1995). "The Evolution of Texaco's Blending Systems from OMEGA to StarBlend," *Interfaces* **25**:5, 64–83.

Robertson, H.H. (1977). "The Accuracy of Error Estimates for Systems of Linear Algebraic Equations," *J. Institute of Mathematics and its Applications* **20**, 409–414.

Rockafellar, R.T. (1970). *Convex Analysis,* Princeton University Press, Princeton, New Jersey.

Rockafellar, R.T. (1984). *Network Flows and Monotropic Optimization,* Wiley, New York.

Rockafellar, R.T. and Wets, R.J. (1989). "Scenario and Policy Aggregation in Optimization Under Uncertainty," *Mathematics of Operations Research* **16**, 119–147.

Rose, D.J. and Willoughby, R.A. (eds.), (1972). *Sparse Matrices and their Applications,* Plenum Press, New York, 67–76.

Rosen, S. (1969). "Electronic Computers: A Historical Survey," *Computing Surveys* **1**, 7–36.

Rosen, J.B. (1960). "The Gradient Projection Method for Nonlinear Programming, Part I–Linear Constraints," *SIAM J. of Applied Mathematics,* **9**, 181–217.

Rosen, J.B. (1964). "Primal Partition Programming for Block Diagonal Matrices," *Numerische Mathematik* **6**, 250–260.

Roush, W.B., Stock, R.H., Cravener, T.L., and D'Alfonso, T.H. (1994). "Using Chance-Constrained Programming for Animal Feed Formulation at Agway," *Interfaces* **24**:2, 53–58.

Roy, J. and Crainic, T.G. (1992). "Improving Freight Routing with a Tactical Planning Model," *Interfaces* **22**:3, 31-44.

Russel, E.J. (1969). "Extensions of Dantzig's Algorithm for Finding an Initial Near-Optimal Basis for the Transportation Problem," *Operations Research* **17**, 187–191.

Ruszczynski, A. (1986). "A Regularized Decomposition Method for Minimizing a Sum of Polyhedral Functions," *Mathematical Programming* **35**, 309–333.

Ryan, D.M., and Osborne, M.R. (1988). "On the Solution of Highly Degenerate Linear Programs," *Mathematical Programming* **41**, 385–392.

Ryder, B.G. (1974). "The PFORT Verifier," *Software Practice and Experience,* **4**, 359–378.

S

Saigal, R. (1992). "A Simple Proof of the Primal Affine Scaling Method," Technical Report 92-60, Department of Industrial Engineering and Operations Research, University of Michigan, Ann Arbor, Michigan, to appear in Annals of Operations Research.

Saigal, R. (1993a). "A Three Step Quadratically Convergent Implementation of the Primal Affine Scaling Method," Technical Report 93-9, Department of Industrial Engineering and Operations Research, University of Michigan, Ann Arbor, Michigan.

Saigal, R. (1993b). "The Primal Power Affine Scaling Method," Technical Report 93-21, Department of Industrial Engineering and Operations Research, University of Michigan, Ann Arbor, Michigan, to appear in *Annals of Operations Research.*

Sakarovitch, M. and Saigal, R. (1967). "An Extension of Generalized Upper Bounding Techniques for Structured Linear Programs," *SIAM J. of Applied Mathematics,* **15**, 4, 906–914.

Salveson, M.E. (1953). "A Computational Technique for the Fabrication Scheduling Problem," Management Sciences Research Project, University of California, Los Angeles, California.

Samelson, H. (1974). *An Introduction to Linear Algebra,* John Wiley and Sons, New York.

Sargent, R.W.H., and Westerberg, A.W., (1964). "'Speed-up' in Chemical Engineering Design," *Transactions of the Institute of Chemical Engineers,* **42**, 190–197.

Saunders, M.A. (1972). "Large-Scale Linear Proramming Using the Cholesky Factorization," Report STAN-CS-72-252, Department of Computer Science, Stanford University, Stanford, CA.

Saunders, M.A. (1976a). "The Complexity of LU Updating in the Simplex Method," in R. S. Anderssen and R. P. Brent (eds.), *The Complexity of Computational Problem Solving,* University Press, Queensland, 214-230.

Saunders, M.A., (1976b). "A Fast, Stable Implementation of the Simplex Method Using Bartels-Golub Updating," in J.R. Bunch and D.J. Rose (eds.), *Sparse Matrix Computations,* Academic Press, London and New York, 213–226.

Saunders, M.A. (1979). "Sparse Least Squares by Conjugate Gradients: a Comparison of Preconditioning Methods," in J.F. Gentlemen (ed.), *Computer Science and Statistics: 12th Annual Symposium on the Interface,* University of Waterloo, Waterloo, Ontario, Canada, 15–20.

Saunders, M.A., (1980). "Large-scale Linear Programming," Notes for the tutorial conference *Practical Optimization,* Systems Optimization Laboratory, Department of Operations Research, Stanford University, Stanford, CA.

Schönauer, W. (1987). *Scientific Computing on Vector Computers,* North-Holland, Amsterdam, the Netherlands.

Schrage, L. (1975). "Implicit Representation of Variable Upper Bounds in Linear Programming," *Mathematical Programming Study* 4, 118-132.

Schrage, L. (1981). "Some Comments on Hidden Structure in Linear Programs," in H. Greenberg and J. Maybee (eds.), *Computer Assisted Analysis and Model Simplification,* Academic Press, London and New York, 389–395.

Scott, D.M. (1985). "A Dynamic Programming Approach to Time-Staged Convex Programs," Technical Report SOL 85-3, Department of Operations Research, Stanford University, Stanford, CA.

Sethi, A.P. and Thompson, G.L. (1984). "The Pivot and Probe Algorithm for Solving a Linear Program," *Mathematical Programming* **29**, 219–233.

Shapiro, J.F., (1979). *Mathematical Programming: Structures and Algorithms*, John Wiley and Sons, New York.

Shapiro, R.D. (1984). *Optimization Models for Planning and Allocation: Text and Cases in Mathematical Programming*, Wiley, New York.

Shanno, D.F. and Bagchi, A. (1990). "A Unified View of Interior Point Methods for Linear Programming," *Annals of Operations Research* **22**, 55–70.

Sharpe, W.F. (1971). "A Linear Programming Approximation for the General Portfolio Selection Problem," *Journal of Financial Quantitative Analysis* **6**, 1263–1275.

Shefi, A. (1969). "Reduction of Linear Inequality Constraints and Determination of all Feasible Extreme Points," Ph.D. thesis, Department of Engineering-Economic Systems, Stanford University, Stanford, CA.

Sherman, A.H. (1978). "Algorithms for Sparse Gaussian Elimination with Partial Pivoting," *ACM Transactions on Mathematical Software,* **4**, 330–338.

Sherman, D.H. and Ladino, G. (1995). "Managing Bank Productivity Using Data Envelopment Analysis (DEA)," *Interfaces* **25**:2, 60–73.

Sherman, J. and Morrison, W.J. (1949). "Adjustment of an Inverse of Matrix Corresponding to Changes in Elements of a Given Column or a Given Row of the Original Matrix," *Annals of Math. Stat.* **20**, 621.

Shor, N.Z. (1970a). "Utilization of the Operation of Space Dilatation in the Minimization of Convex Functions," *Kibernetika* **6**, 6–11. (English Translation in *Cybernetics* **6**, 7–15.)

Shor, N.Z. (1970b). "Convergence Rate of the Gradient Descent Method with Dilatation of the Space," *Kibernetika* **6**, 80–85. (English Translation in *Cybernetics* **6**, 102–108.)

Shor, N.Z. (1971a). "A Minimization Method Using the Operation of Extension of the Space in the Direction of the Difference of Two Successive Gradients," *Kibernetika* **7**, 51–59. (English Translation in *Cybernetics* **7**, 450–459.)

Shor, N.Z. (1971b). "Certain Questions of Convergence of Generalized Gradient Descent," *Kibernetika* **7**, 82–84. (English Translation in *Cybernetics* **7**, 1033-1036.)

Shor, N.Z. (1972a). "Solution of the Minimax Problems by the Method of Generalized Gradient Descent with Dilatation of the Space," *Kibernetika* **8**, 82–88. (English Translation in *Cybernetics* **8**, 88–94.)

Shor, N.Z. (1972b). "A Class of Almost-Differentiable Functions and a Minimization Method for Functions in this Class," *Kibernetika* **8**, 65–70. (English Translation in *Cybernetics* **8**, 599–606.)

Shor, N.Z. (1975). "Convergence of a Gradient Method with Space Dilatation in the Direction of the Difference Between Two Successive Gradients," *Kibernetika* **11**, 48–53. (English Translation in *Cybernetics* **11**, 564–570.)

Shor, N.Z. (1977a). "Cut-off Method with Space Extension in Convex Programming Problems," *Kibernetika* **13**, 94–95. (English Translation in *Cybernetics* **13**, 94–96.)

Shor, N.Z. (1977b). "New Development Trends in Nondifferentaible Optimization," *Kibernetika* **13**, 87–91. (English Translation in *Cybernetics* **13**, 881–886.)

Simonnard, M. (1966). *Linear Programming* Prentice-Hall, Inc., Englewood Cliffs, New Jersey. [English translation by W. S. Jewell].

Sinha, G.P., Chandrasekaran, B.S., Mitter, N., Dutta, G., Singh, S.B., Choudhry, A.R., and Roy, P.N. (1995). "Strategic and Operational Management with Optimization at Tata Steel," *Interfaces* **25**:1, 6–19.

Skeel, R.D. (1979). "Scaling for Numerical Stabilty in Gaussina Elimination," *Journal of the Association for Computing Machinery,* **26**, 494–526.

Skeel, R.D. (1980). "Iterative Refinement Implies Numerical Stability for Gaussian Elimination," *Mathematics of Computation* **35**, 817–832.

Sleator, D.K. (1980). "An $O(nm \log n)$ Algorithm for Maximum Network Flows," Ph.D. thesis, Department of Computer Science, Stanford University, Stanford, CA.

Smale, S., (1976). "A Convergent Process of Price Adjustment and Global Newton Methods," *Journal of Mathematical Economics* **3**, 107–120.

Smale, S. (1982). "On the Average Speed of the Simplex Method of Linear Programming," Department of Mathematics, University of California, Berkeley.

Smale, S. (1983). "The Problem of the Average Speed of the Simplex Method," in A. Bachem, M. Grötschel and B. Korte (eds.), *Mathematical Programming: The State of the Art*, Springer-Verlag, Berlin and New York, , 530–539.

Smith, B.T., Boyle, J.M., Dongarra, J.J., Garbow, B.S., Ikebe, Y., Klema, V.C., and Moler, C.B., (1976). *Matrix Systems Routines—EISPACK Guide, Lecture Notes in Computer Science* **6**, Springer-Verlag, Berlin and New York.

Solow, D. (1984). *Linear Programming: An Introduction to Finite Improvement Algorithms*, North-Holland, Amsterdam, the Netherlands.

Sonnevend, G. (1986). "An 'Analytic Center' for Polyhedrons and New Classes of Global Algorithms for Linear (smooth, convex) Programming, in Prékopa, A., Szelezsan, J., and Strazicky, B. (eds.), *System Modelling and Optimization: Proceedings of the 12th IFIP Conference Held in Budapest, Hungary, September, 1985, Volume 84 of Lecture Notes in Control and Information Science*, Springer-Verlag, Berlin, West Germany, 866–876.

Sorensen, D. (1980). "Newton's method with a model trust region modification," Report ANL-80-106, Argonne National Laboratory, Argonne, Illinois.

Sorensen, D. (1985). "Analysis of Pairwise Pivoting in Gaussian Elimination," *IEEE Transactions on Computers,* **C-34**, 274–278.

Sponk, J. (1981). *Interactive Multiple Goal Programming: Applications to Financial Management*, Martinus Nijhoff, Boston, Massachusetts.

Steinitz, E. (1913). *J. Reine Angew. Math.,* **143**, 128–275.

Steuer, R.E. (1985). *Multiple Criteria Optimization: Theory, Computation, and Application*, Wiley, New York

Stiemke, E. (1915). "Über positive Lösungen homogener linearer Gleichungen," *Math. Ann.* **76**, 340–342.

Stigler, G.J. (1945). "The Cost of Subsistence," *J. Farm Econ.* **27**, 303–314.

Starfield, A.M., Smith, K. A., and Bleloch, A. L. (1990). *How to Model It—Problem Solving for the Computer Age*, McGraw-Hill, New York.

Stewart, G.W. (1973). *Introduction to Matrix Computations*, Academic Press, London and New York.

Strang, G. (1986). *Introduction to Applied Mathematics*, Wellesley-Cambridge Press, Massachusetts.

Strang, G. (1988). *Linear Algebra*, Academic Press, London and New York.

Strassen, V. (1969). "Gaussian Elimination is Not Optimal," *Numerische Mathematik* **13**, 354–356.

Stone, H.S. (1972). *Introduction to Computer Organization and Data Structures*, McGraw-Hill, San Francisco.

T

Tarjan, R.E. (1972). "Depth-First Search and Linear Graph Algorithms," *SIAM J. on Computing,* **1**, 146–160.

Thapa, M.N. (1976). "Operations Research Techniques in Iron and Steel Making," B. Tech. Thesis, Indian Institute of Technology, Bombay, India.

Thapa, M.N. (1983). "Optimization of Unconstrained Functions with Sparse Hessian Matrices—Quasi-Newton Methods," *Mathematical Programming* **25**, 158–182.

Thapa, M.N. (1984a). "Optimal Leveraged Leasing Within LEA2," Final Report for ATEL, Inc.

Thapa, M.N. (1984b). "Optimization of Unconstrained Functions with Sparse Hessian Matrices—Newton-Type Methods," *Mathematical Programming* **29**, 156–186.

Thapa, M.N. (1991). "A Gasoline/Diesel Retail Distribution Model," Final Report for United Refining Company.

Thapa, M.N. (1992). "A Multi-time Period Asphalt Distribution Model," Final Report for United Refining Company.

Thie, P.R. (1988). "An Introduction to Linear Programming and Game Theory," John Wiley and Sons, New York.

Tilanus, C.B., DeGans, O.B., and Lenstra, J.K. (1986). *Quantitative Methods in Management: Case Studies of Failures and Succeses*, Wiley, New York.

Todd, M.J. (1980). "The Monotonic Hirsch Conjecture is False for Dimension at Least 4," *Mathematics of Operations Research* **5**, 599–601.

Todd, M.J. (1982). "An Implementation of the Simplex Method for Linear Programming Problems with Variable Upper Bounds," *Mathematical Programming* **23**, 23–49.

Todd, M.J. (1983). "Large Scale Linear Programming: Geometry, Working Bases and Factorizations," *Mathematical Programming* **26**, 1–20.

Todd, M.J. (1989). "Recent Developments and New Directions in Linear Programming," in Iri, M. and Tanabe, K. (eds.), *Mathematical Programming: Recent Developments and Applications*, Kluwer Academic Press, Dordrecht, The Netherlands, 109–157.

Todd, M.J. (1990a). "The Effects of Degeneracy, Null, and Unbounded Reduction Algorithm for Linear Programming," Technical Report No. 902, School of Operations Research and Industrial College of Engineering, Cornell University, Ithaca, NY 14853.

Todd, M.J. (1990b). "Combining Phase I and Phase II in a Potential Reduction Algorithm for Linear Programming," Technical Report No. 902, School of Operations Research and Industrial College of Engineering, Cornell University, Ithaca, NY 14853.

Todd, M.J. (1994a). "Theory and Practice of Interior-Point Methods," *ORSA Journal on Computing* **6**, 28–31.

Todd, M.J. (1994b). "A Lower Bound on the Number of Iterations of Primal-Dual Interior Point Methods for Linear Programming," Technical Report 1050, School of Operations Research and Industrial College of Engineering, Cornell University, Ithaca, NY 14853.

Todd, M.J. and Burrel, B.P. (1986). "An Extension of Karmarkar's Algorithm for Linear Programming Using Dual Variables," Special issue of *Algorithmica* **1**, Springer-Verlag, Berlin and New York, 409–424.

Toint, Ph.L. (1977). "On Sparse and Symmetric Matrix Updating Subject to a Linear Equation," *Mathematics of Computation* **32**, 839–851.

Toint, Ph.L. (1979). "On the Superlinear Convergence of an Algorithm for Solving a Sparse Minimization Problem," *SIAM J. on Numerical Analysis* **16**,

Tompkins, C.B. (1955). "Projection Methods in Calculation," in H.A. Antosiewicz (ed.), *Proceedings of the Second Symposium in Linear Programming*, Vol. 2, National Bureau of Standards and Directorate of Management Analysis, DCS/Comptroller, USAF, Washington, D.C., 425–488. 1036–1045.

Tompkins, C.B. (1957). "Some Methods of Computational Attack on Programming Problems, Other than the Simplex Method," *Naval Research Logistics Quarterly*, **4**, 95–96.

Tomlin, J.A., (1972a). "Pivoting for Size and Sparsity in Linear Programming Inversion Routines," *J. Institute of Mathematics and its Applications* **10**, 289–295.

Tomlin, J.A., (1972b). "Modifying Triangular Factors of the Basis in the Simplex Method," in D.J. Rose and R.A. Willoughby (eds.), *Sparse Matrices and their Applications*, Plenum Press, New York, .

Tomlin, J.A., (1975a). "LPM1 User's Guide," Systems Optimization Laboratory, Department of Operations Research, Stanford University, Stanford, CA.

Tomlin, J.A., (1975b). "On Scaling Linear Programming Problems," *Mathematical Programming Study* **4**, 146–166.

Tomlin, J.A., (1975c). "An Accuracy Test for Updating Triangular Factors," *Mathematical Programming Study* **4**, 142–145.

Tomlin, J.A. (1987) "An Experimental Approach to Karmarkar's Projective Method for Linear Programming," *Mathematical Programming Study* **31**, 175–191.

Tomlin, J.A. and Welch, J.S. (1986). "Finding Duplicate Rows in a Linear Programming Model," *Operations Research Letters* **5**, 7–11.

Tomlin, J.A. and Welch, J.S. (1983a). "Formal Optimization of some Reduced Linear Programming Problems", *Mathematical Programming* **27**, 232–240.

Tomlin, J.A. and Welch, J.S. (1983b). "A Pathological Case in the Reduction of Linear Programs," *Operations Research Letters* **2**, 53–57.

Tomlin, J.A. and Welch, J.S. (1985). "Integration of a Primal Simplex Network Algorithm with a Large-Scale Mathematical Programming System," *ACM Transactions on Mathematical Programming,* **II**, 1–11.

Tsuchiya, T. and Monteiro, R.D.C. (1996). "Superlinear Convergence of the Affine Scaling Algorithm," *Mathematical Programming* **75**, 77–110.

Tsuchiya, T. and Muramatsu, M. (1995). "Global Convergence of the Long-Step Affine Scaling Algorithm for Degenerate Linear Programming Problems," *SIAM Journal on Optimization* **5**, 525–551. Previously published a Research Memorandum 423, The Institute of Statistical Mathematics, 4-6-7 Minami-Azubu, Minato-ku, Tokyo 106, Japan, 1992.

Tucker, A.W. (1955). "Linear Inequalities and Convex Polyhedral Sets," in H.A. Antosiewicz (ed.), *Proceedings of the Second Symposium in Linear Programming,* Vol. 2, National Bureau of Standards and Directorate of Management Analysis, DCS/ Comptroller, USAF, Washington, D.C., 569–602.

Tucker, A.W. (1956). "Dual Systems of Homogeneous Linear Relations," in H.W. Kuhn and A.W. Tucker (eds.), *Linear Inequalities and Related Systems,* Annals of Mathematics Study No. 38, Princeton University Press, Princeton, New Jersey, 3–18.

U

Ubhaya, V.A. (1974a). "Isotone Optimization, I," JAT12, 146–159.

Ubhaya, V.A. (1974b). "Isotone Optimization, II," JAT12, 315–331.

V

Vajda, S. (1961). *Mathematical Programming,* Addison-Wesley Publishing Company, Reading, Massachusetts.

Vanderbei, R.J. (1991a) "Splitting Dense Columns in Sparse Linear Systems," *Linear Algebra and its Applications* **152**, 107–117.

Vanderbei, R.J. (1991b) "Symmetric Quasi-Definite Matrices" School of Engineering and Applied Science, Department of Civil Engineering and Operations Research, Princeton University, Princeton, New Jersey91–100.

Vanderbei, R.J. (1992) "LOQO User's Manual," School of Engineering and Applied Science, Department of Civil Engineering and Operations Research, Princeton University, Princeton, New Jersey, 92–95.

Vanderbei, R.J. (1994) "Interior Point Methods: Algorithms and Formulations," *ORSA Journal on Computing* **6**, 32–34.

Vanderbei, R.J. and Carpenter, T.J. (1993) "Symmetric Indefinite Systems for Interior Point Methods," *Mathematical Programming* **58**, 1–32.

Vanderbei, R.J. and Lagarias, J.C. (1988) "I.I. Dikin's Convergence Result for the Affine-Scaling Algorithm," in J.C. Lagarias and M.J. Todd (eds.), *Mathematical Developments Arising from Linear Programming*, Contemporary Mathematics, 109–119.

Vanderbei, R.J., Meketon, M.S., and Freedman, B.A. (1986) "A Modification of Karmarkar's Linear Programming Algorithm," *Algorithmica* **1**, 395–407.

Van Loan, C. (1985). "How Near is a Stable Matrix to an Unstable Matrix?" *Contemporary Mathematics* **47**, 465–477.

Van Slyke, R.M. and Wets, R.J. (1969). "L-Shaped Linear Programs with Applications to Optimal Control and Stochastic Programming," *SIAM J. of Applied Mathematics,* **17**, 638–663.

Van Slyke, R.M. and Wets, R.J. (1966). "Programming Under Uncertainty and Stochastic Optimal," *SIAM J. on Control and Optimization* **4**, 179–193.

Van Tassel, D. (1974). *Program style, Design, Efficiency, Debugging and Testing,* Prentice-Hall, Inc., Englewood Cliffs, New Jersey.

Varga, R.S. (1962). *Matrix Iterative Analysis*, Prentice-Hall, Inc., Englewood Cliffs, New Jersey.

Vavasis, S.A. (1993a). "Stable Numerical Algorithms for Equilibrium Systems," *SIAM J. on Matrix Analysis and Applications*

Vavasis, S.A. (1993b). "Stable Finite Elements for Problems with Wild Coefficients," Report TR 93-1364, Department of Computer Science, Cornell University, Ithaca, NY.

Vertinsky, I., Brown, S., Schreier, H., Thompson, W.A., and van Kooten, G.C. (1994). "A Hierarchical-GIS Based Decision Model for Forest Management: The Systems Approach," *Interfaces* **24**:4, 38–53.

Ville, Jean A., (1938). "Sur la théorie général des jeux oú intervient l'habileté des joueurs," *Applications aux jeux de hasard* by Emil Borel and Jean Ville, Tome 4, Fascicule 2, in *Traité du Calculus des Probabilités et de ses Applications*, by Emile Borel, 105–113.

Vliet, A.V., Boender, C.G., and Rinnooy Kan, A.H.G. (1991). "Interactive Optimization of Bulk Sugar Deliveries," *Interfaces* **22**:3, 4–14.

von Neumann, J. (1928). "Zur Theorie de Geselleschaftsspiele," *Math. Ann.* **100**, 295–320. Translated by Sonya Bargmann in H.W. Kuhn and A.W. Tucker (eds.), *Contributions to the Theory of Games*, Vol. IV, Annals of Mathematics Study No. 40, Princeton University Press, Princeton, New Jersey, 13–42.

von Neumann, J. (1937). "Über ein Ökonomisches Gleichungsystem und ein Verallgemeinerung des Brouwerschen Fixpunkstatzes," *Ergebnisse eines Mathematischen Kolloquiums*, No. 8. Translated in *Rev. Econ. Studies,* **13**, 1–9.

von Neumann, J. (1947). "On a Maximization Problem,," (manuscript), Institute for Advanced Study, Princeton, New Jersey, USA.

von Neumann, J. (1948a). "A Numerical Method for the Determination of the Value and Best Strategies of a Zero-Sum Two-Person Game," (manuscript), Institute for Advanced Study, Princeton, New Jersey, USA.

von Neumann, J. (1948b). Private Communication on an Interior Method.

von Neumann, J. (1953). "A Certain Zero-Sum Two-Person Game Equivalent to the Optimal Assignment Problem," in H.W. Kuhn and A.W. Tucker (eds.), *Contributions to the Theory of Games*, Vol. 2, Annals of Mathematics Study No. 28, Princeton University Press, Princeton, New Jersey, 12–15.

von Neumann, J. and Goldstine, H.H. (1947). "Numerical Inverting of Matrices of High Order," *Bull. Amer. Math. Society* **53**, 1021–1089.

von Neumann, J. and Morgenstern, O. (1944). *Theory of Games and Economic Behavior*, Princeton University Press, Princeton, New Jersey.

W

Wegner, P. (1968). *Programming Languages, Information Structures and Machine Organization*, McGraw-Hill, New York.

Wets, R.J. (1984). "Programming Under Uncertainty: The Equivalent Convex Program," *SIAM J. of Applied Mathematics,* **14**, 89–105.

Whiting, P.D. and Hillier, J.A. (1960). "A Method for Finding the Shortest Route Through a Road Network," *Operations Research Quarterly* **11**, 37–40.

Wilkinson, J.H. (1961). "Error Analysis of Direct Methods of Matrix Inversion," *Journal of the Association for Computing Machinery,* **8**, 281–330.

Wilkinson, J.H. (1963). *Rounding Errors in Algebraic Processes*, Prentice-Hall, Inc., Englewood Cliffs, New Jersey.

Wilkinson, J.H. (1965). *The Algebraic Eigenvalue Problem*, Oxford University Press, Oxford and New York.

Wilkinson, J.H. (1971). "Modern Error Analysis," *SIAM Review,* **13**, 548–568.

Wiliinson, J.H. and Reinsch, C. (eds.), (1971). *Handbook for Automatic Computation, Vol. 2, Linear Algebra*, Springer-Verlag, Berlin and New York.

Williams, H.P. (1985). *Model Building in Operations Research*, Wiley, New York.

Wittrock, R.J. (1983). "Advances in a Nested Decomposition Algorithm for Solving Staircase Linear Programs," Technical Report SOL 83-2, Department of Operations Research, Stanford University, Stanford, CA.

Wolfe, P. (1960). "Accelerating the Cutting Plane Method for Nonlinear Programming," *J. Society for Industrial and Applied Mathematics,* **9**, 481-488.

Wolfe, P. (1962). "The Reduced-Gradient Method," unpublished manuscript, *RAND Corporation.*

Wolfe, P. (1963). "A Technique for Resolving Degeneracy in Linear Programming," *SIAM J. of Applied Mathematics,* **11**, 205–211.

Wolfe, P. (1965). "The Composite Simplex Algorithm," *SIAM Review,* **7**, 42–55.

Wolfe, P. (1967). "Methods for Nonlinear Programming," in J. Abadie (ed.), *Nonlinear Programming*, North-Holland, Amsterdam, the Netherlands, 67–86.

Wolfe, P. and Cutler, L. (1963). "Experiments in Linear Programming," in Graves, R.L. and Wolfe, P. (eds.), *Recent Advances in Mathematical Programming*, McGraw-Hill, New York, 177–200.

Wright, M.H. (1976). *Numerical Methods for Nonlinearly Constrained Optimization*, Ph.D. thesis, Department of Computer Science, Stanford University, Stanford, CA.

Wright, M.H. (1992). "Interior Methods for Constrained Optimization," in A. Iserles (ed.), *Acta Numerica*, Cambridge University Press, New York, NY, 341–407.

Wright, M.H. and Glassman, S.C. (1978). "FORTRAN Subroutines to Solve the Linear Least Squares Problem and Compute the Complete Orthogonal Factorization," Technical Report SOL 78-8, Department of Operations Research, Stanford University, Stanford, CA.

Wright, S.J. (1993). "A Superlinear Infeasible-Interior-Point Algorithm for Monotone Nonlinear Complementarity Problems, Technical Report MCS-P344-1292, Mathematical and Computer Science Division, Argonne National Laboratory, Argonne, Illinois.

Wright, W. (1980). "Automatic Identification of Network Rows in Large-Scale Optimization Models," M.S. thesis, Naval Postgraduate School, Monterey, CA.

Y

Yamada, T. and Kitahara, T. (1985). "Qualitative Properties of Systems of Linear Constraints," *Journal of the Operations Research Society of Japan* **28**, 331–343.

Yamashita, H. (1986). "A Polynomially and Quadratically Convergent Method for Linear Programming," Working Paper, Mathematical Systems Institute, Inc., Tokyo, Japan.

Ye, Y. (1987). "Eliminating Columns in the Simplex Method for Linear Programming", Technical Report SOL 87-14, Department of Operations Research, Stanford University, Stanford, CA.

Ye, Y. (1990). "A 'Build-Down' Scheme for Linear Programming," *Mathematical Programming* **46**, 61–72.

Ye, Y., Güler, O., Tapia, R.A., and Zhang, Y. (1993). "A Quadratically Convergent $O(\sqrt{n}L)$-Iteration Algorithm for Linear Programming," *Mathematical Programming* **59**, 151–162.

Ye, Y., Todd, M.J., and Mizuno, S. (1994). "An $O(\sqrt{n}L)$-Iteration Homogeneous and Self-Dual Linear Programming Algorithm," *Mathematics of Operations Research* **19**, 53–67.

Ye, Y. and Tse, E. (1989). "An Extension of Karmarkar's Projective Algorithm for Convex Quadratic Programming," *Mathematical Programming* **44**, 157–180.

Yohe, J.M. (1974). "An Overview of Programming Practices," *Computing Surveys* **6**, 221–245.

Young, D.M. (1971). *Iterative Solution for Large Linear Systems*, Academic Press, London and New York.

Z

Zadeh, N. (1973). "A Bad Network Problem for the Simplex Method and Other Minimum Cost Flow Algorithms," *Mathematical Programming* **5**, 255–266.

Zaki, H. (1990). "A Comparison of Two Algorithms for the Assignment Problem," Technical Report ORL-90-002, Department of Mechanical and Industrial Engineering, University of Illinois at Urbana-Champaign, Urbana, IL.

Zangwill, W.I. (1969). *Nonlinear Programming: A Unified Approach*, Prentice-Hall, Inc., Englewood Cliffs, New Jersey.

Ziemba, W.J. (1970). "Computational Algorithms for Convex Stochastic Programs with Simple Recourse," *Operations Research* **18**, 273–294.

Ziemba, W.J. (1974). "Stochastic Programs with Simple Recourse," in Hammond, P.L. and Zoutendijk, G. (eds.), *Mathematical Programming: Theory and Practice*, North-Holland, Amsterdam, the Netherlands.

Zikan K. and Cottle R.W. (1987). "The Box Method for Linear Programming: Part I—Basic Theory," Technical Report SOL 87-6, Department of Operations Research, Stanford University, Stanford, CA.

Zhang, Y. (1994). "On the Convergence of a Class of Infeasible Interior-Point Methods for the Horizontal Linear-Complementarity Problem," *SIAM Journal on Optimization* **4**, 208–227.

Zhang, Y., Tapia, R.A., and Dennnis, J.E. (1992). "On the Superlinear and Quadratic Convergence of Primal-Dual Interior Point Linear Programming Algorithms," *SIAM Journal on Optimization* **2**, 304–324.

Zhang, Y. and Zhang, D. (1995). "On Polynomiality of the Mehrotra-type Predictor-Corrector Interior Point Algorithms," *Mathematical Programming* **68**, 303–318.

Zlatev, Z. (1980). "On Some Pivotal Strategies in Gaussian Elimination by Sparse Technique," *SIAM J. on Numerical Analysis* **17**, 12–30.

Zlatev, Z., Wasniewski, J., and Schaumburg, K. (1981). "Y12M—Solution of Large and Sparse Systems of Linear Equations," *Lecture Notes in Computer Science* **121**, Springer-Verlag, Berlin and New York.

Index

Not returnable if opened — Please read the following information before opening this package.

If you wish to return this book and the CD-ROM to Springer-Verlag, do not open the CD-ROM envelope, or remove it from the book. Springer-Verlag will not accept any returns if the package has been opened and/or separated from the book.

MIX
Papier aus verantwortungsvollen Quellen
Paper from responsible sources
FSC® C105338

If you have any concerns about our products,
you can contact us on
ProductSafety@springernature.com

In case Publisher is established outside the EU,
the EU authorized representative is:
Springer Nature Customer Service Center GmbH
Europaplatz 3, 69115 Heidelberg, Germany

Printed by Libri Plureos GmbH
in Hamburg, Germany